# STUDENT SOLUTIONS MANUAL
*John Garlow*

# BEGINNING ALGEBRA
## FOURTH EDITION

*K. Elayn Martin-Gay*

PEARSON
Prentice
Hall

Upper Saddle River, NJ 07458

Editor-in-Chief: Chris Hoag
Senior Acquisitions Editor: Paul Murphy
Associate Editor: Elizabeth Covello
Supplement Editor: Christina Simoneau
Executive Managing Editor: Vince O'Brien
Production Editor: Allyson Kloss
Supplement Cover Manager: Paul Gourhan
Supplement Cover Designer: Joanne Alexandris
Manufacturing Buyer: Ilene Kahn

© 2005 Pearson Education, Inc.
Pearson Prentice Hall
Pearson Education, Inc.
Upper Saddle River, NJ 07458

Printed in the United States of America

10  9  8  7  6  5  4  3  2  1

ISBN 0-13-144492-1

Pearson Education Ltd., *London*
Pearson Education Australia Pty. Ltd., *Sydney*
Pearson Education Singapore, Pte. Ltd.
Pearson Education North Asia Ltd., *Hong Kong*
Pearson Education Canada, Inc., *Toronto*
Pearson Educación de Mexico, S.A. de C.V.
Pearson Education—Japan, *Tokyo*
Pearson Education Malaysia, Pte. Ltd.

# Table of Contents

# Chapter 1

**Exercise Set 1.1**

Answers will vary on Exercises 1-19.

**Exercise Set 1.2**

1. $7 > 3$

3. $6.26 = 6.26$

5. $0 < 7$

7. $-2 < 2$

9. $32 < 212$

11. $44,300 > 34,611$

13. True, since $11=11$.

15. False, since 10 is to the left of 11 on the number line.

17. False, since 11 is to the left of 24 on the number line.

19. True, since 7 is to the right of 0 on the number line.

21. $30 \leq 45$

23. $8 < 12$

25. $5 \geq 4$

27. $15 \neq -2$

29. 535 represents an altitude of 535 feet.
    $-8$ represents 8 feet below sea level.

31. $-34,841$ represents a population decrease of 34,841.

33. 350 represents a deposit of $350.
    $-126$ represents a withdrawal of $126.

35. 1993

37. 1993,1994

39. $837 \geq 818$

41. The number 0 belongs to the sets of: whole numbers, integers, rational numbers, and real numbers.

43. The number $-2$ belongs to the sets of: integers, rational numbers, and real numbers.

45. The number 6 belongs to the sets of: natural numbers, whole numbers, integers, rational numbers, and real numbers.

47. The number 2/3 belongs to the sets of: rational numbers and real numbers.

49. The number $-\sqrt{5}$ belongs to the sets of: irrational numbers and real numbers.

**51.** False. Rational numbers may be non-integers.

**53.** True

**55.** True

**57.** True

**59.** False. An irrational number may not be written as a fraction

**59.** False. Irrational numbers are real.

**61.** $-10 > -100$

**63.** $32 > 5.2$

**65.** $\dfrac{18}{3} < \dfrac{24}{3}$

**67.** $-51 < -50$

**69.** $|-5| > -4$ since $5 > -4$

**71.** $|-1| = |1|$ since $1 = 1$

**73.** $|-2| < |-3|$ since $2 < 3$

**75.** $|0| < |-8|$ since $8 < 8$

**77.** $-0.04 > -26.7$

**79.** The sun is brighter since $-26.7 < -0.04$.

**81.** The sun is the brightest since $-26.7$ is to the left of all other numbers listed.

**83.** $20 \le 25$

**85.** $6 > 0$

**87.** $-12 < -10$

**89.** Answers may vary.

**Mental Math 1.3**

**1.** $\dfrac{3}{8}$

**2.** $\dfrac{1}{4}$

**3.** $\dfrac{5}{7}$

**4.** $\dfrac{2}{5}$

**5.** numerator, denominator

**6.** $\dfrac{11}{7}$

**Exercise Set 1.3**

**1.** $33 = 3 \cdot 11$

**3.** $98 = 2 \cdot 7 \cdot 7$

**5.** $20 = 2 \cdot 2 \cdot 5$

**7.** $75 = 3 \cdot 5 \cdot 5$

**9.** $45 = 3 \cdot 3 \cdot 5$

**11.** $\dfrac{2}{4} = \dfrac{2}{2 \cdot 2} = \dfrac{1}{2}$

**13.** $\dfrac{10}{15} = \dfrac{2 \cdot 5}{3 \cdot 5} = \dfrac{2}{3}$

**15.** $\dfrac{3}{7} = \dfrac{3}{7}$

**17.** $\dfrac{18}{30} = \dfrac{2 \cdot 3 \cdot 3}{2 \cdot 3 \cdot 5} = \dfrac{3}{5}$

**19.** $\dfrac{1}{2} \cdot \dfrac{3}{4} = \dfrac{1 \cdot 3}{2 \cdot 2 \cdot 2} = \dfrac{3}{8}$

**21.** $\dfrac{2}{3} \cdot \dfrac{3}{4} = \dfrac{2 \cdot 3}{3 \cdot 2 \cdot 2} = \dfrac{1}{2}$

**23.** $\dfrac{1}{2} \div \dfrac{7}{12} = \dfrac{1}{2} \cdot \dfrac{12}{7} = \dfrac{1 \cdot 2 \cdot 2 \cdot 3}{2 \cdot 7} = \dfrac{2 \cdot 3}{7} = \dfrac{6}{7}$

**25.** $\dfrac{3}{4} \div \dfrac{1}{20} = \dfrac{3}{4} \cdot \dfrac{20}{1} = \dfrac{3 \cdot 2 \cdot 2 \cdot 5}{2 \cdot 2} = \dfrac{3 \cdot 5}{1} = 15$

**27.** $\dfrac{7}{10} \cdot \dfrac{5}{21} = \dfrac{7 \cdot 5}{2 \cdot 5 \cdot 3 \cdot 7} = \dfrac{1}{2 \cdot 3} = \dfrac{1}{6}$

**29.** $2\dfrac{7}{9} \cdot \dfrac{1}{3} = \dfrac{25}{9} \cdot \dfrac{1}{3} = \dfrac{5 \cdot 5 \cdot 1}{3 \cdot 3 \cdot 3} = \dfrac{25}{27}$

**31.** Area $= \dfrac{11}{12} \cdot \dfrac{3}{5} = \dfrac{11 \cdot 3}{2 \cdot 2 \cdot 3 \cdot 5} = \dfrac{11}{2 \cdot 2 \cdot 5}$

              $= \dfrac{11}{20}$ sq. mi

**33.** $\dfrac{4}{5} - \dfrac{1}{5} = \dfrac{4 - 1}{5} = \dfrac{3}{5}$

**35.** $\dfrac{4}{5} + \dfrac{1}{5} = \dfrac{4 + 1}{5} = \dfrac{5}{5} = 1$

**37.** $\dfrac{17}{21} - \dfrac{10}{21} = \dfrac{17 - 10}{21} = \dfrac{7}{21} = \dfrac{7}{3 \cdot 7} = \dfrac{1}{3}$

**39.** $\dfrac{23}{105} + \dfrac{4}{105} = \dfrac{23 + 4}{105} = \dfrac{27}{105} = \dfrac{3 \cdot 3 \cdot 3}{3 \cdot 5 \cdot 7}$

        $= \dfrac{3 \cdot 3}{5 \cdot 7} = \dfrac{9}{35}$

**41.** $\dfrac{7}{10} = \dfrac{7 \cdot 3}{10 \cdot 3} = \dfrac{21}{30}$

**43.** $\dfrac{2}{9} = \dfrac{2 \cdot 2}{9 \cdot 2} = \dfrac{4}{18}$

**45.** $\dfrac{4}{5} = \dfrac{4 \cdot 4}{5 \cdot 4} = \dfrac{16}{20}$

**47.** $\dfrac{2}{3} + \dfrac{3}{7} = \dfrac{2 \cdot 7}{3 \cdot 7} + \dfrac{3 \cdot 3}{7 \cdot 3} = \dfrac{14}{21} + \dfrac{9}{21} = \dfrac{23}{21}$

**49.** $2\dfrac{13}{15} - 1\dfrac{1}{5} = \dfrac{43}{15} - \dfrac{6}{5} = \dfrac{43}{15} - \dfrac{6 \cdot 3}{5 \cdot 3}$

        $= \dfrac{43 - 18}{15} = \dfrac{25}{15} = 1\dfrac{2}{3}$

**51.** $\dfrac{5}{22} - \dfrac{5}{33} = \dfrac{5 \cdot 3}{22 \cdot 3} - \dfrac{5 \cdot 2}{33 \cdot 2} = \dfrac{15}{66} - \dfrac{10}{66} = \dfrac{5}{66}$

**53.** $\dfrac{12}{5} - 1 = \dfrac{12}{5} - \dfrac{5}{5} = \dfrac{12 - 5}{5} = \dfrac{7}{5}$

**55.** $1 - \dfrac{3}{10} - \dfrac{5}{10} = \dfrac{10}{10} - \dfrac{3}{10} - \dfrac{5}{10} = \dfrac{10 - 3 - 5}{10}$

        $= \dfrac{2}{10} = \dfrac{2}{2 \cdot 5} = \dfrac{1}{5}$

     The unknown part is $\dfrac{1}{5}$

**57.** $1 - \dfrac{1}{4} - \dfrac{3}{8} = \dfrac{8}{8} - \dfrac{1 \cdot 2}{4 \cdot 2} - \dfrac{3}{8} = \dfrac{8 - 2 - 3}{8} = \dfrac{3}{8}$

The unknown part is $\dfrac{3}{8}$

**59.** $1 - \dfrac{1}{2} - \dfrac{1}{6} - \dfrac{2}{9} = \dfrac{18}{18} - \dfrac{1 \cdot 9}{2 \cdot 9} - \dfrac{1 \cdot 3}{6 \cdot 3} - \dfrac{2 \cdot 2}{9 \cdot 2}$

$= \dfrac{18 - 9 - 3 - 4}{18} = \dfrac{2}{18} = \dfrac{1}{9}$

The unknown part is $\dfrac{1}{9}$

**61.** $\dfrac{10}{21} + \dfrac{5}{21} = \dfrac{10 + 5}{21} = \dfrac{15}{21} = \dfrac{3 \cdot 5}{3 \cdot 7} = \dfrac{5}{7}$

**63.** $\dfrac{10}{3} - \dfrac{5}{21} = \dfrac{10 \cdot 7}{3 \cdot 7} - \dfrac{5}{3 \cdot 7} = \dfrac{70}{21} - \dfrac{5}{21} = \dfrac{65}{21}$

**65.** $\dfrac{2}{3} \cdot \dfrac{3}{5} = \dfrac{2 \cdot 3}{3 \cdot 5} = \dfrac{2}{5}$

**67.** $\dfrac{3}{4} \div \dfrac{7}{12} = \dfrac{3}{4} \cdot \dfrac{12}{7} = \dfrac{3 \cdot 3 \cdot 4}{4 \cdot 7} = \dfrac{9}{7}$

**69.** $\dfrac{5}{12} + \dfrac{4}{12} = \dfrac{5 + 4}{12} = \dfrac{9}{12} = \dfrac{3 \cdot 3}{3 \cdot 4} = \dfrac{3}{4}$

**71.** $5 + \dfrac{2}{3} = \dfrac{15}{3} + \dfrac{2}{3} = \dfrac{15 + 2}{3} = \dfrac{17}{3}$

**73.** $\dfrac{7}{8} \div 3\dfrac{1}{4} = \dfrac{7}{8} \div \dfrac{13}{4} = \dfrac{7}{8} \cdot \dfrac{4}{13} = \dfrac{7 \cdot 4}{2 \cdot 4 \cdot 13} = \dfrac{7}{26}$

**75.** $\dfrac{7}{18} \div \dfrac{14}{36} = \dfrac{7}{18} \cdot \dfrac{36}{14} = \dfrac{7 \cdot 2 \cdot 18}{18 \cdot 2 \cdot 7} = 1$

**77.** $\dfrac{23}{105} - \dfrac{2}{105} = \dfrac{23 - 2}{105} = \dfrac{21}{105} = \dfrac{21}{21 \cdot 5} = \dfrac{1}{5}$

**79.** $1\dfrac{1}{2} + 3\dfrac{2}{3} = \dfrac{3}{2} + \dfrac{11}{3} = \dfrac{3 \cdot 3}{2 \cdot 3} + \dfrac{11 \cdot 2}{3 \cdot 2} = \dfrac{9 + 22}{6}$

$= \dfrac{31}{6} = 5\dfrac{1}{6}$

**81.** $\dfrac{2}{3} - \dfrac{5}{9} + \dfrac{5}{6} = \dfrac{2 \cdot 2 \cdot 3}{3 \cdot 2 \cdot 3} - \dfrac{5 \cdot 2}{9 \cdot 2} + \dfrac{5 \cdot 3}{6 \cdot 3}$

$= \dfrac{12 - 10 + 15}{18} = \dfrac{17}{18}$

**83.** $5 + 4\dfrac{1}{8} + 4\dfrac{1}{8} + 15\dfrac{3}{4} + 15\dfrac{3}{4} + 10\dfrac{1}{2}$

$= \dfrac{40}{8} + \dfrac{33}{8} + \dfrac{33}{8} + \dfrac{126}{8} + \dfrac{126}{8} + \dfrac{84}{8}$

$= \dfrac{40 + 33 + 33 + 126 + 126 + 84}{8}$

$= \dfrac{442}{8}$

$= 55\dfrac{1}{4}$ feet

**85.** $4\dfrac{3}{4} + 1\dfrac{2}{5} = \dfrac{19}{4} + \dfrac{7}{5} = \dfrac{19 \cdot 5}{4 \cdot 5} + \dfrac{7 \cdot 4}{5 \cdot 4} = \dfrac{95 + 28}{20}$

$= \dfrac{123}{20} = 6\dfrac{3}{20}$ meters

**87.** Answers may vary.

**89.** $5\dfrac{1}{2} - 2\dfrac{1}{8} = \dfrac{11}{2} - \dfrac{17}{8} = \dfrac{11 \cdot 4}{2 \cdot 4} - \dfrac{17}{8} = \dfrac{44 - 17}{8}$

$= \dfrac{27}{8} = 3\dfrac{3}{8}$ miles

**91.** $\dfrac{7}{50}$ are in the physical sciences

**93.** $1 - \dfrac{4}{25} - \dfrac{7}{50} - \dfrac{7}{50} - \dfrac{7}{100} - \dfrac{21}{100} - \dfrac{3}{100}$

$= \dfrac{100}{100} - \dfrac{4 \cdot 4}{25 \cdot 4} - \dfrac{7 \cdot 2}{50 \cdot 2} - \dfrac{7 \cdot 2}{50 \cdot 2} - \dfrac{7}{100}$

$\phantom{=} - \dfrac{21}{100} - \dfrac{3}{100}$

$= \dfrac{100 - 16 - 14 - 14 - 7 - 21 - 3}{100}$

$= \dfrac{25}{100} = \dfrac{1}{4}$

$\dfrac{1}{4}$ are in the biological and

agricultural sciences

**95.** $\dfrac{666}{3678} = \dfrac{111}{613}$ were Old Navy stores.

**97.** Area $= \dfrac{1}{2} \cdot \dfrac{7}{8} \cdot \dfrac{4}{9} = \dfrac{7 \cdot 4}{2 \cdot 2 \cdot 4 \cdot 9} = \dfrac{7}{36}$ sq ft

### Calculator Explorations 1.4

**1.** $5^3 = 125$

**3.** $9^5 = 59{,}049$

**5.** $2(20 - 5) = 30$

**7.** $24(862 - 455) + 89 = 9857$

**9.** $\dfrac{4623 + 129}{36 - 34} = 2376$

### Mental Math 1.4

**1.** multiply

**2.** Add

**3.** Subtract

**4.** divide

### Exercise Set 1.4

**1.** $3^5 = 3 \cdot 3 \cdot 3 \cdot 3 \cdot 3 = 243$

**3.** $3^3 = 3 \cdot 3 \cdot 3 = 27$

**5.** $1^5 = 1 \cdot 1 \cdot 1 \cdot 1 \cdot 1 = 1$

**7.** $5^1 = 5$

**9.** $\left(\dfrac{1}{5}\right)^3 = \left(\dfrac{1}{5}\right)\left(\dfrac{1}{5}\right)\left(\dfrac{1}{5}\right) = \dfrac{1 \cdot 1 \cdot 1}{5 \cdot 5 \cdot 5} = \dfrac{1}{125}$

**11.** $\left(\dfrac{2}{3}\right)^4 = \left(\dfrac{2}{3}\right)\left(\dfrac{2}{3}\right)\left(\dfrac{2}{3}\right)\left(\dfrac{2}{3}\right) = \dfrac{2 \cdot 2 \cdot 2 \cdot 2}{3 \cdot 3 \cdot 3 \cdot 3} = \dfrac{16}{81}$

**13.** $7^2 = 7 \cdot 7 = 49$

**15.** $(4)^2 = (4) \cdot (4) = 16$

**17.** $(1.2)^2 = (1.2) \cdot (1.2) = 1.44$

**19.** $5 + 6 \cdot 2 = 5 + 12 = 17$

**21.** $4 \cdot 8 - 6 \cdot 2 = 32 - 12 = 20$

**23.** $2(8 - 3) = 2(5) = 10$

**25.** $2 + (5 - 2) + 4^2 = 2 + 3 + 4^2 = 2 + 3 + 16 = 21$

**27.** $5 \cdot 3^2 = 5 \cdot 9 = 45$

**29.** $\dfrac{1}{4} \cdot \dfrac{2}{3} - \dfrac{1}{6} = \dfrac{2}{12} - \dfrac{1}{6} = \dfrac{1}{6} - \dfrac{1}{6} = 0$

**31.** $\dfrac{6-4}{9-2} = \dfrac{2}{7}$

**33.** $2\left[5+2(8-3)\right] = 2\left[5+2(5)\right] = 2\left[5+10\right]$
$$= 2\left[15\right] = 30$$

**35.** $\dfrac{19-3\cdot 5}{6-4} = \dfrac{19-15}{6-4} = \dfrac{4}{2} = 2$

**37.** $\dfrac{\left|6-2\right|+3}{8+2\cdot 5} = \dfrac{\left|4\right|+3}{8+2\cdot 5} = \dfrac{4+3}{8+2\cdot 5} = \dfrac{4+3}{8+10}$
$$= \dfrac{7}{18}$$

**39.** $\dfrac{3+3(5+3)}{3^2+1} = \dfrac{3+3(8)}{3^2+1} = \dfrac{3+3(8)}{9+1}$
$$= \dfrac{3+24}{9+1} = \dfrac{27}{10}$$

**41.** $\dfrac{6+\left|8-2\right|+3^2}{18-3} = \dfrac{6+\left|6\right|+3^2}{18-3} = \dfrac{6+6+3^2}{18-3}$
$$= \dfrac{6+6+9}{18-3} = \dfrac{21}{15} = \dfrac{3\cdot 7}{3\cdot 5}$$
$$= \dfrac{7}{5}$$

**43.** No; since in the absence of grouping symbols we always perform multiplications or divisions before additions or subtractions in any expression.

**45.** **a.** $(6+2)\cdot(5+3) = 8\cdot 8 = 64$

**b.** $(6+2)\cdot 5+3 = 8\cdot 5+3 = 40+3$
$$= 43$$

**c.** $6+2\cdot 5+3 = 6+10+3 = 19$

**d.** $6+2\cdot(5+3) = 6+2\cdot 8 = 6+16$
$$= 22$$

**47.** Let $y = 3$
$$3y = 3(3) = 9$$

**49.** Let $x = 1$ and $z = 5$
$$\dfrac{z}{5x} = \dfrac{5}{5(1)} = \dfrac{5}{5} = 1$$

**51.** Let $x = 1$
$$3x-2 = 3(1)-2 = 3-2 = 1$$

**53.** Let $x = 1$ and $y = 3$
$$\left|2x+3y\right| = \left|2(1)+3(3)\right| = \left|2+9\right| = \left|11\right| = 11$$

**55.** Let $y = 3$
$$5y^2 = 5(3)^2 = 5(9) = 45$$

**57.** Let $x = 12$, $y = 8$ and $z = 4$
$$\dfrac{x}{z}+3y = \dfrac{12}{4}+3(8) = 3+24 = 27$$

**59.**          Let $x = 12$ and $y = 8$.
$$x^2-3y+x = (12)^2-3(8)+12$$
$$= 144-24+12$$
$$= 132$$

**61.**     Let $x = 12$, $y = 8$ and $z = 4$

$$\frac{x^2 + z}{y^2 + 2z} = \frac{(12)^2 + 4}{(8)^2 + 2(4)} = \frac{144 + 4}{64 + 8}$$

$$= \frac{148}{72} = \frac{37}{18}$$

**63.** Evaluate $16t^2$ for each value of $t$.

$$t = 1;\ 16(1)^2 = 16(1) = 16$$

$$t = 2;\ 16(2)^2 = 16(4) = 64$$

$$t = 3;\ 16(3)^2 = 16(9) = 144$$

$$t = 4;\ 16(4)^2 = 16(16) = 256$$

| Time $t$ (in seconds) | Distance $16t^2$ (in feet) |
|---|---|
| 1 | 16 |
| 2 | 64 |
| 3 | 144 |
| 4 | 256 |

**65.** Let $x = 5$

$$3x - 6 = 9$$

$$3(5) - 6 \overset{?}{=} 9$$

$$15 - 6 \overset{?}{=} 9$$

$$9 = 9,\ \text{true}$$

5 is a solution of the equation.

**67.** Let $x = 0$

$$2x + 6 = 5x - 1$$

$$2(0) + 6 \overset{?}{=} 5(0) - 1$$

$$0 + 6 \overset{?}{=} 0 - 1$$

$$6 = -1,\ \text{false}$$

0 is not a solution of the equation.

**69.** Let $x = 8$

$$2x - 5 = 5$$

$$2(8) - 5 \overset{?}{=} 5$$

$$16 - 5 \overset{?}{=} 5$$

$$9 = 5,\ \text{false}$$

8 is not a solution of the equation.

**71.** Let $x = 2$

$$x + 6 = x + 6$$

$$2 + 6 \overset{?}{=} 2 + 6$$

$$8 = 8,\ \text{true}$$

2 is a solution of the equation.

**73.** Let $x = 0$

$$x = 5x + 15$$

$$(0) \overset{?}{=} 5(0) + 15$$

$$0 \overset{?}{=} 0 + 15$$

$$0 = 15,\ \text{false}$$

0 is not a solution of the equation.

**75.** $x + 15$

**77.** $x - 5$

**79.** $3x + 22$

**81.** $1 + 2 = 9 \div 3$

**83.** $3 \neq 4 \div 2$

**85.** $5 + x = 20$

**87.** $13 - 3x = 13$

**89.** $\dfrac{12}{x} = \dfrac{1}{2}$

**91.** Answers may vary.

**93.** $(20-4)\cdot 4 \div 2 = (16)\cdot 4 \div 2 = 64 \div 2 = 32$

**95.** Let $l = 8$ and $w = 6$
$2l + 2w = 2(8) + 2(6) = 16 + 12 = 28$ m.

**97.** Let $l = 120$ and $w = 100$
$lw = (120)(100) = 12{,}000$ sq.ft.

**99.** Let $P = 650$, $T = 3$, and $I = 126.75$
$\dfrac{I}{PT} = \dfrac{126.75}{(650)(3)} = \dfrac{126.75}{1950} = 0.065 = 6.5\%$

**101.** Let $m = 228$ .
$\begin{aligned} 4.00 + 0.07m &= 4.00 + 0.07(228) \\ &= 4.00 + 15.96 \\ &= \$19.96 \end{aligned}$

**101.** Answers may vary.

**Mental Math 1.5**

1. negative
2. positive
3. 0
4. negative
5. negative
6. 0

**Exercise Set 1.5**

**1.** $6 + 3 = 9$

**3.** $-6 + (-8) = -14$

**5.** $8 + (-7) = 1$

**7.** $-14 + 2 = -12$

**9.** $-2 + (-3) = -5$

**11.** $-9 + (-3) = -12$

**13.** $-7 + 3 = -4$

**15.** $10 + (-3) = 7$

**17.** $5 + (-7) = -2$

**19.** $-16 + 16 = 0$

**21.** $27 + (-46) = -19$

**23.** $-18 + 49 = 31$

**25.** $-33 + (-14) = -47$

**27.** $6.3 + (-8.4) = -2.1$

**29.** $\left|-8\right| + (-16) = 8 + (-16) = -8$

**31.** $117 + (-79) = 38$

**33.** $-9.6 + (-3.5) = -13.1$

**35.** $-\dfrac{3}{8} + \dfrac{5}{8} = \dfrac{2}{8} = \dfrac{1}{4}$

**37.** $-\dfrac{7}{16}+\dfrac{1}{4}=-\dfrac{7}{16}+\dfrac{1\cdot 4}{4\cdot 4}=-\dfrac{7}{16}+\dfrac{4}{16}=-\dfrac{3}{16}$

**39.** $-\dfrac{7}{10}+\left(-\dfrac{3}{5}\right)=-\dfrac{7}{10}+\left(-\dfrac{3\cdot 2}{5\cdot 2}\right)$

$\qquad\qquad =-\dfrac{7}{10}+\left(-\dfrac{6}{10}\right)=-\dfrac{13}{10}$

**41.** $-15+9+(-2)=-6+(-2)=-8$

**43.** $-21+(-16)+(-22)=-37+(-22)=-59$

**45.** $-23+16+(-2)=-7+(-2)=-9$

**47.** $\left|5+(-10)\right|=\left|-5\right|=5$

**49.** $6+(-4)+9=2+9=11$

**51.** $\left[-17+(-4)\right]+\left[-12+15\right]=\left[-21\right]+\left[3\right]$

$\qquad\qquad\qquad\qquad\qquad\qquad\quad =-18$

**53.** $\left|9+(-12)\right|+\left|-16\right|=\left|-3\right|+16=3+16=19$

**55.** $\quad -1.3+\left[0.5+(-0.3)+0.4\right]$

$\qquad =-1.3+\left[0.2+0.4\right]$

$\qquad =-1.3+\left[0.6\right]$

$\qquad =-0.7$

**57.** $-15+9=-6$

The high temperature in Anoka was $-6°$.

**59.** $-1312+658=-654$

You are 654 feet below sea level.

**61.** $(-1411)+(-567)+(-149)$

$\qquad =(-1978)+(-149)$

$\qquad =-2127$

The total net income
was $-\$2127$ million.

**63.** $5+(-5)+(-2)=0+(-2)=-2$

Her score was 2 under par.

**65.** The opposite of 6 is $-6$.

**67.** The opposite of $-2$ is 2.

**69.** The opposite of 0 is 0.

**71.** Since $\left|-6\right|$ is 6, the opposite of $\left|-6\right|$ is $-6$.

**73.** Answers may vary.

**75.** $-\left|-2\right|=-2$

**77.** $-\left|0\right|=-0=0$

**79.** $-\left|-\dfrac{2}{3}\right|=-\dfrac{2}{3}$

**81.** Answers may vary

**83.** Let $x=-4$

$\qquad x+9=5$

$\qquad (-4)+9\overset{?}{=}5$

$\qquad\qquad 5=5,\text{ true}$

$-4$ is a solution of the equation.

**85.** Let $y = -1$

$$y + (-3) = -7$$

$$(-1) + (-3) \overset{?}{=} -7$$

$$-4 = -7, \text{ false}$$

$-1$ is not a solution of the equation.

**87.** July

**89.** October

**91.** $\left[(-9.1) + 14.4 + 8.8\right] \div 3 = \left[5.3 + 8.8\right] \div 3$

$= \left[14.1\right] \div 3 = 4.7$

The average was $4.7°$ F.

**93.** $-a$ is a <u>negative</u> number.

**95.** $a + a$ is a <u>positive</u> number.

**Exercise Set 1.6**

**1.** $-6 - 4 = -6 + (-4) = -10$

**3.** $4 - 9 = 4 + (-9) = -5$

**5.** $16 - (-3) = 16 + 3 = 19$

**7.** $\dfrac{1}{2} - \dfrac{1}{3} = \dfrac{1}{2} + \left(-\dfrac{1}{3}\right) = \dfrac{1 \cdot 3}{2 \cdot 3} + \left(-\dfrac{1 \cdot 2}{3 \cdot 2}\right)$

$= \dfrac{3}{6} + \left(-\dfrac{2}{6}\right) = \dfrac{1}{6}$

**9.** $-16 - (-18) = -16 + 18 = 2$

**11.** $-6 - 5 = -6 + (-5) = -11$

**13.** $7 - (-4) = 7 + 4 = 11$

**15.** $-6 - (-11) = -6 + 11 = 5$

**17.** $16 - (-21) = 16 + 21 = 37$

**19.** $9.7 - 16.1 = 9.7 + (-16.1) = -6.4$

**21.** $-44 - 27 = -44 + (-27) = -71$

**23.** $-21 - (-21) = -21 + 21 = 0$

**25.** $-2.6 - (-6.7) = -2.6 + 6.7 = 4.1$

**27.** $-\dfrac{3}{11} - \left(-\dfrac{5}{11}\right) = -\dfrac{3}{11} + \dfrac{5}{11} = \dfrac{2}{11}$

**29.** $-\dfrac{1}{6} - \dfrac{3}{4} = -\dfrac{1}{6} + \left(-\dfrac{3}{4}\right) = -\dfrac{1 \cdot 2}{6 \cdot 2} + \left(-\dfrac{3 \cdot 3}{4 \cdot 3}\right)$

$= -\dfrac{2}{12} + \left(-\dfrac{9}{12}\right) = -\dfrac{11}{12}$

**31.** $8.3 - (-0.62) = 8.3 + 0.62 = 8.92$

**33.** $8 - (-5) = 8 + 5 = 13$

**35.** $-6 - (-1) = -6 + 1 = -5$

**37.** $7 - 8 = 7 + (-8) = -1$

**39.** $-8 - 15 = -8 + (-15) = -23$

**41.** Answers may vary

**43.** $-10 - (-8) + (-4) - 20$

$= -10 + 8 + (-4) + (-20)$

$= -2 + (-4) + (-20) = -6 + (-20) = -26$

**45.** $5 - 9 + (-4) - 8 - 8$

$= 5 + (-9) + (-4) + (-8) + (-8)$

$= -4 + (-4) + (-8) + (-8)$

$= -8 + (-8) + (-8) = -16 + (-8) = -24$

**47.** $-6 - (2 - 11) = -6 - (-9) = -6 + 9 = 3$

**49.** $3^3 - 8 \cdot 9 = 27 - 8 \cdot 9$

$\qquad = 27 - 72 = 27 + (-72) = -45$

**51.** $2 - 3(8 - 6) = 2 - 3(2) = 2 - 6 = 2 + (-6) = -4$

**53.** $(3 - 6) + 4^2 = \left[3 + (-6)\right] + 4^2 = \left[-3\right] + 4^2$

$\qquad\qquad = \left[-3\right] + 16 = 13$

**55.** $-2 + \left[(8 - 11) - (-2 - 9)\right]$

$= -2 + \left[(8 + (-11)) - (-2 + (-9))\right]$

$= -2 + \left[(-3) - (-11)\right] = -2 + \left[(-3) + 11\right]$

$= -2 + \left[8\right] = 6$

**57.** $|-3| + 2^2 + \left[-4 - (-6)\right] = 3 + 2^2 + \left[-4 + 6\right]$

$= 3 + 2^2 + \left[2\right] = 3 + 4 + \left[2\right] = 7 + \left[2\right] = 9$

**59.** Let $x = -5$ and $y = 4$.

$x - y = -5 - 4 = -5 + (-4) = -9$

**61.** Let $x = -5$, $y = 4$, and $t = 10$.

$|x| + 2t - 8y = |-5| + 2(10) - 8(4)$

$= 5 + 2(10) - 8(4) = 5 + 20 - 32$

$= 25 - 32 = 25 + (-32) = -7$

**63.** Let $x = -5$ and $y = 4$.

$\dfrac{9 - x}{y + 6} = \dfrac{9 - (-5)}{4 + 6} = \dfrac{9 + 5}{4 + 6} = \dfrac{14}{10} = \dfrac{2 \cdot 7}{2 \cdot 5} = \dfrac{7}{5}$

**65.** Let $x = -5$ and $y = 4$.

$y^2 - x = 4^2 - (-5) = 16 + 5 = 21$

**67.** Let $x = -5$ and $t = 10$.

$\dfrac{|x - (-10)|}{2t} = \dfrac{|-5 - (-10)|}{2(10)} = \dfrac{|-5 + 10|}{2(10)}$

$= \dfrac{|5|}{2(10)} = \dfrac{5}{20} = \dfrac{5}{4 \cdot 5} = \dfrac{1}{4}$

**69.** The change in temperature is the difference between the last temperature and the first temperature.

$-56 - 44 = -56 + (-44) = -100$

The temperature dropped 100°.

**71.** Gains: $+2$

Losses: $-5, -20$

$2 + (-5) + (-20) = -3 + (-20) = -23$

Total loss of 23 yards

**73.** $-475 - 94 = -475 + (-94) = -569$

He was born in 569 B.C.

**75.** Rises: $+120$

Drops: $-250, -178$

$120 + (-250) + (-178)$

$= -130 + (-178) = -308$

The overall vertical change was a drop of 308 feet.

**77.** $19,340 - (-512) = 19,340 + 512$

$\qquad\qquad\qquad = 19,852$

19,852 feet higher

**79.** $y = 180 - 50 = 180 + (-50) = 130$

The supplementary angle is $130°$

**81.** $x = 90 - 60 = 90 + (-60) = 30$

The complementary angle is $30°$

**83.** Let $x = -4$

$x - 9 = 5$

$-4 - 9 \overset{?}{=} 5$

$-13 = 5$, false

$-4$ is not a solution of the equation.

**85.** Let $x = -2$

$-x + 6 = -x - 1$

$-(-2) + 6 \overset{?}{=} -(-2) - 1$

$2 + 6 \overset{?}{=} 2 + (-1)$

$8 = 1$, false

$-2$ is not a solution of the equation.

**87.** Let $x = 2$

$-x - 13 = -15$

$-2 - 13 \overset{?}{=} -15$

$-2 + (-13) \overset{?}{=} -15$

$-15 = -15$, true

2 is a solution of the equation.

**89.** The change in temperature is the difference between the given month's temperature and the previous month's.

F: $-23.7 - (-19.3) = -23.7 + 19.3 = -4.4°$

Mr: $-21.1 - (-23.7) = -21.1 + 23.7 = 2.6°$

Ap: $-9.1 - (-21.1) = -9.1 + 21.1 = 12°$

Ma: $14.4 - (-9.1) = 14.4 + 9.1 = 23.5°$

Jn: $29.7 - 14.4 = 29.7 + (-14.4) = 15.3°$

Jy: $33.6 - 29.7 = 33.6 + (-29.7) = 3.9°$

Au: $33.3 - 33.6 = 33.3 + (-33.6) = -0.3°$

S: $27.0 - 33.3 = 27.0 + (-33.3) = -6.3°$

O: $8.8 - 27.0 = 8.8 + (-27.0) = -18.2°$

N: $-6.9 - 8.8 = -6.9 + (-8.8) = -15.7°$

D: $-17.2 - (-6.9) = -17.2 + 6.9 = -10.3°$

**91.** October

**93.** True: answers may vary.

**95.** True: answers may vary.

**97.** $4.362 - 7.0086 = -2.6466$

**Calculator Explorations 1.7**

**1.** $-38(26 - 27) = 38$

**3.** $134 + 25(68 - 91) = -441$

**5.** $\dfrac{-50(294)}{175 - 265} = 163.\overline{3}$

**7.** $9^5 - 4550 = 54,499$

**9.** $(-125)^2 = 15,625$

**Mental Math 1.7**

**1.** positive

**2.** positive

**3.** negative

**4.** negative

**5.** positive

**6.** negative

**Exercise Set 1.7**

**1.** $-6(4) = -24$

**3.** $2(-1) = -2$

**5.** $-5(-10) = 50$

**7.** $-3 \cdot 4 = -12$

**9.** $-7 \cdot 0 = 0$

**11.** $2(-9) = -18$

**13.** $-\dfrac{1}{2}\left(-\dfrac{3}{5}\right) = \dfrac{1 \cdot 3}{2 \cdot 5} = \dfrac{3}{10}$

**15.** $-\dfrac{3}{4}\left(-\dfrac{8}{9}\right) = \dfrac{3 \cdot 8}{4 \cdot 9} = \dfrac{24}{36} = \dfrac{2 \cdot 12}{3 \cdot 12} = \dfrac{2}{3}$

**17.** $5(-1.4) = -7.0$

**19.** $-0.2(-0.7) = 0.14$

**21.** $-10(80) = -800$

**23.** $4(-7) = -28$

**25.** $(-5)(-5) = 25$

**27.** $\dfrac{2}{3}\left(-\dfrac{4}{9}\right) = -\dfrac{2 \cdot 4}{3 \cdot 9} = -\dfrac{8}{27}$

**29.** $-11(11) = -121$

**31.** $-\dfrac{20}{25}\left(\dfrac{5}{16}\right) = -\dfrac{20 \cdot 5}{25 \cdot 16} = -\dfrac{100}{400} = -\dfrac{1}{4}$

**33.** $(-1)(2)(-3)(-5)$
$= -2(-3)(-5) = 6(-5) = -30$

**35.** $(-2)(5) - (-11)(3) = -10 - (-33)$
$= -10 + 33$
$= 23$

**37.** $(-6)(-1)(-2) - (-5) = -12 + 5 = -7$

**39.** True

**41.** False

**43.** $(-2)^4 = (-2)(-2)(-2)(-2) = 4(-2)(-2)$
$= -8(-2) = 16$

**45.** $-1^5 = -(1)(1)(1)(1)(1) = -1$

**47.** $(-5)^2 = (-5)(-5) = 25$

**49.** $-7^2 = -(7)(7) = -49$

**51.** Reciprocal of $9$ is $\dfrac{1}{9}$ since $9 \cdot \dfrac{1}{9} = 1$

**53.** Reciprocal of $\dfrac{2}{3}$ is $\dfrac{3}{2}$ since $\dfrac{2}{3} \cdot \dfrac{3}{2} = 1$

**55.** Reciprocal of $-14$ is $-\dfrac{1}{14}$

since $-14 \cdot -\dfrac{1}{14} = 1$

**57.** Reciprocal of $-\dfrac{3}{11}$ is $-\dfrac{11}{3}$

since $-\dfrac{3}{11} \cdot -\dfrac{11}{3} = 1$

**59.** Reciprocal of $0.2$ is $\dfrac{1}{0.2}$

since $0.2 \cdot \dfrac{1}{0.2} = 1$

**61.** Reciprocal of $\dfrac{1}{-6.3}$ is $-6.3$

since $\dfrac{1}{-6.3} \cdot -6.3 = 1$

**63.** $\dfrac{18}{-2} = 18 \cdot -\dfrac{1}{2} = -9$

**65.** $\dfrac{-16}{-4} = -16 \cdot -\dfrac{1}{4} = 4$

**67.** $\dfrac{-48}{12} = -48 \cdot \dfrac{1}{12} = -4$

**69.** $\dfrac{0}{-4} = 0 \cdot -\dfrac{1}{4} = 0$

**71.** $-\dfrac{15}{3} = -15 \cdot \dfrac{1}{3} = -5$

**73.** $\dfrac{5}{0}$ is undefined

**75.** $\dfrac{-12}{-4} = -12 \cdot -\dfrac{1}{4} = 3$

**77.** $\dfrac{30}{-2} = 30 \cdot -\dfrac{1}{2} = -15$

**79.** $\dfrac{6}{7} \div -\dfrac{1}{3} = \dfrac{6}{7} \cdot \left(-\dfrac{3}{1}\right) = -\dfrac{6 \cdot 3}{7 \cdot 1} = -\dfrac{18}{7}$

**81.** $-\dfrac{5}{9} \div \left(-\dfrac{3}{4}\right) = -\dfrac{5}{9} \cdot \left(-\dfrac{4}{3}\right) = \dfrac{5 \cdot 4}{9 \cdot 3} = \dfrac{20}{27}$

**83.** $-\dfrac{4}{9} \div \dfrac{4}{9} = -\dfrac{4}{9} \cdot \dfrac{9}{4} = -1$

**85.** $\dfrac{-9(-3)}{-6} = \dfrac{27}{-6} = -\dfrac{9}{2}$

**87.** $\dfrac{12}{9-12} = \dfrac{12}{-3} = -4$

**89.** $\dfrac{-6^2+4}{-2}=\dfrac{-36+4}{-2}=\dfrac{-32}{-2}=16$

**91.** $\dfrac{8+(-4)^2}{4-12}=\dfrac{8+16}{4-12}=\dfrac{24}{-8}=-3$

**93.** $\dfrac{22+(3)(-2)}{-5-2}=\dfrac{22+(-6)}{-5-2}=\dfrac{16}{-7}$

**95.** $\dfrac{-3-5^2}{2(-7)}=\dfrac{-3-25}{2(-7)}=\dfrac{-3+(-25)}{-14}=\dfrac{-28}{-14}=2$

**97.** $\dfrac{6-2(-3)}{4-3(-2)}=\dfrac{6-(-6)}{4-(-6)}=\dfrac{6+6}{4+6}=\dfrac{12}{10}=\dfrac{6}{5}$

**99.** $\dfrac{-3-2(-9)}{-15-3(-4)}=\dfrac{-3-(-18)}{-15-(-12)}$

$\qquad =\dfrac{-3+18}{-15+12}=\dfrac{15}{-3}=-5$

**101.** $\dfrac{|5-9|+|10-15|}{|2(-3)|}=\dfrac{|-4|+|-5|}{|-6|}$

$\qquad =\dfrac{4+5}{6}=\dfrac{9}{6}=\dfrac{3}{2}$

**103.** Let $x=-5$ and $y=-3$.

$3x+2y=3(-5)+2(-3)$

$\qquad =-15+(-6)$

$\qquad =-21$

**105.** Let $x=-5$ and $y=-3$.

$2x^2-y^2=2(-5)^2-(-3)^2=2(25)-9$

$\qquad =50+(-9)=41$

**107.** Let $x=-5$ and $y=-3$.

$x^3+3y=(-5)^3+3(-3)$

$\qquad =-125+(-9)=-134$

**109.** Let $x=-5$ and $y=-3$.

$\dfrac{2x-5}{y-2}=\dfrac{2(-5)-5}{-3-2}=\dfrac{-10-5}{-3-2}=\dfrac{-15}{-5}=3$

**111.** Let $x=-5$ and $y=-3$.

$\dfrac{-3-y}{x-4}=\dfrac{-3-(-3)}{-5-4}=\dfrac{-3+3}{-5-4}=\dfrac{0}{-9}=0$

**113.** $4(-1,272)=-5,088$

The net income will be

$-\$5,088$ million.

**115.** Let $x=7$

$-5x=-35$

$-5(7)\overset{?}{=}-35$

$-35=-35$, true

7 is a solution of the equation.

**117.** Let $x=-20$

$\dfrac{x}{10}=2$

$\dfrac{-20}{10}\overset{?}{=}2$

$-2=2$, false

$-20$ is not a solution of the equation.

**119.** Let $x = 5$

$$-3x - 5 = -20$$

$$-3(5) - 5 \stackrel{?}{=} -20$$

$$-15 - 5 \stackrel{?}{=} -20$$

$$-20 = -20, \text{ true}$$

5 is a solution of the equation.

**121.** Answers may vary

**123.** $-1$ and $1$ are their own reciprocals.

**125.** Positive

**127.** Not possible

**129.** Negative

**131.** $-2 + \dfrac{-15}{3} = \dfrac{-2 \cdot 3}{1 \cdot 3} + \dfrac{-15}{3} = \dfrac{-6 + (-15)}{3}$

$$= \dfrac{-21}{3} = -7$$

**133.** $2\left[-5 + (-3)\right] = 2(-8) = -16$

**Integrated Review-Operations on Real Numbers**

**1.** positive

**2.** positive

**3.** negative

**4.** negative

**5.** positive

**6.** negative

**7.** negative

**8.** positive

**9.** $5(-7) = -35$

**10.** $-3(-10) = 30$

**11.** $\dfrac{-20}{-4} = 5$

**12.** $\dfrac{30}{-6} = -5$

**13.** $7 - (-3) = 7 + 3 = 10$

**14.** $-8 - 10 = -8 + (-10) = -18$

**15.** $-14 - (-12) = -14 + 12 = -2$

**16.** $-3 - (-1) = -3 + 1 = -2$

**17.** $-\dfrac{1}{2}\left(-\dfrac{3}{4}\right) = \dfrac{1 \cdot 3}{2 \cdot 4} = \dfrac{3}{8}$

**18.** $-\dfrac{2}{7}\left(\dfrac{11}{12}\right) = -\dfrac{2 \cdot 11}{7 \cdot 12} = -\dfrac{22}{84} = -\dfrac{11}{42}$

**19.** $\dfrac{-12}{0.2} = -60$

**20.** $\dfrac{-3.8}{-2} = 1.9$

**21.** $-19 + (-23) = -42$

**22.** $18 + (-25) = -7$

**23.** $-15 + 17 = 2$

**24.** $-2 + (-37) = -39$

**25.** $(-8)^2 = (-8)(-8) = 64$

**26.** $-9^2 = -(9)(9) = -81$

**27.** $-3^2 = -(3)(3)(3) = -27$

**28.** $(-2)^4 = (-2)(-2)(-2)(-2) = 16$

**29.** $-1^{10} = -(1)(1)(1)(1)(1)(1)(1)(1)(1)(1) = -1$

**30.** $(-1)^{10} = (-1)(-1)(-1)(-1)(-1)(-1)$
$\qquad\qquad \cdot(-1)(-1)(-1)(-1)$
$\qquad = 1$

**31.** $(-2)^5 = (-2)(-2)(-2)(-2)(-2) = -32$

**32.** $-2^5 = -(2)(2)(2)(2)(2) = -32$

**33.** $(2)(-8)(-3) = (-16)(-3) = 48$

**34.** $3(-2)(5) = (-6)(5) = -30$

**35.** $-6(2) + 20 \div 2 - 4 = -12 + 10 - 4 = -6$

**36.** $-4(-3) + 9 \div 3 - 6 = 12 + 3 - 6 = 9$

**37.** $-3^2 - \left[6 + 5|-2 - 1|\right] = -9 - \left[6 + 5|-3|\right]$
$\quad = -9 - \left[6 + 5(3)\right] = -9 - \left[6 + 15\right] = -9 - 21$
$\quad = -30$

**38.** $-5^2 - \left[4 + 3|-3 - 2|\right] = -25 - \left[4 + 3|-5|\right]$
$\quad = -25 - \left[4 + 3(5)\right] = -25 - \left[4 + 15\right]$
$\quad = -25 - 19 = -44$

**39.** $2(19 - 17)^3 - 3(7 - 9)^2$
$\quad = 2\left[19 + (-17)\right]^3 - 3\left[7 + (-9)\right]^2$
$\quad = 2[2]^3 - 3[-2]^2 = 2[8] - 3[4]$
$\quad = 16 - 12 = 16 + (-12) = 4$

**40.** $3(10 - 9)^2 - 6(20 - 19)^3$
$\quad = 3\left[10 + (-9)\right]^3 - 6\left[20 + (-19)\right]^2$
$\quad = 3[1]^3 - 6[1]^2 = 3[1] - 6[1]$
$\quad = 3 - 6 = 3 + (-6) = -3$

**41.** $\dfrac{19 - 25}{3(-1)} = \dfrac{19 + (-25)}{-3} = \dfrac{-6}{-3} = 2$

**42.** $\dfrac{8(-4)}{-2} = \dfrac{-32}{-2} = 16$

**43.** $\dfrac{-2(3 - 6) - 6(10 - 9)}{-6 - (-5)}$
$\quad = \dfrac{-2\left[3 + (-6)\right] - 6\left[10 + (-9)\right]}{-6 - (-5)}$
$\quad = \dfrac{-2[-3] - 6[1]}{-6 - (-5)} = \dfrac{6 - 6}{-6 - (-5)} = \dfrac{6 + (-6)}{-6 + 5}$
$\quad = \dfrac{0}{-1} = 0$

**44.** $\dfrac{5(7-9)-3(100-97)}{4-5}$

$= \dfrac{5[7+(-9)]-3[100+(-97)]}{4-5}$

$= \dfrac{5[-2]-3[3]}{4-5} = \dfrac{-10-9}{4-5} = \dfrac{-10+(-9)}{4+(-5)}$

$= \dfrac{-19}{-1} = 19$

**45.** $\dfrac{-4(8-10)^3}{-2-1-12} = \dfrac{-4[8+(-10)]^3}{-2-1-12}$

$= \dfrac{-4[-2]^3}{-2-1-12} = \dfrac{-4[-8]}{-2-1-12}$

$= \dfrac{32}{-2+(-1)+(-12)} = \dfrac{32}{-15} = -\dfrac{32}{15}$

**46.** $\dfrac{6(7-10)^2}{6-(-1)-2} = \dfrac{6[7+(-10)]^2}{6-(-1)-2}$

$= \dfrac{6[-3]^2}{6-(-1)-2} = \dfrac{6[9]}{6-(-1)-2}$

$= \dfrac{54}{6+1+(-2)} = \dfrac{54}{5}$

**Exercise Set 1.8**

**1.** $x+16 = 16+x$

**3.** $-4 \cdot y = y \cdot (-4)$

**5.** $xy = yx$

**7.** $2x+13 = 13+2x$

**9.** $(xy) \cdot z = x \cdot (yz)$

**11.** $2+(a+b) = (2+a)+b$

**13.** $4 \cdot (ab) = 4a \cdot (b)$

**15.** $(a+b)+c = a+(b+c)$

**17.** $8+(9+b) = (8+9)+b = 17+b$

**19.** $4(6y) = (4 \cdot 6)y = 24y$

**21.** $\dfrac{1}{5}(5y) = \left(\dfrac{1}{5} \cdot 5\right)y = 1 \cdot y = y$

**23.** $(13+a)+13 = (a+13)+13 = a+(13+13)$
$= a+26$

**25.** $-9(8x) = (-9 \cdot 8)x = -72x$

**27.** $\dfrac{3}{4}\left(\dfrac{4}{3}s\right) = \left(\dfrac{3}{4} \cdot \dfrac{4}{3}\right)s = 1s = s$

**29.** Answers may vary

**31.** $4(x+y) = 4x+4y$

**33.** $9(x-6) = 9x-9 \cdot 6 = 9x-54$

**35.** $2(3x+5) = 2(3x)+2(5) = 6x+10$

**37.** $7(4x-3) = 7(4x)-7(3) = 28x-21$

**39.** $3(6+x) = 3(6)+3x = 18+3x$

**41.** $-2(y-z) = -2y-(-2)z = -2y+2z$

**43.** $-7(3y+5) = -7(3y)+(-7)(5) = -21y-35$

**45.** $5(x+4m+2) = 5x+5(4m)+5(2)$
$= 5x+20m+10$

**47.** $-4(1-2m+n) = -4(1)-(-4)(2m)+(-4)n$
$= -4+8m-4n$

**49.** $-(5x+2) = -1(5x+2) = -1(5x)+(-1)(2)$
$= -5x-2$

**51.** $-(r-3-7p) = -1(r-3-7p)$
$= -1r-(-1)(3)-(-1)(7p)$
$= -r+3+7p$

**53.** $\dfrac{1}{2}(6x+8) = \dfrac{1}{2}(6x)+\dfrac{1}{2}(8)$
$= \left(\dfrac{1}{2}\cdot 6\right)x+\left(\dfrac{1}{2}\cdot 8\right) = 3x+4$

**55.** $-\dfrac{1}{3}(3x-9y) = -\dfrac{1}{3}(3x)-\left(-\dfrac{1}{3}\right)(9y)$
$= \left(-\dfrac{1}{3}\cdot 3\right)x-\left(-\dfrac{1}{3}\cdot 9\right)y = -1\cdot x+3\cdot y$
$= -x+3y$

**57.** $3(2r+5)-7 = 3(2r)+3(5)-7$
$= 6r+15+(-7) = 6r+8$

**59.** $-9(4x+8)+2 = -9(4x)+(-9)(8)+2$
$= -36x-72+2 = -36x-70$

**61.** $-4(4x+5)-5 = -4(4x)+(-4)(5)-5$
$= -16x+(-20)+(-5) = -16x-25$

**63.** $4\cdot 1+4\cdot y = 4(1+y)$

**65.** $11x+11y = 11(x+y)$

**67.** $(-1)\cdot 5+(-1)\cdot x = -1(5+x) = -(5+x)$

**69.** $30a+30b = 30(a+b)$

**71.** Commutative property of multiplication

**73.** Associative property of addition

**75.** Distributive property

**77.** Associative property of multiplication

**79.** Identity element of addition

**81.** Distributive property

**83.** Associative and commutative properties of multiplication

**85.**

| Expression | Opposite | Reciprocal |
|---|---|---|
| 8 | −8 | $\dfrac{1}{8}$ |

**87.**

| Expression | Opposite | Reciprocal |
|---|---|---|
| $x$ | $-x$ | $\dfrac{1}{x}$ |

**89.**

| Expression | Opposite | Reciprocal |
|---|---|---|
| $2x$ | $-2x$ | $\dfrac{1}{2x}$ |

**91.** No

**93.** Yes

**95.** Answers may vary

**Exercise Set 1.9**

**1.** approximately 7.8 million

**3.** 2002

**5.** Red; 23 shades

**7.** $20 - 11 = 9$ more shades

**9.** France

**11.** France, U.S., Spain

**13.** 39 million

**15.** Africa/Middle East, 11 million users

**17.** 187 million users

**19.** Approximately 59 beats per minute

**21.** Approximately 26 beats per minute

**23.** 74,800

**25.** 2000; 76,600

**27.** 2003

**29.** Answers may vary

**31.** 20

**33.** 1985

**35.** 1997

**37.** 18 million

**39.** 63 million

**41.** 1900

**43.** 27 million

**45.** Answers may vary

**47.** 2001, 2002, 2003

**49.** 46 million

**51.** 30° north, 90° west

**53.** 40° north, 104° west

**Chapter 1 Review**

**1.** $8 < 10$

**2.** $7 > 2$

**3.** $-4 > -5$

**4.** $\dfrac{12}{2} > -8$

**5.** $|-7| < |-8|$

**6.** $|-9| > -9$

**7.** $-|-1| \le -1$

**8.** $|-14| = -(-14)$

**9.** $1.2 > 1.02$

**10.** $-\dfrac{3}{2} < -\dfrac{3}{4}$

**11.** $4 \ge -3$

**12.** $6 \ne 5$

**13.** $0.03 < 0.3$

**14.** $50 > 40$

**15. a.** The natural numbers are 1 and 3.
    **b.** The whole numbers are 0, 1, and 3.
    **c.** The integers are $-6$, 0, 1, and 3.
    **d.** The rational numbers are $-6, 0, 1,$

        $1\dfrac{1}{2}$, 3, and 9.62.

    **e.** The irrational number is $\pi$.
    **f.** The real numbers are all numbers
       in the given set.

**16. a.** The natural numbers are 2 and 5.
    **b.** The whole numbers are 2 and 5.
    **c.** The integers are $-3, 2,$ and 5.
    **d.** The rational numbers are $-3, -1.6,$

        $2, 5, \dfrac{11}{2},$ and 15.1.

    **e.** The irrational numbers are $\sqrt{5}$ and $2\pi$.
    **f.** The real numbers are all numbers
       in the given set.

**17.** Friday

**18.** Wednesday

**19.** $36 = 2 \cdot 2 \cdot 3 \cdot 3$

**20.** $120 = 2 \cdot 2 \cdot 2 \cdot 3 \cdot 5$

**21.** $\dfrac{8}{15} \cdot \dfrac{27}{30} = \dfrac{2 \cdot 4 \cdot 3 \cdot 3 \cdot 3}{3 \cdot 5 \cdot 2 \cdot 3 \cdot 5} = \dfrac{12}{25}$

**22.** $\dfrac{7}{8} \div \dfrac{21}{32} = \dfrac{7}{8} \cdot \dfrac{32}{21} = \dfrac{7 \cdot 8 \cdot 4}{8 \cdot 3 \cdot 7} = \dfrac{4}{3}$

**23.** $\dfrac{7}{15} + \dfrac{5}{6} = \dfrac{7 \cdot 2}{15 \cdot 2} + \dfrac{5 \cdot 5}{6 \cdot 5} = \dfrac{14}{30} + \dfrac{25}{30}$

    $= \dfrac{14 + 25}{30} = \dfrac{39}{30} = \dfrac{3 \cdot 13}{3 \cdot 10} = \dfrac{13}{10}$

**24.** $\dfrac{3}{4} - \dfrac{3}{20} = \dfrac{3 \cdot 5}{4 \cdot 5} - \dfrac{3}{20} = \dfrac{15}{20} - \dfrac{3}{20} = \dfrac{15 - 3}{20}$

    $= \dfrac{12}{20} = \dfrac{3 \cdot 4}{5 \cdot 4} = \dfrac{3}{5}$

**25.** $2\dfrac{3}{4} + 6\dfrac{5}{8} = \dfrac{11}{4} + \dfrac{53}{8} = \dfrac{11 \cdot 2}{4 \cdot 2} + \dfrac{53}{8}$

    $= \dfrac{22 + 53}{8} = \dfrac{75}{8} = 9\dfrac{3}{8}$

**26.** $7\dfrac{1}{6} - 2\dfrac{2}{3} = \dfrac{43}{6} - \dfrac{8}{3} = \dfrac{43}{6} - \dfrac{8 \cdot 2}{3 \cdot 2}$

    $= \dfrac{43 - 16}{6} = \dfrac{27}{6} = \dfrac{9 \cdot 3}{2 \cdot 3} = \dfrac{9}{2} = 4\dfrac{1}{2}$

**27.** $5 \div \dfrac{1}{3} = 5 \cdot \dfrac{3}{1} = 15$

**28.** $2 \cdot 8\dfrac{3}{4} = 2 \cdot \dfrac{35}{4} = \dfrac{2 \cdot 35}{2 \cdot 2} = \dfrac{35}{2} = 17\dfrac{1}{2}$

**29.** $1 - \dfrac{112}{6} - \dfrac{1}{4} = \dfrac{12}{12} - \dfrac{1 \cdot 2}{6 \cdot 2} - \dfrac{1 \cdot 3}{4 \cdot 3}$

    $= \dfrac{12 - 2 - 3}{12} = \dfrac{7}{12}$

The unknown part is $\dfrac{7}{12}$

**30.** $P = 2l + 2w$

$$P = 2\left(1\frac{1}{3}\right) + 2\left(\frac{7}{8}\right) = \frac{2}{1} \cdot \frac{4}{3} + \frac{2}{1} \cdot \frac{7}{8}$$

$$= \frac{8}{3} + \frac{14}{8} = \frac{8 \cdot 8}{3 \cdot 8} + \frac{14 \cdot 3}{8 \cdot 3} = \frac{64 + 42}{24}$$

$$= \frac{106}{24} = 4\frac{10}{24} = 4\frac{5}{12} \text{ meters}$$

$A = lw$

$$A = 1\frac{1}{3} \cdot \frac{7}{8} = \frac{4}{3} \cdot \frac{7}{8} = \frac{4 \cdot 7}{3 \cdot 2 \cdot 4}$$

$$= \frac{7}{6} = 1\frac{1}{6} \text{ sq. meters}$$

**31.** $P =$ the sum of the lengths of the sides

$$P = \frac{5}{11} + \frac{8}{11} + \frac{3}{11} + \frac{3}{11} + \frac{2}{11} + \frac{5}{11} = \frac{26}{11}$$

$$= 2\frac{4}{11} \text{ in.}$$

$A =$ the sum of the two areas, each
given by $lw$

$$A = \frac{5}{11} \cdot \frac{5}{11} + \frac{3}{11} \cdot \frac{3}{11} = \frac{25}{121} + \frac{9}{121}$$

$$= \frac{34}{121} \text{ sq. in.}$$

**32.** $7\frac{1}{2} - 6\frac{1}{8} = \frac{15}{2} - \frac{49}{8} = \frac{15 \cdot 4}{2 \cdot 4} - \frac{49}{8}$

$$= \frac{60 - 49}{8} = \frac{11}{8} = 1\frac{3}{8} \text{ ft}$$

**33.** $1\frac{1}{8} + 1\frac{13}{16} = \frac{9}{8} + \frac{29}{16} = \frac{9 \cdot 2}{8 \cdot 2} + \frac{29}{16}$

$$= \frac{18 + 29}{16} = \frac{47}{16} = 2\frac{15}{16} \text{ lb.}$$

**34.** $1\frac{1}{2} + 1\frac{11}{16} + 1\frac{3}{4} + 1\frac{5}{8} + 1\frac{1}{8} + \frac{11}{16}$

$$= \frac{3}{2} + \frac{27}{16} + \frac{7}{4} + \frac{13}{8} + \frac{9}{8} + \frac{11}{16}$$

$$= \frac{3 \cdot 8}{2 \cdot 8} + \frac{27}{16} + \frac{7 \cdot 4}{4 \cdot 4} + \frac{13 \cdot 2}{8 \cdot 2} + \frac{9 \cdot 2}{8 \cdot 2} + \frac{11}{16}$$

$$= \frac{24 + 27 + 28 + 26 + 18 + 11}{16}$$

$$= \frac{134}{16} = 8\frac{3}{8} \text{ lb}$$

**35.** Total weight = weight of girls
+ weight of boys

$$8\frac{3}{8} + 2\frac{15}{16} = \frac{67}{8} + \frac{47}{16} = \frac{67 \cdot 2}{8 \cdot 2} + \frac{47}{16}$$

$$= \frac{134 + 47}{16} = \frac{181}{16} = 11\frac{5}{16} \text{ lb.}$$

**36.** Jioke

**37.** Odera

**38.** $1\frac{13}{16} - \frac{11}{16} = \frac{29}{16} - \frac{11}{16} = \frac{29 - 11}{16}$

$$= \frac{18}{16} = 1\frac{2}{6} = 1\frac{1}{8} \text{ lb}$$

**39.** $5\frac{1}{2} - 1\frac{5}{8} = \frac{11}{2} - \frac{13}{8} = \frac{11 \cdot 4}{2 \cdot 4} - \frac{13}{8}$

$$= \frac{44 - 13}{8} = \frac{31}{8} = 3\frac{7}{8} \text{ lb}$$

**40.** $4\frac{5}{32} - 1\frac{1}{8} = \frac{133}{32} - \frac{9}{8} = \frac{133}{32} - \frac{9 \cdot 4}{8 \cdot 4}$

$$= \frac{133 - 36}{32} = \frac{97}{32} = 3\frac{1}{32} \text{ lb}$$

**41.** $2^4 = 2 \cdot 2 \cdot 2 \cdot 2 = 16$

**42.** $5^2 = 5 \cdot 5 = 25$

**43.** $\left(\dfrac{2}{7}\right)^2 = \dfrac{2}{7} \cdot \dfrac{2}{7} = \dfrac{4}{49}$

**44.** $\left(\dfrac{3}{4}\right)^3 = \dfrac{3}{4} \cdot \dfrac{3}{4} \cdot \dfrac{3}{4} = \dfrac{27}{64}$

**45.** $6 \cdot 3^2 + 2 \cdot 8 = 6 \cdot 9 + 2 \cdot 8 = 54 + 16 = 70$

**46.** $68 - 5 \cdot 2^3 = 68 - 5 \cdot 8 = 68 - 40 = 28$

**47.** $3(1 + 2 \cdot 5) + 4 = 3(1 + 10) + 4 = 3(11) + 4$
$$= 33 + 4 = 37$$

**48.** $8 + 3(2 \cdot 6 - 1) = 8 + 3(12 - 1) = 8 + 3(11)$
$$= 8 + 33 = 41$$

**49.** $\dfrac{4 + |6 - 2| + 8^2}{4 + 6 \cdot 4} = \dfrac{4 + |4| + 64}{4 + 24} = \dfrac{4 + 4 + 64}{4 + 24}$
$$= \dfrac{72}{28} = \dfrac{4 \cdot 18}{4 \cdot 7} = \dfrac{18}{7}$$

**50.** $5[3(2 + 5) - 5] = 5[3(7) - 5] = 5[21 - 5]$
$$= 5[16] = 80$$

**51.** $20 - 12 = 2 \cdot 4$

**52.** $\dfrac{9}{2} > -5$

**53.** Let $x = 6$ and $y = 2$.
$$2x + 3y = 2(6) + 3(2) = 12 + 6 = 18$$

**54.** Let $x = 6$, $y = 2$, and $z = 8$.
$$x(y + 2z) = 6[2 + 2(8)] = 6[2 + 16]$$
$$= 6[18] = 108$$

**55.** Let $x = 6$, $y = 2$, and $z = 8$.
$$\dfrac{x}{y} + \dfrac{z}{2y} = \dfrac{6}{2} + \dfrac{8}{2(2)} = \dfrac{6}{2} + \dfrac{8}{4} = 3 + 2 = 5$$

**56.** Let $x = 6$ and $y = 2$.
$$x^2 - 3y^2 = (6)^2 - 3(2)^2 = 36 - 3(4)$$
$$= 36 - 12 = 36 + (-12) = 24$$

**57.** Let $a = 37$ and $b = 80$.
$$180 - a - b = 180 - 37 - 80$$
$$= 180 + (-37) + (-80) = 143 + (-80) = 63°$$

**58.** Let $x = 3$.
$$7x - 3 = 18$$
$$7(3) - 3 \overset{?}{=} 18$$
$$21 - 3 \overset{?}{=} 18$$
$$18 = 18, \text{ true}$$
3 is a solution to the equation.

**59.** Let $x = 1$.
$$3x^2 + 4 = x - 1$$
$$3(1)^2 + 4 \overset{?}{=} 1 - 1$$
$$3 + 7 \overset{?}{=} 0$$
$$10 = 0, \text{ false}$$
1 is not a solution to the equation.

**60.** The additive inverse of $-9$ is $9$.

**61.** The additive inverse of $\dfrac{2}{3}$ is $-\dfrac{2}{3}$.

**62.** The additive inverse of $|-2|$ is $-2$
since $|-2| = 2$.

**63.** The additive inverse of $-|-7|$ is $7$
since $-|-7| = -7$.

**64.** $-15 + 4 = -11$

**65.** $-6 + (-11) = -17$

**66.** $\dfrac{1}{16} + \left(-\dfrac{1}{4}\right) = \dfrac{1}{16} + \left(-\dfrac{1 \cdot 4}{4 \cdot 4}\right)$
$\qquad = \dfrac{1}{16} + \left(-\dfrac{4}{16}\right) = -\dfrac{3}{16}$

**67.** $-8 + |-3| = -8 + 3 = -5$

**68.** $-4.6 + (-9.3) = -13.9$

**69.** $-2.8 + 6.7 = 3.9$

**70.** $-282 + 728 = 446$ feet

**71.** $6 - 20 = 6 + (-20) = -14$

**72.** $-3.1 - 8.4 = -3.1 + (-8.4) = -11.5$

**73.** $-6 - (-11) = -6 + 11 = 5$

**74.** $4 - 15 = 4 + (-15) = -11$

**75.** $-21 - 16 + 3(8 - 2)$
$= -21 + (-16) + 3\left[8 + (-2)\right]$
$= -21 + (-16) + 3[6] = -21 + (-16) + 18$
$= -37 + 18 = -19$

**76.** $\dfrac{11 - (-9) + 6(8 - 2)}{2 + 3 \cdot 4} = \dfrac{11 + 9 + 6\left[8 + (-2)\right]}{2 + 3 \cdot 4}$
$= \dfrac{11 + 9 + 6[6]}{2 + 3 \cdot 4} = \dfrac{11 + 9 + 36}{2 + 12} = \dfrac{56}{14} = 4$

**77.** Let $x = 3$, $y = -6$, and $z = -9$.
$2x^2 - y + z = 2(3)^2 - (-6) + (-9)$
$= 2(9) + 6 + (-9) = 18 + 6 + (-9)$
$= 24 + (-9) = 15$

**78.** Let $x = 3$ and $y = -6$.
$\dfrac{y - x + 5x}{2x} = \dfrac{y + 4x}{2x} = \dfrac{-6 + 4(3)}{2(3)}$
$= \dfrac{-6 + 12}{6} = \dfrac{6}{6} = 1$

**79.** The multiplicative inverse of $-6$ is $-\dfrac{1}{6}$
since $-6 \cdot -\dfrac{1}{6} = 1$.

**80.** The multiplicative inverse of $\dfrac{3}{5}$ is $\dfrac{5}{3}$
since $\dfrac{3}{5} \cdot \dfrac{5}{3} = 1$.

**81.** $6(-8) = -48$

**82.** $(-2)(-14) = 28$

**83.** $\dfrac{-18}{-6} = 3$

**84.** $\dfrac{42}{-3} = -14$

**85.** $\dfrac{4 \cdot (-3) + (-8)}{2 + (-2)} = \dfrac{-12 + (-8)}{2 + (-2)} = \dfrac{-20}{0}$

The expression is undefined.

**86.** $\dfrac{3(-2)^2 - 5}{-14} = \dfrac{3(4) - 5}{-14} = \dfrac{12 - 5}{-14} = \dfrac{7}{-14} = -\dfrac{1}{2}$

**87.** $\dfrac{-6}{0}$ is undefined

**88.** $\dfrac{0}{-2} = 0$

**89.** $-4^2 - (-3 + 5) \div (-1) \cdot 2$
$= -16 - (2) \div (-1) \cdot 2 = -16 + 2 \cdot 2$
$= -16 + 4 = -12$

**90.** $-5^2 - (2 - 20) \div (-3) \cdot 3$
$= -25 - (-18) \div (-3) \cdot 3 = -25 - 6 \cdot 3$
$= -25 - 18 = -43$

**91.** Let $x = -5$ and $y = -2$.
$x^2 - y^4 = (-5)^2 - (-2)^4 = 25 - 16 = 9$

**92.** Let $x = -5$ and $y = -2$.
$x^2 - y^3 = (-5)^2 - (-2)^3 = 25 - (-8)$
$= 25 + 8 = 33$

**93.** $\dfrac{-9 + (-7) + 1}{3} = \dfrac{-15}{3} = -5$

Her average score per round
was 5 under par.

**94.** $\dfrac{-1 + 0 + (-3) + 0}{4} = \dfrac{-4}{4} = -1$

His average score per round
was 1 under par.

**95.** Commutative property of addition

**96.** Multiplicative identity property

**97.** Distributive property

**98.** Additive inverse property

**99.** Associative property of addition

**100.** Commutative property of multiplication

**101.** Distributive property

**102.** Associative property of multiplication

**103.** Multiplicative inverse property

**104.** Additive identity property

**105.** Commutative property of addition

**106.** 128 million

**107.** 19 million

**108.** 2000

**109.** The number of subscribers is increasing.

**110.** cross country skiing, 181 calories

**111.** sleeping, 15 calories

**112.** 52 more calories

**113.** 10 more calories

**Chapter 1 Test**

**1.** $|-7| > 5$

**2.** $(9+5) \geq 4$

**3.** $-13 + 8 = -5$

**4.** $-13 - (-2) = -13 + 2 = -11$

**5.** $12 \div 4 \cdot 3 - 6 \cdot 2 = 3 \cdot 3 - 12 = 9 - 12 = -3$

**6.** $(13)(-3) = -39$

**7.** $(-6)(-2) = 12$

**8.** $\dfrac{|-16|}{-8} = \dfrac{16}{-8} = -2$

**9.** $\dfrac{-8}{0}$ is undefined

**10.** $\dfrac{|-6| + 2}{5 - 6} = \dfrac{6 + 2}{5 + (-6)} = \dfrac{8}{-1} = -8$

**11.** $\dfrac{1}{2} - \dfrac{5}{6} = \dfrac{1 \cdot 3}{2 \cdot 3} - \dfrac{5}{6} = \dfrac{3 - 5}{6} = \dfrac{-2}{6} = -\dfrac{1}{3}$

**12.** $-1\dfrac{1}{8} + 5\dfrac{3}{4} = -\dfrac{9}{8} + \dfrac{23}{4} = -\dfrac{9}{8} + \dfrac{2 \cdot 23}{2 \cdot 4}$
$\qquad = \dfrac{-9 + 46}{8} = \dfrac{37}{8} = 4\dfrac{5}{8}$

**13.** $(2-6) \div \dfrac{-2-6}{-3-1} - \dfrac{1}{2} = (2-6) \div \dfrac{-8}{-4} - \dfrac{1}{2}$
$\qquad = -4 \div 2 - \dfrac{1}{2} = -2 - \dfrac{1}{2} = -2\dfrac{1}{2}$

**14.** $3(-4)^2 - 80 = 3(16) - 80 = 48 + (-80) = -32$

**15.** $6[5 + 2(3-8) - 3]$
$\qquad = 6\{5 + 2[3 + (-8)] + (-3)\}$
$\qquad = 6\{5 + 2[-5] + (-3)\}$
$\qquad = 6\{5 + (-10) + (-3)\}$
$\qquad = 6\{-5 + (-3)\} = 6\{-8\} = -48$

**16.** $\dfrac{-12 + 3 \cdot 8}{4} = \dfrac{-12 + 24}{4} = \dfrac{12}{4} = 3$

**17.** $\dfrac{(-2)(0)(-3)}{-6} = \dfrac{0(-3)}{-6} = \dfrac{0}{-6} = 0$

**18.** $-3 > -7$

**19.** $4 > -8$

**20.** $2 < |-3|$

**21.** $|-2| = -1 - (-3)$

**22.** $2221 < 10{,}993$

**23. a.** The natural numbers are 1 and 7.

   **b.** The whole numbers are 0, 1 and 7.

   **c.** The integers are $-5, -1, 0, 1,$ and 7.

   **d.** The rational numbers are

      $-5, -1, \dfrac{1}{4}, 0, 1, 7,$ and 11.6.

   **e.** The irrational numbers are $\sqrt{7}$ and $3\pi$.

   **f.** The real numbers are all numbers in the given set.

**24.** Let $x = 6$ and $y = -2$.

$$x^2 + y^2 = (6)^2 + (-2)^2$$
$$= 36 + 4$$
$$= 40$$

**25.** Let $x = 6$, $y = -2$ and $z = -3$.

$$x + yz = 6 + (-2)(-3)$$
$$= 6 + 6$$
$$= 12$$

**26.** Let $x = 6$ and $y = -2$.

$$2 + 3x - y = 2 + 3(6) - (-2)$$
$$= 2 + 18 + 2$$
$$= 20 + 2$$
$$= 22$$

**27.** Let $x = 6$, $y = -2$ and $z = -3$.

$$\frac{y + z - 1}{x} = \frac{-2 + (-3) - 1}{6}$$
$$= \frac{-5 + (-1)}{6}$$
$$= \frac{-6}{6}$$
$$= -1$$

**28.** Associative property of addition

**29.** Commutative property of multiplication

**30.** Distributive property

**31.** Multiplicative inverse property

**32.** The opposite of $-9$ is 9.

**33.** The reciprocal of $-\dfrac{1}{3}$ is $-3$.

**34.** Second down

**35.** Gains: 5, 29

    Losses: $-10, -2$

    Total gain or loss

$$= 5 + (-10) + (-2) + 29 = (-5) + (-2) + 29$$
$$= -7 + 29 = 22 \text{ yards gained.}$$

    Yes, they scored a touchdown.

**36.** Since $-14 + 31 = 17,$ the temperature at noon was $17°$

**37.** $356 + 460 + (-166) = 650$

    The net income was $650 million

**38.** Change in value per share $= -1.50$

    Change in total value $= 28(-1.50) = -420$

    Total loss of $420

**39.** 2000, 27 billion pounds

**40.** 1940, 7 billion pounds

**41.** 1970, 1980, 1990, 2000

**42.** 1970

**43.** Indiana, 25.2 million tons

**44.** Texas, 5 million tons

**45.** 16 million tons

**46.** $8 - 5 = 3$
3 million tons more

# Chapter 2

**Mental Math 2.1**

1. $-7$

2. $3$

3. $1$

4. $-1$

5. $17$

6. $1.2$

7. $\dfrac{1}{8}$

8. $-\dfrac{5}{3}$

9. $-\dfrac{2}{3}$

10. Like terms

11. Like terms

12. Unlike terms

13. Like terms

14. Like terms

15. Unlike terms

**Exercise Set 2.1**

1. $7y + 8y = (7 + 8)y = 15y$

3. $-9n - 6n = (-9 - 6)n = -15n$

5. $3.5t - 4.5t = (3.5 - 4.5)t = -1t$ or $-t$

7. $8w - w + 6w = (8 - 1 + 6)w = 13w$

9. $3b - 5 - 10b - 4 = 3b - 10b - 5 - 4$
$$= (3 - 10)b - 9 = -7b - 9$$

11. $m - 4m + 2m - 6 = (1 - 4 + 2)m - 6$
$$= -m - 6$$

13. $5(y - 4) = 5(y) - 5(4) = 5y - 20$

15. $7(d - 3) + 10 = 7d - 21 + 10 = 7d - 11$

17. $-(3x - 2y + 1) = -3x + 2y - 1$

19. $5(x + 2) - (3x - 4) = 5x + 10 - 3x + 4$
$$= 2x + 14$$

21. Answers may vary

23. $(4x - 10) + (6x + 7) = 4x - 10 + 6x + 7$
$$= 10x - 3$$

25. $(3x - 8) - (7x + 1) = 3x - 8 - 7x - 1 = -4x - 9$

27. $7x^2 + 8x^2 - 10x^2 = (7 + 8 - 10)x^2 = 5x^2$

29. $6x - 5x + x - 3 + 2x = 6x - 5x + x + 2x - 3$
$$= (6 - 5 + 1 + 2)x - 3 = 4x - 3$$

**31.** $-5+8(x-6)=-5+8x-48=8x-53$

**33.** $6.2x-4+x-1.2=6.2x+x-4-1.2$
$=(6.2+1)x-5.2=7.2x-5.2$

**35.** $2k-k-6=(2-1)k-6=k-6$

**37.** $0.5(m+2)+0.4m=0.5m+1-0.4m$
$=0.9m+1$

**39.** $-4(3y-4)=-12y+16$

**41.** $3(2x-5)-5(x-4)=6x-15-5x+20$
$=x+5$

**43.** $6x+0.5-4.3x-0.4x+3$
$=6x-4.3x-0.4x+0.5+3$
$=(6-4.3-0.4)x+3.5$
$=1.3x+3.5$

**45.** $-2(3x-4)+7x-6=-6x+8+7x-6$
$=x+2$

**47.** $-9x+4x+18-10x=-9x+4x-10x+18$
$=(-9+4-10)x+18=-15x+18$

**49.** $5k-(3k-10)=5k-3k+10=2k+10$

**51.** $(3x+4)-(6x-1)=3x+4-6x+1$
$=-3x+5$

**53.**

| twice a number | decreased by | 4 |
|:---:|:---:|:---:|
| ↓ | ↓ | ↓ |
| $2x$ | $-$ | $4$ |

**55.**

| three-fourths of a number | increased by | 12 |
|:---:|:---:|:---:|
| ↓ | ↓ | ↓ |
| $\dfrac{3}{4}x$ | $+$ | $12$ |

**57.**

| 5 times a number | added to | $-2$ | added to | 7 times a number |
|:---:|:---:|:---:|:---:|:---:|
| ↓ | ↓ | ↓ | ↓ | ↓ |
| $5x$ | $+$ | $-2$ | $+$ | $7x$ |

$5x+(-2)+7x=12x-2$

**59.** $(m-9)-(5m-6)=m-9-5m+6=-4m-3$

**61.**

| 8 times | the sum of a number and 6 |
|:---:|:---:|
| ↓ | ↓ |
| $8$ | $(x+6)$ |

$8(x+6)=8x+48$

**63.**

| double a number | minus | the sum of the number and 10 |
|:---:|:---:|:---:|
| ↓ | ↓ | ↓ |
| $2x$ | $-$ | $(x+10)$ |

$2x-(x+10)=2x-x-10=x-10$

**65.**

| 7 | multiplied by | the quotient of a number and 6 |
|:---:|:---:|:---:|
| ↓ | ↓ | ↓ |
| $7$ | $\times$ | $\dfrac{x}{6}$ |

$7\left(\dfrac{x}{6}\right)=\dfrac{7x}{6}$

**67.**

| 2 | added to | 3 times a number | added to | −9 | added to | 4 times a number |
|---|---|---|---|---|---|---|
| ↓ ↓ | | ↓ | ↓ | ↓ ↓ | | ↓ |
| 2 | + | $3x$ | + | −9 | + | $4x$ |

$2 + 3x + (-9) + 4x = 7x - 7$

**69.** $5x + (4x - 1) + 5x + (4x - 1)$

$= 5x + 4x - 1 + 5x + 4x - 1$

$= (18x - 2)$ ft.

**71.** $y - x^2 = 3 - (-1)^2 = 3 - 1 = 2$

**73.** $a - b^2 = 2 - (-5)^2 = 2 - 25 = -23$

**75.** $yz - y^2 = (-5)(0) - (-5)^2 = 0 - 25 = -25$

**77.** $1 \text{ cone} + 1 \text{ cylinder} \overset{?}{=} 3 \text{ cubes}$

$1 \text{ cube} + 2 \text{ cubes} \overset{?}{=} 3 \text{ cubes}$

$3 \text{ cubes} = 3 \text{ cubes: Balanced}$

**79.** $2 \text{ cylinders} + 1 \text{ cube} \overset{?}{=} 3 \text{ cones} + 2 \text{ cubes}$

$2 \cdot 2 \text{ cubes} + 1 \text{ cube} \overset{?}{=} 3 \text{ cubes} + 2 \text{ cubes}$

$4 \text{ cubes} + 1 \text{ cube} \overset{?}{=} 3 \text{ cubes} + 2 \text{ cubes}$

$5 \text{ cubes} = 5 \text{ cubes: Balanced}$

**81.** $12(x + 2) + (3x - 1) = 12x + 24 + 3x - 1$

$= 15x + 23$

The total length is $(15x + 23)$ inches

**83.** $5b^2c^3 + 8b^3c^2 - 7b^3c^2 = 5b^2c^3 + b^3c^2$

**85.** $3x - (2x^2 - 6x) + 7x^2$

$= 3x - 2x^2 + 6x + 7x^2$

$= 5x^2 + 9x$

**87.** $-(2x^2y + 3z) + 3z - 5x^2y$

$= -2x^2y - 3z + 3z - 5x^2y$

$= -7x^2y$

**Mental Math 2.2**

**1.** 2,   **2.** 3,   **3.** 12,   **4.** 18,   **5.** 17,   **6.** 21

**Exercise Set 2.2**

**1.** $x + 7 = 10$

$x + 7 - 7 = 10 - 7$

$x = 3$

Check: $x + 7 = 10$

$3 + 7 \overset{?}{=} 10$

$10 = 10$

The solution is 3

**3.** $x - 2 = -4$

$x - 2 + 2 = -4 + 2$

$x = -2$

Check: $x - 2 = -4$

$-2 - 2 \overset{?}{=} -4$

$-4 = -4$

The solution is $-2$

**5.** $-2 = t - 5$

$-2 + 5 = t - 5 + 5$

$3 = t$

Check: $-2 = t - 5$

$-2 \overset{?}{=} 3 - 5$

$-2 = -2$

The solution is 3

**7.**    $r - 8.6 = -8.1$

$r - 8.6 + 8.6 = -8.1 + 8.6$

$r = 0.5$

Check: $x - 8.6 = -8.1$

$0.5 - 8.6 \overset{?}{=} -8.1$

$-8.1 = -8.1$

The solution is 0.5

**9.**    $\dfrac{3}{4} = \dfrac{1}{3} + f$

$\dfrac{3}{4} - \dfrac{1}{3} = \dfrac{1}{3} - \dfrac{1}{3} + f$

$\dfrac{9}{12} - \dfrac{4}{12} = f$

$\dfrac{5}{12} = f$

Check: $\dfrac{3}{4} = \dfrac{1}{3} + f$

$\dfrac{3}{4} \overset{?}{=} \dfrac{1}{3} + \dfrac{5}{12}$

$\dfrac{3}{4} \overset{?}{=} \dfrac{4}{12} + \dfrac{5}{12}$

$\dfrac{3}{4} \overset{?}{=} \dfrac{9}{12}$

$\dfrac{3}{4} = \dfrac{3}{4}$

The solution is $\dfrac{5}{12}$

**11.**    $5b - 0.7 = 6b$

$5b - 5b - 0.7 = 6b - 5b$

$-0.7 = b$

Check: $5b - 0.7 = 6b$

$5(-0.7) - 0.7 \overset{?}{=} 6(-0.7)$

$-3.5 - 0.7 \overset{?}{=} -4.2$

$-4.2 = -4.2$

The solution is $-0.7$

**13.**    $7x - 3 = 6x$

$7x - 6x - 3 = 6x - 6x$

$x - 3 = 0$

$x - 3 + 3 = 0 + 3$

$x = 3$

Check: $7x - 3 = 6x$

$7(3) - 3 \overset{?}{=} 6(3)$

$21 - 3 \overset{?}{=} 18$

$18 = 18$

The solution is 3

**15.** Answers may vary.

**17.**    $-8 = p - 4$

$-8 + 4 = p - 4 + 4$

$-4 = p$

Check: $-8 = p - 4$

$-8 \overset{?}{=} (-4) - 4$

$-8 = -8$

The solution is $-4$

**19.** $7x + 2x = 8x - 3$

$9x = 8x - 3$

$9x - 8x = 8x - 8x - 3$

$x = -3$

Check: $7x + 2x = 8x - 3$

$7(-3) + 2(-3) \stackrel{?}{=} 8(-3) - 3$

$-21 - 6 \stackrel{?}{=} -24 - 3$

$-27 = -27$

The solution is $-3$

**21.**     $2y + 10 = 5y - 4y$

$2y + 10 = y$

$2y - y + 10 = y - y$

$y + 10 = 0$

$y + 10 - 10 = 0 - 10$

$y = -10$

Check: $2y + 10 = 5y - 4y$

$2(-10) + 10 \stackrel{?}{=} 5(-10) - 4(-10)$

$-20 + 10 \stackrel{?}{=} -50 + 40$

$-10 = -10$

The solution is $-10$

**23.**     $3x - 6 = 2x + 5$

$3x - 2x - 6 = 2x - 2x + 5$

$x - 6 = 5$

$x - 6 + 6 = 5 + 6$

$x = 11$

Check: $3x - 6 = 2x + 5$

$3(11) - 6 \stackrel{?}{=} 2(11) + 5$

$33 - 6 \stackrel{?}{=} 22 + 5$

$27 = 27$

The solution is 11

**25.**     $5x - \dfrac{1}{6} = 6x - \dfrac{5}{6}$

$5x - 5x - \dfrac{1}{6} = 6x - 5x - \dfrac{5}{6}$

$-\dfrac{1}{6} = x - \dfrac{5}{6}$

$-\dfrac{1}{6} + \dfrac{5}{6} = x - \dfrac{5}{6} + \dfrac{5}{6}$

$\dfrac{4}{6} = x$

$\dfrac{2}{3} = x$

Check: $5x - \dfrac{1}{6} = 6x - \dfrac{5}{6}$

$5\left(\dfrac{2}{3}\right) - \dfrac{1}{6} \stackrel{?}{=} 6\left(\dfrac{2}{3}\right) - \dfrac{5}{6}$

$\dfrac{10}{3} - \dfrac{1}{6} \stackrel{?}{=} \dfrac{12}{3} - \dfrac{5}{6}$

$\dfrac{19}{6} = \dfrac{19}{6}$

The solution is $\dfrac{2}{3}$

**27.** $8y + 2 - 6y = 3 + y - 10$

$$2y + 2 = y - 7$$
$$2y - y + 2 = y - y - 7$$
$$y + 2 = -7$$
$$y + 2 - 2 = -7 - 2$$
$$y = -9$$

Check: $8y + 2 - 6y = 3 + y - 10$

$$8(-9) + 2 - 6(-9) \overset{?}{=} 3 + (-9) - 10$$
$$-72 + 2 + 54 \overset{?}{=} -16$$
$$-16 = -16$$

The solution is $-9$

**29.** $-6.5 - 4x - 1.6 - 3x = -6x + 9.8$

$$-8.1 - 7x = -6x + 9.8$$
$$-8.1 - 7x + 7x = -6x + 7x + 9.8$$
$$-8.1 = x + 9.8$$
$$-8.1 - 9.8 = x + 9.8 - 9.8$$
$$-17.9 = x$$

Check: $-6.5 - 4x - 1.6 - 3x = -6x + 9.8$

$$-6.5 - 4(-17.9) - 1.6 - 3(-17.9)$$
$$\overset{?}{=} -6(-17.9) + 9.8$$
$$-6.5 + 71.6 - 1.6 - 3 \overset{?}{=} 107.4 + 9.8$$
$$117.2 = 117.2$$

The solution is $-17.9$

**31.** $\dfrac{3}{8}x - \dfrac{1}{6} = -\dfrac{5}{8}x - \dfrac{2}{3}$

$$\frac{3}{8}x + \frac{5}{8}x - \frac{1}{6} = -\frac{5}{8}x + \frac{5}{8}x - \frac{2}{3}$$
$$\frac{8}{8}x - \frac{1}{6} = -\frac{2}{3}$$
$$x - \frac{1}{6} + \frac{1}{6} = -\frac{2}{3} + \frac{1}{6}$$
$$x = -\frac{4}{6} + \frac{1}{6}$$
$$x = -\frac{3}{6}$$
$$x = -\frac{1}{2}$$

Check: $\dfrac{3}{8}x - \dfrac{1}{6} = -\dfrac{5}{8}x - \dfrac{2}{3}$

$$\frac{3}{8}\left(-\frac{1}{2}\right) - \frac{1}{6} \overset{?}{=} -\frac{5}{8}\left(-\frac{1}{2}\right) - \frac{2}{3}$$
$$-\frac{3}{16} - \frac{1}{6} \overset{?}{=} \frac{5}{16} - \frac{2}{3}$$
$$-\frac{9}{48} - \frac{8}{48} \overset{?}{=} \frac{15}{48} - \frac{32}{48}$$
$$-\frac{17}{48} = -\frac{17}{48}$$

The solution is $-\dfrac{1}{2}$

**33.** $2(x - 4) = x + 3$

$$2x - 8 = x + 3$$
$$2x - x - 8 = x - x + 3$$
$$x - 8 = 3$$
$$x - 8 + 8 = 3 + 8$$
$$x = 11$$

Check: $2(x-4) = x+3$

$$2(11-4) \overset{?}{=} 11+3$$

$$2(7) \overset{?}{=} 14$$

$$14 = 14$$

The solution is 11

**35.** $\quad 7(6+w) = 6(2+w)$

$$42+7w = 12+6w$$

$$42+7w-6w = 12+6w-6w$$

$$42+w = 12$$

$$42-42+w = 12-42$$

$$w = -30$$

Check: $7(6+w) = 6(2+w)$

$$7(6-30) \overset{?}{=} 6(2-30)$$

$$7(-24) \overset{?}{=} 6(-28)$$

$$-168 = -168$$

The solution is $-30$

**37.** $10-(2x-4) = 7-3x$

$$10-2x+4 = 7-3x$$

$$14-2x = 7-3x$$

$$14-2x+3x = 7-3x+3x$$

$$14+x = 7$$

$$14-14+x = 7-14$$

$$x = -7$$

Check: $10-(2x-4) = 7-3x$

$$10-(2(-7)-4) \overset{?}{=} 7-3(-7)$$

$$10-(-14-4) \overset{?}{=} 7+21$$

$$10-(-18) \overset{?}{=} 28$$

$$28 = 28$$

The solution is $-7$

**39.** $\quad -5(n-2) = 8-4n$

$$-5n+10 = 8-4n$$

$$-5n+5n+10 = 8-4n+5n$$

$$10 = 8+n$$

$$10-8 = 8-8+n$$

$$2 = n$$

Check: $-5(n-2) = 8-4n$

$$-5(2-2) \overset{?}{=} 8-4(2)$$

$$-5(0) \overset{?}{=} 8-8$$

$$0 = 0$$

The solution is 2

**41.** $\quad -3\left(x-\dfrac{1}{4}\right) = -4x$

$$-3x+\dfrac{3}{4} = -4x$$

$$-3x+4x+\dfrac{3}{4} = -4x+4x$$

$$x+\dfrac{3}{4} = 0$$

$$x+\dfrac{3}{4}-\dfrac{3}{4} = 0-\dfrac{3}{4}$$

$$x = -\dfrac{3}{4}$$

Check: $-3\left(x - \dfrac{1}{4}\right) = -4x$

$-3\left(-\dfrac{3}{4} - \dfrac{1}{4}\right) \stackrel{?}{=} -4\left(-\dfrac{3}{4}\right)$

$-3(-1) \stackrel{?}{=} 3$

$3 = 3$

The solution is $-\dfrac{3}{4}$

**43.** $3(n-5) - (6-2n) = 4n$

$3n - 15 - 6 + 2n = 4n$

$5n - 21 = 4n$

$5n - 4n - 21 = 4n - 4n$

$n - 21 = 0$

$n - 21 + 21 = 0 + 21$

$n = 21$

Check: $3(n-5) - (6-2n) = 4n$

$3(21-5) - (6-2(21)) \stackrel{?}{=} 4(21)$

$3(16) - (6-42) \stackrel{?}{=} 84$

$48 - (-36) \stackrel{?}{=} 84$

$84 = 84$

The solution is 21

**45.** $-2(x+6) + 3(2x-5) = 3(x-4) + 10$

$-2x - 12 + 6x - 15 = 3x - 12 + 10$

$4x - 27 = 3x - 2$

$4x - 3x - 27 = 3x - 3x - 2$

$x - 27 = -2$

$x - 27 + 27 = -2 + 27$

$x = 25$

Check:

$-2(x+6) + 3(2x-5) = 3(x-4) + 10$

$-2(25+6) + 3(2(25)-5) \stackrel{?}{=} 3(25-4) + 10$

$-2(31) + 3(50-5) \stackrel{?}{=} 3(21) + 10$

$-62 + 3(45) \stackrel{?}{=} 63 + 10$

$-62 + 135 \stackrel{?}{=} 63 + 10$

$73 = 73$

The solution is 25

**47.** $7(m-2) - 6(m+1) = -20$

$7m - 14 - 6m - 6 = -20$

$m - 20 = -20$

$m - 20 + 20 = -20 + 20$

$m = 0$

Check: $7(m-2) - 6(m+1) = -20$

$7(0-2) - 6(0+1) \stackrel{?}{=} -20$

$7(-2) - 6(+1) \stackrel{?}{=} -20$

$-14 - 6 \stackrel{?}{=} -20$

$-20 = -20$

The solution is 0

**49.** $0.8t + 0.2(t - 0.4) = 1.75$

$0.8t + 0.2t - 0.08 = 1.75$

$1t - 0.08 = 1.75$

$t - 0.08 + 0.08 = 1.75 + 0.08$

$t = 1.83$

Check: $0.8t + 0.2(t - 0.4) = 1.75$

$0.8(1.83) + 0.2(1.83 - 0.4) \overset{?}{=} 1.75$

$1.464 + 0.366 - 0.08 \overset{?}{=} 1.75$

$1.75 = 1.75$

The solution is 1.83

**51.** The other number is $20 - p$.

**53.** The length of the other piece is $(10 - x)$ feet

**55.** The supplement of the angle $x°$ is $(180 - x)°$.

**57.** Catarella received $(n + 284)$ votes.

**59.** The length of I-90 is $(m + 178.5)$ miles.

**61.** The area of the Sahara Desert is $7x$ square miles.

**63.** The reciprocal of $\dfrac{5}{8}$ is $\dfrac{8}{5}$ since $\dfrac{5}{8} \cdot \dfrac{8}{5} = 1$

**65.** The reciprocal of $2$ is $\dfrac{1}{2}$ since $2 \cdot \dfrac{1}{2} = 1$

**67.** The reciprocal of $-\dfrac{1}{9}$ is $-9$

since $-\dfrac{1}{9} \cdot -9 = 1$

**69.** $\dfrac{3x}{3} = x$

**71.** $-5\left(-\dfrac{1}{5}y\right) = y$

**73.** $\dfrac{3}{5}\left(\dfrac{5}{3}x\right) = x$

**75.** $180 - [x + (2x + 7)] = 180 - [x + 2x + 7]$

$= 180 - [3x + 7] = 180 - 3x - 7 = 173 - 3x$

The third angle is $(173 - 3x)°$.

**77.** $200 + 150 + 400 + x = 1000$

$750 + x = 1000$

$750 - 750 + x = 1000 - 750$

$x = 250$

The fluid needed by the patient is 250 ml

**79.** Answers may vary

**81.** $8.13 + 5.85y = 20.05y - 8.91$

Check $y = 1.2$

$8.13 + 5.85(1.2) = 20.05(1.2) - 8.91$

$8.13 + 7.02 = 24.06 - 8.91$

$15.15 = 15.15$

Solution

**83.** $7(z - 1.7) + 9.5 = 5(z + 3.2) - 9.2$

Check $z = 4.8$

$7(4.8 - 1.7) + 9.5 = 5(4.8 + 3.2) - 9.2$

$7(3.1) + 9.5 = 5(8.0) - 9.2$

$21.7 + 9.5 = 40.0 - 9.2$

$31.2 \neq 30.8$

Not a solution

**Mental Math 2.3**

**1.** $3a = 27$

   $a = 9$

**2.** $9c = 54$

   $c = 6$

**3.** $5b = 10$

   $b = 2$

**4.** $7t = 14$

   $t = 2$

**5.** $6x = -30$

   $x = -5$

**6.** $8r = -64$

   $r = -8$

**Exercise Set 2.3**

**1.** $-5x = 20$

   $\dfrac{-5x}{-5} = \dfrac{20}{-5}$

   $x = -4$

   Check: $-5x = 20$

   $-5(-4) \overset{?}{=} 20$

   $20 = 20$

   The solution is $-4$

**3.** $3x = 0$

   $\dfrac{3x}{3} = \dfrac{0}{3}$

   $x = 0$

Check: $3x = 0$

$3(0) \overset{?}{=} 0$

$0 = 0$

The solution is 0

**5.** $-x = -12$

   $\dfrac{-x}{-1} = \dfrac{-12}{-1}$

   $x = 12$

Check: $-x = -12$

$-(12) \overset{?}{=} -12$

$-12 = -12$

The solution is 12

**7.** $\dfrac{2}{3}x = -8$

   $\dfrac{3}{2}\left(\dfrac{2}{3}x\right) = \dfrac{3}{2}(-8)$

   $x = -12$

Check: $\dfrac{2}{3}x = -8$

$\dfrac{2}{3}(-12) \overset{?}{=} -8$

$-8 = -8$

The solution is $-12$

**9.** $\dfrac{1}{6}d = \dfrac{1}{2}$

   $6\left(\dfrac{1}{6}d\right) = 6\left(\dfrac{1}{2}\right)$

   $d = 3$

Check: $\dfrac{1}{6}d = \dfrac{1}{2}$

$\dfrac{1}{6}(3) \overset{?}{=} \dfrac{1}{2}$

$\dfrac{1}{2} = \dfrac{1}{2}$

The solution is 3

**11.**    $\dfrac{a}{2} = 1$

$2\left(\dfrac{a}{2}\right) = 2(1)$

$a = 2$

Check: $\dfrac{a}{2} = 1$

$\dfrac{2}{2} \overset{?}{=} 1$

$1 = 1$

The solution is 2

**13.**    $\dfrac{k}{-7} = 0$

$-7\left(\dfrac{k}{-7}\right) = -7(0)$

$k = 0$

Check: $\dfrac{k}{-7} = 0$

$\dfrac{0}{-7} \overset{?}{=} 0$

$0 = 0$

The solution is 0

**15.**   $1.7x = 10.71$

$\dfrac{1.7x}{1.7} = \dfrac{10.71}{1.7}$

$x = 6.3$

Check: $1.7x = 10.71$

$1.7(6.3) \overset{?}{=} 10.71$

$10.71 = 10.71$

The solution is 6.3

**17.**    $-x + 2 = 22$

$-x + 2 - 2 = 22 - 2$

$-x = 20$

$x = -20$

Check: $-x + 2 = 22$

$-(-20) + 2 \overset{?}{=} 22$

$20 + 2 \overset{?}{=} 22$

$22 = 22$

The solution is $-20$

**19.**    $6a + 3 = 3$

$6a + 3 - 3 = 3 - 3$

$6a = 0$

$\dfrac{6a}{6} = \dfrac{0}{6}$

$a = 0$

Check: $6a + 3 = 3$

$6(0) + 3 \overset{?}{=} 3$

$0 + 3 \overset{?}{=} 3$

$3 = 3$

The solution is 0

**21.**     $6x + 10 = -20$

$6x + 10 - 10 = -20 - 10$

$6x = -30$

$\dfrac{6x}{6} = \dfrac{-30}{6}$

$x = -5$

Check:  $6x + 10 = -20$

$6(-5) + 10 \overset{?}{=} -20$

$-30 + 10 \overset{?}{=} -20$

$-20 = -20$

The solution is $-5$

**23.**     $-2x + \dfrac{1}{2} = \dfrac{7}{2}$

$-2x + \dfrac{1}{2} - \dfrac{1}{2} = \dfrac{7}{2} - \dfrac{1}{2}$

$-2x = \dfrac{6}{2}$

$-2x = 3$

$\dfrac{-2x}{-2} = \dfrac{3}{-2}$

$x = -\dfrac{3}{2}$

Check:  $-2x + \dfrac{1}{2} = \dfrac{7}{2}$

$-2\left(-\dfrac{3}{2}\right) + \dfrac{1}{2} \overset{?}{=} \dfrac{7}{2}$

$3 + \dfrac{1}{2} \overset{?}{=} \dfrac{7}{2}$

$\dfrac{6}{2} + \dfrac{1}{2} \overset{?}{=} \dfrac{7}{2}$

$\dfrac{7}{2} = \dfrac{7}{2}$

The solution is $-\dfrac{3}{2}$

**25.**  $6z - 8 - z + 3 = 0$

$5z - 5 = 0$

$5z - 5 + 5 = 0 + 5$

$5z = 5$

$\dfrac{5z}{5} = \dfrac{5}{5}$

$z = 1$

Check:  $6z - 8 - z + 3 = 0$

$6(1) - 8 - (1) + 3 \overset{?}{=} 0$

$6 - 8 - 1 + 3 \overset{?}{=} 0$

$0 = 0$

The solution is 1

**27.**  $10 - 3x - 6 - 9x = 7$

$4 - 12x = 7$

$4 - 4 - 12x = 7 - 4$

$-12x = 3$

$\dfrac{-12x}{-12} = \dfrac{3}{-12}$

$x = -\dfrac{1}{4}$

Check: $10 - 3x - 6 - 9x = 7$

$$10 - 3\left(-\frac{1}{4}\right) - 6 - 9\left(-\frac{1}{4}\right) \overset{?}{=} 7$$

$$10 + \frac{3}{4} - 6 + \frac{9}{4} \overset{?}{=} 7$$

$$4 + \frac{12}{4} \overset{?}{=} 7$$

$$4 + 3 \overset{?}{=} 7$$

$$7 = 7$$

The solution is $-\dfrac{1}{4}$

**29.** $0.4x - 0.6x - 5 = 1$

$$-0.2x - 5 = 1$$

$$-0.2x - 5 + 5 = 1 + 5$$

$$-0.2x = 6$$

$$\frac{-0.2x}{-0.2} = \frac{6}{-0.2}$$

$$x = -30$$

Check: $0.4x - 0.6x - 5 = 1$

$$0.4(-30) - 0.6(-30) - 5 \overset{?}{=} 1$$

$$-12 + 18 - 5 \overset{?}{=} 1$$

$$1 = 1$$

The solution is $-30$

**31.** $42 = 7x$

$$\frac{42}{7} = \frac{7x}{7}$$

$$6 = x$$

Check: $42 = 7x$

$$42 \overset{?}{=} 7(6)$$

$$42 = 42$$

The solution is 6

**33.** $4.4 = -0.8x$

$$\frac{4.4}{-0.8} = \frac{-0.8x}{-0.8}$$

$$-5.5 = x$$

Check: $4.4 = -0.8x$

$$4.4 \overset{?}{=} -0.8(-5.5)$$

$$4.4 = 4.4$$

The solution is $-5.5$

**35.** $-\dfrac{3}{7}p = -2$

$$-\frac{7}{3}\left(-\frac{3}{7}p\right) = -\frac{7}{3}(-2)$$

$$p = \frac{14}{3}$$

Check: $-\dfrac{3}{7}p = -2$

$$-\frac{3}{7}\left(\frac{14}{3}\right) \overset{?}{=} -2$$

$$-2 = -2$$

The solution is $\dfrac{14}{3}$

**37.** $2x - 4 = 16$

$$2x - 4 + 4 = 16 + 4$$

$$2x = 20$$

$$\frac{2x}{2} = \frac{20}{2}$$

$$x = 10$$

Check: $2x - 4 = 16$

$$2(10) - 4 \overset{?}{=} 16$$

$$20 - 4 \overset{?}{=} 16$$

$$16 = 16$$

The solution is 10

**39.**
$$5 - 0.3k = 5$$
$$5 - 5 - 0.3k = 5 - 5$$
$$-0.3k = 0$$
$$\frac{-0.3k}{-0.3} = \frac{0}{-0.3}$$
$$k = 0$$

Check: $5 - 0.3k = 5$
$$5 - 0.3(0) \overset{?}{=} 5$$
$$5 - 0 \overset{?}{=} 5$$
$$5 = 5$$
The solution is 0

**41.**
$$-\frac{4}{3}x = 12$$
$$-\frac{3}{4}\left(-\frac{4}{3}x\right) = -\frac{3}{4}(12)$$
$$x = -9$$

Check: $-\frac{4}{3}x = 12$
$$-\frac{4}{3}(-9) \overset{?}{=} 12$$
$$12 = 12$$
The solution is $-9$

**43.**
$$10 = 2x - 1$$
$$10 + 1 = 2x - 1 + 1$$
$$11 = 2x$$
$$\frac{11}{2} = \frac{2x}{2}$$
$$\frac{11}{2} = x$$

Check: $10 = 2x - 1$
$$10 \overset{?}{=} 2\left(\frac{11}{2}\right) - 1$$
$$10 \overset{?}{=} 11 - 1$$
$$10 = 10$$
The solution is $\frac{11}{2}$

**45.**
$$\frac{x}{3} + 2 = -5$$
$$\frac{x}{3} + 2 - 2 = -5 - 2$$
$$\frac{x}{3} = -7$$
$$3\left(\frac{x}{3}\right) = 3(-7)$$
$$x = -21$$

Check: $\frac{x}{3} + 2 = -5$
$$\frac{-21}{3} + 2 \overset{?}{=} -5$$
$$-7 + 2 \overset{?}{=} -5$$
$$-5 = -5$$
The solution is $-21$

**47.**
$$1 = 0.4x - 0.6x - 5$$
$$1 = -0.2x - 5$$
$$1 + 5 = -0.2x - 5 + 5$$
$$6 = -0.2x$$
$$\frac{6}{-0.2} = \frac{-0.2x}{-0.2}$$
$$-30 = x$$

Check: $1 = 0.4x - 0.6x - 5$

$1 \overset{?}{=} 0.4(-30) - 0.6(-30) - 5$

$1 \overset{?}{=} -12 + 18 - 5$

$1 = 1$

The solution is $-30$

**49.** $z - 5z = 7z - 9 - z$

$-4z = 6z - 9$

$-4z - 6z = 6z - 6z - 9$

$-10z = -9$

$\dfrac{-10z}{-10} = \dfrac{-9}{-10}$

$z = \dfrac{9}{10}$

Check: $z - 5z = 7z - 9 - z$

$\dfrac{9}{10} - 5\left(\dfrac{9}{10}\right) \overset{?}{=} 7\left(\dfrac{9}{10}\right) - 9 - \dfrac{9}{10}$

$\dfrac{9}{10} - \dfrac{45}{10} \overset{?}{=} \dfrac{63}{10} - 9 - \dfrac{9}{10}$

$-\dfrac{36}{10} \overset{?}{=} \dfrac{54}{10} - \dfrac{90}{10}$

$-\dfrac{36}{10} = -\dfrac{36}{10}$

The solution is $\dfrac{9}{10}$

**51.** $6 - 2x + 8 = 10$

$-2x + 14 = 10$

$-2x + 14 - 14 = 10 - 14$

$-2x = -4$

$\dfrac{-2x}{-2} = \dfrac{-4}{-2}$

$x = 2$

Check: $6 - 2x + 8 = 10$

$6 - 2(2) + 8 \overset{?}{=} 10$

$6 - 4 + 8 \overset{?}{=} 10$

$10 = 10$

The solution is 2

**53.** $-3a + 6 + 5a = 7a - 8a$

$2a + 6 = -a$

$2a + a + 6 = -a + a$

$3a + 6 = 0$

$3a + 6 - 6 = -6$

$3a = -6$

$\dfrac{3a}{3} = \dfrac{-6}{3}$

$a = -2$

Check: $-3a + 6 + 5a = 7a - 8a$

$-3(-2) + 6 + 5(-2) \overset{?}{=} 7(-2) - 8(-2)$

$6 + 6 - 10 \overset{?}{=} -14 + 16$

$2 = 2$

The solution is $-2$

**55.** Answers may vary

**57.** Answers may vary

**59.** Sum = first integer + second integer.

Sum $= x + (x + 2)$

$= x + x + 2 = 2x + 2$

**61.** Sum = first integer + third integer.

Sum $= x + (x + 2)$

$= x + x + 2 = 2x + 2$

**63.** $5x + 2(x-6) = 5x + 2x - 12 = 7x - 12$

**65.** $-(x-1) + x = -x + 1 + x = 1$

**67.** $(-3)^2 > -3^2$

**69.** $(-2)^3 = -2^3$

**71.** $-|-6| < 6$

**73.** $9x = 2100$

$$\frac{9x}{9} = \frac{2100}{9}$$

$$x = \frac{700}{3}$$

Each dose should be $\dfrac{700}{3}$ mg

**75.** $-3.6x = 10.62$

$$\frac{-3.6x}{-3.6} = \frac{10.62}{-3.6}$$

$$x = -2.95$$

**77.** $7x - 5.06 = -4.92$

$$7x - 5.06 + 5.06 = -4.92 + 5.06$$

$$7x = 0.14$$

$$\frac{7x}{7} = \frac{0.14}{7}$$

$$x = 0.02$$

**Calculator Explorations 2.4**

**1.** Solution $(-24 = -24)$

**3.** Not a solution $(19.4 \neq 10.4)$

**5.** Solution $(17,061 = 17,061)$

**Exercise Set 2.4**

**1.** $-2(3x - 4) = 2x$

$$-6x + 8 = 2x$$

$$-6x + 6x + 8 = 2x + 6x$$

$$8 = 8x$$

$$\frac{8}{8} = \frac{8x}{8}$$

$$1 = x$$

**3.** $4(2n - 1) = (6n + 4) + 1$

$$8n - 4 = 6n + 4 + 1$$

$$8n - 4 = 6n + 5$$

$$8n - 6n - 4 = 6n - 6n + 5$$

$$2n - 4 = 5$$

$$2n - 4 + 4 = 5 + 4$$

$$2n = 9$$

$$\frac{2n}{2} = \frac{9}{2}$$

$$n = \frac{9}{2}$$

**5.** $5(2x - 1) - 2(3x) = 1$

$$10x - 5 - 6x = 1$$

$$4x - 5 = 1$$

$$4x - 5 + 5 = 1 + 5$$

$$4x = 6$$

$$\frac{4x}{4} = \frac{6}{4}$$

$$x = \frac{3}{2}$$

**7.** $6(x-3)+10=-8$

$6x-18+10=-8$

$6x-8=-8$

$6x-8+8=-8+8$

$6x=0$

$\dfrac{6x}{6}=\dfrac{0}{6}$

$x=0$

**9.** $\dfrac{3}{4}x-\dfrac{1}{2}=1$

$4\left(\dfrac{3}{4}x-\dfrac{1}{2}\right)=4(1)$

$3x-2=4$

$3x-2+2=4+2$

$3x=6$

$\dfrac{3x}{3}=\dfrac{6}{3}$

$x=2$

**11.** $x+\dfrac{5}{4}=\dfrac{3}{4}x$

$4\left(x+\dfrac{5}{4}\right)=4\left(\dfrac{3}{4}x\right)$

$4x+5=3x$

$4x-3x+5=3x-3x$

$x+5=0$

$x+5-5=0-5$

$x=-5$

**13.** $\dfrac{x}{2}-1=\dfrac{x}{5}+2$

$10\left(\dfrac{x}{2}-1\right)=10\left(\dfrac{x}{5}+2\right)$

$5x-10=2x+20$

$5x-2x-10=2x-2x+20$

$3x-10=20$

$3x-10+10=20+10$

$3x=30$

$\dfrac{3x}{3}=\dfrac{30}{3}$

$x=10$

**15.** $\dfrac{6(3-z)}{5}=-z$

$5\left[\dfrac{6(3-z)}{5}\right]=5(-z)$

$6(3-z)=-5z$

$18-6z=-5z$

$18-6z+6z=-5z+6z$

$18=z$

**17.** $\dfrac{2(x+1)}{4}=3x-2$

$4\left[\dfrac{2(x+1)}{4}\right]=4(3x-2)$

$2(x+1)=12x-8$

$2x+2=12x-8$

$2x-12x+2=12x-12x-8$

$-10x+2=-8$

$-10x+2-2=-8-2$

$-10x=-10$

$\dfrac{-10x}{-10}=\dfrac{-10}{-10}$

$x=1$

**19.** $0.50x+0.15(70)=0.25(142)$

$100\left[0.50x+0.15(70)\right]=100\left[0.25(142)\right]$

$50x+15(70)=25(142)$

$50x+1050=3550$

45

$$50x + 1050 - 1050 = 3550 - 1050$$
$$50x = 2500$$
$$\frac{50x}{50} = \frac{2500}{50}$$
$$x = 50$$

**21.**
$$0.12(y-6) + 0.06y = 0.08y - 0.07(10)$$
$$100[0.12(y-6) + 0.06y] = 100[0.08y - 0.07(10)]$$
$$12(y-6) + 6y = 8y - 7(10)$$
$$12y - 72 + 6y = 8y - 70$$
$$18y - 72 = 8y - 70$$
$$18y - 8y - 72 = 8y - 8y - 70$$
$$10y - 72 = -70$$
$$10y - 72 + 72 = -70 + 72$$
$$10y = 2$$
$$\frac{10y}{10} = \frac{2}{10}$$
$$y = \frac{1}{5} = 0.2$$

**23.**
$$5x - 5 = 2(x+1) + 3x - 7$$
$$5x - 5 = 2x + 2 + 3x - 7$$
$$5x - 5 = 5x - 5$$
$$5x - 5x - 5 = 5x - 5x - 5$$
$$-5 = -5$$
Every real number is a solution.

**25.**
$$\frac{x}{4} + 1 = \frac{x}{4}$$
$$4\left(\frac{x}{4} + 1\right) = 4\left(\frac{x}{4}\right)$$
$$x + 4 = x$$
$$x - x + 4 = x - x$$
$$4 = 0$$
There is no solution.

**27.**
$$3x - 7 = 3(x+1)$$
$$3x - 7 = 3x + 3$$
$$3x - 3x - 7 = 3x - 3x + 3$$
$$-7 = 3$$
There is no solution.

**29.** Answers may vary

**31.** Answers may vary

**33.**
$$4x + 3 = 2x + 11$$
$$4x - 2x + 3 = 2x - 2x + 11$$
$$2x + 3 = 11$$
$$2x + 3 - 3 = 11 - 3$$
$$2x = 8$$
$$\frac{2x}{2} = \frac{8}{2}$$
$$x = 4$$

**35.**
$$-2y - 10 = 5y + 18$$
$$-2y - 5y - 10 = 5y - 5y + 18$$
$$-7y - 10 = 18$$
$$-7y - 10 + 10 = 18 + 10$$
$$-7y = 28$$
$$\frac{-7y}{-7} = \frac{28}{-7}$$
$$y = -4$$

**37.**
$$0.6x - 0.1 = 0.5x + .2$$
$$0.6x - 0.5x - 0.1 = 0.5x - 0.5x + 0.2$$
$$0.1x - 0.1 = 0.2$$
$$0.1x - 0.1 + 0.1 = 0.2 + 0.1$$
$$0.1x = 0.3$$
$$\frac{0.1x}{0.1} = \frac{0.3}{0.1}$$
$$x = 3$$

**39.**
$$2y + 2 = y$$
$$2y - y + 2 = y - y$$
$$y + 2 = 0$$
$$y + 2 - 2 = -2$$
$$y = -2$$

**41.**
$$3(5c - 1) - 2 = 13c + 3$$
$$15c - 3 - 2 = 13c + 3$$
$$15c - 5 = 13c + 3$$
$$15c - 13c - 5 = 13c - 13c + 3$$
$$2c - 5 = 3$$
$$2c - 5 + 5 = 3 + 5$$
$$2c = 8$$
$$\frac{2c}{2} = \frac{8}{2}$$
$$c = 4$$

**43.**
$$x + \frac{7}{6} = 2x - \frac{7}{6}$$
$$6\left(x + \frac{7}{6}\right) = 6\left(2x - \frac{7}{6}\right)$$
$$6x + 7 = 12x - 7$$
$$6x - 12x + 7 = 12x - 12x - 7$$
$$-6x + 7 = -7$$
$$-6x + 7 - 7 = -7 - 7$$
$$-6x = -14$$
$$\frac{-6x}{-6} = \frac{-14}{-6}$$
$$x = \frac{14}{6}$$
$$x = \frac{7}{3}$$

**45.**
$$2(x - 5) = 7 + 2x$$
$$2x - 10 = 7 + 2x$$
$$2x - 2x - 10 = 7 + 2x - 2x$$
$$-10 = 7$$
There is no solution.

**47.**
$$\frac{2(z + 3)}{3} = 5 - z$$
$$3\left[\frac{2(z + 3)}{3}\right] = 3(5 - z)$$
$$2z + 6 = 15 - 3z$$
$$2z + 3z + 6 = 15 - 3z + 3z$$
$$5z + 6 = 15$$
$$5z + 6 - 6 = 15 - 6$$
$$5z = 9$$
$$\frac{5z}{5} = \frac{9}{5}$$
$$z = \frac{9}{5}$$

**49.**
$$\frac{4(y - 1)}{5} = -3y$$
$$5\left[\frac{4(y - 1)}{5}\right] = 5(-3y)$$
$$4y - 4 = -15y$$
$$4y + 15y - 4 = -15y + 15y$$
$$19y - 4 = 0$$
$$19y - 4 + 4 = 4$$
$$19y = 4$$
$$\frac{19y}{19} = \frac{4}{19}$$
$$y = \frac{4}{19}$$

**51.**    $8 - 2(a-1) = 7 + a$

$$8 - 2a + 2 = 7 + a$$
$$-2a + 10 = 7 + a$$
$$-2a + 2a + 10 = 7 + a + 2a$$
$$10 = 7 + 3a$$
$$10 - 7 = 7 - 7 + 3a$$
$$3 = 3a$$
$$\frac{3}{3} = \frac{3a}{3}$$
$$1 = a$$

**53.** $2(x+3) - 5 = 5x - 3(1+x)$

$$2x + 6 - 5 = 5x - 3 - 3x$$
$$2x + 1 = 2x - 3$$
$$2x - 2x + 1 = 2x - 2x - 3$$
$$1 = -3$$

There is no solution.

**55.**    $\dfrac{5x - 7}{3} = x$

$$3\left[\frac{5x-7}{3}\right] = 3(x)$$
$$5x - 7 = 3x$$
$$5x - 3x - 7 = 3x - 3x$$
$$2x - 7 = 0$$
$$2x - 7 + 7 = 7$$
$$2x = 7$$
$$\frac{2x}{2} = \frac{7}{2}$$
$$x = \frac{7}{2}$$

**57.**    $\dfrac{9 + 5v}{2} = 2v - 4$

$$2\left(\frac{9+5v}{2}\right) = 2(2v - 4)$$
$$9 + 5v = 4v - 8$$
$$9 + 5v - 4v = 4v - 4v - 8$$
$$9 + v = -8$$
$$9 - 9 + v = -8 - 9$$
$$v = -17$$

**59.** $-3(t-5) + 2t = 5t - 4$

$$-3t + 15 + 2t = 5t - 4$$
$$-t + 15 = 5t - 4$$
$$-t - 5t + 15 = 5t - 5t - 4$$
$$-6t + 15 = -4$$
$$-6t + 15 - 15 = -4 - 15$$
$$-6t = -19$$
$$\frac{-6t}{-6} = \frac{-19}{-6}$$
$$t = \frac{19}{6}$$

**61.**    $0.02(6t - 3) = 0.12(t - 2) + 0.18$

$$0.12t - 0.06 = 0.12t - 0.24 + 0.18$$
$$0.12t - 0.06 = 0.12t - 0.06$$
$$0.12t - 0.12t - 0.06 = 0.12t - 0.12t - 0.06$$
$$-0.06 = -0.06$$

Every real number is a solution.

**63.**
$$0.06 - 0.01(x+1) = -0.02(2-x)$$
$$100\left[0.06 - 0.01(x+1)\right] = 100\left[-0.02(2-x)\right]$$
$$6 - (x+1) = -2(2-x)$$
$$6 - x - 1 = -4 + 2x$$
$$5 - x = -4 + 2x$$
$$5 - x - 2x = -4 + 2x - 2x$$
$$5 - 3x = -4$$
$$5 - 5 - 3x = -4 - 5$$
$$-3x = -9$$
$$\frac{-3x}{-3} = \frac{-9}{-3}$$
$$x = 3$$

**65.**
$$\frac{3(x-5)}{2} = \frac{2(x+5)}{3}$$
$$6\left[\frac{3(x-5)}{2}\right] = 6\left[\frac{2(x+5)}{3}\right]$$
$$9(x-5) = 4(x+5)$$
$$9x - 45 = 4x + 20$$
$$9x - 4x - 45 = 4x - 4x + 20$$
$$5x - 45 = 20$$
$$5x - 45 + 45 = 20 + 45$$
$$5x = 65$$
$$\frac{5x}{5} = \frac{65}{5}$$
$$x = 13$$

**67.**
$$2x + 7 = x + 6$$
$$2x - x + 7 = x - x + 6$$
$$x + 7 = 6$$
$$x + 7 - 7 = 6 - 7$$
$$x = -1$$
The number is $-1$

**69.**
$$3x - 6 = 2x + 8$$
$$3x - 2x - 6 = 2x - 2x + 8$$
$$x - 6 = 8$$
$$x - 6 + 6 = 8 + 6$$
$$x = 14$$
The number is 14

**71.**
$$\frac{1}{3}x = \frac{5}{6}$$
$$6\left(\frac{1}{3}x\right) = 6\left(\frac{5}{6}\right)$$
$$2x = 5$$
$$\frac{2x}{2} = \frac{5}{2}$$
$$x = \frac{5}{2}$$

The number is $\dfrac{5}{2}$

**73.**
$$\frac{x}{4} + \frac{1}{2} = \frac{3}{4}$$
$$4\left(\frac{x}{4} + \frac{1}{2}\right) = 4\left(\frac{3}{4}\right)$$
$$x + 2 = 3$$
$$x + 2 - 2 = 3 - 2$$
$$x = 1$$
The number is 1

**75.**
$$10 - 5x = 3x$$
$$10 - 5x + 5x = 3x + 5x$$
$$10 = 8x$$
$$\frac{10}{8} = \frac{8x}{8}$$
$$\frac{5}{4} = x$$

The number is $\dfrac{5}{4}$

**77.** Since the perimeter is the sum of the lengths of the sides,
$$x + x + x + 2x + 2x = 28$$
$$7x = 28$$
$$\frac{7x}{7} = \frac{28}{7}$$
$$x = 4$$
$$2x = 2(4) = 8$$

The lengths are 4 cm and 8 cm.

**79.** $\left|2^3 - 3^2\right| - |5 - 7|$
$$= |8 - 9| - |-2|$$
$$= |-1| - 2$$
$$= 1 - 2$$
$$= -1$$

**81.** $\dfrac{5}{4 + 3 \cdot 7} = \dfrac{5}{4 + 21} = \dfrac{5}{25} = \dfrac{1}{5}$

**83.** $x + (2x - 3) + (3x - 5) = x + 2x - 3 + 3x - 5$
$$= 6x - 8$$

The perimeter is $(6x - 8)$ meters

**85.** Midway

**87.**
$$65 + x = 2x - 90$$
$$65 + x - x = 2x - x - 90$$
$$65 = x - 90$$
$$65 + 90 = x - 90 + 90$$
$$155 = x$$

The number of places named
Five Points is 155.

**89.** $1000(7x - 10) = 50(412 + 100x)$
$$7000x - 10,000 = 20,600 + 5000x$$
$$7000x - 5000x - 10,000$$
$$\qquad = 20,600 + 5000x - 5000x$$
$$2000x - 10,000 = 20,600$$
$$2000x - 10,000 + 10,000$$
$$\qquad = 20,600 + 10,000$$
$$2000x = 30,600$$
$$\frac{2000x}{2000} = \frac{30,600}{2000}$$
$$x = 15.3$$

**91.**
$$0.035x + 5.112 = 0.010x + 5.107$$
$$1000(0.035x + 5.112) = 1000(0.010x + 5.107)$$
$$35x + 5112 = 10x + 5107$$
$$35x - 10x + 5112 = 10x - 10x + 5107$$
$$25x + 5112 = 5107$$
$$25x + 5112 - 5112 = 5107 - 5112$$
$$25x = -5$$
$$\frac{25x}{25} = \frac{-5}{25}$$
$$x = -\frac{1}{5} = -0.2$$

**93.**
$$x(x-3) = x^2 + 5x + 7$$
$$x^2 - 3x = x^2 + 5x + 7$$
$$x^2 - x^2 - 3x = x^2 - x^2 + 5x + 7$$
$$-3x = 5x + 7$$
$$-3x - 5x = 5x - 5x + 7$$
$$-8x = 7$$
$$\frac{-8x}{-8} = \frac{7}{-8}$$
$$x = -\frac{7}{8}$$

**95.**
$$2z(z+6) = 2z^2 + 12z - 8$$
$$2z^2 + 12z = 2z^2 + 12z - 8$$
$$2z^2 - 2z^2 + 12z = 2z^2 - 2z^2 + 12z - 8$$
$$12z = 12z - 8$$
$$12z - 12z = 12z - 12z - 8$$
$$0 = -8$$

There is no solution

**Integrated Review-Solving Linear Equations**

**1.**
$$x - 10 = -4$$
$$x - 10 + 10 = -4 + 10$$
$$x = 6$$

**2.**
$$y + 14 = -3$$
$$y + 14 - 14 = -3 - 14$$
$$y = -17$$

**3.** $9y = 108$
$$\frac{9y}{9} = \frac{108}{9}$$
$$y = 12$$

**4.** $-3x = 78$
$$\frac{-3x}{-3} = \frac{78}{-3}$$
$$x = -26$$

**5.**
$$-6x + 7 = 25$$
$$-6x + 7 - 7 = 25 - 7$$
$$-6x = 18$$
$$\frac{-6x}{-6} = \frac{18}{-6}$$
$$x = -3$$

**6.**
$$5y - 42 = -47$$
$$5y - 42 + 42 = -47 + 42$$
$$5y = -5$$
$$\frac{5y}{5} = \frac{-5}{5}$$
$$y = -1$$

**7.**
$$\frac{2}{3}x = 9$$
$$\frac{3}{2}\left(\frac{2}{3}x\right) = \frac{3}{2}(9)$$
$$x = \frac{27}{2} = 13.5$$

**8.**
$$\frac{4}{5}z = 10$$
$$\frac{5}{4}\left(\frac{4}{5}z\right) = \frac{5}{4}(10)$$
$$z = \frac{25}{2} = 12.5$$

**9.**     $\dfrac{r}{-4} = -2$

$-4\left(\dfrac{r}{-4}\right) = -4(-2)$

$r = 8$

**10.**     $\dfrac{y}{-8} = 8$

$-8\left(\dfrac{y}{-8}\right) = -8(8)$

$y = -64$

**11.**     $6 - 2x + 8 = 10$

$-2x + 14 = 10$

$-2x + 14 - 14 = 10 - 14$

$-2x = -4$

$\dfrac{-2x}{-2} = \dfrac{-4}{-2}$

$x = 2$

**12.**  $-5 - 6y + 6 = 19$

$-6y + 1 = 19$

$-6y + 1 - 1 = 19 - 1$

$-6y = 18$

$\dfrac{-6y}{-6} = \dfrac{18}{-6}$

$y = -3$

**13.**     $2x - 7 = 2x - 27$

$2x - 2x - 7 = 2x - 2x - 27$

$-7 = -27$  a contradiction

There is no solution

**14.**     $3 + 8y = 8y - 2$

$3 + 8y - 8y = 8y - 8y - 2$

$3 = -2$  a contradiction

There is no solution

**15.**  $-3a + 6 + 5a = 7a - 8a$

$2a + 6 = -a$

$2a - 2a + 6 = -a - 2a$

$6 = -3a$

$\dfrac{6}{-3} = \dfrac{-3a}{-3}$

$-2 = a$

**16.**  $4b - 8 - b = 10b - 3b$

$3b - 8 = 7b$

$3b - 3b - 8 = 7b - 3b$

$-8 = 4b$

$\dfrac{-8}{4} = \dfrac{4b}{4}$

$-2 = b$

**17.**     $-\dfrac{2}{3}x = \dfrac{5}{9}$

$-\dfrac{3}{2}\left(-\dfrac{2}{3}x\right) = -\dfrac{3}{2}\left(\dfrac{5}{9}\right)$

$x = -\dfrac{5}{6}$

**18.**     $-\dfrac{3}{8}y = -\dfrac{1}{16}$

$-\dfrac{8}{3}\left(-\dfrac{3}{8}y\right) = -\dfrac{8}{3}\left(-\dfrac{1}{16}\right)$

$y = \dfrac{1}{6}$

**19.**
$$10 = -6n + 16$$
$$10 - 16 = -6n + 16 - 16$$
$$-6 = -6n$$
$$\frac{-6}{-6} = \frac{-6n}{-6}$$
$$1 = n$$

**20.**
$$-5 = -2m + 7$$
$$-5 - 7 = -2m + 7 - 7$$
$$-12 = -2m$$
$$\frac{-12}{-2} = \frac{-2m}{-2}$$
$$6 = m$$

**21.** $3(5c - 1) - 2 = 13c + 3$
$$15c - 3 - 2 = 13c + 3$$
$$15c - 5 = 13c + 3$$
$$15c - 13c - 5 = 13c - 13c + 3$$
$$2c - 5 = 3$$
$$2c - 5 + 5 = 3 + 5$$
$$2c = 8$$
$$\frac{2c}{2} = \frac{8}{2}$$
$$c = 4$$

**22.** $4(3t + 4) - 20 = 3 + 5t$
$$12t + 16 - 20 = 3 + 5t$$
$$12t - 4 = 3 + 5t$$
$$12t - 5t - 4 = 3 + 5t - 5t$$
$$7t - 4 = 3$$
$$7t - 4 + 4 = 3 + 4$$
$$7t = 7$$
$$\frac{7t}{7} = \frac{7}{7}$$
$$t = 1$$

**23.**
$$\frac{2(z + 3)}{3} = 5 - z$$
$$3\left[\frac{2(z + 3)}{3}\right] = 3(5 - z)$$
$$2z + 6 = 15 - 3z$$
$$2z + 3z + 6 = 15 - 3z + 3z$$
$$5z + 6 = 15$$
$$5z + 6 - 6 = 15 - 6$$
$$5z = 9$$
$$\frac{5z}{5} = \frac{9}{5}$$
$$z = \frac{9}{5}$$

**24.**
$$\frac{3(w + 2)}{4} = 2w + 3$$
$$4\left[\frac{3(w + 2)}{4}\right] = 4(2w + 3)$$
$$3w + 6 = 8w + 12$$
$$3w - 8w + 6 = 8w - 8w + 12$$
$$-5w + 6 = 12$$
$$-5w + 6 - 6 = 12 - 6$$
$$-5w = 6$$
$$\frac{-5w}{-5} = \frac{6}{-5}$$
$$w = -\frac{6}{5}$$

**25.**
$$-2(2x - 5) = -3x + 7 - x + 3$$
$$-4x + 10 = -4x + 10$$
$$-4x + 4x + 10 = -4x + 4x + 10$$
$$10 = 10$$
Every real number is a solution

53

**26.**     $-4(5x-2) = -12x+4-8x+4$
$-20x+8 = -20x+8$
$-20x+20x+8 = -20x+20x+8$
$8 = 8$

Every real number is a solution

**27.**      $0.02(6t-3) = 0.04(t-2)+0.02$
$100[0.02(6t-3)] = 100[0.04(t-2)+0.02]$
$2(6t-3) = 4(t-2)+2$
$12t-6 = 4t-8+2$
$12t-6 = 4t-6$
$12t-4t-6 = 4t-4t-6$

$8t-6 = -6$
$8t-6+6 = -6+6$
$8t = 0$
$\dfrac{8t}{8} = \dfrac{0}{8}$
$t = 0$

**28.**      $0.03(m+7) = 0.02(5-m)+0.03$
$100[0.03(m+7)] = 100[0.02(5-m)+0.03]$
$3(m+7) = 2(5-m)+3$
$3m+21 = 10-2m+3$
$3m+21 = 13-2m$
$3m+2m+21 = 13-2m+2m$
$5m+21 = 13$
$5m+21-21 = 13-21$
$5m = -8$
$\dfrac{5m}{5} = \dfrac{-8}{5}$
$m = -\dfrac{8}{5} = -1.6$

**29.**      $-3y = \dfrac{4(y-1)}{5}$
$5(-3y) = 5\left[\dfrac{4(y-1)}{5}\right]$
$-15y = 4y-4$
$-15y-4y = 4y-4y-4$
$-19y = -4$
$\dfrac{-19y}{-19} = \dfrac{-4}{-19}$
$y = \dfrac{4}{19}$

**30.**      $-4x = \dfrac{5(1-x)}{6}$
$6(-4x) = 6\left[\dfrac{5(1-x)}{6}\right]$
$-24x = 5-5x$
$-24x+5x = 5-5x+5x$
$-19x = 5$
$\dfrac{-19x}{-19} = \dfrac{5}{-19}$
$x = -\dfrac{5}{19}$

**31.**    $\dfrac{5}{3}x - \dfrac{7}{3} = x$

$$3\left(\dfrac{5}{3}x - \dfrac{7}{3}\right) = 3(x)$$

$$5x - 7 = 3x$$

$$5x - 5x - 7 = 3x - 5x$$

$$-7 = -2x$$

$$\dfrac{-7}{-2} = \dfrac{-2x}{-2}$$

$$\dfrac{7}{2} = x$$

**32.**    $\dfrac{7}{5}n + \dfrac{3}{5} = -n$

$$5\left(\dfrac{7}{5}n + \dfrac{3}{5}\right) = 5(-n)$$

$$7n + 3 = -5n$$

$$7n - 7n + 3 = -5n - 7n$$

$$3 = -12n$$

$$\dfrac{3}{-12} = \dfrac{-12n}{-12}$$

$$-\dfrac{1}{4} = n$$

**Exercise Set 2.5**

**1.**      Let $x$ = the number.

$$2(x - 8) = 3(x + 3)$$

$$2x - 16 = 3x + 9$$

$$2x - 2x - 16 = 3x - 2x + 9$$

$$-16 = x + 9$$

$$-16 - 9 = x + 9 - 9$$

$$-25 = x$$

The number $= -25$

**3.**      Let $x$ = the number.

$$2x(3) = 5x - \dfrac{3}{4}$$

$$6x = 5x - \dfrac{3}{4}$$

$$6x - 5x = 5x - 5x - \dfrac{3}{4}$$

$$x = -\dfrac{3}{4}$$

The number $= -\dfrac{3}{4}$

**5.**    Let $x$ = the number of the left page
and $x + 1 =$ the number of the right page.

$$x + x + 1 = 469$$

$$2x + 1 = 469$$

$$2x + 1 - 1 = 469 - 1$$

$$2x = 468$$

$$\dfrac{2x}{2} = \dfrac{468}{2}$$

$$x = 234$$

$$x + 1 = 235$$

The page numbers are 234 and 239.

**7.**    Let $x$ = the code for Belgium,
$x + 1 =$ the code for France,
$x + 2 =$ the code for Spain.

$$x + x + 1 + x + 2 = 99$$

$$3x + 3 = 99$$

$$3x + 3 - 3 = 99 - 3$$

$$3x = 96$$

$$\dfrac{3x}{3} = \dfrac{96}{3}$$

$$x = 32$$

$$x + 1 = 33$$
$$x + 2 = 34$$

The codes are;

Belgium:32; France:33; Spain:34

9.  Let $x =$ length of the first piece,
    $2x =$ length of the second piece,
    and $5x =$ length of the third piece.

$$x + 2x + 5x = 40$$
$$8x = 40$$
$$\frac{8x}{8} = \frac{40}{8}$$
$$x = 5$$
$$2x = 2(5) = 10$$
$$5x = 5(5) = 25$$

The lengths are 5, 10, and 25 inches

11. Let $x =$ the salary of the governor of
    Nebraska and $2x =$ the salary of
    the governor of Washington.

$$x + 2x = 195,000$$
$$3x = 195,000$$
$$\frac{3x}{3} = \frac{195,000}{3}$$
$$x = 65,000$$
$$2x = 2(65,000) = 130,000$$

The governor of Nebraska makes $65,000
and the governor of Washington makes
$130,000.

13. Let $x =$ the number of miles driven,
    $0.29 =$ the charge per mile, and
    $24.95 =$ the charge per day.

$$2(24.95) + 0.29x = 100$$
$$49.90 + 0.29x = 100$$
$$49.90 - 49.90 + 0.29x = 100 - 49.90$$
$$0.29x = 50.10$$
$$\frac{0.29x}{0.29} = \frac{50.10}{0.29}$$
$$x = 172$$

172 miles were driven

15. Let $x =$ the number of miles driven,
    $0.80 =$ the charge per mile,
    $3.00 =$ the taxi charge
    $4.50 =$ the toll charges.

$$0.80x + 3.00 + 4.50 = 27.50$$
$$0.80x + 7.50 = 27.50$$
$$0.80x + 7.50 - 7.50 = 27.50 - 7.50$$
$$0.80x = 20.00$$
$$\frac{0.80x}{0.80} = \frac{20.00}{0.80}$$
$$x = 25$$

You can travel 25 miles

17. Let $x =$ the measure of each of the two
    equal angle, and $2x + 30 =$ the measure
    of the third.

$$x + x + 2x + 30 = 180$$
$$4x + 30 = 180$$
$$4x + 30 - 30 = 180 - 30$$
$$4x = 150$$
$$\frac{4x}{4} = \frac{150}{4}$$
$$x = 37.5$$
$$2x + 30 = 2(37.5) + 30 = 105$$

The 3 angles are 37.5°, 37.5°, 105°

**19.** Let $x$ = the measure of each of the two equal angles $A$ and $D$, and $2x$ = the measure of each of the other two equal angles $C$ and $B$.

$$x + x + 2x + 2x = 360$$
$$6x = 360$$
$$\frac{6x}{6} = \frac{360}{6}$$
$$x = 60$$
$$2x = 2(60) = 120$$

The angles are $A = 60°, D = 60°$
$C = 120°, B = 120°$

**21.** Let $x$ = length of the shorter piece and $2x + 2$ = length of the longer piece.

$$x + 2x + 2 = 17$$
$$3x + 2 = 17$$
$$3x + 2 - 2 = 17 - 2$$
$$3x = 15$$
$$\frac{3x}{3} = \frac{15}{3}$$
$$x = 5$$
$$2x + 2 = 2(5) + 2 = 12$$

The shorter piece = 5 ft. and the longer piece = 12 ft

**23.** Let $x$ = the number of millions of prescriptions in 1997 and $x + 5.5$ = the number in 2001.

$$x + x + 5.5 = 35.7$$
$$2x + 5.5 = 35.7$$
$$2x + 5.5 - 5.5 = 35.7 - 5.5$$

$$2x = 30.2$$
$$\frac{2x}{2} = \frac{30.2}{2}$$
$$x = 15.1$$
$$x + 5.5 = 15.1 + 5.5 = 20.6$$

15.1 million prescriptions were written in 1997 and 20.6 were written in 2001

**25.** Let $x$ = the measure of the smaller angle and $3x$ = the measure of the other.

$$x + 3x = 180$$
$$4x = 180$$
$$\frac{4x}{4} = \frac{180}{4}$$
$$x = 45$$
$$3x = 3(45) = 135$$

The 2 angles are $45°, 135°$.

**27.** Let $x$ = the measure of the smallest angle, $x + 2$ = the measure of the second, and $x + 4$ = the measure of the third.

$$x + x + 2 + x + 4 = 180$$
$$3x + 6 = 180$$
$$3x + 6 - 6 = 180 - 6$$
$$3x = 174$$
$$\frac{3x}{3} = \frac{174}{3}$$
$$x = 58$$
$$x + 2 = 58 + 2 = 60$$
$$x + 4 = 58 + 4 = 62$$

The 3 angles are $58°, 60°, 62°$

**29.** Let $x$ = the number of miles driven,
$0.20$ = the charge per mile, and
$39$ = the charge per day.

$$39 + 0.20x = 95$$
$$39 - 39 + 0.20x = 95 - 39$$
$$0.20x = 56$$
$$\frac{0.20x}{0.20} = \frac{56}{0.20}$$
$$x = 280$$

280 miles were driven

**31.** Let $x$ = the Oakland score and
$x + 27$ = the Tampa Bay score.

$$x + x + 27 = 69$$
$$2x + 27 = 69$$
$$2x + 27 - 27 = 69 - 27$$
$$2x = 42$$
$$\frac{2x}{2} = \frac{42}{2}$$
$$x = 21$$
$$x + 27 = 21 + 27 = 48$$

The score was
Tampa Bay 48, Oakland 21.

**33.** Let $x$ = number of medals won by
Germany, $x + 1$ = number of medals
won by Australia, and $x + 2$ = number
of medals won by China.

$$x + x + 1 + x + 2 = 174$$
$$3x + 3 = 174$$
$$3x + 3 - 3 = 174 - 3$$
$$3x = 171$$
$$\frac{3x}{3} = \frac{171}{3}$$
$$x = 57$$

$x + 1 = 57 + 1 = 58$, $x + 2 = 57 + 2 = 59$
Germany won 57 medals.
Australia won 58 medals.
China won 59 medals.

**35.** Let $x$ = the number of electoral votes for
for Texas and $x + 21$ = the number
for California.

$$x + x + 21 = 89$$
$$2x + 21 = 89$$
$$2x + 21 - 21 = 89 - 21$$
$$2x = 68$$
$$\frac{2x}{2} = \frac{68}{2}$$
$$x = 34$$
$$x + 21 = 34 + 21 = 55$$

Texas will have 34 votes.
California will have 55 votes.

**37.** Let $x$ = number of moons won of
Neptune, $x + 13$ = number of moons
of Uranus, and $2x + 2$ = number
of moons of Saturn.

$$x + x + 13 + 2x + 2 = 47$$
$$4x + 15 = 47$$
$$4x + 15 - 15 = 47 - 15$$
$$4x = 32$$
$$\frac{4x}{4} = \frac{32}{4}$$
$$x = 8$$

$x + 13 = 8 + 13 = 21$, $2x + 2 = 2(8) + 2 = 18$
Neptune: 8 moons; Uranus: 21 moons;
Saturn: 18 moons

**39.**      Let $x$ = the number.

$$3(x+5) = 2x - 1$$
$$3x + 15 = 2x - 1$$
$$3x - 2x + 15 = 2x - 2x - 1$$
$$x + 15 = -1$$
$$x + 15 - 15 = -1 - 15$$
$$x = -16$$

The number $= -16$

**41.** Let $x$ = the number of votes for Randall and $x + 13,288$ = the number for Brown.

$$x + x + 13,288 = 119,436$$
$$2x + 13,288 = 119,436$$
$$2x + 13,288 - 13,288 = 119,436 - 13,288$$
$$2x = 106,148$$
$$\frac{2x}{2} = \frac{106,148}{2}$$
$$x = 53,074$$

$x + 13,288 = 53,074 + 13,288 = 66,362$

Randall had 53,074 votes.

Brown had 66,362 votes.

**43.** Illinois

**45.** Let $x$ = the amount spent in millions by Florida and $x + 2.2$ = the amount spent in millions by Texas.

$$x + x + 2.2 = 56.6$$
$$2x + 2.2 = 56.6$$
$$2x + 2.2 - 2.2 = 56.6 - 2.2$$

$$2x = 54.4$$
$$\frac{2x}{2} = \frac{54.4}{2}$$
$$x = 27.2$$

$x + 2.2 = 27.2 + 2.2 = 29.4$

Florida spent \$27.2 million.

Texas spent \$29.4 million.

**47.** Answers may vary

**49.** $-2 + (-8) = -10$

**51.** $-11 + 2 = -9$

**53.** $-12 - 3 = -12 + (-3) = -15$

**55.** Let $x$ = the measure of the smallest angle, $2x$ = the measure of the second, and $3x$ = the measure of the third.

$$x + 2x + 3x = 180$$
$$6x = 180$$
$$\frac{6x}{6} = \frac{180}{6}$$
$$x = 30$$
$$2x = 2(30) = 60$$
$$3x = 3(30) = 90$$

The 3 angles are $30°$, $60°$, $90°$

**57.** Answers may vary

**59.** Answers may vary

**61. c.**

**Exercise Set 2.6**

**1.** Let $A = 45$ and $b = 15$

$$A = bh$$
$$45 = 15h$$
$$\frac{45}{15} = \frac{15h}{15}$$
$$3 = h$$

**3.** Let $S = 102$, $l = 7$, and $w = 3$

$$S = 4lw + 2wh$$
$$102 = 4(7)(3) + 2(3)h$$
$$102 = 84 + 6h$$
$$102 - 84 = 84 - 84 + 6h$$
$$18 = 6h$$
$$\frac{18}{6} = \frac{6h}{6}$$
$$3 = h$$

**5.** Let $A = 180$, $B = 11$, and $b = 7$

$$A = \frac{1}{2}(B + b)h$$
$$180 = \frac{1}{2}(11 + 7)h$$
$$2(180) = 2\left[\frac{1}{2}(18)h\right]$$
$$360 = 18h$$
$$\frac{360}{18} = \frac{18h}{18}$$
$$20 = h$$

**7.** Let $P = 30$, $a = 8$, and $b = 10$

$$P = a + b + c$$
$$30 = 8 + 10 + c$$
$$30 = 18 + c$$
$$30 - 18 = 18 - 18 + c$$
$$12 = c$$

**9.** Let $C = 15.7$, and $\pi = 3.14$

$$C = 2\pi r$$
$$15.7 = 2(3.14)r$$
$$15.7 = 6.28r$$
$$\frac{15.7}{6.28} = \frac{6.28r}{6.28}$$
$$2.5 = r$$

**11.** Let $I = 3750$, $P = 25,000$, and $R = 0.05$

$$I = PRT$$
$$3750 = 25,000(0.05)T$$
$$3750 = 1250T$$
$$\frac{3750}{1250} = \frac{1250T}{1250}$$
$$3 = T$$

**13.** Let $V = 565.2$, $r = 6$, and $\pi = 3.14$

$$V = \frac{1}{3}\pi r^2 h$$
$$565.2 = \frac{1}{3}(3.14)(6)^2 h$$
$$565.2 = 37.68h$$
$$\frac{565.2}{37.68} = \frac{37.68h}{37.68}$$
$$15 = h$$

**15.**   $f = 5gh$

$$\frac{f}{5g} = \frac{5gh}{5g}$$

$$\frac{f}{5g} = h$$

**17.**   $V = LWH$

$$\frac{V}{LH} = \frac{LWH}{LH}$$

$$\frac{V}{LH} = W$$

**19.**   $3x + y = 7$

$$3x - 3x + y = 7 - 3x$$

$$y = 7 - 3x$$

**21.**   $A = P + PRT$

$$A - P = P - P + PRT$$

$$A - P = PRT$$

$$\frac{A - P}{PT} = \frac{PRT}{PT}$$

$$\frac{A - P}{PT} = R$$

**23.**   $V = \frac{1}{3}Ah$

$$3V = 3\left(\frac{1}{3}Ah\right)$$

$$3V = Ah$$

$$\frac{3V}{h} = \frac{Ah}{h}$$

$$\frac{3V}{h} = A$$

**25.**   $P = a + b + c$

$$P - b - c = a + b - b + c - c$$

$$P - b - c = a$$

**27.**   $S = 2\pi rh + 2\pi r^2$

$$S - 2\pi r^2 = 2\pi rh + 2\pi r^2 - 2\pi r^2$$

$$S - 2\pi r^2 = 2\pi rh$$

$$\frac{S - 2\pi r^2}{2\pi r} = \frac{2\pi rh}{2\pi r}$$

$$\frac{S - 2\pi r^2}{2\pi r} = h$$

**29.**   Let $A = 52{,}400$ and $l = 400$

$$A = lw$$

$$52{,}400 = 400w$$

$$\frac{52{,}400}{400} = \frac{400w}{400}$$

$$131 = w$$

The width is 131 ft

**31.**   Let $F = 14$

$$14 = \frac{9}{5}C + 32$$

$$5(14) = 5\left(\frac{9}{5}\right)C + 5(32)$$

$$70 = 9C + 160$$

$$70 - 160 = 9C + 160 - 160$$

$$-90 = 9C$$

$$\frac{-90}{9} = \frac{9C}{9}$$

$$-10 = C$$

The equivalent temperature is $-10°\,$C.

**33.** Let $d = 25,000$ and $r = 4000$

$$d = rt$$
$$25,000 = 4000t$$
$$\frac{25,000}{4000} = \frac{4000t}{4000}$$
$$6.25 = t$$

It will take 6.25 hours

**35.** Let $P = 260$ and $w = \frac{2}{3}l$

$$P = 2l + 2w$$
$$260 = 2l + 2\left(\frac{2}{3}l\right)$$
$$260 = \frac{10}{3}l$$
$$3(260) = 3\left(\frac{10}{3}l\right)$$
$$780 = 10l$$
$$\frac{780}{10} = \frac{10l}{10}$$
$$78 = l$$
$$w = \frac{2}{3}l = \frac{2}{3}(78) = 52$$

The width is 52 ft and the length is 78 ft.

**37.** Let $P = 102$, $a =$ the length of the shortest side, $b = 2a$, and $c = a + 30$

$$P = a + b + c$$
$$102 = a + 2a + a + 30$$
$$102 = 4a + 30$$
$$102 - 30 = 4a + 30 - 30$$
$$72 = 4a$$

$$\frac{72}{4} = \frac{4a}{4}$$
$$18 = a$$
$$b = 2a = 2(18) = 36$$
$$c = a + 30 = 18 + 30 = 48$$

The lengths are 18 ft, 36 ft, and 48 ft.

**39.** Let $t = 2.5$ and $r = 55$

$$d = rt$$
$$d = 55(2.5)$$
$$d = 137.5$$

They are 137.5 miles apart.

**41.** Let $l = 8$, $w = 3$, and $h = 6$

$$V = lwh$$
$$V = 8(3)(6) = 144$$

Let $x =$ number of piranha and volume per fish $= 1.5$

$$144 = 1.5x$$
$$\frac{144}{1.5} = \frac{1.5x}{1.5}$$
$$96 = x$$

96 piranha can be placed in the tank.

**43.** Let $h = 60$, $B = 130$, and $b = 70$

$$A = \frac{1}{2}(B + b)h$$
$$A = \frac{1}{2}(130 + 70)60 = \frac{1}{2}(200)(60) = 6000$$

Let $x =$ number of bags of fertilizer and the area per bag $= 4000$.

$$4000x = 6000$$

$$\frac{4000x}{4000} = \frac{6000}{4000}$$

$$x = 1.5$$

Two bags must be purchased.

**45.** Let $d = 16$, so $r = 8$

$$A = \pi r^2 = \pi (8)^2 = 64\pi$$

Let $d = 10$, so $r = 5$

$$A = 2\pi r^2 = 2\pi (5)^2 = 50\pi$$

One 16 inch pizza has more area and therefore gives more pizza for the price.

**47.**    Let $d = 42.8$ and

$$r = 552 \,\frac{\text{km}}{\text{hr}} \times \frac{1 \text{ hr}}{60 \text{ min}} = 9.2 \,\frac{\text{km}}{\text{min}}$$

$$d = rt$$

$$42.8 = 9.2t$$

$$\frac{42.8}{9.2} = \frac{9.2t}{9.2}$$

$$4.65 = t$$

It would take 4.65 minutes

**49.** Let $x =$ the length of a side of the square and $x + 5 =$ the length of a side of the triangle.

$$P(\text{triangle}) = P(\text{square}) + 7$$

$$3(x + 5) = 4x + 7$$

$$3x + 15 = 4x + 7$$

$$3x - 3x + 15 = 4x - 3x + 7$$

$$15 = x + 7$$

$$15 - 7 = x + 7 - 7$$

$$8 = x$$

$$x + 5 = 8 + 5 = 13.$$

The side of the triangle is 13 in.

**51.**    Let $d = 135$ and $r = 60$

$$d = rt$$

$$135 = 60t$$

$$\frac{135}{60} = \frac{60t}{60}$$

$$2.25 = t$$

It would take 2.25 hours.

**53.** Let $A = 1,813,500$ and $w = 150$

$$A = lw$$

$$1,813,500 = l(150)$$

$$\frac{1,813,500}{150} = \frac{150l}{150}$$

$$12,090 = l$$

The length is 12,090 ft

**55.** Let $F = 122$

$$122 = \frac{9}{5}C + 32$$

$$5(122) = 5\left(\frac{9}{5}\right)C + 5(32)$$

$$610 = 9C + 160$$

$$610 - 160 = 9C + 160 - 160$$

$$450 = 9C$$

$$\frac{450}{9} = \frac{9C}{9}$$

$$50 = C$$

The equivalent temperature is $50°\,$C.

**57.** Let $l = 199$, $w = 78.5$, and $h = 33$

$V = lwh$

$V = 199(78.5)(33) = 515,509.5$

The volume must be 515,509.5 cu in.

**59.** Let $\pi = 3.14$ and $d = 9.5$ so $r = 4.75$

$V = \frac{4}{3}\pi r^3 = \frac{4}{3}(3.14)(4.75)^3 = 449$

The volume is 449 cu in.

**61.** Let $C = 167$

$F = \frac{9}{5}C + 32 = \frac{9}{5}(167) + 32$

$\phantom{F} = 300.6 + 32 = 332.6$

The equivalent temperature is $332.6°$F.

**63.** $\dfrac{9}{x+5}$

**65.** $3(x+4)$

**67.** $3(x-12)$

**69.** Let $C = -78.5$

$F = \frac{9}{5}C + 32 = \frac{9}{5}(-78.5) + 32$

$\phantom{F} = -141.3 + 32 = -109.3$

The equivalent temperature is $-109.3°$F.

**71.** Let $d = 93,000,000$ and $r = 186,000$

$$d = rt$$

$$93,000,000 = 186,000t$$

$$\frac{93,000,000}{186,000} = \frac{186,000t}{186,000}$$

$$500 = t$$

It will take 500 seconds or $8\frac{1}{3}$ minutes.

**73.** Let $t = 365$ and $r = 20$

$d = rt = 20(365) = 7300$ inches

$\dfrac{7300 \text{ inches}}{1} \cdot \dfrac{1 \text{ foot}}{12 \text{ inch}} \approx 608.33$ feet

It moves about 608.33 feet.

**75.** Let $h = 15$, and $d = 2$ then $r = 1$.

$h = \dfrac{15 \text{ feet}}{1} \cdot \dfrac{12 \text{ inches}}{1 \text{ foot}} = 180$ inches

Use $\pi = 3.14$.

$V = \pi r^2 h$

$V = (3.14)(1)^2 (180)$

$V = 565.2$

The volume of the column is 565 cu in.

**77.** Let $F = 78$

$$78 = \frac{9}{5}C + 32$$

$$5(78) = 5\left(\frac{9}{5}\right)C + 5(32)$$

$$390 = 9C + 160$$

$$390 - 160 = 9C + 160 - 160$$

$$230 = 9C$$

$$\frac{230}{9} = \frac{9C}{9}$$

$$25\frac{5}{9} = C$$

The equivalent temperature is $25\frac{5}{9}°$ C.

**79.** The original parallelogram has an area

$V = bh$

The altered box, has a base $2b$,

a height $2h$ and a new area.

$A = 2b(2h)$

$A = 4bh.$

The area is multiplied by 4.

**Exercise Set 2.7**

**1.** Let $x$ = the decrease in price.

$x = 0.25(256) = 64$

The decrease in price is \$64.

The sale price is $256 - 64 = \$192.$

**3.** Let $x$ = the original price and

$0.25x$ = the discount

$x - 0.25x = 78$

$0.75x = 78$

$\dfrac{0.75x}{0.75} = \dfrac{78}{0.75}$

$x = 104$

The original price was \$104

**5.** Increase $= 380,000 - 220,000 = 160,000$

Let $x$ = the percent increase

$x(220,000) = 160,000$

$\dfrac{220,000x}{220,000} = \dfrac{160,000}{220,000}$

$x = 0.727$

The percent increase was 73%

**7.** Let $x$ = the time traveled by the jet plane

$Rate \cdot Time = Distance$

| | | | |
|---|---|---|---|
| Jet | 500 | $x$ | $500x$ |
| Prop | 200 | $x + 2$ | $200(x+2)$ |

$d = d$

$500x = 200(x + 2)$

$500x = 200x + 400$

$500x - 200x = 200x - 200x + 400$

$300x = 400$

$\dfrac{300x}{300} = \dfrac{400}{300}$

$x = \dfrac{4}{3}$

The jet traveled for $\dfrac{4}{3}$ hours

$d = rt$

$d = 500\left(\dfrac{4}{3}\right) = 666\dfrac{2}{3}$ miles

**9.** Let $x$ = the time to get to Disneyland

and $7.2 - x$ = the time to return

$Rate \cdot Time = Distance$

| | | | |
|---|---|---|---|
| Going | 50 | $x$ | $50x$ |
| Returning | 40 | $7.2 - x$ | $40(7.2 - x)$ |

$d = d$

$50x = 40(7.2 - x)$

$50x = 288 - 40x$

$50x + 40x = 288 - 40x + 40x$

$90x = 288$

$\dfrac{90x}{90} = \dfrac{288}{90}$

$x = 3.2$

It took 3.2 hours to get to Disneyland

$d = rt$

$d = 50(3.2) = 160$ miles

**11.** Let $x$ = the amount of pure acid.

*No.of gallons · Strength = Amt of Acid*

| 100% | $x$ | 1.00 | $x$ |
|------|-----|------|-----|
| 40% | 2 | 0.4 | $2(0.4)$ |
| 70% | $x+2$ | 0.7 | $0.7(x+2)$ |

$$x + 2(0.4) = 0.7(x+2)$$
$$x + 0.8 = 0.7x + 1.4$$
$$x - 0.7x + 0.8 = 0.7x - 0.7x + 1.4$$
$$0.3x + 0.8 = 1.4$$
$$0.3x + 0.8 - 0.8 = 1.4 - 0.8$$
$$0.3x = 0.6$$
$$\frac{0.3x}{0.3} = \frac{0.6}{0.3}$$
$$x = 2$$

Mix 2 gallons of pure acid

**13.** Let $x$ = the pounds of cashew nuts.

*Cost / lb · No.of lbs = Cost*

| Peanuts | 3 | 20 | $3(20)$ |
|---------|---|-----|---------|
| Cashews | 5 | $x$ | $5x$ |
| Mix | 3.50 | $x+20$ | $3.50(x+20)$ |

$$3(20) + 5x = 3.50(x+20)$$
$$60 + 5x = 3.5x + 70$$
$$60 + 5x - 3.5x = 3.5x - 3.5x + 70$$
$$60 + 1.5x = 70$$
$$60 - 60 + 1.5x = 70 - 60$$

$$1.5x = 10$$
$$\frac{1.5x}{1.5} = \frac{10}{1.5}$$
$$x = 6\frac{2}{3}$$
$$x = 2$$

Add $6\frac{2}{3}$ pounds of cashews

**15.** No. The 30% solution will only dilute the 50% solution, not make it more concentrated.

**17.** Let $x$ = the amount invested at 9% for one year.

*Principal · Rate = Interest*

| 9% | $x$ | .09 | $.09x$ |
|-------|------------|-----|------------------|
| 8% | $25,000 - x$ | .08 | $.08(25,000 - x)$ |
| Total | 25,000 | | 2,135 |

$$.09x + .08(25,000 - x) = 2,135$$
$$.09x + 2,000 - .08x = 2,135$$
$$.01x + 2,000 = 2,135$$
$$.01x + 2,000 - 2,000 = 2,135 - 2,000$$
$$.01x = 135$$
$$\frac{.01x}{.01} = \frac{135}{.01}$$
$$x = 13,500$$

$25,000 - x = 25,000 - 13,500 = 11,500$

She invested $11,500 @ 8% and $13,500 @ 9%

**19.** Let $x$ = the amount invested at 11% for one year.

*Principal · Rate = Interest*

| 11% | $x$ | .11 | $.11x$ |
|---|---|---|---|
| 4% | $10,000 - x$ | $-.04$ | $-.04(10,000 - x)$ |
| Total | 10,000 | | 650 |

$.11x - .04(10,000 - x) = 650$

$.11x - 400 + .04x = 650$

$.15x - 400 = 650$

$.15x - 400 + 400 = 650 + 400$

$.15x = 1050$

$\dfrac{.15x}{.15} = \dfrac{1050}{.15}$

$x = 7,000$

$10,000 - x = 10,000 - 7,000 = 3,000$

He invested $7,000 @ 11% and $3,000 @ 4%

**21.** Let $x$ = the amount paid.

Profit $= .2x$

Amount paid + profit = selling price

$x + .2x = 4,680$

$1.2x = 4,680$

$\dfrac{1.2x}{1.2} = \dfrac{4,680}{1.2}$

$x = 3,900$

He paid $3,900

**23.** Let $x$ = the amount invested at 10% for one year.

*Principal · Rate = Interest*

| 10% | $x$ | .10 | $.10x$ |
|---|---|---|---|
| 8% | $54,000 - x$ | .08 | $.08(54,000 - x)$ |

$.10x = .08(54,000 - x)$

$.10x = 4320 - .08x$

$.10x + .08x = 4320 - .08x + .08x$

$.18x = 4320$

$\dfrac{.18x}{.18} = \dfrac{4320}{.18}$

$x = 24,000$

$54,000 - x = 54,000 - 24,000 = 30,000$

Invest $30,000 @ 8% and $24,000 @ 10%

**25.** Let $x$ = the time they talk

*Rate · Time = Distance*

| Cade | 5 | $x$ | $5x$ |
|---|---|---|---|
| Kathleen | 4 | $x$ | $4x$ |
| Total | | | 20 |

$5x + 4x = 20$

$9x = 20$

$\dfrac{9x}{9} = \dfrac{20}{9}$

$x = 2\dfrac{2}{9}$

They can talk for $2\dfrac{2}{9}$ hours

**27.** Let $x$ = the amount of 20% alloy.

*No.of Oz · Strength = Amt of Copper*

|       | No.of Oz | Strength | Amt of Copper |
|-------|----------|----------|---------------|
| 50%   | 200      | 0.5      | $200(0.5)$    |
| 20%   | $x$      | 0.2      | $0.2x$        |
| Mix   | $x+200$  | 0.3      | $0.3(x+200)$  |

$$200(0.5)+0.2x = 0.3(x+200)$$
$$100+0.2x = 0.3x+60$$
$$100+0.2x-0.2x = 0.3x-0.2x+60$$
$$100 = 0.1x+60$$
$$100-60 = 0.1x+60-60$$
$$40 = 0.1x$$
$$\frac{40}{0.1} = \frac{0.1x}{0.1}$$
$$400 = x$$

Mix with 400 ounces of 20% alloy.

**29.** Increase $= 70-40 = 30$

Let $x$ = the percent increase

$$x(40) = 30$$
$$\frac{40x}{40} = \frac{30}{40}$$
$$x = 0.75$$

The percent increase was 75%

**31.** Let $x$ = the amount invested at 9% for one year.

*Principal · Rate = Interest*

|       | Principal | Rate | Interest   |
|-------|-----------|------|------------|
| 9%    | $x$       | .09  | $.09x$     |
| 6%    | 3000      | .06  | $.06(3000)$|
| Total |           |      | 585        |

$$.09x+.06(3000) = 585$$
$$.09x+180 = 585$$
$$.09x+180-180 = 585-180$$

$$.09x = 405$$
$$\frac{.09x}{.09} = \frac{405}{.09}$$
$$x = 4500$$

Should invest $4500 @ 9%.

**33.** Let $x$ = the number of cards(in millions) issued in 2001.

Increase $= 1.17x$

No. in 2001 + increase = no. in 2003

$$x+1.17x = 500$$
$$2.17x = 500$$
$$\frac{2.17x}{2.17} = \frac{500}{2.17}$$
$$x = 230$$

230 million cards were issued in 2001

**35.** Let $x$ = the rate of hiker1

*Rate · Time = Distance*

|        | Rate    | Time | Distance     |
|--------|---------|------|--------------|
| Hiker1 | $x$     | 2    | $2x$         |
| Hiker2 | $x+1.1$ | 2    | $2(x+1.1)$   |
| Total  |         |      | 11           |

$$2x+2(x+1.1) = 11$$
$$2x+2x+2.2 = 11$$
$$4x+2.2 = 11$$
$$4x+2.2-2.2 = 11-2.2$$
$$4x = 8.8$$
$$\frac{4x}{4} = \frac{8.8}{4}$$
$$x = 2.2$$

$x+1.1 = 2.2+1.1 = 3.3$

Hiker1: 2.2 mph; Hiker2: 3.3 mph.

**37.** Increase $= 21.0 - 20.7 = 0.3$

Let $x =$ the percent increase

$x(20.7) = 0.3$

$$\frac{20.7x}{20.7} = \frac{0.3}{20.7}$$

$x = 0.0145$

The percent increase was 1.45%

**39.** Let $x =$ the time it takes them to meet

$$Rate \cdot Time = Distance$$

|  | | | |
|--------|---|---|-----|
| Nedra | 3 | $x$ | $3x$ |
| Latonya | 4 | $x$ | $4x$ |
| Total | | | 12 |

$3x + 4x = 12$

$7x = 12$

$$\frac{7x}{7} = \frac{12}{7}$$

$$x = 1\frac{5}{7}$$

They meet in $1\dfrac{5}{7}$ hours

**41.** $3 + (-7) = -4$

**43.** $\dfrac{3}{4} - \dfrac{3}{16} = \dfrac{3}{4}\left(\dfrac{4}{4}\right) - \dfrac{3}{16} = \dfrac{12}{16} - \dfrac{3}{16} = \dfrac{9}{16}$

**45.** $-5 - (-1) = -5 + 1 = -4$

**47.** $-5 > -7$

**49.** $|-5| = -(-5)$

**51.** $\qquad\qquad R = C$

$24x = 100 + 20x$

$24x - 20x = 100 + 20x - 20x$

$4x = 100$

$$\frac{4x}{x} = \frac{100}{4}$$

$x = 25$

Should sell 25 skateboards to break even.

**53.** $\qquad\qquad R = C$

$7.50x = 4.50x + 2400$

$7.50x - 4.50x = 4.50x - 4.50x + 2400$

$3x = 2400$

$$\frac{3x}{3} = \frac{2400}{3}$$

$x = 800$

Should sell 800 books to break even.

**55.** Answers may vary

**57.** $x(300) = 23$

$$x = \frac{23}{300} \approx 0.077 = 7.7\%$$

This is about 7.7% of the daily value.

**59.** Let $x =$ percent of calories from fat.

$x(280) = 9(6)$

$$x = \frac{54}{280} \approx 0.193 = 19.3\%$$

One serving contains 19.3% of calories from fat.

**61.** Answers may vary

**Mental Math 2.8**

**1.** $5x > 10$

   $x > 2$

**2.** $4x < 20$

   $x < 5$

**3.** $2x \geq 16$

   $x \geq 8$

**4.** $9x \leq 63$

   $x \leq 7$

**Exercise Set 2.8**

**1.** $x \geq 2, \ [2, \infty)$

**3.** $(-\infty, -5), \ x < -5$

**5.** $x \leq -1, \ (-\infty, -1]$

**7.** $x < \dfrac{1}{2}, \ \left(-\infty, \dfrac{1}{2}\right)$

**9.** $y \geq 5, \ [5, \infty)$

**11.** $2x < -6$

   $x < -3, \ (-\infty, -3)$

**13.** $x - 2 \geq -7$

   $x \geq -5, \ [-5, \infty)$

**15.** $-8x \leq 16$

   $\dfrac{-8x}{-8} \geq \dfrac{16}{-8}$

   $x \geq -2, \ [-2, \infty)$

**17.** $3x - 5 > 2x - 8$

   $x - 5 > -8$

   $x > -3, \ (-3, \infty)$

**19.** $4x - 1 \leq 5x - 2x$

   $4x - 1 \leq 3x$

   $x - 1 \leq 0$

   $x \leq 1, \ (-\infty, 1]$

**21.** $x - 7 < 3(x + 1)$

   $x - 7 < 3x + 3$

   $-2x - 7 < 3$

   $-2x < 10$

   $x > -5, \ (-5, \infty)$

[number line marked at -5, open bracket]

[number line marked at 16, open bracket]

**23.** $-6x + 2 \le 2(5 - x)$

$-6x + 2 \le 10 - 2x$

$-4x + 2 \le 10$

$-4x \le 8$

$x \ge -2, \ [-2, \infty)$

[number line marked at -2]

**25.** $4(3x - 1) \le 5(2x - 4)$

$12x - 4 \le 10x - 20$

$2x - 4 \le -20$

$2x \le -16$

$x \le -8, \ (-\infty, -8]$

[number line marked at -8]

**27.** $3(x + 2) - 6 > -2(x - 3) + 14$

$3x + 6 - 6 > -2x + 6 + 14$

$3x > -2x + 20$

$5x > 20$

$x > 4, \ (4, \infty)$

[number line marked at 4, open bracket]

**29.** $-2x \le -40$

$x \ge 20, \ [20, \infty)$

[number line marked at 20]

**31.** $-9 + x > 7$

$x > 16, \ (16, \infty)$

**33.** $3x - 7 < 6x + 2$

$-3x - 7 < 2$

$-3x < 9$

$x > -3, \ (-3, \infty)$

[number line marked at -3, open bracket]

**35.** $5x - 7x \ge x + 2$

$-2x \ge x + 2$

$-3x \ge 2$

$x \le -\dfrac{2}{3}, \ \left(-\infty, -\dfrac{2}{3}\right]$

[number line marked at -2/3]

**37.** $\dfrac{3}{4}x > 2$

$x > \dfrac{8}{3}, \ \left(\dfrac{8}{3}, \infty\right)$

[number line marked at 8/3, open bracket]

**39.** $3(x - 5) < 2(2x - 1)$

$3x - 15 < 4x - 2$

$-x - 15 < -2$

$-x < 13$

$x > -13, \ (-13, \infty)$

[number line marked at -13, open bracket]

**41.** $4(2x+1) < 4$

$8x + 4 < 4$

$8x < 0$

$x < 0, \ (-\infty, 0)$

**43.** $-5x + 4 \geq -4(x-1)$

$-5x + 4 \geq -4x + 4$

$-x + 4 \geq 4$

$-x \geq 0$

$x \leq 0, \ (-\infty, 0]$

**45.** $-2(x-4) - 3x < -(4x+1) + 2x$

$-2x + 8 - 3x < -4x - 1 + 2x$

$-5x + 8 < -2x - 1$

$-3x + 8 < -1$

$-3x < -9$

$x > 3, \ (3, \infty)$

**47.** $-3x + 6 \geq 2x + 6$

$-5x + 6 \geq 6$

$-5x \geq 0$

$x \leq 0, \ (-\infty, 0]$

**49.** Answers may vary

**51.** $-1 < x < 3, \ (-1, 3)$

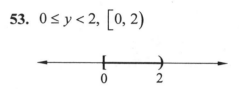

**53.** $0 \leq y < 2, \ [0, 2)$

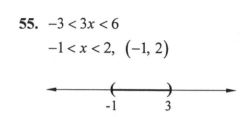

**55.** $-3 < 3x < 6$

$-1 < x < 2, \ (-1, 2)$

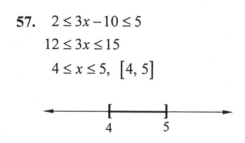

**57.** $2 \leq 3x - 10 \leq 5$

$12 \leq 3x \leq 15$

$4 \leq x \leq 5, \ [4, 5]$

**59.** $-4 < 2(x-3) \leq 4$

$-4 < 2x - 6 \leq 4$

$2 < 2x \leq 10$

$1 < x \leq 5, \ (1, 5]$

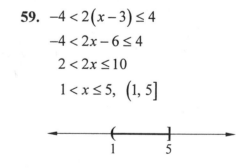

**61.** $-2 < 3x - 5 < 7$

$3 < 3x < 12$

$1 < x < 4, \ (1, 4)$

**63.** $-6 < 3(x-2) \le 8$

$-6 < 3x - 6 \le 8$

$0 < 3x \le 14$

$0 < x \le \dfrac{14}{3}, \quad \left(0, \dfrac{14}{3}\right]$

**65.** Answers may vary

**67.** $2x + 6 > -14$

$2x > -20$

$x > -10$

**69.** Let $x$ = the number of people invited.

$34x + 50 \le 3000$

$34x \le 2950$

$x \le 86.8$

They must invite no more than 86 people.

**71.** Let $x$ = the length.

$2l + 2w = P$

$2x + 2(15) \le 100$

$2x + 30 \le 100$

$2x \le 70$

$x \le 35$

The length can be no greater than 35 cm..

**73.** Let $x$ = the rate for \$5000 for one year.

*Principal · Rate = Interest*

| 11% | 10,000 | .11 | .11(10,000) |
|---|---|---|---|
| ? | 5000 | $x$ | $5000x$ |
| Total | | | 1600 |

$1100 + 5000x \ge 1600$

$5000x \ge 500$

$x \ge .1$

Should invest the \$5000 at 10% or more.

**75.** Let $x$ = his score on the third game.

$\dfrac{146 + 201 + x}{3} \ge 180$

$3\left(\dfrac{146 + 201 + x}{3}\right) \ge 3(180)$

$347 + x \ge 540$

$x \ge 193$

He must score at least 193.

**77.** $x < 200$       recommended

$200 \le x \le 240$   borderline

$x > 240$         high

**79.** Let $x$ = the unknown number.

$-5 < 2x + 1 < 7$

$-6 < 2x < 6$

$-3 < x < 3$

All numbers between $-3$ and 3.

**81.**
$$C = \frac{5}{9}(F - 32)$$
$$-39 \le \frac{5}{9}(F - 32) \le 45$$
$$-351 \le 5(F - 32) \le 405$$
$$-351 \le 5F - 160 \le 405$$
$$-191 \le 5F \le 565$$
$$-38.2° \le F \le 113°$$

**83.** $2^3 = (2)(2)(2) = 8$

**85.** $1^{12} = (1)(1)(1)(1)(1)(1)(1)(1)(1)(1)(1)(1) = 1$

**87.** $\left(\frac{4}{7}\right)^2 = \left(\frac{4}{7}\right)\left(\frac{4}{7}\right) = \frac{16}{49}$

**89.** \$3225

**91.** 1986-1990

**93.**
$$C = 3.14d$$
$$2.9 \le 3.14d \le 3.1$$
$$0.924 \le d \le 0.987$$
The diameter must be between
0.924 cm and 0.987 cm.

**95.** $x(x + 4) > x^2 - 2x + 6$
$$x^2 + 4x > x^2 - 2x + 6$$
$$4x > -2x + 6$$
$$6x > 6$$
$$x > 1, \quad (1, \infty)$$

**97.** $x^2 + 6x - 10 < x(x - 10)$
$$x^2 + 6x - 10 < x^2 - 10x$$
$$6x - 10 < -10x$$
$$16x - 10 < 0$$
$$16x < 10$$
$$x < \frac{10}{16} = \frac{5}{8}$$
$$x < \frac{5}{8}, \quad \left(-\infty, \frac{5}{8}\right)$$

**Chapter 2 Review**

**1.** $5x - x + 2x = 6x$

**2.** $0.2z - 4.6x - 7.4z = -4.6x - 7.2z$

**3.** $\frac{1}{2}x + 3 + \frac{7}{2}x - 5 = \frac{8}{2}x - 2 = 4x - 2$

**4.** $\frac{4}{5}y + 1 + \frac{6}{5}y + 2 = \frac{10}{5}y + 3 = 2y + 3$

**5.** $2(n - 4) + n - 10 = 2n - 8 + n - 10 = 3n - 18$

**6.** $3(w + 2) - (12 - w) = 3w + 6 - 12 + w$
$$= 4w - 6$$

**7.** $(x + 5) - (7x - 2) = x + 5 - 7x + 2 = -6x + 7$

**8.** $(y - 0.7) - (1.4y - 3) = y - 0.7 - 1.4y + 3$
$$= -0.4y + 2.3$$

**9.** $3x - 7$

**10.** $3x + 2(x + 2.8) = 3x + 2x + 5.6 = 5x + 5.6$

**11.** $8x + 4 = 9x$
$4 = x$

**12.** $5y - 3 = 6y$
$-3 = y$

**13.** $3x - 5 = 4x + 1$
$-5 = x + 1$
$-6 = x$

**14.** $2x - 6 = x - 6$
$x - 6 = -6$
$x = 0$

**15.** $4(x + 3) = 3(1 + x)$
$4x + 12 = 3 + 3x$
$x + 12 = 3$
$x = -9$

**16.** $6(3 + n) = 5(n - 1)$
$18 + 6n = 5n - 5$
$18 + n = -5$
$n = -23$

**17.** $\quad x - 5 = 3$
$x - 5 + \underline{5} = 3 + \underline{5}$
$x = 8$

**18.** $\quad x + 9 = -2$
$x + 9 - \underline{9} = -2 - \underline{9}$
$x = -11$

**19.** $10 - x$

**20.** $(x - 5)$ in.

**21.** $180 - (x + 5) = 180 - x - 5 = (175 - x)^\circ$

**22.** $\qquad \dfrac{3}{4}x = -9$
$\dfrac{4}{3}\left(\dfrac{3}{4}x\right) = \dfrac{4}{3}(-9)$
$x = -12$

**23.** $\dfrac{x}{6} = \dfrac{2}{3}$
$x = 4$

**24.** $-3x + 1 = 19$
$-3x = 18$
$x = -6$

**25.** $5x + 25 = 20$
$5x = -5$
$x = -1$

**26.** $5x + x = 9 + 4x - 1 + 6$
$6x = 4x + 14$
$2x = 14$
$x = 7$

**27.** $-y + 4y = 7 - y - 3 - 8$
$3y = -y - 4$
$4y = -4$
$y = -1$

**28.** Let $x =$ the first even integer.
Sum $= x + (x + 2) + (x + 4) = 3x + 6$

**29.**     $\dfrac{2}{7}x - \dfrac{5}{7} = 1$

$7\left(\dfrac{2}{7}x\right) - 7\left(\dfrac{5}{7}\right) = 7(1)$

$2x - 5 = 7$

$2x = 12$

$x = 6$

**30.** $\dfrac{5}{3}x + 4 = \dfrac{2}{3}x$

$\dfrac{3}{3}x + 4 = 0$

$x = -4$

**31.** $-(5x + 1) = -7x + 3$

$-5x - 1 = -7x + 3$

$2x - 1 = 3$

$2x = 4$

$x = 2$

**32.** $-4(2x + 1) = -5x + 5$

$-8x - 4 = -5x + 5$

$-3x - 4 = 5$

$-3x = 9$

$x = -3$

**33.** $-6(2x - 5) = -3(9 + 4x)$

$-12x + 30 = -27 - 12x$

$30 = -27$

There is no solution.

**34.** $3(8y - 1) = 6(5 + 4y)$

$24y - 3 = 30 + 24y$

$-3 = 30$

There is no solution.

**35.** $\dfrac{3(2 - z)}{5} = z$

$3(2 - z) = 5z$

$6 - 3z = 5z$

$6 = 8z$

$\dfrac{6}{8} = z$

$\dfrac{3}{4} = z$

**36.** $\dfrac{4(n + 2)}{5} = -n$

$4(n + 2) = -5n$

$4n + 8 = -5n$

$8 = -9n$

$-\dfrac{8}{9} = n$

**37.** $5(2n - 3) - 1 = 4(6 + 2n)$

$10n - 15 - 1 = 24 + 8n$

$10n - 16 = 24 + 8n$

$2n - 16 = 24$

$2n = 40$

$n = 20$

**38.** $-2(4y - 3) + 4 = 3(5 - y)$

$-8y + 6 + 4 = 15 - 3y$

$-8y + 10 = 15 - 3y$

$10 = 15 + 5y$

$-5 = 5y$

$-1 = y$

**39.** $9z - z + 1 = 6(z - 1) + 7$

$8z + 1 = 6z - 6 + 7$

$8z + 1 = 6z + 1$

$2z + 1 = 1$

$2z = 0$

$z = 0$

**40.** $5t - 3 - t = 3(t + 4) - 15$

$4t - 3 = 3t + 12 - 15$

$4t - 3 = 3t - 3$

$t - 3 = -3$

$t = 0$

**41.** $-n + 10 = 2(3n - 5)$

$-n + 10 = 6n - 10$

$10 = 7n - 10$

$20 = 7n$

$\dfrac{20}{7} = n$

**42.** $-9 - 5a = 3(6a - 1)$

$-9 - 5a = 18a - 3$

$-9 = 23a - 3$

$-6 = 23a$

$-\dfrac{6}{23} = a$

**43.** $\dfrac{5(c + 1)}{6} = 2c - 3$

$5(c + 1) = 6(2c - 3)$

$5c + 5 = 12c - 18$

$-7c + 5 = -18$

$-7c = -23$

$c = \dfrac{23}{7}$

**44.** $\dfrac{2(8 - a)}{3} = 4 - 4a$

$2(8 - a) = 3(4 - 4a)$

$16 - 2a = 12 - 12a$

$10a + 16 = 12$

$10a = -4$

$a = \dfrac{-4}{10}$

$a = -\dfrac{2}{5}$

**45.** $200(70x - 3560) = -179(150x - 19,300)$

$14,000x - 712,000 = -26,850x + 3,454,700$

$40,850x - 712,000 = 3,454,700$

$40,850x = 4,166,700$

$x = 102$

**46.** $1.72y - 0.04y = 0.42$

$1.68y = 0.42$

$y = 0.25$

**47.** Let $x =$ the unknown number.

$\dfrac{x}{3} = x - 2$

$x = 3(x - 2)$

$x = 3x - 6$

$-2x = -6$

$x = 3$

The number is 3.

**48.** Let $x$ = the unknown number.

$$2(x+6) = -x$$
$$2x+12 = -x$$
$$12 = -3x$$
$$-4 = x$$

The number is $-4$.

**49.** Let $x$ = the length of a side of the base and $3x + 68$ = the height.

$$x + 3x + 68 = 1380$$
$$4x + 68 = 1380$$
$$4x = 1312$$
$$x = 328$$
$$3x + 68 = 3(328) + 68 = 1052$$

The height is 1052 ft.

**50.** Let $x$ = the length of the shorter piece and $2x$ = the length of the other.

$$x + 2x = 12$$
$$3x = 12$$
$$x = 4$$
$$2x = 8$$

The lengths are 4 feet and 8 feet.

**51.** Let $x$ = the first area code and $3x + 34$ = the other.

$$x + 3x + 34 = 1262$$
$$4x + 34 = 1262$$
$$4x = 1228$$
$$x = 307$$
$$3x + 34 = 3(307) + 34 = 955$$

The codes are 307 and 955.

**52.** Let $x$ = the first integer, $x + 2$ = the second integer, and $x + 4$ = the third.

$$x + x + 2 + x + 4 = -114$$
$$3x + 6 = -114$$
$$3x = -120$$
$$x = -40$$
$$x + 2 = -38, \text{ and } x + 4 = -36$$

The integers are $-40, -38, -36$.

**53.** Let $P = 46$ and $l = 14$.

$$P = 2l + 2w$$
$$46 = 2(14) + 2w$$
$$46 = 28 + 2w$$
$$18 = 2w$$
$$9 = w$$

**54.** Let $V = 192$, $l = 8$, and $w = 6$.

$$V = lwh$$
$$192 = 8(6)h$$
$$192 = 48h$$
$$4 = h$$

**55.**
$$y = mx + b$$
$$y - b = mx$$
$$\frac{y-b}{x} = m$$

**56.**
$$r = vst - 9$$
$$r + 9 = vst$$
$$\frac{r+9}{vt} = s$$

**57.** $2y - 5x = 7$
$$-5x = -2y + 7$$
$$x = \frac{-2y+7}{-5} = \frac{2y-7}{5}$$

**58.** $3x - 6y = -2$

$$-6y = -3x - 2$$

$$y = \frac{-3x - 2}{-6} = \frac{3x + 2}{6}$$

**59.** $C = \pi D$

$$\frac{C}{D} = \pi$$

**60.** $C = 2\pi r$

$$\frac{C}{2r} = \pi$$

**61.** Let $V = 900$, $l = 20$, and $h = 3$.

$$V = lwh$$

$$900 = 20w(3)$$

$$900 = 60w$$

$$15 = w$$

Width $= 15$ meters

**62.** Let $F = 104$

$$C = \frac{5}{9}(F - 32) = \frac{5}{9}(104 - 32) = \frac{5}{9}(72)$$

$$= 40$$

The temperature was $40°$ C.

**63.** Let $d = 10,000$ and $r = 125$

$$d = rt$$

$$10,000 = 125t$$

$$80 = t$$

It will take 1 hour and 20 minutes.

**64.** Let $x =$ the amount invested at 10.5% for one year.

*Principal $\cdot$ Rate $=$ Interest*

| 10.5% | $x$ | .105 | $.105x$ |
|---|---|---|---|
| 8.5% | $50,000 - x$ | .085 | $.085(50,000 - x)$ |
| Total | $50,000$ | | $4,550$ |

$$.105x + .085(50,000 - x) = 4,550$$

$$.105x + 4,250 - .085x = 4,550$$

$$.02x + 4,250 = 4,550$$

$$.02x = 300$$

$$x = 15,000$$

$$50,000 - x = 50,000 - 15,000 = 35,000$$

She invested \$35,000 @ 8.5% and
\$15,000 @ 10.5%

**65.** Let $x =$ the number of dimes, $2x =$ the number of quarters, and
$500 - x - 2x =$ the number of nickles.

*No. of Coins $\cdot$ Value $=$ Amt of Money*

| | | | |
|---|---|---|---|
| Dimes | $x$ | .1 | $.1x$ |
| Quarters | $2x$ | .25 | $.25(2x)$ |
| Nickles | $500 - 3x$ | .05 | $.05(500 - 3x)$ |
| Total | $500$ | | $88$ |

$$.1x + .25(2x) + .05(500 - 3x) = 88$$

$$.1x + .5x + 25 - .15x = 88$$

$$.45x + 25 = 88$$

$$.45x = 63$$

$$x = 140$$

$$500 - 3x = 500 - 3(140) = 500 - 420 = 80$$

There were 80 nickles in the pay phone.

**66.** Let $x$ = the time traveled by the Amtrak

$$Rate \cdot Time = Distance$$

| Amtrak | 60 | $x$ | $60x$ |
|--------|-----|--------|------------|
| Freight | 45 | $x+1.5$ | $45(x+1.5)$ |

$$d = d$$
$$60x = 45(x+1.5)$$
$$60x = 45x + 67.5$$
$$15x = 67.5$$
$$x = 4.5$$

It will take 4.5 hours

**67.** Let $x$ = the time to cycle up

and $5 - x$ = the time to cycle down

$$Rate \cdot Time = Distance$$

| Up | 8 | $x$ | $8x$ |
|------|-----|--------|-----------|
| Down | 12 | $5-x$ | $12(5-x)$ |

$$d = d$$
$$8x = 12(5-x)$$
$$8x = 60 - 12x$$
$$20x = 60$$
$$x = 3$$

It took 3 hours to cycle up.

$$d = rt$$

$$d = 8(3) = 24 \text{ miles up.}$$

Total distance was 48 miles.

**68.** $x \le -2, \ (-\infty, -2]$

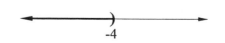

**69.** $x > 0, \ (0, \infty)$

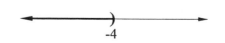

**70.** $-1 < x < 1, \ (-1, 1)$

**71.** $0.5 \le y < 1.5, \ [0.5, 1.5)$

**72.** $-2x \ge -20$

$$\frac{-2x}{-2} \le \frac{-20}{-2}$$

$$x \le 10, \ (-\infty, 10]$$

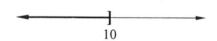

**73.** $-3x > 12$

$$\frac{-3x}{-3} < \frac{12}{-3}$$

$$x < -4, \ (-\infty, -4)$$

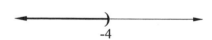

**74.** $5x - 7 > 8x + 5$

$$-3x - 7 > 5$$

$$-3x > 12$$

$$\frac{-3x}{-3} < \frac{12}{-3}$$

$$x < -4, \ (-\infty, -4)$$

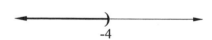

**75.** $x + 4 \geq 6x - 16$

$-5x + 4 \geq -16$

$-5x \geq -20$

$\dfrac{-5x}{-5} \leq \dfrac{-20}{-5}$

$x \leq 4, \ (-\infty, 4]$

**76.** $2 \leq 3x - 4 < 6$

$6 \leq 3x < 10$

$2 \leq x < \dfrac{10}{3}$

$2 \leq x < \dfrac{10}{3}, \ \left[2, \dfrac{10}{3}\right)$

**77.** $-3 < 4x - 1 < 2$

$-2 < 4x < 3$

$-\dfrac{1}{2} < x < \dfrac{3}{4}, \ \left(-\dfrac{1}{2}, \dfrac{3}{4}\right)$

**78.** $-2(x - 5) > 2(3x - 2)$

$-2x + 10 > 6x - 4$

$-8x + 10 > -4$

$-8x > -14$

$\dfrac{-8x}{-8} < \dfrac{-14}{-8}$

$x < \dfrac{7}{4}, \ \left(-\infty, \dfrac{7}{4}\right)$

7/4

**79.** $4(2x - 5) \leq 5x - 1$

$8x - 20 \leq 5x - 1$

$3x - 20 \leq -1$

$3x \leq 19$

$x \leq \dfrac{19}{3}, \ \left(-\infty, \dfrac{19}{3}\right]$

19/3

**80.** Let $x$ = the amount of sales then

$0.05x$ = her commission.

$175 + 0.05x \geq 300$

$0.05x \geq 125$

$x \geq 2500$

Sales must be at least $2500.

**81.** Let $x$ = her score on the fourth round.

$\dfrac{76 + 82 + 79 + x}{4} < 80$

$237 + x < 320$

$x < 83$

Her score must be less than 83.

**Chapter 2 Test**

**1.** $2y - 6 - y - 4 = y - 10$

**2.** $2.7x + 6.1 + 3.2x - 4.9 = 5.9x + 1.2$

**3.** $4(x - 2) - 3(2x - 6) = 4x - 8 - 6x + 18$

$= -2x + 10$

**4.**  $7 + 2(5y - 3) = 7 + 10y - 6$
$$= 10y + 1$$

**5.**  $-\dfrac{4}{5}x = 4$
$$x = -5$$

**6.**  $4(n - 5) = -(4 - 2n)$
$$4n - 20 = -4 + 2n$$
$$2n - 20 = -4$$
$$2n = 16$$
$$n = 8$$

**7.**  $5y - 7 + y = -(y + 3y)$
$$6y - 7 = -4y$$
$$-7 = -10y$$
$$\dfrac{7}{10} = y$$

**8.**  $4z + 1 - z = 1 + z$
$$3z + 1 = 1 + z$$
$$2z + 1 = 1$$
$$2z = 0$$
$$z = 0$$

**9.**  $\dfrac{2(x + 6)}{3} = x - 5$
$$2(x + 6) = 3(x - 5)$$
$$2x + 12 = 3x - 15$$
$$12 = x - 15$$
$$27 = x$$

**10.**  $\dfrac{1}{2} - x + \dfrac{3}{2} = x - 4$
$$2\left(\dfrac{1}{2} - x + \dfrac{3}{2}\right) = 2(x - 4)$$
$$1 - 2x + 3 = 2x - 8$$
$$-2x + 4 = 2x - 8$$
$$-4x + 4 = -8$$
$$-4x = -12$$
$$x = 3$$

**11.**  $-0.3(x - 4) + x = 0.5(3 - x)$
$$10\left[-0.3(x - 4) + x\right] = 10\left[0.5(3 - x)\right]$$
$$-3(x - 4) + 10x = 5(3 - x)$$
$$-3x + 12 + 10x = 15 - 5x$$
$$7x + 12 = 15 - 5x$$
$$12x + 12 = 15$$
$$12x = 3$$
$$x = \dfrac{3}{12} = \dfrac{1}{4} = 0.25$$

**12.**  $-4(a + 1) - 3a = -7(2a - 3)$
$$-4a - 4 - 3a = -14a + 21$$
$$-7a - 4 = -14a + 21$$
$$7a - 4 = 21$$
$$7a = 25$$
$$a = \dfrac{25}{7}$$

**13.**  $-2(x - 3) = x + 5 - 3x$
$$-2x + 6 = -2x + 5$$
$$6 = 5 \quad \text{a contradiction}$$
There is no solution.

**14.** Let $x$ = the number.

$$x + \frac{2}{3}x = 35$$
$$3x + 2x = 105$$
$$5x = 105$$
$$x = 21$$

The number is 21.

**15.** Let $l = 35$, and $w = 20$.

$$2A = 2lw = 2(35)(20) = 1400$$

Let $x$ = the number of gallons needed at 200 square feet per gallon.

$$1400 = 200x$$
$$7 = x$$

7 gallons are needed.

**16.** Let $x$ = the amount invested at 10% for one year.

*Principal · Rate = Interest*

| | | | |
|---|---|---|---|
| 10% | $x$ | .10 | $.1x$ |
| 12% | $2x$ | .12 | $.12(2x)$ |
| Total | | | 2890 |

$$.1x + .12(2x) = 2890$$
$$.1x + .24x = 2890$$
$$.34x = 2890$$
$$x = 8500$$
$$2x = 2(8500) = 17,000$$

He invested $8500 @ 10% and $17,000 @ 12%.

**17.** Let $x$ = the time they travel

*Rate · Time = Distance*

| | | | |
|---|---|---|---|
| Train1 | 50 | $x$ | $50x$ |
| Train2 | 64 | $x$ | $64x$ |
| Total | | | 285 |

$$50x + 64x = 285$$
$$114x = 285$$
$$x = 2\frac{1}{2}$$

They must travel for $2\frac{1}{2}$ hours

**18.** Let $y = -14$, $m = -2$, and $b = -2$.

$$y = mx + b$$
$$-14 = -2x - 2$$
$$-12 = -2x$$
$$6 = x$$

**19.** $V = \pi r^2 h$

$$\frac{V}{\pi r^2} = \frac{\pi r^2 h}{\pi r^2}$$
$$\frac{V}{\pi r^2} = h$$

**20.** $3x - 4y = 10$

$$-4y = -3x + 10$$
$$y = \frac{-3x + 10}{-4}$$
$$y = \frac{3x - 10}{4}$$

**21.**  $3x - 5 \geq 7x + 3$

$-4x - 5 \geq 3$

$-4x \geq 8$

$\dfrac{-4x}{-4} \leq \dfrac{8}{-4}$

$x \leq -2, \ \left(-\infty, -2\right]$

**22.**  $x + 6 > 4x - 6$

$-3x + 6 > -6$

$-3x > -12$

$\dfrac{-3x}{-3} < \dfrac{-12}{-3}$

$x < 4, \ \left(-\infty, 4\right)$

**23.**  $-2 < 3x + 1 < 8$

$-3 < 3x < 7$

$-1 < x < \dfrac{7}{3}, \ \left(-1, \dfrac{7}{3}\right)$

**24.**  $\dfrac{2(5x + 1)}{3} > 2$

$2(5x + 1) > 6$

$10x + 2 > 6$

$10x > 4$

$x > \dfrac{4}{10} = \dfrac{2}{5}, \ \left(\dfrac{2}{5}, \infty\right)$

## Chapter 2 Cumulative Review

**1.  a.** The natural numbers are 11 and 112.

   **b.** The whole numbers are 0, 11 and 112.

   **c.** The integers are $-3, -2, 0, 11$ and 112.

   **d.** The rational numbers are $-3, -2, 0, \frac{1}{4}, 11$ and 112.

   **e.** The irrational number is $\sqrt{2}$.

   **f.** The real numbers are all numbers in the given set.

**2.  a.** The natural numbers are 2, 7 and 8.

   **b.** The whole numbers are 0, 2, 7 and 8.

   **c.** The integers are $-185, 0, 2, 7$ and 8.

   **d.** The rational numbers are $-185, -\frac{1}{5}, 0, 2, 7$ and 8.

   **e.** The irrational number is $\sqrt{3}$.

   **f.** The real numbers are all numbers in the given set.

**3.  a.** $|4| = 4$   **b.** $|-5| = 5$   **c.** $|0| = 0$

**4.  a.** $|5| = 5$   **b.** $|-8| = 8$   **c.** $|-2/3| = 2/3$

**5.  a.** $40 = 2 \cdot 2 \cdot 2 \cdot 5$   **b.** $63 = 3 \cdot 3 \cdot 7$

**6.  a.** $44 = 2 \cdot 2 \cdot 11$   **b.** $90 = 2 \cdot 3 \cdot 3 \cdot 5$

**7.**  $\dfrac{2}{5} = \dfrac{2}{5} \cdot \dfrac{4}{4} = \dfrac{8}{20}$

**8.**  $\dfrac{2}{3} = \dfrac{2}{3} \cdot \dfrac{8}{8} = \dfrac{16}{24}$

**9.** $3\left[4(5+2)-10\right]=3\left[4(7)-10\right]$
$$=3\left[28-10\right]$$
$$=3\left[18\right]$$
$$=54$$

**10.** $5\left[16-4(2+1)\right]=5\left[16-4(3)\right]$
$$=5\left[16-12\right]$$
$$=5\left[4\right]$$
$$=20$$

**11.** Let $x=2$.
$$3x+10=8x$$
$$3(2)+10\overset{?}{=}8(2)$$
$$6+10\overset{?}{=}16$$
$$16=16$$
2 is a solution of the equation.

**12.** Let $x=3$.
$$5x-2=4x$$
$$5(3)-2\overset{?}{=}4(3)$$
$$15-2\overset{?}{=}12$$
$$13\neq 12$$
3 is not a solution of the equation.

**13.** $-1+(-2)=-3$

**14.** $(-2)+(-8)=-10$

**15.** $-4+6=2$

**16.** $-3+10=7$

**17. a.** $-(-10)=10$    **b.** $-\left(-\dfrac{1}{2}\right)=\dfrac{1}{2}$

    **c.** $-(-2x)=2x$    **d.** $-|-6|=-(6)=-6$

**18. a.** $-(-5)=5$    **b.** $-\left(-\dfrac{2}{3}\right)=\dfrac{2}{3}$

    **c.** $-(-a)=a$    **d.** $-|-3|=-(3)=-3$

**19. a.** $5.3-(-4.6)=5.3+4.6=9.9$

    **b.** $-\dfrac{3}{10}-\dfrac{5}{10}=-\dfrac{3}{10}+\left(-\dfrac{5}{10}\right)=\dfrac{-3-5}{10}$
$$=-\dfrac{8}{10}=-\dfrac{4}{5}$$

    **c.** $-\dfrac{2}{3}-\left(-\dfrac{4}{5}\right)=-\dfrac{2}{3}\cdot\dfrac{5}{5}+\dfrac{4}{5}\cdot\dfrac{3}{3}$
$$=-\dfrac{10}{15}+\dfrac{12}{15}=\dfrac{2}{15}$$

**20. a.** $-2.7-8.4=-2.7+(-8.4)=-11.1$

    **b.** $-\dfrac{4}{5}-\left(-\dfrac{3}{5}\right)=-\dfrac{4}{5}+\dfrac{3}{5}=\dfrac{-4+3}{5}=-\dfrac{1}{5}$

    **c.** $\dfrac{1}{4}-\left(-\dfrac{1}{2}\right)=\dfrac{1}{4}+\dfrac{1}{2}\cdot\dfrac{2}{2}=\dfrac{1}{4}+\dfrac{2}{4}=\dfrac{3}{4}$

**21. a.** $x=90-38=90+(-38)=52$
    The complementary angle is $52°$

    **b.** $y=180-62=180+(-62)=118$
    The supplementary angle is $118°$

**22. a.** $x=90-72=90+(-72)=18$
    The complementary angle is $18°$

    **b.** $y=180-47=180+(-47)=133$
    The supplementary angle is $133°$

**23.** **a.** $(-1.2)(0.05) = -0.06$

**b.** $\dfrac{2}{3}\left(-\dfrac{7}{10}\right) = -\dfrac{2 \cdot 7}{3 \cdot 10} = -\dfrac{14}{30} = -\dfrac{7}{15}$

**24.** **a.** $(4.5)(-0.08) = -0.36$

**b.** $-\dfrac{3}{4}\left(-\dfrac{8}{17}\right) = \dfrac{3 \cdot 8}{4 \cdot 17} = \dfrac{24}{68} = \dfrac{6}{17}$

**25.** **a.** $\dfrac{-24}{-4} = 6$    **b.** $\dfrac{-36}{3} = -12$

**c.** $\dfrac{2}{3} \div \left(-\dfrac{5}{4}\right) = \dfrac{2}{3}\left(-\dfrac{4}{5}\right) = -\dfrac{8}{15}$

**26.** **a.** $\dfrac{-32}{8} = -4$    **b.** $\dfrac{-108}{-12} = 9$

**c.** $-\dfrac{5}{7} \div \left(-\dfrac{9}{2}\right) = -\dfrac{5}{7}\left(-\dfrac{2}{9}\right) = \dfrac{10}{63}$

**27.** **a.** $x + 5 = 5 + x$

**b.** $3 \cdot x = x \cdot 3$

**28.** **a.** $y + 1 = 1 + y$

**b.** $y \cdot 4 = 4 \cdot y$

**29.** **a.** $8 \cdot 2 + 8 \cdot x = 8(2 + x)$

**b.** $7s + 7t = 7(s + t)$

**30.** **a.** $4 \cdot y + 4 \cdot \frac{1}{3} = 4\left(y + \frac{1}{3}\right)$

**b.** $0.10x + 0.10y = 0.10(x + y)$

**31.** $(2x - 3) - (4x - 2) = 2x - 3 - 4x + 2$
$$= -2x - 1$$

**32.** $(-5x + 1) - (10x + 3) = -5x + 1 - 10x - 3$
$$= -15x - 2$$

**33.** $\dfrac{1}{2} = x - \dfrac{3}{4}$

$$4\left(\dfrac{1}{2}\right) = 4(x) - 4\left(\dfrac{3}{4}\right)$$
$$2 = 4x - 3$$
$$5 = 4x$$
$$\dfrac{5}{4} = x$$

**34.** $\dfrac{5}{6} + x = \dfrac{2}{3}$

$$6\left(\dfrac{5}{6}\right) + 6(x) = 6\left(\dfrac{2}{3}\right)$$
$$5 + 6x = 4$$
$$6x = -1$$
$$x = -\dfrac{1}{6}$$

**35.** $6(2a - 1) - (11a + 6) = 7$
$$12a - 6 - 11a - 6 = 7$$
$$a - 12 = 7$$
$$a = 19$$

**36.** $-3x + 1 - (-4x - 6) = 10$
$$-3x + 1 + 4x + 6 = 10$$
$$x + 7 = 10$$
$$x = 3$$

**37.** $\dfrac{y}{7} = 20$
$$y = 140$$

**38.** $\dfrac{x}{4} = 18$
$$x = 72$$

**39.** $4(2x-3)+7=3x+5$

$$8x-12+7=3x+5$$
$$8x-5=3x+5$$
$$5x-5=5$$
$$5x=10$$
$$x=2$$

**40.** $6x+5=4(x+4)-1$

$$6x+5=4x+16-1$$
$$6x+5=4x+15$$
$$2x+5=15$$
$$2x=10$$
$$x=5$$

**41.** Let $x=$ a number.

$$2(x+4)=4x-12$$
$$2x+8=4x-12$$
$$8=2x-12$$
$$20=2x$$
$$10=x$$

The number is 10

**42.** Let $x=$ a number.

$$x+4=3x-8$$
$$4=2x-8$$
$$12=2x$$
$$6=x$$

The number is 6

**43.** $V=lwh$

$$\frac{V}{wh}=\frac{lwh}{wh}$$
$$\frac{V}{wh}=l$$

**44.** $C=2\pi r$

$$\frac{C}{2\pi}=\frac{2\pi r}{2\pi}$$
$$\frac{C}{2\pi}=r$$

**45.** $x+4\le-6$

$$x\le-10,\ \left(-\infty,-10\right]$$

**46.** $x-3>2$

$$x>5,\ \left(5,\infty\right)$$

# Chapter 3

## Mental Math 3.1

**1.** $x + y = 10$

Answers may vary; Ex. $(5,5), (7,3)$

**2.** $x + y = 6$

Answers may vary; Ex. $(0,6), (6,0)$

**3.** $x = 3$

Answers may vary; Ex. $(3,5), (3,0)$

**4.** $y = -2$

Answers may vary; Ex. $(0,-2), (1,-2)$

## Exercise Set 3.1

Use the following graph for Exercises 1, 3, 5, 7, 9, 11.

**1-11.**

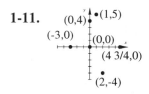

**1.** Point $(1,5)$ lies in quadrant I.

**3.** Point $(-3,0)$ lies on the $x$-axis.

**5.** Point $(2,-4)$ lies in quadrant IV.

**7.** Point $\left(4\frac{3}{4},0\right)$ lies on the $x$-axis.

**9.** Point $(0,0)$ lies on the origin.

**11.** Point $(0,4)$ lies on the $y$-axis.

**13.** $A:(0,0)$, $B:\left(3\frac{1}{2},0\right)$, $C:(3,2)$,

$D:(-1,3)$, $E:(-2,-2)$, $F:(0,-1)$,

$G:(2,-1)$

**15.**

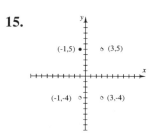

Rectangle is 9 units by 4 units.

Perimeter is $9 + 4 + 9 + 4 = 26$ units.

**17.a.** $(1994,1.11), (1995,1.15), (1996,1.23),$

$(1997,1.23), (1998,1.06), (1999,1.17),$

$(2000,1.51), (2001,1.46), (2002,1.29)$

**b.**

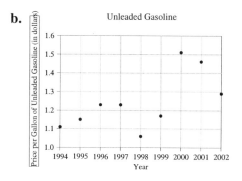

**19.a.** $(1998,690), (1999,733), (2000,818),$

$(2001,789), (2002,827)$

**b.**

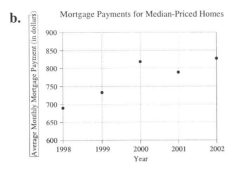

**c.** The average monthly mortgage payment increases each year.

**2.a.** $(2313,2),(2085,1),(2711,21),(2869,39),$
$(2920,42),(4038,99),(1783,0),(2493,9)$

**b.**

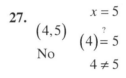

**c.** The farther from the equator, the more snowfall.

**23.**
$(3,1)$   $2x + y = 7$
Yes   $2(3)+(1)\overset{?}{=}7$
$7 = 7$

$(7,0)$   $2x + y = 7$
No   $2(7)+(0)\overset{?}{=}7$
$14 \neq 7$

$(0,7)$   $2x + y = 7$
Yes   $2(0)+(7)\overset{?}{=}7$
$7 = 7$

**25.**
$(-1,-5)$   $y = -5x$
No   $(-5)\overset{?}{=}-5(-1)$
$-5 \neq 5$

$(0,0)$   $y = -5x$
Yes   $(0)\overset{?}{=}-5(0)$
$0 = 0$

$(2,-10)$   $y = -5x$
Yes   $(-10)\overset{?}{=}-5(2)$
$-10 = -10$

**27.**
$(4,5)$   $x = 5$
No   $(4)\overset{?}{=}5$
$4 \neq 5$

$(5,4)$   $x = 5$
Yes   $(5)\overset{?}{=}5$
$5 = 5$

$(5,0)$   $x = 5$
Yes   $(5)\overset{?}{=}5$
$5 = 5$

**29.**
$(5,2)$   $x + 2y = 9$
Yes   $(5)+2(2)\overset{?}{=}9$
$9 = 9$

$$x + 2y = 9$$

$(0, 9)$

No

$$(0) + 2(9) \overset{?}{=} 9$$

$$18 \neq 9$$

**31.**

$$2x - y = 11$$

$(3, -4)$

No

$$2(3) - (-4) \overset{?}{=} 11$$

$$10 \neq 11$$

$$2x - y = 11$$

$(9, 8)$

No

$$2(9) - (8) \overset{?}{=} 11$$

$$10 \neq 11$$

**33.**

$$x = \frac{1}{3} y$$

$(0, 0)$

Yes

$$0 \overset{?}{=} \frac{1}{3}(0)$$

$$0 = 0$$

$$x = \frac{1}{3} y$$

$(3, 9)$

Yes

$$3 \overset{?}{=} \frac{1}{3}(9)$$

$$3 = 3$$

**35.** $x - 4y = 4$

$$y = -2, \ x - 4(-2) = 4$$

$$x + 8 = 4$$

$$x = -4; \ (-4, -2)$$

$$x = 4, \ 4 - 4y = 4$$

$$-4y = 0$$

$$y = 0; \ (4, 0)$$

**37.** $3x + y = 9$

$$x = 0, \ 3(0) + y = 9$$

$$0 + y = 9$$

$$y = 9; \ (0, 9)$$

$$y = 0, \ 3x + 0 = 9$$

$$3x = 9$$

$$x = 3; \ (3, 0)$$

**39.** $y = -7$

$$x = 11, \ y = -7; \ (11, -7)$$

$$y = -7, \ x = \text{any value}$$

**41.** $-2x + 7y = -3$

$$y = 1, \ -2x + 7(1) = -3$$

$$x = 5; \ (5, 1)$$

$$x = 1, \ -2(1) + 7y = -3$$

$$y = -\frac{1}{7}; \ \left(1, -\frac{1}{7}\right)$$

**43.** $x + 3y = 6$

$$x = 0, \ 0 + 3y = 6, \ y = 2; \ (0, 2)$$

$$y = 0, \ x + 3(0) = 6, \ x = 6; \ (6, 0)$$

$$y = 1, \ x + 3(1) = 6, \ x = 3; \ (3, 1)$$

| x | y |
|---|---|
| 0 | 2 |
| 6 | 0 |
| 3 | 1 |

**45.** $2x - y = 12$

$x = 0, \ 2(0) - y = 12, \ y = -12; \ (0, -12)$

$y = -2, \ 2x - (-2) = 12$

$\qquad 2x + 2 = 12$

$\qquad 2x = 10$

$\qquad x = 5; \ (5, -2)$

$x = -3, \ 2(-3) - y = 12,$

$\qquad -6 - y = 12$

$\qquad -y = 18$

$\qquad y = -18; \ (-3, -18)$

| $x$ | $y$ |
|-----|-----|
| 0 | $-12$ |
| 5 | $-2$ |
| $-3$ | $-18$ |

**47.** $2x + 7y = 5$

$x = 0, \ 2(0) + 7y = 5,$

$\qquad 7y = 5$

$\qquad y = \dfrac{5}{7}; \ \left(0, \dfrac{5}{7}\right)$

$y = 0, \ 2x + 7(0) = 5$

$\qquad 2x = 5$

$\qquad x = \dfrac{5}{2}; \ \left(\dfrac{5}{2}, 0\right)$

$y = 1, \ 2x + 7(1) = 5,$

$\qquad 2x + 7 = 5$

$\qquad 2x = -2$

$\qquad x = -1; \ (-1, 1)$

| $x$ | $y$ |
|-----|-----|
| 0 | $\frac{5}{7}$ |
| $\frac{5}{2}$ | 0 |
| $-1$ | 1 |

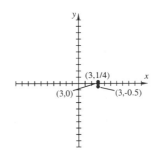

**49.** $x = 3$

| $x$ | $y$ |
|-----|-----|
| 3 | 0 |
| 3 | $-0.5$ |
| 3 | $\frac{1}{4}$ |

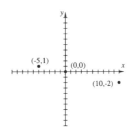

**51.** $x = -5y$

$y = 0, \ x = -5(0) = 0$

$y = 1, \ x = -5(1) = -5$

$x = 10, \ 10 = -5y, \ y = -2$

| $x$ | $y$ |
|-----|-----|
| 0 | 0 |
| $-5$ | 1 |
| 10 | $-2$ |

**53.** Answers may vary.

**55.a.** $y = 80x + 5000$

$\qquad x = 100, \ y = 80(100) + 5000 = 13,000$

$\qquad x = 200, \ y = 80(200) + 5000 = 21,000$

$\qquad x = 300, \ y = 80(300) + 5000 = 29,000$

| $x$ | 100 | 200 | 300 |
|-----|-----|-----|-----|
| $y$ | $13,000$ | $21,000$ | $29,000$ |

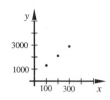

**b.** Let $y = 8600$

$8600 = 80x + 5000$

$3600 = 80x$

$45 = x$

45 desks can be produced

**57.a.** $y = -3.95x + 24.93$

$x = 1, \ y = -3.95(1) + 24.93 = 20.98$

$x = 3, \ y = -3.95(3) + 24.93 = 13.08$

$x = 5, \ y = -3.95(5) + 24.93 = 5.18$

| $x$ | 1 | 3 | 5 |
|---|---|---|---|
| $y$ | 20.98 | 13.08 | 5.18 |

**b.** Let $y = 17$

$17 = -3.95x + 24.93$

$17 - 24.93 = -3.95x + 24.93 - 24.93$

$-7.93 = -3.95x$

$\dfrac{-7.93}{-3.95} = \dfrac{-3.95x}{-3.95}$

$2 \approx x$

The year $1995 + 2 = 1997$.

**59.** In 1995, there were 670 Target Stores.

**61.** year 6: 66 stores; year 7: 60 stores; year 8: 55 stores

**63.** When $a = b$.

**65.** $x + y = 5$

$y = 5 - x$

**67.** $2x + 4y = 5$

$4y = -2x + 5$

$y = -\dfrac{1}{2}x + \dfrac{5}{4}$

**69.** $10x = -5y$

$-2x = y$

**71.** $x - 3y = 6$

$-3y = -x + 6$

$y = \dfrac{1}{3}x - 2$

**73.** Quadrant IV

**75.** Quadrants II or III

**77.**

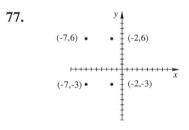

**a.** The fourth vertex is $(-2, 6)$

The rectangle is 9 units by 5 units.

**b.** The perimeter is $9 + 5 + 9 + 5 = 28$ units.

**c.** The area is $9 \times 5 = 45$ square units.

## Graphing Calculator Explorations 3.2

**1.** $y = -3x + 7$

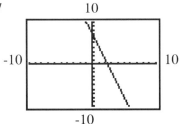

**3.** $y = 2.5x - 7.9$

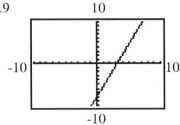

**5.** $y = -\dfrac{3}{10}x + \dfrac{32}{5}$

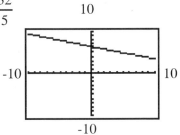

## Exercise Set 3.2

**1.** Yes; It can be written in the form $Ax + By = C$.

**3.** Yes; It can be written in the form $Ax + By = C$.

**5.** No; $x$ is squared.

**7.** Yes; It can be written in the form $Ax + By = C$.

**9.** $x + y = 4$

| $x$ | $y$ |
|-----|-----|
| $-2$ | $6$ |
| $0$ | $4$ |
| $2$ | $2$ |

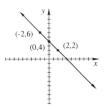

**11.** $x - y = -2$

| $x$ | $y$ |
|-----|-----|
| $-2$ | $0$ |
| $0$ | $2$ |
| $2$ | $4$ |

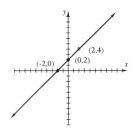

**13.** $x - 2y = 6$

| $x$ | $y$ |
|-----|-----|
| $-4$ | $-5$ |
| $0$ | $-3$ |
| $4$ | $-1$ |

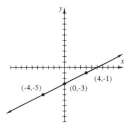

**15.** $y = 6x + 3$

| $x$ | $y$ |
|-----|-----|
| $-1$ | $-3$ |
| $0$ | $3$ |
| $1$ | $9$ |

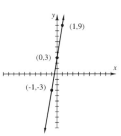

**17.** $x - 2y = -6$

| $x$ | $y$ |
|-----|-----|
| $-4$ | $1$ |
| $0$ | $3$ |
| $4$ | $5$ |

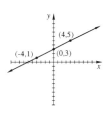

**19.** $y = 6x$

| $x$ | $y$ |
|-----|-----|
| $-1$ | $-6$ |
| $0$ | $0$ |
| $1$ | $6$ |

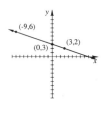

**21.** $3y - 10 = 5x$

| $x$ | $y$ |
|-----|-----|
| $-4$ | $-\dfrac{10}{3}$ |
| $0$ | $\dfrac{10}{3}$ |
| $1$ | $5$ |

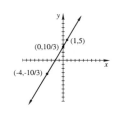

**23.** $x + 3y = 9$

| $x$ | $y$ |
|-----|-----|
| $-9$ | $6$ |
| $0$ | $3$ |
| $3$ | $2$ |

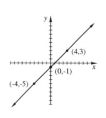

**25.** $y - x = -1$

| $x$ | $y$ |
|-----|-----|
| $-4$ | $-5$ |
| $0$ | $-1$ |
| $4$ | $3$ |

**27.** $x = -3y$

| $x$ | $y$ |
|-----|-----|
| $-6$ | $2$ |
| $0$ | $0$ |
| $6$ | $-2$ |

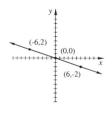

**29.** $5x - y = 10$

| $x$ | $y$ |
|-----|-----|
| $1$ | $-5$ |
| $2$ | $0$ |
| $3$ | $5$ |

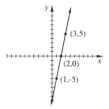

**31.** $y = \dfrac{1}{2}x + 2$

| $x$ | $y$ |
|-----|-----|
| $-4$ | $0$ |
| $0$ | $2$ |
| $4$ | $4$ |

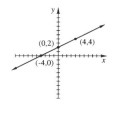

**33.** $y = 5x$      $y = 5x + 4$

| $x$ | $y$ |
|-----|-----|
| $-1$ | $-5$ |
| $0$ | $0$ |
| $1$ | $5$ |

| $x$ | $y$ |
|-----|-----|
| $-1$ | $-1$ |
| $0$ | $4$ |
| $1$ | $9$ |

**35.** $y = -2x$      $y = -2x - 3$

| $x$ | $y$ |
|-----|-----|
| $-2$ | $4$ |
| $0$ | $0$ |
| $2$ | $-4$ |

| $x$ | $y$ |
|-----|-----|
| $-2$ | $1$ |
| $0$ | $-3$ |
| $2$ | $-7$ |

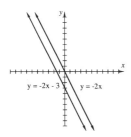

**45.**

$y = 16x + 144$

The average weekly earnings in 2007 should be $288.

**37.**  $y = \dfrac{1}{2}x$        $y = \dfrac{1}{2}x + 2$

| $x$ | $y$ |
|----|----|
| $-4$ | $-2$ |
| $0$ | $0$ |
| $4$ | $2$ |

| $x$ | $y$ |
|----|----|
| $-4$ | $0$ |
| $0$ | $2$ |
| $4$ | $4$ |

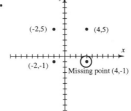

Answers may vary.

**47.**

**39.** c

**41.** d

**43.**  $y = 180x + 450$

In 2009, $x = 13$

$y = 180(13) + 450$

$y = 2790$

The total sales in 2009
should be $2790 billion.

**49.**  $3(x-2) + 5x = 6x - 16$

$3x - 6 + 5x = 6x - 16$

$8x - 6 = 6x - 16$

$2x - 6 = -16$

$2x = -10$

$x = -5$

**51.**            $3x + \dfrac{2}{5} = \dfrac{1}{10}$

$10(3x) + 10\left(\dfrac{2}{5}\right) = 10\left(\dfrac{1}{10}\right)$

$30x + 4 = 1$

$30x = -3$

$x = -\dfrac{1}{10}$

**53.** $x - y = -3$

$y = 0, \; x - 0 = -3, \; x = -3$

$x = 0, \; 0 - y = -3, \; y = 3$

| $x$ | $y$ |
|---|---|
| 0 | 3 |
| -3 | 0 |

**55.** $y = 2x$

$y = 0, \; 0 = 2x, \; 0 = x$

$x = 0, \; y = 2(0) = 0$

| $x$ | $y$ |
|---|---|
| 0 | 0 |
| 0 | 0 |

**57.** $y = x + 5$

| $x$ | $y$ |
|---|---|
| -3 | 2 |
| 0 | 5 |
| 3 | 8 |

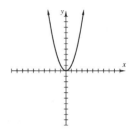

**59.** $2x + 3y = 6$

| $x$ | $y$ |
|---|---|
| 0 | 2 |
| 3 | 0 |

**61.** Answers may vary.

**63.** If $(a,b)$ is a solution of $x + y = 5$, then $(b,a)$ is also a solution. Explanations may vary.

**65.** $y = x^2$

| $x$ | $y$ |
|---|---|
| 0 | 0 |
| 1 | 1 |
| -1 | 1 |
| 2 | 4 |
| -2 | 4 |

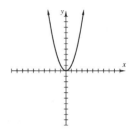

**Graphing Calculator Explorations 3.3**

**1.** $x = 3.78y$

$y = \dfrac{x}{3.78}$

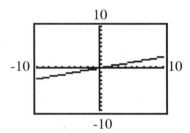

**3.** $3x + 7y = 21$

$7y = -3x + 21$

$y = -\dfrac{3}{7}x + 3$

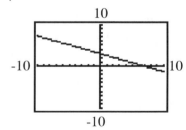

**5.** $-2.2x + 6.8y = 15.5$

$6.8y = 2.2x + 15.5$

$y = \dfrac{2.2}{6.8}x + \dfrac{15.5}{6.8}$

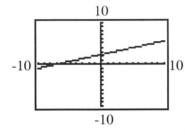

**9.** 0

**11.** $x - y = 3$

$y = 0, \ x - 0 = 3, \ x = 3$

$x = 0, \ 0 - y = 3, \ y = -3$

$x$-intercept: $(3, 0)$; $y$-intercept: $(0, -3)$

| $x$ | $y$ |
|---|---|
| 3 | 0 |
| 0 | $-3$ |

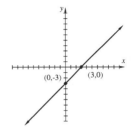

## Mental Math 3.3

**1.** False

**2.** False

**3.** True

**4.** True

**5.** False

**6.** True

## Exercise Set 3.3

**1.** $x$-intercept: $(-1, 0)$; $y$-intercept: $(0, 1)$

**3.** $x$-intercept: $(-2, 0)$

**5.** $x$-intercepts: $(-1, 0), (1, 0)$

    $y$-intercept: $(0, 1), (0, -2)$

**7.** Infinite

**13.** $x = 5y$

$y = 0, \ x = 5(0) = 0$

$x = 0, \ 0 = 5y, \ y = 0$

$x$-intercept: $(0, 0)$; $y$-intercept: $(0, 0)$

$y = 1, \ x = 5(1) = 5$

| $x$ | $y$ |
|---|---|
| 0 | 0 |
| 5 | 1 |

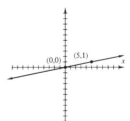

**15.** $-x + 2y = 6$

$y = 0, \ -x + 2(0) = 6, \ x = -6$

$x = 0, \ -0 + 2y = 6, \ y = 3$

$x$-intercept: $(-6, 0)$; $y$-intercept: $(0, 3)$

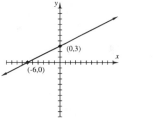

| $x$ | $y$ |
|-----|-----|
| −6  | 0   |
| 0   | 3   |

**23.** $y + 7 = 0$

   $y = -7$

   for all values of $x$.

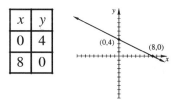

**17.** $2x - 4y = 8$

   $y = 0,\ 2x - 4(0) = 8,\ x = 4$

   $x = 0,\ 2(0) - 4y = 8,\ y = -2$

   $x$-intercept: $(4, 0)$;  $y$-intercept: $(0, -2)$

| $x$ | $y$ |
|-----|-----|
| 4   | 0   |
| 0   | −2  |

**25.** $x + 2y = 8$

   $x$-intercept: $(8, 0)$;  $y$-intercept: $(0, 4)$

| $x$ | $y$ |
|-----|-----|
| 0   | 4   |
| 8   | 0   |

**27.** $x - 7 = 3y$

   $x$-intercept: $(7, 0)$;  $y$-intercept: $\left(0, -\dfrac{7}{3}\right)$

**19.** $x = -1$

   for all values of $y$.

| $x$ | $y$             |
|-----|-----------------|
| 0   | $-\dfrac{7}{3}$ |
| 7   | 0               |

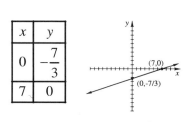

**29.** $x = -3$

   for all values of $y$.

**21.** $y = 0$

   for all values of $x$.

**31.** $3x + 5y = 7$

   $x$-intercept: $\left(\dfrac{7}{3}, 0\right)$;  $y$-intercept: $\left(0, \dfrac{7}{5}\right)$

| $x$ | $y$ |
|-----|-----|
| $0$ | $\dfrac{7}{5}$ |
| $\dfrac{7}{3}$ | $0$ |

| $x$ | $y$ |
|-----|-----|
| $\dfrac{5}{6}$ | $0$ |
| $0$ | $-5$ |

**33.** $x = y$

$x$-intercept: $(0,0)$; $y$-intercept: $(0,0)$

Second point $(4,4)$

| $x$ | $y$ |
|-----|-----|
| $4$ | $4$ |
| $0$ | $0$ |

**39.** $-x + 10y = 11$

$x$-intercept: $(-11,0)$; $y$-intercept: $\left(0, \dfrac{11}{10}\right)$

| $x$ | $y$ |
|------|-----|
| $-11$ | $0$ |
| $0$ | $\dfrac{11}{10}$ |

**35.** $x + 8y = 8$

$x$-intercept: $(8,0)$; $y$-intercept: $(0,1)$

| $x$ | $y$ |
|-----|-----|
| $8$ | $0$ |
| $0$ | $1$ |

**41.**     $y = 4.5$

for all values of $x$.

**43.** $y = \dfrac{1}{2}x$

$x$-intercept: $(0,0)$; $y$-intercept: $(0,0)$

Second point, $(6,3)$

| $x$ | $y$ |
|-----|-----|
| $0$ | $0$ |
| $6$ | $3$ |

**37.** $5 = 6x - y$

$x$-intercept: $\left(\dfrac{5}{6},0\right)$; $y$-intercept: $(0,-5)$

**45.** $x + 4 = 0$

$x = -4$

for all values of $y$.

**47.** $3x - 4y = -12$

$x$-intercept: $(-4, 0)$; $y$-intercept: $(0, 3)$

| $x$ | $y$ |
|-----|-----|
| $-4$ | $0$ |
| $0$ | $-3$ |

(-4,0)   (0,3)

**49.** $2x + 3y = 6$

$x$-intercept: $(3, 0)$; $y$-intercept: $(0, 2)$

| $x$ | $y$ |
|-----|-----|
| $3$ | $0$ |
| $0$ | $2$ |

(0,2)   (3,0)

**51.** $y = 3$

C

**53.** $x = -1$

E

**55.** $y = 2x + 3$

B

**57.** $\dfrac{-6 - 3}{2 - 8} = \dfrac{-9}{-6} = \dfrac{3}{2}$

**59.** $\dfrac{-8 - (-2)}{-3 - (-2)} = \dfrac{-6}{-1} = 6$

**61.** $\dfrac{0 - 6}{5 - 0} = \dfrac{-6}{5} = -\dfrac{6}{5}$

**63.** $y = 78.1x + 569.9$

**a.** $(0, 569.9)$

**b.** In 1999, the average price of a digital camera was $569.90.

**65.** $y = -37.2x + 264.4$

**a.** $y = 0$,       $0 = -37.2x + 264.4$

$37.2x = 264.4$

$x = 7.1$

$(7.1, 0)$.

**b.** 7.1 years after 1995 $(2002+)$

no music cassettes will be shipped.

**c.** Answers may vary.

**67.** $3x + 6y = 1200$

**a.** $x = 0$, $3(0) + 6y = 1200$, $y = 200$

$(0, 200)$ corresponds to no chairs

and 200 desks being manufactured.

**b.** $y = 0$, $3x + 6(0) = 1200$, $x = 400$

$(400, 0)$ corresponds to 400 chairs

and no desks being manufactured.

**c.**

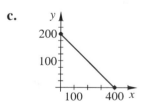

**d.** $y = 50,\ 3x + 6(50) = 1200$

$$3x + 300 = 1200$$
$$3x = 900$$
$$x = 300$$

300 chairs can be made.

**69.** Parallel to $y = -1$ is horizontal.

$y$-intercept is $(0, -4)$, so $y = -4$

for all values of $x.$  $y = -4$

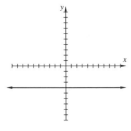

**71.** Answers may vary

**73.** Answers may vary

**Graphing Calculator Explorations 3.4**

**1.** $y_1 = 3.8x,\ y_2 = 3.8x - 3,\ y_3 = 3.8x + 9$

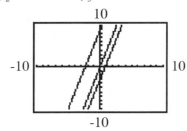

**3.** $y_1 = \dfrac{1}{4}x,\ y_2 = \dfrac{1}{4}x + 5,\ y_3 = \dfrac{1}{4}x - 8$

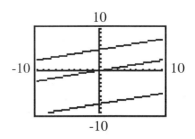

**Mental Math 3.4**

**1.** Upward

**2.** Downward

**3.** Horizontal

**4.** Vertical

**Exercise Set 3.4**

**1.** $(-1, 2)$ and $(2, -2)$

$$m = \frac{y_2 - y_1}{x_2 - x_1} = \frac{-2 - 2}{2 - (-1)} = -\frac{4}{3}$$

**3.** $(2, -1)$ and $(2, 3)$

$$m = \frac{y_2 - y_1}{x_2 - x_1} = \frac{3 - (-1)}{2 - 2} = \frac{4}{0} \text{ is undefined.}$$

**5.** $(-3, -2)$ and $(-1, 3)$

$$m = \frac{y_2 - y_1}{x_2 - x_1} = \frac{3 - (-2)}{-1 - (-3)} = \frac{5}{2}$$

**7.** $(0, 0)$ and $(7, 8)$

$$m = \frac{y_2 - y_1}{x_2 - x_1} = \frac{8 - 0}{7 - 0} = \frac{8}{7}$$

**9.** $(-1, 5)$ and $(6, -2)$

$$m = \frac{y_2 - y_1}{x_2 - x_1} = \frac{-2 - 5}{6 - (-1)} = -\frac{7}{7} = -1$$

**11.** $(1, 4)$ and $(5, 3)$

$$m = \frac{y_2 - y_1}{x_2 - x_1} = \frac{3 - 4}{5 - 1} = -\frac{1}{4}$$

**13.** $(-4, 3)$ and $(-4, 5)$

$$m = \frac{y_2 - y_1}{x_2 - x_1} = \frac{5 - 3}{-4 - (-4)} = \frac{2}{0} \text{ is undefined.}$$

**15.** $(-2, 8)$ and $(1, 6)$

$$m = \frac{y_2 - y_1}{x_2 - x_1} = \frac{6 - 8}{1 - (-2)} = -\frac{2}{3}$$

**17.** $(1, 0)$ and $(1, 1)$

$$m = \frac{y_2 - y_1}{x_2 - x_1} = \frac{1 - 0}{1 - 1} = \frac{1}{0} \text{ is undefined.}$$

**19.** $(5, 1)$ and $(-2, 1)$

$$m = \frac{y_2 - y_1}{x_2 - x_1} = \frac{1 - 1}{-2 - 5} = \frac{0}{-7} = 0$$

**21.** Line 1

**23.** Line 2

**25.** $(0, 0)$ and $(1, 1)$

$$m = \frac{y_2 - y_1}{x_2 - x_1} = \frac{1 - 0}{1 - 0} = 1$$

**D**

**27.** A vertical line has undefined slope.

**B**

**29.** $(2, 0)$ and $(4, -1)$

$$m = \frac{y_2 - y_1}{x_2 - x_1} = \frac{-1 - 0}{4 - 2} = -\frac{1}{2}$$

**E**

**31.** $x = 6$ is a vertical line, so it has an undefined slope.

**33.** $y = -4$ is a horizontal line, so it has a slope $m = 0$.

**35.** $x = -3$ is a vertical line, so it has an undefined slope.

**37.** $y = 0$ is a horizontal line, so it has a slope $m = 0$.

**39.** $(-3, -3)$ and $(0, 0)$

$$m = \frac{y_2 - y_1}{x_2 - x_1} = \frac{0 - (-3)}{0 - (-3)} = \frac{3}{3} = 1$$

**a.** $m = 1$

**b.** $m = -1$

**41.** $(-8, -4)$ and $(3, 5)$

$$m = \frac{y_2 - y_1}{x_2 - x_1} = \frac{5 - (-4)}{3 - (-8)} = \frac{9}{11}$$

**a.** $m = \frac{9}{11}$

**b.** $m = -\frac{11}{9}$

**43.** $(0,6)$ and $(-2,0)$

$$m_1 = \frac{y_2 - y_1}{x_2 - x_1} = \frac{0-6}{-2-0} = \frac{-6}{-2} = 3$$

$(0,5)$ and $(1,8)$

$$m_2 = \frac{y_2 - y_1}{x_2 - x_1} = \frac{8-5}{1-0} = \frac{3}{1} = 3$$

$m_1 = m_2$, parallel

**45.** $(2,6)$ and $(-2,8)$

$$m_1 = \frac{y_2 - y_1}{x_2 - x_1} = \frac{8-6}{-2-2} = \frac{2}{-4} = -\frac{1}{2}$$

$(0,3)$ and $(1,5)$

$$m_2 = \frac{y_2 - y_1}{x_2 - x_1} = \frac{5-3}{1-0} = \frac{2}{1} = 2$$

$$m_1 m_2 = \left(-\frac{1}{2}\right)(2) = -1,\ \text{perpendicular}$$

**47.** $(3,6)$ and $(7,8)$

$$m_1 = \frac{y_2 - y_1}{x_2 - x_1} = \frac{8-6}{7-3} = \frac{2}{4} = \frac{1}{2}$$

$(0,6)$ and $(2,7)$

$$m_2 = \frac{y_2 - y_1}{x_2 - x_1} = \frac{7-6}{2-0} = \frac{1}{2}$$

$m_1 = m_2$, parallel

**49.** $(2,-3)$ and $(6,-5)$

$$m_1 = \frac{y_2 - y_1}{x_2 - x_1} = \frac{-5-(-3)}{6-2} = \frac{-2}{4} = -\frac{1}{2}$$

$(5,-2)$ and $(-3,-4)$

$$m_2 = \frac{y_2 - y_1}{x_2 - x_1} = \frac{-4-(-2)}{-3-5} = \frac{-2}{-8} = \frac{1}{4}$$

$m_1 \neq m_2$ and $m_1 m_2 \neq -1$, neither

**51.** $(-4,-3)$ and $(-1,0)$

$$m_1 = \frac{y_2 - y_1}{x_2 - x_1} = \frac{0-(-3)}{-1-(-4)} = \frac{3}{3} = 1$$

$(4,-4)$ and $(0,0)$

$$m_2 = \frac{y_2 - y_1}{x_2 - x_1} = \frac{0-(-4)}{0-4} = \frac{4}{-4} = -1$$

$m_1 m_2 = (1)(-1) = -1$, perpendicular

**53.** $(-7,-5)$ and $(-2,-6)$

$$m_1 = \frac{y_2 - y_1}{x_2 - x_1} = \frac{-6-(-5)}{-2-(-7)} = \frac{-1}{5} = -\frac{1}{5}$$

parallel; $m_2 = m_1 = -\dfrac{1}{5}$

**55.** $(0,0)$ and $(1,-3)$

$$m_1 = \frac{y_2 - y_1}{x_2 - x_1} = \frac{-3-0}{1-0} = \frac{-3}{1} = -3$$

perpendicular; $m_2 =$ (negative

reciprocal of $m_1$) $= \dfrac{1}{3}$

**57.** $(3,3)$ and $(-3,-3)$

$$m_1 = \frac{y_2 - y_1}{x_2 - x_1} = \frac{-3-3}{-3-3} = \frac{-6}{-6} = 1$$

parallel; $m_2 = m_1 = 1$

**59.** $\text{pitch} = \dfrac{6}{10} = \dfrac{3}{5}$

**61.** $\text{grade} = \dfrac{\text{rise}}{\text{run}} = \dfrac{2}{16} = 0.125 = 12.5\%$

**63.** $\text{grade} = \dfrac{\text{rise}}{\text{run}} = \dfrac{2580}{6450} = 0.40 = 40\%$

**65.** $\text{slope} = \dfrac{\text{rise}}{\text{run}} = \dfrac{0.25}{12} = 0.02$

**67.** $(2003, 148)$ and $(2007, 168)$

$m = \dfrac{y_2 - y_1}{x_2 - x_1} = \dfrac{168 - 148}{2007 - 2003} = \dfrac{20}{4} = 5$

$= 5$ million users per year.
Every year there will be 5 million
more cell phone users.

**69.** $(5000, 1800)$ and $(20,000, 7200)$

$m = \dfrac{y_2 - y_1}{x_2 - x_1} = \dfrac{7200 - 1800}{20,000 - 5000} = \dfrac{5400}{15,000}$

$= 0.36$ dollars per mile
It costs $0.36 per mile to own and
operate a compact car.

**71.** $x + y = 10$

$y = -x + 10$

**73.** $x + 2y = -12$

$2y = -x - 12$

$y = -\dfrac{1}{2}x - 6$

**75.** $5x - y = 17$

$y = 5x - 17$

**77.** 28.3 miles per gallon

**79.** 1992 the average was 27.6 miles per
gallon.

**81.** The greatest slope was from 1992 to 1993.

**83.** $\text{pitch} = \dfrac{\text{rise}}{\text{run}}$

$\dfrac{1}{3} = \dfrac{x}{18}$

$3x = 18$

$x = 6$

**85. a.** $(1994, 782)$ and $(2001, 1132)$

**b.** $m = \dfrac{y_2 - y_1}{x_2 - x_1} = \dfrac{1132 - 782}{2001 - 1994} = \dfrac{350}{7} = 50$

**c.** For the years 1994 through 2001, the
price per acre of U.S. farmland rose
$50 every year.

**87.** $(1, 1)$, $(-4, 4)$ and $(-3, 0)$

$m_1 = \dfrac{0 - 1}{-3 - 1} = \dfrac{1}{4}$, $m_2 = \dfrac{0 - 4}{-3 - (-4)} = -4$

$m_1 m_2 = -1$, so the sides are
perpendicular.

**89.** $(2.1, 6.7)$ and $(-8.3, 9.3)$

$m = \dfrac{y_2 - y_1}{x_2 - x_1} = \dfrac{9.3 - 6.7}{-8.3 - 2.1} = \dfrac{2.6}{-10.4} = -0.25$

**91.** $(2.3, 0.2)$ and $(7.9, 5.1)$

$m = \dfrac{y_2 - y_1}{x_2 - x_1} = \dfrac{5.1 - 0.2}{7.9 - 2.3} = \dfrac{4.9}{5.6} = 0.875$

**93.** $y = -\dfrac{1}{3}x + 2$

$y = -2x + 2$

$y = -4x + 2$

The line becomes steeper.

**Integrated Review-Summary on Slope and Graphing Linear Equations**

**1.** $(0,0)$ and $(2,4)$

$$m = \frac{y_2 - y_1}{x_2 - x_1} = \frac{4-0}{2-0} = \frac{4}{2} = 2$$

**2.** Horizontal line, $m = 0$

**3.** $(0,1)$ and $(3,-1)$

$$m = \frac{y_2 - y_1}{x_2 - x_1} = \frac{-1-1}{3-0} = -\frac{2}{3}$$

**4.** Vertical line, slope is undefined.

**5.** $y = -2x$

$m = -2, b = 0$

**6.** $x + y = 3$

$y = -x + 3$

$m = -1, b = 3$

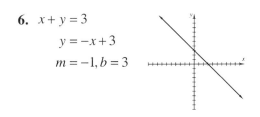

**7.** $x = -1$ for all values of $y$.

Vertical line

Slope is undefined.

**8.** $y = 4$ for all values of $x$

Horizontal line

$m = 0$

**9.** $x - 2y = 6$

$-2y = -x + 6$

$y = \dfrac{1}{2}x - 3$

$m = \dfrac{1}{2}, b = -3$

**10.** $y = 3x + 2$

$m = 3, b = 2$

**11.** $5x + 3y = 15$

| $x$ | $y$ |
|-----|-----|
| 0   | 5   |
| 3   | 0   |

**12.** $2x - 4y = 8$

| $x$ | $y$ |
|---|---|
| 0 | -2 |
| 4 | 0 |

**13.** $(-1, 3)$ and $(1, -3)$

$$m_1 = \frac{y_2 - y_1}{x_2 - x_1} = \frac{-3 - 3}{1 - (-1)} = \frac{-6}{2} = -3$$

$(2, -1)$ and $(4, -7)$

$$m_2 = \frac{y_2 - y_1}{x_2 - x_1} = \frac{-7 - (-1)}{4 - 2} = \frac{-6}{2} = -3$$

$m_1 = m_2$, parallel

**14.** $(-6, -6)$ and $(-1, -2)$

$$m_1 = \frac{y_2 - y_1}{x_2 - x_1} = \frac{-2 - (-6)}{-1 - (-6)} = \frac{4}{5}$$

$(-4, 3)$ and $(3, -3)$

$$m_2 = \frac{y_2 - y_1}{x_2 - x_1} = \frac{-3 - 3}{3 - (-4)} = \frac{-6}{7} = -\frac{6}{7}$$

$m_1 \neq m_2$ and $m_1 m_2 \neq -1$, neither

**15.** $y = 110x + 1407$

　**a.** $(0, 1407)$

　**b.** In 2000, there were 1407 million admissions to movie theatres in the U.S.

　**c.** $m = 110$

　**d.** For the years 2000 through 2002, the number of movie theater admissions has increased at a rate of 110 million per year.

**Graphing Calculator Explorations 3.5**

**1.** $y_1 = x$, $y_2 = 6x$, $y_3 = -6x$

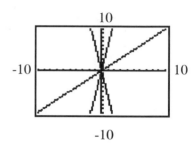

**3.** $y_1 = \frac{1}{2}x + 2$, $y_2 = \frac{3}{4}x + 2$, $y_3 = x + 2$

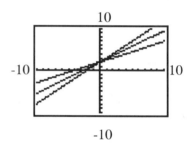

**5.** $y_1 = -7x + 5$, $y_2 = 7x + 5$

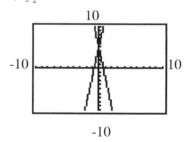

**Mental Math 3.5**

**1.** $y = 2x - 1$

　$m = 2$, $(0, -1)$

**2.** $y = -7x + 3$

$m = -7,\ (0, 3)$

**3.** $y = x + \dfrac{1}{3}$

$m = 1,\ \left(0, \dfrac{1}{3}\right)$

**4.** $y = -x - \dfrac{2}{9}$

$m = -1,\ \left(0, -\dfrac{2}{9}\right)$

**5.** $y = \dfrac{5}{7}x - 4$

$m = \dfrac{5}{7},\ (0, -4)$

**6.** $y = -\dfrac{1}{4}x + \dfrac{3}{5}$

$m = -\dfrac{1}{4},\ \left(0, \dfrac{3}{5}\right)$

**Exercise Set 3.5**

**1.** $2x + y = 4$

$y = -2x + 4$

$y = mx + b$

$m = -2,\ b = 4,\ (0, 4)$

**3.** $x + 9y = 1$

$9y = -x + 1$

$y = -\dfrac{1}{9}x + \dfrac{1}{9}$

$y = mx + b$

$m = -\dfrac{1}{9},\ b = \dfrac{1}{9},\ \left(0, \dfrac{1}{9}\right)$

**5.** $4x - 3y = 12$

$-3y = -4x + 12$

$y = \dfrac{4}{3}x - 4$

$y = mx + b$

$m = \dfrac{4}{3},\ b = -4,\ (0, -4)$

**7.** $x + y = 0$

$y = -x$

$y = mx + b$

$m = -1,\ b = 0,\ (0, 0)$

**9.** $y = -3$

$y = mx + b$

$m = 0,\ b = -3,\ (0, -3)$

**11.** $-x + 5y = 20$

$5y = x + 20$

$y = \dfrac{1}{5}x + 4$

$y = mx + b$

$m = \dfrac{1}{5},\ b = 4,\ (0, 4)$

**13. B**

**15. D**

**17.** $x - 3y = -6,\ -3y = -x - 6,$

$y = \dfrac{1}{3}x + 2,\ m_1 = \dfrac{1}{3}$

$3x - y = 0,\ -y = -3x,\ y = 3x,\ m_2 = 3$

$m_1 \neq m_2$ and $m_1 m_2 \neq -1$, neither

**19.** $2x - 7y = 1$, $-7y = -2x + 1$,

$y = \frac{2}{7}x - \frac{1}{7}$, $m_1 = \frac{2}{7}$

$2y = 7x - 2$, $y = \frac{7}{2}x - 1$, $m_2 = \frac{7}{2}$

$m_1 \neq m_2$ and $m_1 m_2 \neq -1$, neither

**21.** $10 + 3x = 5y$, $2 + \frac{3}{5}x = y$, $m_1 = \frac{3}{5}$

$5x + 3y = 1$, $3y = -5x + 1$, $y = -\frac{5}{3}x + \frac{1}{3}$

$m_2 = -\frac{5}{3}$

$m_1 m_2 = \left(\frac{3}{5}\right)\left(-\frac{5}{3}\right) = -1$, perpendicular

**23.** $6x = 5y + 1$, $6x - 1 = 5y$, $\frac{6}{5}x - \frac{1}{5} = y$,

$m_1 = \frac{6}{5}$

$-12x + 10y = 1$, $10y = 12x + 1$,

$y = \frac{12}{10}x + \frac{1}{10}$, $y = \frac{6}{5}x + \frac{1}{10}$, $m_2 = \frac{6}{5}$

$m_1 = m_2$, parallel

**25.** Answers may vary.

**27.** $m = -1$, $b = 1$

$y = mx + b$

$y = -x + 1$

**29.** $m = 2$, $b = \frac{3}{4}$

$y = mx + b$

$y = 2x + \frac{3}{4}$

**31.** $m = \frac{2}{7}$, $b = 0$

$y = mx + b$

$y = \frac{2}{7}x$

**33.** $y = \frac{2}{3}x + 5$

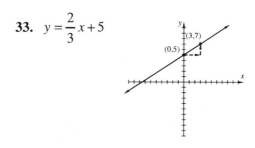

**35.** $y = -\frac{3}{5}x - 2$

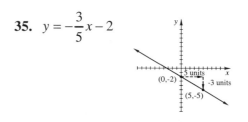

**37.** $y = 2x + 1$

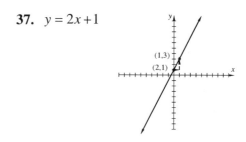

**39.** $y = -5x$

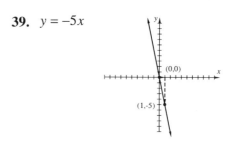

**41.** $4x + y = 6$

$y = -4x + 6$

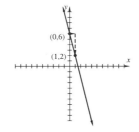

**43.** $x - y = -2$

$y = x + 2$

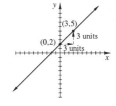

**45.** $3x + 5y = 10$

$5y = -3x + 10$

$y = -\dfrac{3}{5}x + 2$

**47.** $4x - 7y = -14$

$-7y = -4x - 14$

$y = \dfrac{4}{7}x + 2$

**49.** $y - (-6) = 2(x - 4)$

$y + 6 = 2x - 8$

$y = 2x - 14$

**51.** $y - 1 = -6\big(x - (-2)\big)$

$y - 1 = -6(x + 2)$

$y - 1 = -6x - 12$

$y = -6x - 11$

**53.** $(0,0)$ and $(1,1)$

$m = \dfrac{y_2 - y_1}{x_2 - x_1} = \dfrac{1 - 0}{1 - 0} = 1$

**D**

**55.** A vertical line has undefined slope.

**B**

**57.** $(2,0)$ and $(4,-1)$

$m = \dfrac{y_2 - y_1}{x_2 - x_1} = \dfrac{-1 - 0}{4 - 2} = -\dfrac{1}{2}$

**E**

**59. a.** $(0,21)$ and $(22,45)$

$m = \dfrac{y_2 - y_1}{x_2 - x_1} = \dfrac{45 - 21}{22 - 0} = \dfrac{24}{22} = \dfrac{12}{11}$

**b.** $m = \dfrac{12}{11}; \ (0,21)$

$y = mx + b$

$y = \dfrac{12}{11}x + 21$

**61.** $2y + 4x = 12$

$2y = -4x + 12$

$y = -2x + 6$

$m_1 = -2$

parallel; $m_2 = -2$

point $(0,5)$, $b_2 = 5$
$$y = m_2 x + b_2$$
$$y = -2x + 5$$

**63. a.** The temperature $100°$ Celsius is equivalent to $212°$ Fahrenheit.

**b.** $68°$ F

**c.** $27°$ C

**d.** $(0,32)$ and $(100,212)$

$$m = \frac{F_2 - F_1}{C_2 - C_1} = \frac{212 - 32}{100 - 0} = \frac{180}{100} = \frac{9}{5}$$

$$m = \frac{9}{5}, \ (0,32)$$

$$F = mC + b$$

$$F = \frac{9}{5}C + 32$$

**Mental Math 3.6**

**1.** $y - 8 = 3(x - 4)$

$m = 3$

Answers may vary. Example:$(4,8)$

**2.** $y - 1 = 5(x - 2)$

$m = 5$

Answers may vary. Example:$(2,1)$

**3.** $y + 3 = -2(x - 10)$

$m = -2$

Answers may vary. Example:$(10, -3)$

**4.** $y + 6 = -7(x - 2)$

$m = -7$

Answers may vary. Example:$(2, -6)$

**5.** $y = \frac{2}{5}(x + 1)$

$m = \frac{2}{5}$

Answers may vary. Example:$(-1,0)$

**6.** $y = \frac{3}{7}(x + 4)$

$m = \frac{3}{7}$

Answers may vary. Example:$(-4,0)$

**Exercise Set 3.6**

**1.** $m = 6; (2,2)$

$$y - y_1 = m(x - x_1)$$
$$y - 2 = 6(x - 2)$$
$$y - 2 = 6x - 12$$
$$6x - y = 10$$

**3.** $m = -8; (-1, -5)$

$$y - y_1 = m(x - x_1)$$
$$y - (-5) = -8(x - (-1))$$
$$y + 5 = -8x - 8$$
$$8x + y = -13$$

**5.** $m = \frac{1}{2}; (5, -6)$

$$y - y_1 = m(x - x_1)$$
$$y - (-6) = \frac{1}{2}(x - 5)$$
$$2(y + 6) = x - 5$$
$$2y + 12 = x - 5$$
$$-x + 2y = -17$$
$$x - 2y = 17$$

**7.** $(3,2)$ and $(5,6)$

$$m = \frac{y_2 - y_1}{x_2 - x_1} = \frac{6-2}{5-3} = \frac{4}{2} = 2$$

$m = 2;\ (3,2)$

$$y - y_1 = m(x - x_1)$$
$$y - 2 = 2(x - 3)$$
$$y - 2 = 2x - 6$$
$$-2x + y = -4$$
$$2x - y = 4$$

**9.** $(-1,3)$ and $(-2,-5)$

$$m = \frac{y_2 - y_1}{x_2 - x_1} = \frac{-5-3}{-2-(-1)} = \frac{-8}{-1} = 8$$

$m = 8;\ (-1,3)$

$$y - y_1 = m(x - x_1)$$
$$y - 3 = 8(x - (-1))$$
$$y - 3 = 8x + 8$$
$$-8x + y = 11$$
$$8x - y = -11$$

**11.** $(2,3)$ and $(-1,-1)$

$$m = \frac{y_2 - y_1}{x_2 - x_1} = \frac{-1-3}{-1-2} = \frac{-4}{-3} = \frac{4}{3}$$

$m = \dfrac{4}{3};\ (2,3)$

$$y - y_1 = m(x - x_1)$$
$$y - 3 = \frac{4}{3}(x - 2)$$
$$3(y - 3) = 4(x - 2)$$
$$3y - 9 = 4x - 8$$
$$-4x + 3y = 1$$
$$4x - 3y = -1$$

**13.** Vertical line, point $(0,2)$

$x = c$

$x = 0$

**15.** Horizontal line, point $(-1,3)$

$y = c$

$y = 3$

**17.** Vertical line, point $\left(-\dfrac{7}{3}, -\dfrac{2}{5}\right)$

$x = c$

$x = -\dfrac{7}{3}$

**19.** $y = 5$ is horizontal.

Parallel to $y = 5$ is horizontal; $y = c$.

Point $(1,2)$

$y = 2$

**21.** $x = -3$ is vertical.

Perpendicular to $x = -3$ is horizontal; $y = c$.

Point $(-2,5)$

$y = 5$

**23.** $x = 0$ is vertical.

Parallel to $x = 0$ is vertical; $x = c$.

Point $(6,-8)$

$x = 6$

**25.**   $m = -\dfrac{1}{2}; \left(0, \dfrac{5}{3}\right)$

$y = mx + b$

$y = -\dfrac{1}{2}x + \dfrac{5}{3}$

$6y = -3x + 10$

$3x + 6y = 10$

**27.**   $m = 1; (-7, 9)$

$y - y_1 = m(x - x_1)$

$y - 9 = 1[x - (-7)]$

$y - 9 = x + 7$

$x - y = -16$

**29.**   $(10, 7)$ and $(7, 10)$

$m = \dfrac{y_2 - y_1}{x_2 - x_1} = \dfrac{10 - 7}{7 - 10} = \dfrac{3}{-3} = -1$

$m = -1; (10, 7)$

$y - y_1 = m(x - x_1)$

$y - 7 = -1(x - 10)$

$y - 7 = -x + 10$

$x + y = 17$

**31.**   $x$-axis is horizontal.

Parallel to $y$-axis is horizontal; $y = c$.

Point $(6, 7)$

$y = 7$

**33.**   $m = -\dfrac{4}{7}; (-1, -2)$

$y - y_1 = m(x - x_1)$

$y - (-2) = -\dfrac{4}{7}[x - (-1)]$

$y + 2 = -\dfrac{4}{7}x - \dfrac{4}{7}$

$7y + 14 = -4x - 4$

$4x + 7y = -18$

**35.**   $(-8, 1)$ and $(0, 0)$

$m = \dfrac{y_2 - y_1}{x_2 - x_1} = \dfrac{0 - 1}{0 - (-8)} = -\dfrac{1}{8}$

$m = -\dfrac{1}{8}; (0, 0)$

$y - y_1 = m(x - x_1)$

$y - 0 = -\dfrac{1}{8}(x - 0)$

$8y = -x$

$x + 8y = 0$

**37.**   $m = 3; (0, 0)$

$y = mx + b$

$y = 3x + 0$

$3x - y = 0$

**39.**   $(-6, -6)$ and $(0, 0)$

$m = \dfrac{y_2 - y_1}{x_2 - x_1} = \dfrac{0 - (-6)}{0 - (-6)} = \dfrac{6}{6} = 1$

$m = 1; (0, 0)$

$y - y_1 = m(x - x_1)$

$y - 0 = 1(x - 0)$

$y = x$

$x - y = 0$

**41.**   $m = -5, b = 7$

$y = mx + b$

$y = -5x + 7$

$5x + y = 7$

**43.**   $(-1, 5)$ and $(0, -6)$

$$m = \frac{y_2 - y_1}{x_2 - x_1} = \frac{-6 - 5}{0 - (-1)} = \frac{-11}{1} = -11$$

$m = -11;\ (0, -6)$

$y = mx + b$

$y = -11x - 6$

$11x + y = -6$

**45.** Undefined slope is vertical., point $\left(-\dfrac{3}{4}, 1\right)$

$x = c$

$x = -\dfrac{3}{4}$

**47.** *y*-axis is vertical.

Perpendicular to *y*-axis is horizontal; $y = c$.

Point $(-2, -3)$

$y = -3$

**49.**  $m = 7;\ (1, 3)$

$y - y_1 = m(x - x_1)$

$y - 3 = 7(x - 1)$

$y - 3 = 7x - 7$

$7x - y = 4$

**51.** **a.** $(0, 4760)$ and $(3, 6680)$

$$m = \frac{y_2 - y_1}{x_2 - x_1} = \frac{6680 - 4760}{3 - 0} = \frac{1920}{3} = 640$$

$m = 640;\ (0, 4760)$

$y = mx + b$

$y = 640x + 4760$

**b.** If $x = 10$,

then $y = 640(10) + 4760 = 11{,}160$

Expect 11,160 vehicles.

**53.** **a.** $(1, 32)$ and $(3, 96)$

$$m = \frac{y_2 - y_1}{x_2 - x_1} = \frac{96 - 32}{3 - 1} = \frac{64}{2} = 32$$

$m = 32;\ (1, 32)$

$s - s_1 = m(t - t_1)$

$s - 32 = 32(t - 1)$

$s - 32 = 32t - 32$

$s = 32t$

**b.** If $t = 4$, then $s = 32(4) = 128$ ft/sec.

**55.** **a.** $(0, 70.3)$ and $(10, 79.6)$

$$m = \frac{y_2 - y_1}{x_2 - x_1} = \frac{79.6 - 70.3}{10 - 0} = \frac{9.3}{10} = 0.93$$

$m = 0.93;\ (0, 70.3)$

$y - y_1 = m(x - x_1)$

$y - 70.3 = 0.93(x - 0)$

$y - 70.3 = 0.93x$

$y = 0.93x + 70.3.$

**b.** If $x = 17$,

then $y = 0.93(17) + 70.3 = 86.11$

Expect 86.11 person per square mile.

**57.  a.** $(0,191)$ and $(5,260)$

**b.** $m = \dfrac{y_2 - y_1}{x_2 - x_1} = \dfrac{260 - 191}{5 - 0} = \dfrac{69}{5} = 13.8$

$m = 13.8; \ (0,191)$

$y - y_1 = m(x - x_1)$

$y - 191 = 13.8(x - 0)$

$y - 191 = 13.8x$

$y = 13.8x + 191$

**c.**  If $x = 4$,

then $y = 13.8(4) + 191 = 246.2$

Expect \$246.2 million in sales.

**59.**  $(10,63)$ and $(15,94)$

$m = \dfrac{y_2 - y_1}{x_2 - x_1} = \dfrac{94 - 63}{15 - 10} = \dfrac{31}{5}$

$m = \dfrac{31}{5}; \ (10,63)$

$y - y_1 = m(x - x_1)$

$y - 63 = \dfrac{31}{5}(x - 10)$

$5y - 315 = 31(x - 10)$

$5y - 315 = 31x - 310$

$31x - 5y = -5$

**61.  a.** $(3,10{,}000)$ and $(5,8000)$

$m = \dfrac{y_2 - y_1}{x_2 - x_1} = \dfrac{8000 - 10{,}000}{5 - 3}$

$= \dfrac{-2000}{2} = -1000$

$m = -1000; \ (5,8000)$

$S - S_1 = m(p - p_1)$

$S - 8000 = -1000(p - 5)$

$S - 8000 = -1000p + 5000$

$S = -1000p + 13{,}000$

**b.**  If $p = 3.50$,

then $S = -1000(3.5) + 13{,}000 = 9500$

Expect \$9500 in daily sales.

**63.**  If $x = 2$, then

$x^2 - 3x + 1 = (2)^2 - 3(2) + 1 = 4 - 6 + 1 = -1$

**65.**  If $x = -1$, then

$x^2 - 3x + 1 = (-1)^2 - 3(-1) + 1 = 1 + 3 + 1 = 5$

**67.**  No

**69.**  Yes

**71.**  Answers may vary.

**73.**  $y = 3x - 1, \ m_1 = 3$

**a.**  Parallel: $m_2 = m_1 = 3; \ (-1,2)$

$y - y_1 = m_2(x - x_1)$

$y - 2 = 3(x - (-1))$

$y - 2 = 3x + 3$

$-3x + y = 5$

$3x - y = -5$

**b.** Perpendicular: $m_2 = -\dfrac{1}{m_1} = -\dfrac{1}{3}$;

$(-1, 2)$

$y - y_1 = m_2 (x - x_1)$

$y - 2 = -\dfrac{1}{3}(x - (-1))$

$3(y - 2) = -1(x + 1)$

$3y - 6 = -x - 1$

$x + 3y = 5$

**75.** $3x + 2y = 7,\ y = -\dfrac{3}{2}x + \dfrac{7}{2},\ m_1 = -\dfrac{3}{2}$

**a.** Parallel: $m_2 = m_1 = -\dfrac{3}{2}$; $(3, -5)$

$y - y_1 = m_2 (x - x_1)$

$y - (-5) = -\dfrac{3}{2}(x - 3)$

$2(y + 5) = -3(x - 3)$

$2y + 10 = -3x + 9$

$3x + 2y = -1$

**b.** Perpendicular: $m_2 = -\dfrac{1}{m_1} = \dfrac{2}{3}$; $(3, -5)$

$y - y_1 = m_2 (x - x_1)$

$y - (-5) = \dfrac{2}{3}(x - 3)$

$3(y + 5) = 2(x - 3)$

$3y + 15 = 2x - 6$

$2x - 3y = 21$

**Exercise Set 3.7**

**1.** $\{(2, 4), (0, 0), (-7, 10), (10, -7)\}$
Domain: $\{-7, 0, 2, 10\}$
Range: $\{-7, 0, 4, 10\}$

**3.** $\{(0, -2), (1, -2), (5, -2),\}$
Domain: $\{0, 1, 5\}$
Range: $\{-2\}$

**5.** Every point has a unique $x$-value: it is a function.

**7.** Two points have the same $x$-value: it is not a function.

**9.** No

**11.** Yes

**13.** Yes

**15.** No

**17.** No

**19.** Yes

**21.** Yes; $y = x + 1$ is a non-vertical line.

**23.** Yes; $y - x = 7$ is a non-vertical line.

**25.** Yes; $y = 6$ is a non-vertical line.

**27.** No; $x = -2$ is a vertical line.

**29.** No; does not pass the vertical line test.

**31.** 5:20 A.M.

**33.** Answers may vary

**35.** $4.75 per hour

**37.** 1996

**39.** Yes; answers may vary

**41.** $f(x) = 2x - 5$

$f(-2) = 2(-2) - 5 = -4 - 5 = -9$

$f(0) = 2(0) - 5 = -5$

$f(3) = 2(3) - 5 = 6 - 5 = 1$

**43.** $f(x) = x^2 + 2$

$f(-2) = (-2)^2 + 2 = 4 + 2 = 6$

$f(0) = (0)^2 + 2 = 2$

$f(3) = (3)^2 + 2 = 9 + 2 = 11$

**45.** $f(x) = x^3$

$f(-2) = (-2)^3 = (-8) = -8$

$f(0) = (0)^3 = 0$

$f(3) = (3)^3 = 27$

**47.** $f(x) = |x|$

$f(-2) = |-2| = 2$

$f(0) = |0| = 0$

$f(3) = |3| = 3$

**49.** $h(x) = 5x$

$h(-1) = 5(-1) = -5, (-1, -5)$

$h(0) = 5(0) = 0, (0, 0)$

$h(4) = 5(4) = 20, (4, 20)$

**51.** $h(x) = 2x^2 + 3$

$h(-1) = 2(-1)^2 + 3 = 2 + 3 = 5, (-1, 5)$

$h(0) = 2(0)^2 + 3 = 3, (0, 3)$

$h(4) = 2(4)^2 + 3 = 2 \cdot 16 + 3 = 32 + 3$

$= 35, (4, 35)$

**53.** $h(x) = -x^2 - 2x + 3$

$h(-1) = -(-1)^2 - 2(-1) + 3 = 4, (-1, 4)$

$h(0) = -(0)^2 - 2(0) + 3 = 3, (0, 3)$

$h(4) = -(4)^2 - 2(4) + 3 = -21, (4, -21)$

**55.** $h(x) = 6$

$h(-1) = 6, (-1, 6)$

$h(0) = 6, (0, 6)$

$h(4) = 6, (4, 6)$

**57.** $(-\infty, \infty)$

**59.** $x + 5 \neq 0 \Rightarrow x \neq -5$, therefore

$(-\infty, -5) \cup (-5, \infty)$

**61.** $(-\infty, \infty)$

**63.** D; $(-\infty, \infty)$, R; $x \geq -4$, $[-4, \infty)$

**65.** D; $(-\infty, \infty)$, R; $(-\infty, \infty)$

**67.** D; $(-\infty, \infty)$, R; $\{2\}$

**69.** $(-2, 1)$

**71.** $(-3, -1)$

**73.** $H(x) = 2.59x + 47.24$

**a.** $H(46) = 2.59(46) + 47.24 = 166.38$ cm

**b.** $H(39) = 2.59(39) + 47.24 = 148.25$ cm

**75.** Answers may vary

**77.**    $y = x + 7$

$f(x) = x + 7$

**79.**  $g(x) = -3x + 12$

**a.** $g(s) = -3(s) + 12 = -3s + 12$

**b.** $g(r) = -3(r) + 12 = -3r + 12$

**81.** $f(x) = x^2 - 12$

**a.** $f(12) = (12)^2 - 12 = 132$

**b.** $f(a) = (a)^2 - 12 = a^2 - 12$

## Chapter 3 Review

**1.-6.**

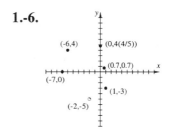

**7. A.** $(8.00, 1), (7.50, 10), (6.50, 25),$
$(5.00, 50), (2.00, 100)$

**B.**

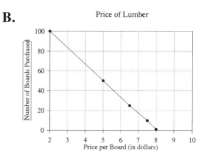

**8. a.** $(1996, 10.5), (1997, 10), (1998, 9.8),$
$(1999, 9.9), (2000, 9.6), (2001, 9.8)$

**B.**

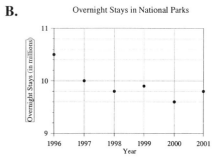

**9.**  $(0,56)$   No

$$7x - 8y = 56$$
$$7(0) - 8(56) \overset{?}{=} 56$$
$$-448 \neq 56$$

$(8,0)$   Yes

$$7x - 8y = 56$$
$$7(8) - 8(0) \overset{?}{=} 56$$
$$56 = 56$$

**10.**  $(-5,0)$   Yes

$$-2x + 5y = 10$$
$$-2(-5) + 5(0) \overset{?}{=} 10$$
$$10 = 10$$

$(1,1)$   No

$$-2x + 5y = 10$$
$$-2(1) + 5(1) \overset{?}{=} 10$$
$$3 \neq 10$$

**11.**  $(13,5)$   Yes

$$x = 13$$
$$(13) \overset{?}{=} 13$$
$$13 = 13$$

117

$$x = 13$$
$$(13,13)$$
$$(13)\overset{?}{=}13$$
Yes
$$13 = 13$$

**12.**
$$y = 2$$
$$(7,2)$$
$$(2)\overset{?}{=}2$$
Yes
$$2 = 2$$

$$y = 2$$
$$(2,7)$$
$$(7)\overset{?}{=}2$$
No
$$7 \neq 2$$

**13.** $-2 + y = 6x, \ x = 7$
$$-2 + y = 6(7)$$
$$-2 + y = 42$$
$$y = 44$$
$$(7, 44)$$

**14.** $y = 3x + 5, \ y = -8$
$$-8 = 3x + 5$$
$$-13 = 3x$$
$$-\frac{13}{3} = x$$
$$\left(-\frac{13}{3}, -8\right)$$

**15.** $9 = -3x + 4y$
$$y = 0, \ 9 = -3x + 4(0), \ 9 = -3x, \ -3 = x$$
$$y = 3, \ 9 = -3x + 4(3), \ 9 = -3x + 12$$
$$-3 = -3x, \ 1 = x$$
$$x = 9, \ 9 = -3(9) + 4y, \ 9 = -27 + 4y$$
$$36 = 4y, \ 9 = y$$

| $x$ | $y$ |
|----|----|
| $-3$ | $0$ |
| $1$ | $3$ |
| $9$ | $9$ |

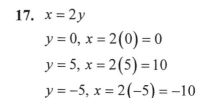

**16.** $y = 5$ for all values of $x$.

| $x$ | $y$ |
|----|----|
| $7$ | $5$ |
| $-7$ | $5$ |
| $0$ | $5$ |

**17.** $x = 2y$
$$y = 0, \ x = 2(0) = 0$$
$$y = 5, \ x = 2(5) = 10$$
$$y = -5, \ x = 2(-5) = -10$$

| $x$ | $y$ |
|----|----|
| $0$ | $0$ |
| $10$ | $5$ |
| $-10$ | $-5$ |

**18.a.** $y = 5x + 2000$
$$x = 1, \ y = 5(1) + 2000 = 2005$$
$$x = 100, \ y = 5(100) + 2000 = 2500$$
$$x = 1000, \ y = 5(1000) + 2000 = 7000$$

| $x$ | 1 | 100 | 1000 |
|----|----|----|----|
| $y$ | 2005 | 2500 | 7000 |

**b.**   Let $y = 6430$

$$6430 = 5x + 2000$$

$$4430 = 5x$$

$$886 = x$$

886 CD holders can be produced

**19.**   $x - y = 1$

| $x$ | $y$ |
|-----|-----|
| 1 | 0 |
| 0 | −1 |

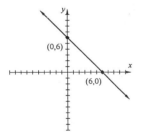

**20.**   $x + y = 6$

| $x$ | $y$ |
|-----|-----|
| 6 | 0 |
| 0 | 6 |

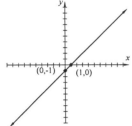

**21.**   $x - 3y = 12$

| $x$ | $y$ |
|-----|-----|
| 12 | 0 |
| 0 | −4 |

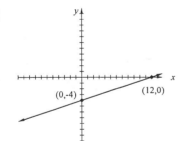

**22.**   $5x - y = -8$

| $x$ | $y$ |
|-----|-----|
| −2 | −2 |
| 0 | 8 |

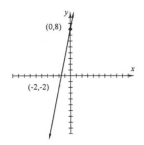

**23.**   $x = 3y$

| $x$ | $y$ |
|-----|-----|
| 0 | 0 |
| 6 | 2 |

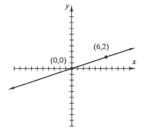

**24.**   $y = -2x$

| $x$ | $y$ |
|-----|-----|
| 0 | 0 |
| 4 | −8 |

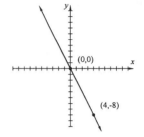

**25.**   $2x - 3y = 6$

| $x$ | $y$ |
|-----|-----|
| 0 | −2 |
| 3 | 0 |

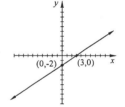

**26.**   $4x - 3y = 12$

| $x$ | $y$ |
|-----|-----|
| 0 | −4 |
| 3 | 0 |

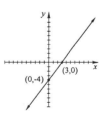

**27.**   $y = 3x + 111$

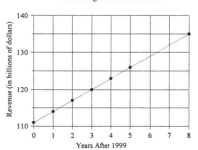

U.S. Long-Distance Revenue

Expect a revenue of $135 billion in 2007.

**28.** $x$-intercept: $(4,0)$

   $y$-intercept: $(0,-2)$

**29.** $y$-intercept: $(0,-3)$

**30.** $x$-intercepts: $(-2,0),(2,0)$

   $y$-intercepts: $(0,2),(0,-2)$

**31.** $x$-intercepts: $(-1,0),(2,0),(3,0)$

   $y$-intercept: $(0,-2)$

**32.** $x-3y=12$

| $x$ | $y$ |
|-----|-----|
| 0   | $-4$ |
| 12  | 0   |

**33.** $-4x+y=8$

| $x$ | $y$ |
|-----|-----|
| 0   | 8   |
| $-2$ | 0   |

**34.** $y=-3$ for all $x$

| $x$ | $y$ |
|-----|-----|
| 0   | $-3$ |

**35.** $x=5$ for all $y$

| $x$ | $y$ |
|-----|-----|
| 5   | 0   |

**36.** $y=-3x$

   Find a second point.

| $x$ | $y$ |
|-----|-----|
| 0   | 0   |
| 3   | $-9$ |

**37.** $x=5y$

   Find a second point.

| $x$ | $y$ |
|-----|-----|
| 0   | 0   |
| 5   | 1   |

**38.** $x-2=0$

   $x=2$ for all $y$

| $x$ | $y$ |
|-----|-----|
| 2   | 0   |

**39.** $y+6=0$

   $y=-6$ for all $x$

| $x$ | $y$ |
|-----|-----|
| 0   | $-6$ |

**40.** $(-1,2)$, and $(3,-1)$

$$m=\frac{y_2-y_1}{x_2-x_1}=\frac{-1-2}{3-(-1)}=-\frac{3}{4}$$

**41.** $(-2,-2)$, and $(3,-1)$

$$m = \frac{y_2 - y_1}{x_2 - x_1} = \frac{-1-(-2)}{3-(-2)} = \frac{1}{5}$$

**42.** $m = 0$

   **D**

**43.** $m = -1$

   **B**

**44.** Slope is undefined.

   **C**

**45.** $m = 3$

   **A**

**46.** $m = \dfrac{2}{3}$

   **E**

**47.** $(2,5)$, and $(6,8)$

$$m = \frac{y_2 - y_1}{x_2 - x_1} = \frac{8-5}{6-2} = \frac{3}{4}$$

**48.** $(4,7)$, and $(1,2)$

$$m = \frac{y_2 - y_1}{x_2 - x_1} = \frac{2-7}{1-4} = \frac{-5}{-3} = \frac{5}{3}$$

**49.** $(1,3)$, and $(-2,-9)$

$$m = \frac{y_2 - y_1}{x_2 - x_1} = \frac{-9-3}{-2-1} = \frac{-12}{-3} = 4$$

**50.** $(-4,1)$, and $(3,-6)$

$$m = \frac{y_2 - y_1}{x_2 - x_1} = \frac{-6-1}{3-(-4)} = \frac{-7}{7} = -1$$

**51.** Vertical; slope is undefined

**52.** Horizontal; slope is zero

**53.** Horizontal; slope is zero

**54.** Vertical; slope is undefined

**55.** $(-3,1)$ and $(1,-2)$

$$m_1 = \frac{y_2 - y_1}{x_2 - x_1} = \frac{-2-1}{1-(-3)} = \frac{-3}{4} = -\frac{3}{4}$$

$(2,4)$ and $(6,1)$

$$m_2 = \frac{y_2 - y_1}{x_2 - x_1} = \frac{1-4}{6-2} = \frac{-3}{4} = -\frac{3}{4}$$

$m_1 = m_2$, parallel

**56.** $(-7,6)$ and $(0,4)$

$$m_1 = \frac{y_2 - y_1}{x_2 - x_1} = \frac{4-6}{0-(-7)} = \frac{-2}{7} = -\frac{2}{7}$$

$(-9,-3)$ and $(1,5)$

$$m_2 = \frac{y_2 - y_1}{x_2 - x_1} = \frac{5-(-3)}{1-(-9)} = \frac{8}{10} = \frac{4}{5}$$

$m_1 \neq m_2$ and $m_1 m_2 \neq -1$, neither

**57.** $(9,10)$ and $(8,-7)$

$$m_1 = \frac{y_2 - y_1}{x_2 - x_1} = \frac{-7-10}{8-9} = \frac{-17}{-1} = 17$$

$(-1,-3)$ and $(2,-8)$

$$m_2 = \frac{y_2 - y_1}{x_2 - x_1} = \frac{-8-(-3)}{2-(-1)} = \frac{-5}{3} = -\frac{5}{3}$$

$m_1 \neq m_2$ and $m_1 m_2 \neq -1$, neither

**58.** $(-1, 3)$ and $(3, -2)$

$$m_1 = \frac{y_2 - y_1}{x_2 - x_1} = \frac{-2-3}{3-(-1)} = \frac{-5}{4} = -\frac{5}{4}$$

$(-2, -2)$ and $(3, 2)$

$$m_2 = \frac{y_2 - y_1}{x_2 - x_1} = \frac{2-(-2)}{3-(-2)} = \frac{4}{5}$$

$$m_1 m_2 = \left(-\frac{5}{4}\right)\left(\frac{4}{5}\right) = -1, \text{ perpendicular}$$

**59.** Every 1 year, monthly day care increases by \$17.75.

**60.** Every 1 year, 7.7 billion more dollars are spent on technology.

**61.** $3x + y = 7$

$$y = -3x + 7$$
$$y = mx + b$$
$$m = -3, \ y\text{-intercept} = (0, 7)$$

**62.** $x - 6y = -1$

$$-6y = -x - 1$$
$$y = \frac{1}{6}x + \frac{1}{6}$$
$$y = mx + b$$
$$m = \frac{1}{6}, \ y\text{-intercept} = \left(0, \frac{1}{6}\right)$$

**63.** $y = 2$

$$y = mx + b$$
$$m = 0, \ y\text{-intercept} = (0, 2)$$

**64.** $x = -5$

$$y = mx + b$$

$m$ is undefined,

There is no $y$-intercept.

**65.** $x - y = -6, \ -y = -x - 6,$

$$y = x + 6, \ m_1 = 1$$
$$x + y = 3, \ y = -x + 3, \ m_2 = -1$$
$$m_1 m_2 = (1)(-1) = -1, \text{ perpendicular}$$

**66.** $3x + y = 7, \ y = -3x + 7, \ m_1 = -3$

$$-3x - y = 10, \ -y = 3x + 10,$$
$$y = -3x - 10, \ m_2 = -3$$
$$m_1 = m_2, \text{ parallel}$$

**67.** $y = 4x + \dfrac{1}{2}, \ m_1 = 4$

$$4x + 2y = 1, \ 2y = -4x + 1,$$
$$y = -2x + \frac{1}{2}, \ m_2 = -2$$
$$m_1 \neq m_2 \text{ and } m_1 m_2 \neq -1, \text{ neither}$$

**68.** $m = -5, \ b = \dfrac{1}{2}$

$$y = mx + b$$
$$y = -5x - \frac{1}{2}$$

**69.** $m = \dfrac{2}{3}, \ b = 6$

$$y = mx + b$$
$$y = \frac{2}{3}x + 6$$

**70.** $y = -3x$

$y = mx + b$

$m = -3, b = 0$

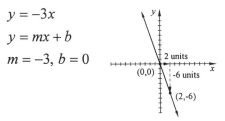

**71.** $y = 3x - 1$

$y = mx + b$

$m = 3, b = -1$

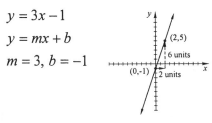

**72.** $-x + 2y = 8$

$2y = x + 8$

$y = \dfrac{1}{2}x + 4$

$y = mx + b$

$m = \dfrac{1}{2}, b = 4$

**73.** $5x - 3y = 15$

$-3y = -5x + 15$

$y = \dfrac{5}{3}x - 5$

$y = mx + b$

$m = \dfrac{5}{3}, b = -5$

**74.** $y = -2x + 1$

$m = 2, b = 1$

**D**

**75.** $y = -4x$

$m = -4, b = 0$

**C**

**76.** $y = 2x$

$m = 2, b = 0$

**A**

**77.** $y = 2x - 1$

$m = 2, b = -1$

**B**

**78.** $\quad m = 4; (2, 0)$

$y - y_1 = m(x - x_1)$

$y - 0 = 4(x - 2)$

$y = 4x - 8$

$4x - y = 8$

**79.** $\quad m = -3; (0, -5)$

$y = mx + b$

$y = -3x - 5$

$3x + y = -5$

**80.** $\quad m = \dfrac{1}{2}; \left(0, -\dfrac{7}{2}\right)$

$y = mx + b$

$y = \dfrac{1}{2}x - \dfrac{7}{2}$

$2y = x - 7$

$x - 2y = 7$

**81.** Horizontal line, point $(-2, -3)$

$y = c$

$y = -3$

**82.** Horizontal line, point $(0, 04)$

$y = c$

$y = 0$

**83.** $\quad m = -6;\ (2, -1)$

$$y - y_1 = m(x - x_1)$$
$$y - (-1) = -6(x - 2)$$
$$y + 1 = -6x + 12$$
$$6x + y = 11$$

**84.** $\quad m = 12;\ \left(\dfrac{1}{2}, 5\right)$

$$y - y_1 = m(x - x_1)$$
$$y - 5 = 12\left(x - \dfrac{1}{2}\right)$$
$$y - 5 = 12x - 6$$
$$12x - y = 1$$

**85.** $(0, 6)$ and $(6, 0)$

$$m = \dfrac{y_2 - y_1}{x_2 - x_1} = \dfrac{0 - 6}{6 - 0} = \dfrac{-6}{6} = -1$$
$$m = -1;\ (0, 6)$$
$$y - y_1 = m(x - x_1)$$
$$y - 6 = -1(x - 0)$$
$$y - 6 = -x$$
$$x + y = 6$$

**86.** $(0, -4)$ and $(-8, 0)$

$$m = \dfrac{y_2 - y_1}{x_2 - x_1} = \dfrac{0 - (-4)}{-8 - 0} = \dfrac{4}{-8} = -\dfrac{1}{2}$$
$$m = -\dfrac{1}{2};\ (0, -4)$$
$$y - y_1 = m(x - x_1)$$
$$y - (-4) = -\dfrac{1}{2}(x - 0)$$

$$y + 4 = -\dfrac{1}{2}x$$
$$2y + 8 = -x$$
$$x + 2y = -8$$

**87.** Vertical line, point $(5, 7)$

$$x = c$$
$$x = 5$$

**88.** Horizontal line, point $(-6, 8)$

$$y = c$$
$$y = 8$$

**89.** $y = 8$ is horizontal.

Perpendicular to $y = 8$ is vertical; $x = c$.

Point $(6, 0)$

$$x = 6$$

**90.** $x = -2$ is vertical.

Perpendicular to $x = -2$ is horizontal;

$y = c$, point $(10, 12)$

$$y = 12$$

**91.** $y = -3x + 7,\ m_1 = -3$

**a.** Parallel: $m_2 = m_1 = -3;\ (5, 0)$

$$y - y_1 = m_2(x - x_1)$$
$$y - 0 = -3(x - 5)$$
$$y = -3x + 15$$
$$3x + y = 15$$

**b.** Perpendicular: $m_2 = -\dfrac{1}{m_1} = \dfrac{1}{3}$; $(5,0)$

$$y - y_1 = m_2(x - x_1)$$
$$y - 0 = \frac{1}{3}(x - 5)$$
$$3(y - 0) = 1(x - 5)$$
$$3y - 0 = x - 5$$
$$x - 3y = 5$$

**92.** Two points have the same $x$-value: it is not a function.

**93.** Every point has a unique $x$-value: it is a function.

**94.** Yes; $7x - 6y = 1$ is a non-vertical line.

**95.** Yes; $y = 7$ is a non-vertical line.

**96.** No; $x = 2$ is a vertical line.

**97.** Yes; for each value of $x$ there is only one value of $y$.

**98.** No; some values of $x$ give 2 values of $y$.

**99.** No

**100.** Yes

**101.** $f(x) = -2x + 6$

    **a.** $f(0) = -2(0) + 6 = 6$

    **b.** $f(-2) = -2(-2) + 6 = 4 + 6 = 10$

    **c.** $f\left(\dfrac{1}{2}\right) = -2\left(\dfrac{1}{2}\right) + 6 = -1 + 6 = 5$

**102.** $h(x) = -5 - 3x$

    **a.** $h(2) = -5 - 3(2) = -11$

    **b.** $h(-3) = -5 - 3(-3) = 4$

    **c.** $h(0) = -5 - 3(0) = -5$

**103.** $g(x) = x^2 + 12x$

    **a.** $g(3) = (3)^2 + 12(3) = 45$

    **b.** $g(-5) = (-5)^2 + 12(-5) = -35$

    **c.** $g(0) = (0)^2 + 12(0) = 0$

**104.** $h(x) = 6 - |x|$

    **a.** $h(-1) = 6 - |-1| = 6 - 1 = 5$

    **b.** $h(1) = 6 - |1| = 6 - 1 = 5$

    **c.** $h(-4) = 6 - |-4| = 6 - 4 = 2$

**105.** $(-\infty, \infty)$

**106.** $x - 2 \ne 0 \Rightarrow x \ne 2$, therefore
$(-\infty, 2) \cup (2, \infty)$

**107.** D; $[-3, 5]$,  R; $[-4, 2]$

**108.** D; $(-\infty, \infty)$,  R; $x \ge 0$, $[0, \infty)$

**109.** D; $\{3\}$,  R; $(-\infty, \infty)$

**110.** D; $(-\infty, \infty)$,  R; $x \le 2$, $(-\infty, 2]$

**Chapter 3  Test**

**1. a.** $(1980, 38), (1984, 47), (1988, 51),$
$(1992, 54), (1996, 59), (2000, 55)$

**b.**

New Mothers Returning to Work

**2.** $2x + y = 8$

| x | y |
|---|---|
| 4 | 0 |
| 0 | 8 |

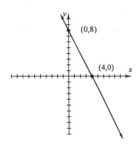

**3.** $5x - 7y = 10$

| x | y |
|---|---|
| 2 | 0 |
| -5 | -5 |

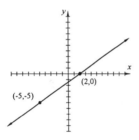

**4.** $y = -1$

for all values of $x$

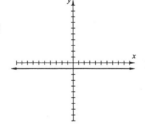

**5.** $x - 3 = 0$

$x = 3$

for all values of $y$

**6.** $(-1, -1)$, and $(4, 1)$

$$m = \frac{y_2 - y_1}{x_2 - x_1} = \frac{1 - (-1)}{4 - (-1)} = \frac{2}{5}$$

**7.** Horizontal line: $m = 0$

**8.** $(6, -5)$, and $(-1, 2)$

$$m = \frac{y_2 - y_1}{x_2 - x_1} = \frac{2 - (-5)}{-1 - 6} = \frac{7}{-7} = -1$$

**9.** $-3x + y = 5$

$y = 3x + 5$

$y = mx + b$

$m = 3$

**10.** $x = 6$ is a vertical line.

The slope is undefined.

**11.** $7x - 3y = 2$

$-3y = -7x + 2$

$$y = \frac{7}{3}x - \frac{2}{3}$$

$y = mx + b$

$$m = \frac{7}{3}, b = -\frac{2}{3}, \left(0, -\frac{2}{3}\right)$$

**12.** $y = 2x - 6, \ m_1 = 2$

$-4x = 2y, \ -2x = y,$

$y = -2x, \ m_2 = -2$

$m_1 \neq m_2$ and $m_1 m_2 \neq -1,$ neither

**13.** $\qquad m = -\dfrac{1}{4}; \ (2,2)$

$y - y_1 = m(x - x_1)$

$y - 2 = -\dfrac{1}{4}(x - 2)$

$4(y - 2) = -(x - 2)$

$4y - 8 = -x + 2$

$x + 4y = 10$

**14.** $(0,0)$ and $(6, -7)$

$m = \dfrac{y_2 - y_1}{x_2 - x_1} = \dfrac{-7 - 0}{6 - 0} = -\dfrac{7}{6}$

$m = -\dfrac{7}{6}; \ (0,0)$

$y - y_1 = m(x - x_1)$

$y - 0 = -\dfrac{7}{6}(x - 0)$

$6y = -7x$

$7x + 6y = 0$

**15.** $(2, -5)$ and $(1, 3)$

$m = \dfrac{y_2 - y_1}{x_2 - x_1} = \dfrac{3 - (-5)}{1 - 2} = \dfrac{8}{-1} = -8$

$m = -8; \ (1, 3)$

$y - y_1 = m(x - x_1)$

$y - 3 = -8(x - 1)$

$y - 3 = -8x + 8$

$8x + y = 11$

**16.** $x = 7$ is vertical.

Parallel to $x = 7$ is vertical;

$x = c,$ point $(-5, -1)$

$x = -5$

**17.** $\qquad m = \dfrac{1}{8}, \ b = 12$

$y = mx + b$

$y = \dfrac{1}{8}x + 12$

$8y = x + 96$

$x - 8y = -96$

**18.** Yes

**19.** No

**20.** $h(x) = x^3 - x$

**a.** $h(-1) = (-1)^3 - (-1) = -1 + 1 = 0$

**b.** $h(0) = (0)^3 - (0) = 0$

**c.** $h(4) = (4)^3 - (4) = 64 - 4 = 60$

**21.** $x + 1 \neq 0 \Rightarrow x \neq -1,$ therefore

$(-\infty, -1) \cup (-1, \infty)$

**22.** D; $(-\infty, \infty),$ R; $x \leq 4, \ (-\infty, 4]$

**23.** D; $(-\infty, \infty),$ R; $(-\infty, \infty)$

**24.** 9 p.m.

**25.** 4 p.m.

**26.** January 1st and December 1st

**27.** June 1st and end of July

**28.** Yes; it passes the verical line test.

**29.** Yes; every location has exactly 1 sunset time per day.

**Cumulative Review Chapter 3**

**1.** **a.** $2 < 3$    **b.** $7 > 4$    **c.** $72 > 27$

**2.** $\dfrac{56}{64} = \dfrac{7 \cdot 8}{8 \cdot 8} = \dfrac{7}{8}$

**3.** $\dfrac{2}{15} \cdot \dfrac{5}{13} = \dfrac{2 \cdot 5}{3 \cdot 5 \cdot 13} = \dfrac{2}{39}$

**4.** $\dfrac{10}{3} + \dfrac{5}{21} = \dfrac{10 \cdot 7}{3 \cdot 7} + \dfrac{5}{21} = \dfrac{70 + 5}{21} = \dfrac{75}{21}$

$= \dfrac{3 \cdot 25}{3 \cdot 7} = \dfrac{25}{7} = 3\dfrac{4}{7}$

**5.** $\dfrac{3 + |4 - 3| + 2^2}{6 - 3} = \dfrac{3 + |1| + 2^2}{6 - 3} = \dfrac{3 + 1 + 4}{6 - 3}$

$= \dfrac{8}{3}$

**6.** $16 - 3 \cdot 3 + 2^4 = 16 - 3 \cdot 3 + 16$

$= 16 - 9 + 16$

$= 23$

**7.** **a.** $-8 + (-11) = -19$

**b.** $-5 + 35 = 30$

**c.** $0.6 + (-1.1) = -0.5$

**d.** $-\dfrac{7}{10} + \left(-\dfrac{1}{10}\right) = -\dfrac{8}{10} = -\dfrac{4}{5}$

**e.** $11.4 + (-4.7) = 6.7$

**f.** $-\dfrac{3}{8} + \dfrac{2}{5} = -\dfrac{3 \cdot 5}{8 \cdot 5} + \dfrac{2 \cdot 8}{5 \cdot 8} = \dfrac{-15 + 16}{40} = \dfrac{1}{40}$

**8.** $|9 + (-20)| + |-10| = |-11| + |-10|$

$= 11 + 10$

$= 21$

**9.** **a.** $-14 - 8 + 10 - (-6)$

$= -14 + (-8) + 10 + 6$

$= -6$

**b.** $1.6 - (-10.3) + (-5.6)$

$= 1.6 + 10.3 + (-5.6)$

$= 6.3$

**10.** $-9 - (3 - 8) = -9 - (-5) = -9 + 5 = -4$

**11.** Let $x = -2$ and $y = -4$.

**a.** $5x - y = 5(-2) - (-4) = -10 + 4$

$= -6$

**b.** $x^4 - y^2 = (-2)^4 - (-4)^2 = 16 - 16 = 0$

**c.** $\dfrac{3x}{2y} = \dfrac{3(-2)}{2(-4)} = \dfrac{-6}{-8} = \dfrac{3}{4}$

**12.** $\dfrac{x}{-10} = 2$

Let $x = -20$.

$\dfrac{-20}{-10} \overset{?}{=} 2$

$2 = 2$   True

2 is a solution to the equation.

**13.** **a.** $10 + (x + 12) = 10 + x + 12 = x + 22$

**b.** $-3(7x) = -21x$

**14.** $(12 + x) - (4x - 7) = 12 + x - 4x + 7$

$= 19 - 3x$

**15.**   **a.** $-3$    **b.** $22$    **c.** $1$    **d.** $-1$    **e.** $\dfrac{1}{7}$

**16.**   $-5(x-7) = -5x - (-5)(7) = -5x + 35$

**17.**   $y + 0.6 = -1.0$
$$y = -1.6$$

**18.**   $5(3+z) - (8z+9) = -4$
$$15 + 5z - 8z - 9 = -4$$
$$-3z + 6 = -4$$
$$-3z = -10$$
$$z = \frac{10}{3}$$

**19.**   $-\dfrac{2}{3}x = -\dfrac{5}{2}$
$$6\left(-\frac{2}{3}x\right) = 6\left(-\frac{5}{2}\right)$$
$$-4x = -15$$
$$x = \frac{15}{4}$$

**20.**   $\dfrac{x}{4} - 1 = -7$
$$4\left(\frac{x}{4}\right) - 4(1) = 4(-7)$$
$$x - 4 = -28$$
$$x = -24$$

**21.**   Sum = first integer + second integer
              + third integer.
$$\text{Sum} = x + (x+1) + (x+2)$$
$$= x + x + 1 + x + 2$$
$$= 3x + 3$$

**22.**   $\dfrac{x}{3} - 2 = \dfrac{x}{3}$
$$3\left(\frac{x}{3}\right) - 3(2) = 3\left(\frac{x}{3}\right)$$
$$x - 6 = x$$
$$-6 = 0$$
This is false. There is no solution

**23.**   $\dfrac{2(a+3)}{3} = 6a + 2$
$$2(a+3) = 18a + 6$$
$$2a + 6 = 18a + 6$$
$$-16a + 6 = 6$$
$$-16a = 0$$
$$a = 0$$

**24.**   $x + 2y = 6$
$$x - x + 2y = 6 - x$$
$$2y = 6 - x$$
$$\frac{2y}{2} = \frac{6-x}{2}$$
$$y = \frac{6-x}{2}$$

**25.**   Let $x =$ the number of Democratic representatives and $x + 15 =$ the number of Republican representatives.
$$x + x + 15 = 431$$
$$2x + 15 = 431$$
$$2x+ = 416$$
$$x = 208$$
$$x + 15 = 223$$
There were 208 Democratic representatives and 223 Republican.

**26.**  $5(x+4) \geq 4(2x+3)$

$5x + 20 \geq 8x + 12$

$-3x + 20 \geq 12$

$-3x \geq -8$

$\dfrac{-3x}{-3} \leq \dfrac{-8}{-3}$

$x \leq \dfrac{8}{3}, \ \left(-\infty, \dfrac{8}{3}\right]$

**27.**  The perimeter of a rectangle is given by the formula $P = 2l + 2w.$ Let $l =$ the length of the garden.  $P = 2l + 2w$

$140 = 2l + 2w$

$140 = 2l + 2(30)$

$140 = 2l + 60$

$80 = 2l$

$40 = l$

The length of the garden is 40 feet.

**28.**  $-3 < 4x - 1 \leq 2$

$-2 < 4x \leq 3$

$-\dfrac{1}{2} < x \leq \dfrac{3}{4}, \ \left(-\dfrac{1}{2}, \dfrac{3}{4}\right]$

**29.**  $y = mx + b$

$y - b = mx + b - b$

$y - b = mx$

$\dfrac{y-b}{m} = \dfrac{mx}{m}$

$\dfrac{y-b}{m} = x$

**30.**  $y = -5x$

| $x$ | $y$ |
|-----|-----|
| 0   | 0   |
| −1  | 5   |
| 2   | −10 |

**31.**  Let $x =$ the amount of 70% acid.

*No. of liters $\cdot$ Strength $=$ Amt of Acid*

| 70% | $x$ | 0.7 | $0.7x$ |
|-----|-----|-----|--------|
| 40% | $12 - x$ | 0.4 | $0.4(12 - x)$ |
| 50% | 12  | 0.5 | $0.5(12)$ |

$0.7x + 0.4(12 - x) = 0.5(12)$

$0.7x + 4.8 - 0.4x = 6$

$0.3x + 4.8 = 6$

$0.3x = 1.2$

$x = 4$

$12 - x = 12 - 4 = 8$

Mix 4 liters of 70% acid with 8 liters of 40% acid.

**32.**  $y = -3x + 5$

| $x$ | $y$ |
|-----|-----|
| −1  | 8   |
| 0   | 0   |
| 1   | 2   |

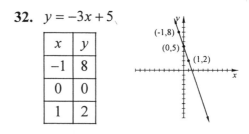

**33.**  $x \geq -1, \ [-1, \infty)$

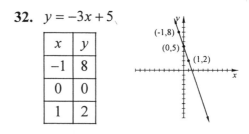

**34.** $2x + 4y = -8$

$x$-intercept, $y = 0$

$2x + 4(0) = -8 \Rightarrow x = -4 : (-4, 0)$

$y$-intercept, $x = 0$

$2(0) + 4y = -8 \Rightarrow y = -2 : (0, -2)$

**35.** $-1 \le 2x - 3 < 5$

$2 \le 2x < 8$

$1 \le x < 4, \ [1, 4)$

**36.** $x = 2$

$x = 2$ for all

values of $y$

**37.a.** $\qquad x - 2y = 6$

$(6, 0) \qquad (6) - 2(0) \overset{?}{=} 6$

Yes $\qquad\qquad 6 = 6$

**b.** $\qquad x - 2y = 6$

$(0, 3) \qquad (0) - 2(3) \overset{?}{=} 6$

No $\qquad\qquad -6 \ne 6$

$\qquad\qquad x - 2y = 6$

**c.** $\left(1, -\dfrac{5}{2}\right) \quad (1) - 2\left(-\dfrac{5}{2}\right) \overset{?}{=} 6$

Yes $\qquad\qquad 1 + 5 \overset{?}{=} 6$

$\qquad\qquad 6 = 6$

**38.** $(0, 5)$ and $(-5, 4)$

$m = \dfrac{y_2 - y_1}{x_2 - x_1} = \dfrac{4 - 5}{-5 - 0} = \dfrac{-1}{-5} = \dfrac{1}{5}$

**39.a.** linear

**b.** linear

**c.** Not linear

**d.** Linear

**40.** $x = -10$ is a vertical line.

The slope is undefined.

**41.** $y = -1$ is horizontal, slope is 0.

**42.** $2x - 5y = 10$

$-5y = -2x + 10$

$y = \dfrac{2}{5}x - 2$

$y = mx + b$

$m = \dfrac{2}{5}, b = -2$

The slope is $\dfrac{2}{5}$.

The $y$-intercept is $(0, -2)$

**43.** $(-1, 7)$ and $(2, 2)$

$m_1 = \dfrac{y_2 - y_1}{x_2 - x_1} = \dfrac{2 - 7}{2 - (-1)} = \dfrac{-5}{3} = -\dfrac{5}{3}$

perpendicular; $m_2 =$ (negative

reciprocal of $m_1$) $= \dfrac{3}{5}$

**44.** $(2, 3)$ and $(0, 0)$

$$m = \frac{y_2 - y_1}{x_2 - x_1} = \frac{0 - 3}{0 - 2} = \frac{-3}{-2} = \frac{3}{2}$$

Point: $(0, 0)$

$$y - y_1 = m(x - x_1)$$

$$y - 0 = \frac{3}{2}(x - 0)$$

$$2y = 3x$$

$$3x - 2y = 0$$

# Chapter 4

## Graphing Calculator Explorations 4.1

**1.** $\begin{cases} y = -2.68x + 1.21 \\ y = 5.22x - 1.68 \end{cases}$

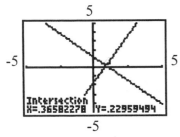

The solution of the system is $(0.37, 0.23)$.

**3.** $\begin{cases} 4.3x - 2.9y = 5.6 \\ 8.1x + 7.6y = -14.1 \end{cases}$

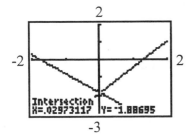

The solution of the system is $(0.03, -1.89)$.

## Mental Math 4.1

**1.** One solution, $(-1, 3)$

**2.** No solution

**3.** Infinite number of solutions.

**4.** One solution, $(3, 4)$

**5.** No solution

**6.** Infinite number of solutions.

**7.** One solution, $(3, 2)$

**8.** One solution, $(0, -3)$

## Exercise Set 4.1

**1. a.** Let $x = 2$ and $y = 4$.

$$x + y = 8 \qquad\qquad 3x + 2y = 21$$
$$2 + 4 \overset{?}{=} 8 \qquad 3(2) + 2(4) \overset{?}{=} 21$$
$$6 = 8 \qquad\qquad 6 + 8 \overset{?}{=} 21$$
$$\text{False} \qquad\qquad 14 = 21$$
$$\text{False}$$

$(2, 4)$ is not a solution of the system.

**b.** Let $x = 5$ and $y = 3$.

$$x + y = 8 \qquad\qquad 3x + 2y = 21$$
$$5 + 3 \overset{?}{=} 8 \qquad 3(5) + 2(3) \overset{?}{=} 21$$
$$8 = 8 \qquad\qquad 15 + 6 \overset{?}{=} 21$$
$$\text{True} \qquad\qquad 21 = 21$$
$$\text{True}$$

$(5, 3)$ is a solution of the system

**c.** Let $x = 1$ and $y = 9$.

$$x + y = 8 \qquad\qquad 3x + 2y = 21$$
$$1 + 9 \overset{?}{=} 8 \qquad 3(1) + 2(9) \overset{?}{=} 21$$
$$10 = 8 \qquad\qquad 3 + 18 \overset{?}{=} 21$$
$$\text{False} \qquad\qquad 21 = 21$$
$$\text{True}$$

$(1, 9)$ is not a solution of the system

**3..** **a.** Let $x = 2$ and $y = -1$.

$$3x - y = 5 \qquad\qquad x + 2y = 11$$

$$3(2) - (-1) \overset{?}{=} 5 \qquad\qquad 2 + 2(-1) \overset{?}{=} 11$$

$$6 + 1 \overset{?}{=} 5 \qquad\qquad 2 - 2 \overset{?}{=} 11$$

$$7 = 5 \qquad\qquad 0 = 11$$

$$\text{False} \qquad\qquad \text{False}$$

$(2, -1)$ is not a solution of the system.

**b.** Let $x = 3$ and $y = 4$.

$$3x - y = 5 \qquad\qquad x + 2y = 11$$

$$3(3) - 4 \overset{?}{=} 5 \qquad\qquad 3 + 2(4) \overset{?}{=} 11$$

$$9 - 4 \overset{?}{=} 5 \qquad\qquad 3 + 8 \overset{?}{=} 11$$

$$5 = 5 \qquad\qquad 11 = 11$$

$$\text{True} \qquad\qquad \text{True}$$

$(3, 4)$ is a solution of the system.

**c.** Let $x = 0$ and $y = -5$.

$$3x - y = 5 \qquad\qquad x + 2y = 11$$

$$3(0) - (-5) \overset{?}{=} 5 \qquad\qquad 0 + 2(-5) \overset{?}{=} 11$$

$$0 + 5 \overset{?}{=} 5 \qquad\qquad 0 - 10 \overset{?}{=} 11$$

$$5 = 5 \qquad\qquad -10 = 11$$

$$\text{True} \qquad\qquad \text{False}$$

$(0, -5)$ is not a solution of the system

**5.** **a.** Let $x = -3$ and $y = -6$.

$$2y = 4x \qquad\qquad 2x - y = 0$$

$$2(-6) \overset{?}{=} 4(-3) \qquad\qquad 2(-3) - (-6) \overset{?}{=} 0$$

$$-12 = -12 \qquad\qquad -6 + 6 \overset{?}{=} 0$$

$$\text{True} \qquad\qquad 0 = 0$$

$$\text{True}$$

$(-3, -6)$ is a solution of the system

**b.** Let $x = 0$ and $y = 0$.

$$2y = 4x \qquad\qquad 2x - y = 0$$

$$2(0) \overset{?}{=} 4(0) \qquad\qquad 2(0) - (0) \overset{?}{=} 0$$

$$0 = 0 \qquad\qquad 0 - 0 \overset{?}{=} 0$$

$$\text{True} \qquad\qquad 0 = 0$$

$$\text{True}$$

$(0, 0)$ is a solution of the system.

**c.** Let $x = 1$ and $y = 2$.

$$2y = 4x \qquad\qquad 2x - y = 0$$

$$2(2) \overset{?}{=} 4(1) \qquad\qquad 2(1) - (2) \overset{?}{=} 0$$

$$4 = 4 \qquad\qquad 2 - 2 \overset{?}{=} 0$$

$$\text{True} \qquad\qquad 0 = 0$$

$$\text{True}$$

$(1, 2)$ is a solution of the system.

**7.** Answers may vary

**9.** $\begin{cases} y = x + 1 \\ y = 2x - 1 \end{cases}$

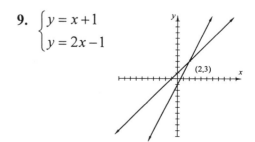

The solution of the system is $(2, 3)$.

consistent and independent

**11.** $\begin{cases} 2x + y = 0 \\ 3x + y = 1 \end{cases}$

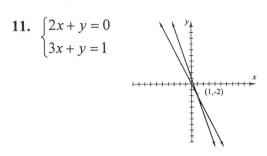

The solution of the system is $(1, -2)$.

consistent and independent

**13.** $\begin{cases} y = -x - 1 \\ y = 2x + 5 \end{cases}$

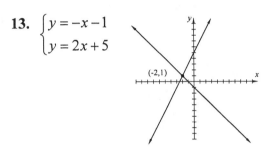

The solution of the system is $(-2, 1)$.

consistent and independent

**15.** $\begin{cases} 2x - y = 6 \\ y = 2 \end{cases}$

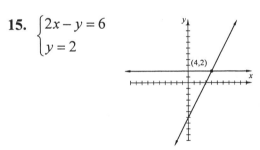

The solution of the system is $(4, 2)$.

consistent and independent

**17.** $\begin{cases} x + y = 5 \\ x + y = 6 \end{cases}$

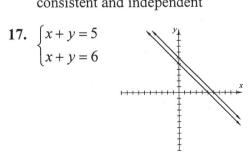

There is no solution.

inconsistent and independent

**19.** $\begin{cases} y - 3x = -2 \\ 6x - 2y = 4 \end{cases}$

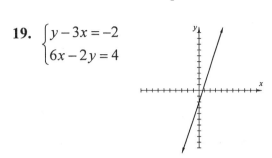

The solution of the system is $(2, 0)$.

consistent and independent

**21.** $\begin{cases} x - 2y = 2 \\ 3x + 2y = -2 \end{cases}$

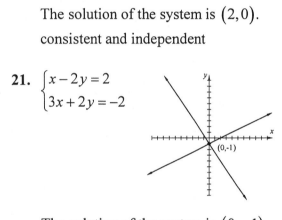

The solution of the system is $(0, -1)$.

consistent and independent

**23.** $\begin{cases} \dfrac{1}{2}x + y = -1 \\ x = 4 \end{cases}$

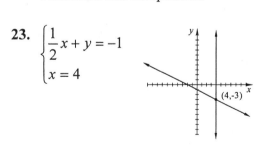

The solution of the system is $(4, -3)$.

consistent and independent

**25.** $\begin{cases} y = x - 2 \\ y = 2x + 3 \end{cases}$

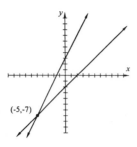

(-5,-7)

The solution of the system is $(-5, -7)$.

consistent and independent

**27.** $\begin{cases} x + y = 7 \\ x - y = 3 \end{cases}$

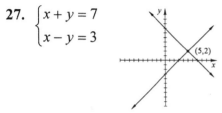

(5,2)

The solution of the system is $(5, 2)$.

consistent and independent

**29.** Answers may vary

**31.** Intersecting, one solution

**33.** Parallel, no solution

**35.** Identical lines, infinite number of solutions

**37.** Intersecting, one solution

**39.** Intersecting, one solution

**41.** Identical lines, infinite number of solutions

**43.** Parallel, no solution

**45.** $5(x - 3) + 3x = 1$

$5x - 15 + 3x = 1$

$8x - 15 = 1$

$8x = 16$

$x = 2$

The solution is 2.

**47.** $4\left(\dfrac{y+1}{2}\right) + 3y = 0$

$2(y + 1) + 3y = 0$

$2y + 2 + 3y = 0$

$5y + 2 = 0$

$5y = -2$

$y = -\dfrac{2}{5}$

The solution is $-\dfrac{2}{5}$.

**49.** $8a - 2(3a - 1) = 6$

$8a - 6a + 2 = 6$

$2a + 2 = 6$

$2a = 4$

$a = 2$

The solution is 2.

**51.** Answers may vary

**53.** 1984, 1988

**55.** 1996

**57.** Answers may vary

**59.** Answers may vary.
Possible answer

$\begin{cases} 3x - y = 1 \\ x - 2y = -8 \end{cases}$

**61.** Answers may vary.

**63.a.** Each table includes the point $(4, 9)$.
Therefore $(4, 9)$ is a solution of the
system.

**b.**

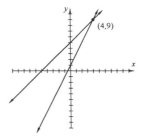

(4,9)

**c.** Yes

Possible answer

$$\begin{cases} 3x - 2y = -2 \\ 6x + 4y = 4 \end{cases}$$

**Mental Math 4.2**

1. When solving, you obtain $x = 1$. $(1, 4)$

2. When solving, you obtain $0 = 34$.
   No solution.

3. When solving, you obtain $0 = 0$.
   Infinite number of solutions.

4. When solving, you obtain $y = 0$. $(5, 0)$

5. When solving, you obtain $x = 0$. $(0, 0)$

6. When solving, you obtain $0 = 0$.
   Infinite number of solutions.

**Exercise Set 4.2**

1. $\begin{cases} x + y = 3 \\ x = 2y \end{cases}$

   Substitute $2y$ for $x$ in the first equation.
   $$2y + y = 3$$
   $$3y = 3$$
   $$y = 1$$

Let $y = 1$ in the second equation.
$$x = 2(1)$$
$$x = 2$$
The solution is $(2, 1)$.

3. $\begin{cases} x + y = 6 \\ y = -3x \end{cases}$

   Substitute $-3x$ for $y$ in the first equation.
   $$x + (-3x) = 6$$
   $$-2x = 6$$
   $$x = -3$$
   Let $x = -3$ in the second equation.
   $$y = -3(-3)$$
   $$y = 9$$
   The solution is $(-3, 9)$.

5. $\begin{cases} 3x + 2y = 16 \\ x = 3y - 2 \end{cases}$

   Substitute $3y - 2$ for $x$ in the first equation.
   $$3(3y - 2) + 2y = 16$$
   $$9y - 6 + 2y = 16$$
   $$11y = 22$$
   $$y = 2$$

   Let $y = 2$ in the second equation.
   $$x = 3(2) - 2$$
   $$x = 4$$
   The solution is $(4, 2)$.

**7.** $\begin{cases} 3x - 4y = 10 \\ x = 2y \end{cases}$

Substitute $2y$ for $x$ in the first equation.

$3(2y) - 4y = 10$

$6y - 4y = 10$

$2y = 10$

$y = 5$

Let $y = 5$ in the second equation.

$x = 2(5)$

$x = 10$

The solution is $(10, 5)$.

**9.** $\begin{cases} y = 3x + 1 \\ 4y - 8x = 12 \end{cases}$

Substitute $3x + 1$ for $y$ in the second equation.

$4(3x + 1) - 8x = 12$

$12x + 4 - 8x = 12$

$4x = 8$

$x = 2$

Let $x = 2$ in the first equation.

$y = 3(2) + 1$

$y = 7$

The solution is $(2, 7)$.

**11.** $\begin{cases} x + 2y = 6 \\ 2x + 3y = 8 \end{cases}$

Solve the first equation for $x$.

$x = 6 - 2y$

Substitute $6 - 2y$ for $x$ in the second equation.

$2(6 - 2y) + 3y = 8$

$12 - 4y + 3y = 8$

$-y = -4$

$y = 4$

Let $y = 4$ in $x = 6 - 2y$.

$x = 6 - 2(4)$

$y = -2$

The solution is $(-2, 4)$.

**13.** $\begin{cases} 2x - 5y = 1 \\ 3x + y = -7 \end{cases}$

Solve the second equation for $y$.

$y = -7 - 3x$

Substitute $-7 - 3x$ for $y$ in the first equation.

$2x - 5(-7 - 3x) = 1$

$2x + 35 + 15x = 1$

$17x = -34$

$x = -2$

Let $x = -2$ in $y = -7 - 3x$.

$y = -7 - 3(-2)$

$y = -1$

The solution is $(-2, -1)$.

**15.** $\begin{cases} 2y = x+2 \\ 6x - 12y = 0 \end{cases}$

Solve the first equation for $x$.

$x = 2y - 2$

Substitute $2y - 2$ for $x$ in the second equation.

$6(2y-2) - 12y = 0$

$12y - 12 - 12y = 0$

$-12 = 0$

The system has no solution.

**17.** $\begin{cases} \dfrac{1}{3}x - y = 2 \\ x - 3y = 6 \end{cases}$

Solve the second equation for $x$.

$x = 6 + 3y$

Substitute $6 + 3y$ for $x$ in the first equation.

$\dfrac{1}{3}(6+3y) - y = 2$

$2 + y - y = 2$

$2 = 2$

The equations in the original system are equivalent and there are an infinite number of solutions.

**19.** $\begin{cases} 4x + y = 11 \\ 2x + 5y = 1 \end{cases}$

Solve the first equation for $y$.

$y = 11 - 4x$

Substitute $11 - 4x$ for $y$ in the second equation.

$2x + 5(11 - 4x) = 1$

$2x + 55 - 20x = 1$

$-18x = -54$

$x = 3$

Let $x = 3$ in $y = 11 - 4x$.

$y = 11 - 4(3)$

$y = -1$

The solution is $(3, -1)$.

**21.** $\begin{cases} 2x - 3y = -9 \\ 3x = y + 4 \end{cases}$

Solve the second equation for $y$.

$y = 3x - 4$

Substitute $3x - 4$ for $y$ in the first equation.

$2x - 3(3x - 4) = -9$

$2x - 9x + 12 = -9$

$-7x = -21$

$x = 3$

Let $x = 3$ in $y = 3x - 4$.

$y = 3(3) - 4$

$y = 5$

The solution is $(3, 5)$.

**23.** $\begin{cases} 6x - 3y = 5 \\ x + 2y = 0 \end{cases}$

Solve the second equation for $x$.

$x = -2y$

Substitute $-2y$ for $x$ in the first equation.

$6(-2y) - 3y = 5$

$-12y - 3y = 5$

$$-15y = 5$$

$$y = -\frac{1}{3}$$

Let $y = -\frac{1}{3}$ in $x = -2y$.

$$x = -2\left(-\frac{1}{3}\right)$$

$$x = \frac{2}{3}$$

The solution is $\left(\frac{2}{3}, -\frac{1}{3}\right)$.

**25.** $\begin{cases} 3x - y = 1 \\ 2x - 3y = 10 \end{cases}$

Solve the first equation for $y$.

$$y = 3x - 1$$

Substitute $3x - 1$ for $y$ in the second equation.

$$2x - 3(3x - 1) = 10$$

$$2x - 9x + 3 = 10$$

$$-7x = 7$$

$$x = -1$$

Let $x = -1$ in $y = 3x - 1$.

$$y = 3(-1) - 1$$

$$y = -4$$

The solution is $(-1, -4)$.

**27.** $\begin{cases} -x + 2y = 10 \\ -2x + 3y = 18 \end{cases}$

Solve the first equation for $x$.

$$x = 2y - 10$$

Substitute $2y - 10$ for $x$ in the second equation.

$$-2(2y - 10) + 3y = 18$$

$$-4y + 20 + 3y = 18$$

$$-y = -2$$

$$y = 2$$

Let $y = 2$ in $x = 2y - 10$.

$$x = 2(2) - 10$$

$$x = -6$$

The solution is $(-6, 2)$.

**29.** $\begin{cases} 5x + 10y = 20 \\ 2x + 6y = 10 \end{cases}$

Solve the first equation for $x$.

$$x + 2y = 4$$

$$x = 4 - 2y$$

Substitute $4 - 2y$ for $x$ in the second equation.

$$2(4 - 2y) + 6y = 10$$

$$8 - 4y + 6y = 10$$

$$2y = 2$$

$$y = 1$$

Let $y = 1$ in $x = 4 - 2y$.

$$x = 4 - 2(1)$$

$$x = 2$$

The solution is $(2, 1)$.

**31.** $\begin{cases} 3x + 6y = 9 \\ 4x + 8y = 16 \end{cases}$

Solve the first equation for $x$.

$$x + 2y = 3$$

$$x = 3 - 2y$$

Substitute $3 - 2y$ for $x$ in the second equation.

$$4(3 - 2y) + 8y = 16$$
$$12 - 8y + 8y = 16$$
$$12 = 16$$

The system has no solution.

**33.** $\begin{cases} y = 2x + 9 \\ y = 7x + 10 \end{cases}$

Substitute $2x + 9$ for $y$ in the second equation.

$$2x + 9 = 7x + 10$$
$$-5x = 1$$
$$x = -\frac{1}{5}$$

Let $x = -\frac{1}{5}$ in the first equation.

$$y = 2\left(-\frac{1}{5}\right) + 9$$
$$y = \frac{43}{5}$$

The solution is $\left(-\frac{1}{5}, \frac{43}{5}\right)$.

**35.** Answers may vary.

**37.** $\begin{cases} -5y + 6y = 3x + 2(x - 5) - 3x + 5 \\ \qquad\quad y = 3x + 2x - 10 - 3x + 5 \\ \qquad\quad y = 2x - 5 \\ \\ 4(x + y) - x + y = -12 \\ \quad 4x + 4y - x + y = -12 \\ \qquad\qquad 3x + 5y = -12 \end{cases}$

Substitute $2x - 5$ for $y$ in the second equation.

$$3x + 5(2x - 5) = -12$$
$$3x + 10x - 25 = -12$$
$$13x = 13$$
$$x = 1$$

Let $x = 1$ in $y = 2x - 5$.

$$y = 2(1) - 5$$
$$y = -3$$

The solution is $(1, -3)$.

**39.** $\qquad 3x + 2y = 6$
$$-2(3x + 2y) = -2(6)$$
$$-6x - 4y = -12$$

**41.** $\qquad -4x + y = 3$
$$3(-4x + y) = 3(3)$$
$$-12x + 3y = 9$$

**43.** $\quad 3n + 6m$
$$\underline{2n - 6m}$$
$$5n$$

**45.** $\quad -5a - 7b$
$$\underline{5a - 8b}$$
$$-15b$$

**47.a.** $\begin{cases} y = 2.5x + 450 \\ y = 11.85x + 337 \end{cases}$

Substitute $11.85x + 337$ for $y$ in the first equation.

$$11.85x + 337 = 2.5x + 450$$

$$9.35x = 113$$
$$x = 12.09$$

Let $x = 12.09$ in $y = 2.5x + 450$

$$y = 2.5(12.09) + 450$$
$$y = 480.225$$

The solution is $(12, 480)$.

**b.** In $1970 + 12 = 1982$, 480,000 men and 480,000 women received bachelor degrees.

**c.**

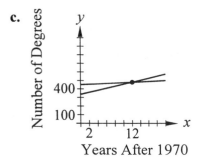

Years After 1970

**49.** $\begin{cases} y = 5.1x + 14.56 \\ y = -2x - 3.9 \end{cases}$

Substitute $-2x - 3.9$ for $y$ in the first equation.

$$-2x - 3.9 = 5.1x + 14.56$$
$$-7.1x = 18.46$$
$$x = -2.6$$

Let $x = -2.6$ in $y = -2x - 3.9$

$$y = -2(-2.6) - 3.9$$
$$y = 1.3$$

The solution is $(-2.6, 1.3)$.

**51.** $\begin{cases} 3x + 2y = 14.05 \\ 5x + y = 18.5 \end{cases}$

Solve the first equation for $y = -5x + 18.5$

Substitute $-5x + 18.5$ for $y$ in the first equation.

$$3x + 2(-5x + 18.5) = 14.05$$
$$3x - 10x + 37 = 14.05$$
$$-7x = -22.95$$
$$x = 3.279$$

Let $x = 3.279$ in $y = -5x + 18.5$

$$y = -5(3.279) + 18.5$$
$$y = 2.105$$

The solution is $(3.28, 2.11)$.

**Exercise Set 4.3**

**1.** $\begin{cases} 3x + y = 5 \\ 6x - y = 4 \end{cases}$

$$\begin{array}{r} 3x + y = 5 \\ 6x - y = 4 \\ \hline 9x \quad\;\; = 9 \\ x \quad\; = 1 \end{array}$$

Let $x = 1$ in the first equation.

$$3(1) + y = 5$$
$$3 + y = 5$$
$$y = 2$$

The solution of the system is $(1, 2)$

**3.** $\begin{cases} x - 2y = 8 \\ -x + 5y = -17 \end{cases}$

$$\begin{array}{r} x - 2y = \quad 8 \\ -x + 5y = -17 \\ \hline 3y = -9 \\ y = -3 \end{array}$$

Let $y = -3$ in the first equation.

$x - 2(-3) = 8$

$x + 6 = 8$

$x = 2$

The solution of the system is $(2, -3)$.

**5.** $\begin{cases} x + y = 6 \\ x - y = 6 \end{cases}$

$x + y = 6$

$\underline{x - y = 6}$

$2x \quad = 12$

$x \quad = 6$

Let $x = 6$ in the first equation.

$6 + y = 6$

$y = 0$

The solution of the system is $(6, 0)$

**7.** $\begin{cases} 3x + y = 4 \\ 9x + 3y = 6 \end{cases}$

Multiply the first equation by $-3$.

$-9x - 3y = -12$

$\underline{9x + 3y = \quad 6}$

$0 = -6$

The system has no solution.

**9.** $\begin{cases} 3x - 2y = 7 \\ 5x + 4y = 8 \end{cases}$

Multiply the first equation by 2.

$6x - 4y = 14$

$\underline{5x + 4y = \ 8}$

$11x \quad = 22$

$x \quad = 2$

Let $x = 2$ in the first equation.

$3(2) - 2y = 7$

$6 - 2y = 7$

$-2y = 1$

$y = -\dfrac{1}{2}$

The solution of the system is $\left(2, -\dfrac{1}{2}\right)$.

**11.** $\begin{cases} \dfrac{2}{3}x + 4y = -4 \\ 5x + 6y = 18 \end{cases}$

Multiply the first equation by 3 and the second equation by $-2$.

$2x + 12y = -12$

$\underline{-10x - 12y = \ 36}$

$-8x \quad = -48$

$x \quad = 6$

Let $x = 6$ in the first equation.

$\dfrac{2}{3}(6) + 4y = -4$

$4 + 4y = -4$

$4y = -8$

$y = -2$

The solution of the system is $(6, -2)$.

**13.** $\begin{cases} 4x - 6y = 8 \\ 6x - 9y = 12 \end{cases}$

Multiply the first equation by 3 and the second equation by $-2$.

$12x - 18y = 24$

$\underline{-12x + 18y = \ 24}$

$0 = 0$

The equations in the original system are equivalent and there are an infinite number of solutions..

**15.** $\begin{cases} 3x + y = -11 \\ 6x - 2y = -2 \end{cases}$

Multiply the first equation by 2.

$$6x + 2y = -22$$
$$\underline{6x - 2y = -2}$$
$$12x \quad = -24$$
$$x \quad = -2$$

Let $x = -2$ in the first equation.

$$3(-2) + y = -11$$
$$-6 + y = -11$$
$$y = -5$$

The solution of the system is $(-2, -5)$.

**17.** $\begin{cases} 3x + 2y = 11 \\ 5x - 2y = 29 \end{cases}$

$$3x + \quad y = 11$$
$$\underline{5x - 2y = 29}$$
$$8x \quad = 40$$
$$x \quad = 5$$

Let $x = 5$ in the first equation.

$$3(5) + 2y = 11$$
$$15 + 2y = 11$$
$$2y = -4$$
$$y = -2$$

The solution of the system is $(5, -2)$

**19.** $\begin{cases} x + 5y = 18 \\ 3x + 2y = -11 \end{cases}$

Multiply the first equation by $-3$.

$$-3x - 15y = -54$$
$$\underline{3x + 2y = -11}$$
$$-13y = -65$$
$$y = 5$$

Let $y = 5$ in the first equation.

$$x + 5(5) = 18$$
$$x + 25 = 18$$
$$x = -7$$

The solution of the system is $(-7, 5)$.

**21.** $\begin{cases} 2x - 5y = 4 \\ 3x - 2y = 4 \end{cases}$

Multiply the first equation by $-3$ and the second equation by 2.

$$-6x + 15y = -12$$
$$\underline{6x - 4y = \quad 8}$$
$$11y = -4$$
$$y = -\frac{4}{11}$$

Let $y = -\frac{4}{11}$ in the first equation.

$$2x - 5\left(-\frac{4}{11}\right) = 4$$
$$2x + \frac{20}{11} = 4$$
$$11(2x) + 11\left(\frac{20}{11}\right) = 11(4)$$
$$22x + 20 = 44$$

$$22x = 24$$

$$x = \frac{12}{11}$$

The solution of the system is $\left(\frac{12}{11}, -\frac{4}{11}\right)$.

**23.** $\begin{cases} 2x + 3y = 0 \\ 4x + 6y = 3 \end{cases}$

Multiply the first equation by $-2$.

$$-4x - 6y = 0$$
$$\underline{4x + 6y = 3}$$
$$0 = 3$$

The system has no solution.

**25.** $\begin{cases} \dfrac{x}{3} + \dfrac{y}{6} = 1 \\ \dfrac{x}{2} - \dfrac{y}{4} = 0 \end{cases}$

Multiply the first equation by 6 and the second equation by 4.

$\begin{cases} 2x + y = 6 \\ \underline{2x - y = 0} \end{cases}$  Simplified system

$$4x = 6$$
$$x = \frac{3}{2}$$

Multiply the second equation of the simplified system by $-1$.

$\begin{cases} 2x + y = 6 \\ \underline{-2x + y = 0} \end{cases}$

$$2y = 6$$
$$y = 3$$

The solution of the system is $\left(\frac{3}{2}, 3\right)$.

**27.** $\begin{cases} x - \dfrac{y}{3} = -1 \\ -\dfrac{x}{2} + \dfrac{y}{8} = \dfrac{1}{4} \end{cases}$

Multiply the first equation by 3 and the second equation by 8.

$\begin{cases} 3x - y = -3 \\ \underline{-4x + y = 2} \end{cases}$  Simplified system

$$-x = -1$$
$$x = 1$$

Multiply the first equation of the simplified system by 4 and the second equation by 3.

$\begin{cases} 12x - 4y = -12 \\ \underline{-12x + 3y = 6} \end{cases}$

$$-y = -6$$
$$y = 6$$

The solution of the system is $(1, 6)$.

**29.** $\begin{cases} \dfrac{x}{3} - y = 2 \\ -\dfrac{x}{2} + \dfrac{3y}{2} = -3 \end{cases}$

Multiply the first equation by 3 and the second equation by 2.

$\begin{cases} x - 3y = 6 \\ \underline{-x + 3y = -6} \end{cases}$  Simplified system

$$0 = 0$$

The equations in the original system are equivalent and there are an infinite number of solutions..

**31.** $\begin{cases} 8x + 11y = -16 \\ 2x + 3y = -4 \end{cases}$

Multiply the second equation by $-4$.

$\phantom{-}8x + 11y = -16$
$\underline{-8x - 12y = \phantom{-}16}$
$\phantom{-8x-12}-y = 0$
$\phantom{-8x-12-}y = 0$

Let $y = 0$ in the first equation.

$8x + 11(0) = -16$
$\phantom{8x+11}8x = -16$
$\phantom{8x+11}x = -2$

The solution of the system is $(-2, 0)$.

**33.** Answers may vary.

**35.** $\begin{cases} 2x - 3y = -11 \\ y = 4x - 3 \end{cases}$

Substitute $4x - 3$ for $y$ in the first equation.

$2x - 3(4x - 3) = -11$
$2x - 12x + 9 = -11$
$-10x + 9 = -11$
$-10x = -20$
$x = 2$

Let $x = 2$ in the second equation.

$y = 4(2) - 3$
$y = 5$

The solution of the system is $(2, 5)$.

**37.** $\begin{cases} x + 2y = 1 \\ 3x + 4y = -1 \end{cases}$

Multiply the first equation by $-2$.

$-2x - 4y = -2$
$\underline{\phantom{-}3x + 4y = -1}$
$\phantom{-2x-4y-}x = -3$

Let $x = -3$ in the first equation.

$-3 + 2y = 1$
$\phantom{-3+}2y = 4$
$\phantom{-3+}y = 2$

The solution is $(-3, 2)$

**39.** $\begin{cases} 2y = x + 6 \\ 3x - 2y = -6 \end{cases}$

Subtract $x$ from both sides
of the first equation.

$\begin{cases} -x + 2y = 6 \\ \underline{3x - 2y = -6} \end{cases}$
$\phantom{-x+2}2x \phantom{-2y} = 0$
$\phantom{-x+2x}x = 0$

Let $x = 0$ in the first equation.

$2y = 0 + 6$
$\phantom{2}y = 3$

The solution of the system is $(0, 3)$.

**41.** $\begin{cases} y = 2x - 3 \\ y = 5x - 18 \end{cases}$

Substitute $5x - 18$ for $y$ in the first equation.

$5x - 18 = 2x - 3$
$3x - 18 = -3$

$$3x = 15$$
$$x = 5$$

Let $x = 5$ in the second equation.

$$y = 5(5) - 18$$
$$y = 7$$

The solution of the system is $(5, 7)$.

**43.** $\begin{cases} x + \dfrac{1}{6}y = \dfrac{1}{2} \\ 3x + 2y = 3 \end{cases}$

Multiply the first equation by $-12$.

$\begin{cases} -12x - 2y = -6 \\ \phantom{-1}3x + 2y = 3 \end{cases}$

$$-9x \phantom{+ 2y} = -3$$
$$x = \frac{1}{3}$$

Substitute $\dfrac{1}{3}$ for $x$

in the second equation.

$$3\left(\frac{1}{3}\right) + 2y = 3$$
$$1 + 2y = 3$$
$$2y = 2$$
$$y = 1$$

The solution of the system is $\left(\dfrac{1}{3}, 1\right)$.

**45.** $\begin{cases} \dfrac{x+2}{2} = \dfrac{y+11}{3} \\ \dfrac{x}{2} = \dfrac{2y+16}{6} \end{cases}$

Multiply the first equation by 6
and the second equation by $-6$.

$\begin{cases} 3(x+2) = 2(y+11) \\ 3x + 6 = 2y + 22 \\ \\ -3x = -2y - 16 \end{cases}$

Add the two equations.

$$6 = 6$$

There are an infinite number of solutions.

**47.** $\begin{cases} 2x + 3y = 14 \\ 3x - 4y = -69.1 \end{cases}$

Multiply the first equation by 3 and
the second equation by $-2$.

$$6x + 9y = \phantom{0}42$$
$$-6x + 8y = 138.2$$
$$17y = 180.2$$
$$y = 10.6$$

Let $y = 10.6$ in the first equation.

$$2x + 3(10.6) = 14$$
$$2x + 31.8 = 14$$
$$2x = -17.8$$
$$x = -8.9$$

The solution of the system is $(-8.9, 10.6)$.

**49.** Let $x =$ a number.

$$2x + 6 = x - 3$$

**51.** Let $x =$ a number.

$$20 - 3x = 2$$

**53.** Let $x =$ a number.

$$4(x + 6) = 2x$$

**55.a.**
$$\begin{cases} 0.35x + y = 9.3 \\ \underline{0.56x - y = -2.5} \end{cases}$$
$$0.91x \qquad = 6.8$$
$$x = 7.47$$

The number of skiers will equal the number of snowboarders in
$1996 + 8 = 2004$.

**b.** Let $x = 8$ in the first equation.
$$0.35(8) + y = 9.3$$
$$2.8 + y = 9.3$$
$$y = 6.5$$

There will be 6.5 million skiers.
Let $x = 8$ in the second equation.
$$0.56(8) - y = -2.5$$
$$4.48 - y = -2.5$$
$$y = 6.98$$

There will be 6.98 million snowboarders. The numbers differ because of rounding in part a.

**57.**
$$\begin{cases} x + y = 5 \\ 3x + 3y = b \end{cases}$$
Multiply the first equation by $-3$.
$$-3x - 3y = -15$$
$$\underline{3x + 3y = \quad b}$$
$$0 = b - 15$$

**a.** The system has an infinite number of solutions if this statement is true.
$$b = 15$$

**b.** The system has no solution if this statement is false. $b = $ any real number except 15.

**59. b.** Answers may vary.

**c.** Answers may vary

**Integrated Review-Solving Systems of Equations**

**1.**
$$\begin{cases} 2x - 3y = -11 \\ y = 4x - 3 \end{cases}$$
Substitute $4x - 3$ for $y$ in the first equation.
$$2x - 3(4x - 3) = -11$$
$$2x - 12x + 9 = -11$$
$$-10x = -20$$
$$x = 2$$
Let $x = 2$ in the second equation.
$$y = 4(2) - 3$$
$$y = 5$$
The solution is $(2, 5)$.

**2.**
$$\begin{cases} 4x - 5y = 6 \\ y = 3x - 10 \end{cases}$$
Substitute $3x - 10$ for $y$ in the first equation.
$$4x - 5(3x - 10) = 6$$
$$4x - 15x + 50 = 6$$
$$-11x = -44$$
$$x = 4$$
Let $x = 4$ in the second equation.
$$y = 3(4) - 10$$
$$y = 2$$
The solution is $(4, 2)$.

**3.** $\begin{cases} x + y = 3 \\ x - y = 7 \end{cases}$

$\phantom{xx}2x \phantom{xxx}= 10$

$\phantom{xx}x \phantom{xxxx}= 5$

Let $x = 5$ in the first equation.

$5 + y = 3$

$\phantom{xx}y = -2$

The solution of the system is $(5, -2)$

**4.** $\begin{cases} x - y = 20 \\ x + y = -8 \end{cases}$

$\phantom{xx}2x \phantom{xxx}= 12$

$\phantom{xx}x \phantom{xxxx}= 6$

Let $x = 6$ in the second equation.

$6 + y = -8$

$\phantom{xx}y = -14$

The solution of the system is $(6, -14)$

**5.** $\begin{cases} x + 2y = 1 \\ 3x + 4y = -1 \end{cases}$

Solve the first equation for $x$.

$x = 1 - 2y$

Substitute $1 - 2y$ for $x$ in the

second equation.

$3(1 - 2y) + 4y = -1$

$\phantom{xx}3 - 6y + 4y = -1$

$\phantom{xxxxx}-2y = -4$

$\phantom{xxxxxxx}y = 2$

Let $y = 2$ in $x = 1 - 2y$.

$x = 1 - 2(2)$

$x = -3$

The solution is $(-3, 2)$.

**6.** $\begin{cases} x + 3y = 5 \\ 5x + 6y = -2 \end{cases}$

Solve the first equation for $x$.

$x = 5 - 3y$

Substitute $5 - 3y$ for $x$ in the

second equation.

$5(5 - 3y) + 6y = -2$

$25 - 15y + 6y = -2$

$\phantom{xxxxx}-9y = -27$

$\phantom{xxxxxxx}y = 3$

Let $y = 3$ in $x = 5 - 3y$.

$x = 5 - 3(3)$

$x = -4$

The solution is $(-4, 3)$.

**7.** $\begin{cases} y = x + 3 \\ 3x - 2y = -6 \end{cases}$

Substitute $x + 3$ for $y$ in the

second equation.

$3x - 2(x + 3) = -6$

$\phantom{xx}3x - 2x - 6 = -6$

$\phantom{xxxxxxx}x = 0$

Let $x = 0$ in the first equation.

$y = 0 + 3$

$y = 3$

The solution is $(0, 3)$.

**8.** $\begin{cases} y = -2x \\ 2x - 3y = -16 \end{cases}$

Substitute $-2x$ for $y$ in the

second equation.

$2x - 3(-2x) = -16.$

$$2x + 6x = -16$$
$$8x = -16$$
$$x = -2$$

Let $x = -2$ in the first equation.

$$y = -2(-2)$$
$$y = 4$$

The solution is $(-2, 4)$.

9. $\begin{cases} y = 2x - 3 \\ y = 5x - 18 \end{cases}$

Substitute $5x - 18$ for $y$ in the first equation.

$$5x - 18 = 2x - 3$$
$$3x = 15$$
$$x = 5$$

Let $x = 5$ in the second equation.

$$y = 5(5) - 18$$
$$y = 7$$

The solution is $(5, 7)$.

10. $\begin{cases} y = 6x - 5 \\ y = 4x - 11 \end{cases}$

Substitute $6x - 5$ for $y$ in the second equation.

$$6x - 5 = 4x - 11$$
$$2x = -6$$
$$x = -3$$

Let $x = -3$ in the first equation.

$$y = 6(-3) - 5$$
$$y = -23$$

The solution is $(-3, -23)$.

11. $\begin{cases} x + \dfrac{1}{6}y = \dfrac{1}{2} \\ 3x + 2y = 3 \end{cases}$

Multiply the first equation by 6.

$\begin{cases} 6x + y = 3 \\ 3x + 2y = 3 \end{cases}$   Simplified system

Multiply the first equation of the simplified system by $-2$.

$\begin{cases} -12x - 2y = -6 \\ \underline{\phantom{-1}3x + 2y = \phantom{-}3} \end{cases}$

$$-9x \phantom{+2y} = -3$$
$$x \phantom{+2y} = \dfrac{1}{3}$$

Multiply the second equation of the simplified system by $-2$.

$\begin{cases} 6x + \phantom{4}y = \phantom{-}3 \\ \underline{-6x - 4y = -6} \end{cases}$

$$-3y = -3$$
$$y = 1$$

The solution of the system is $\left(\dfrac{1}{3}, 1\right)$.

12. $\begin{cases} x + \dfrac{1}{3}y = \dfrac{5}{12} \\ 8x + 3y = 4 \end{cases}$

Multiply the first equation by 12.

$\begin{cases} 12x + 4y = 5 \\ 8x + 3y = 4 \end{cases}$   Simplified system

Multiply the first equation of the simplified system by 2 and the second equation by $-3$.

$$\begin{cases} 24x + 8y = 10 \\ -24x - 9y = -12 \end{cases}$$

$$-y = -2$$
$$y = 2$$

Multiply the first equation of the simplified system by 3 and the second equation by 4.

$$\begin{cases} 36x + 12y = 15 \\ -32x - 12y = -16 \end{cases}$$

$$4x = -1$$
$$x = -\frac{1}{4}$$

The solution of the system is $\left(-\frac{1}{4}, 2\right)$.

**13.** $\begin{cases} x - 5y = 1 \\ -2x + 10y = 3 \end{cases}$

Multiply the first equation by 2.

$$2x - 10y = 2$$
$$-2x + 10y = 3$$
$$\overline{\qquad\qquad}$$
$$0 = 5$$

The system has no solution.

**14.** $\begin{cases} -x + 2y = 3 \\ 3x - 6y = -9 \end{cases}$

Multiply the first equation by 3.

$$-3x + 6y = 9$$
$$3x - 6y = -9$$
$$\overline{\qquad\qquad}$$
$$0 = 0$$

The equations in the original system are equivalent and there are an infinite number of solutions..

**15.** $\begin{cases} 0.2x - 0.3y = -0.95 \\ 0.4x + 0.1y = 0.55 \end{cases}$

Multiply both equations by 10.

$\begin{cases} 2x - 3y = -9.5 \\ 4x + y = 5.5 \end{cases}$   Simplified system

Multiply the first equation of the simplified system by $-2$.

$$\begin{cases} -4x + 6y = 19 \\ 4x + y = 5.5 \end{cases}$$

$$7y = 24.5$$
$$y = 3.5$$

Multiply the second equation of the simplified system by 3.

$$\begin{cases} 2x - 3y = -9.5 \\ 12x + 3y = 16.5 \end{cases}$$

$$14x = 7$$
$$x = 0.5$$

The solution of the system is $(0.5, 3.5)$.

**16.** $\begin{cases} 0.08x - 0.04y = -0.11 \\ 0.02x - 0.06y = -0.09 \end{cases}$

Multiply both equations by 100.

$\begin{cases} 8x - 4y = -11 \\ 2x - 6y = -9 \end{cases}$   Simplified system

Multiply the second equation of the simplified system by $-4$.

$$\begin{cases} 8x - 4y = -11 \\ -8x + 24y = 36 \end{cases}$$

$$20y = 25$$
$$y = 1.25$$

Multiply the first equation of the simplified system by $-3$ and the second equation by 2.

$$\begin{cases} -24x + 12y = 33 \\ 4x - 12y = -18 \end{cases}$$
$$\begin{aligned} -20x \quad\quad &= 15 \\ x \quad\quad &= -0.75 \end{aligned}$$

The solution of the system is $(-0.75, 1.25)$.

**17.** $\begin{cases} x = 3y - 7 \\ 2x - 6y = -14 \end{cases}$

Substitute $3y - 7$ for $x$ in the second equation.

$$2(3y - 7) - 6y = -14$$
$$6y - 14 - 6y = -14$$
$$0 = 0$$

The equations in the original system are equivalent and there are an infinite number of solutions..

**18.** $\begin{cases} y = \dfrac{x}{2} - 3 \\ 2x - 4y = 0 \end{cases}$

Substitute $\dfrac{x}{2} - 3$ for $y$ in the second equation.

$$2x - 4\left(\frac{x}{2} - 3\right) = 0$$
$$2x - 2x + 12 = 0$$
$$12 = 0$$

There is no solution.

**19.** Answers may vary.

**20.** Answers may vary.

**Exercise Set 4.4**

**1. c**

**3. b**

**5. a**

**7.** Let $x$ = the first number and $y$ = the second number.
$$\begin{cases} x + y = 15 \\ x - y = 7 \end{cases}$$

**9.** Let $x$ = the amount invested in the larger account and $y$ = the amount invested in the smaller account.
$$\begin{cases} x + y = 6500 \\ x = y + 800 \end{cases}$$

**11.** Let $x$ = the first number and $y$ = the second number.
$$\begin{cases} x + y = 83 \\ x - y = 17 \end{cases}$$
$$\begin{aligned} 2x \quad\quad &= 100 \\ x \quad\quad &= 50 \end{aligned}$$
Let $x = 50$ in the first equation.
$$50 + y = 83$$
$$y = 33$$
The numbers are 50 and 33.

**13.** Let $x$ = the first number and
$y$ = the second number.

$$\begin{cases} x + 2y = 8 \\ 2x + y = 25 \end{cases}$$

Multiply the first equation by $-2$.

$$-2x - 4y = -16$$
$$\underline{2x + \ y = \ \ 25}$$
$$-3y = 9$$
$$y = -3$$

Let $y = -3$ in the first equation.

$$x + 2(-3) = 8$$
$$x - 6 = 8$$
$$x = 14$$

The numbers are 14 and $-3$.

**15.** Let $x$ = points scored by Smith and
$y$ = points scored by Jackson.

$$\begin{cases} x + y = 990 \\ y = x + 12 \end{cases}$$

Substitute $x + 12$ for $y$ in the
first equation.

$$x + (x + 12) = 990$$
$$x + x + 12 = 990$$
$$2x = 978$$
$$x = 489$$

Let $x = 489$ in the second equation.

$$y = 489 + 12$$
$$y = 501$$

Smith scored 489 points and
Jackson scored 501 points.

**17.** Let $x$ = the price of an adult's ticket and
$y$ = the price of a child's ticket.

$$\begin{cases} 3x + 4y = 159 \\ 2x + 3y = 112 \end{cases}$$

Multiply the first equation by $-2$ and
the second equation by 3.

$$-6x - 8y = -318$$
$$\underline{6x + 9y = \ \ 336}$$
$$y = \ \ 18$$

Let $y = 18$ in the first equation.

$$3x + 4(18) = 159$$
$$3x + 72 = 159$$
$$3x = 87$$
$$x = 29$$

An adult's ticket is $29 and
a child's ticket is $18.

**19.** Let $x$ = the number of quarters and
$y$ = the number of nickels.

$$\begin{cases} x + y = 80 \\ 0.25x + 0.05y = 14.6 \end{cases}$$

Solve the first equation for $y$.

$$y = 80 - x$$

Substitute $80 - x$ for $y$ in the
second equation.

$$0.25x + 0.05(80 - x) = 14.6$$
$$0.25x + 4 - 0.05x = 14.6$$
$$0.20x = 10.6$$
$$x = 53$$

Let $x = 53$ in $y = 80 - x$.

$$y = 80 - 53$$
$$y = 27$$

There are 53 quarters and 27 nickels.

**21.** Let $x$ = price of IBM stock and

$y$ = price of GA Financial stock.

$$\begin{cases} 50x + 40y = 5278 \\ y = x - 59.12 \end{cases}$$

Substitute $x - 59.12$ for $y$ in the

first equation.

$$50x + 40(x - 59.12) = 5278$$

$$50x + 40x - 2364.8 = 5278$$

$$90x = 7642.8$$

$$x = 84.92$$

Let $x = 84.92$ in the second equation.

$$y = 84.92 - 59.12$$

$$y = 25.8$$

IBM was $84.92 and

GA Financial was $25.80

**23.**

| $d$ | = | $r$ | $\cdot$ | $t$ |
|---|---|---|---|---|
| Downstream | 18 | | $x + y$ | 2 |
| Upstream | 18 | | $x - y$ | $4\frac{1}{2}$ |

$$\begin{cases} 2(x + y) = 18 \\ \dfrac{9}{2}(x - y) = 18 \end{cases}$$

Multiply the first equation by $\dfrac{1}{2}$ and

the second equation by $\dfrac{2}{9}$.

$$\begin{cases} x + y = 9 \\ x - y = 4 \end{cases}$$

$$\underline{\phantom{xxxxxxx}}$$

$$2x \quad = 13$$

$$x \quad = 6.5$$

Multiply the second equation of the

simplified system by $-1$.

$$\begin{cases} x + y = 9 \\ -x + y = -4 \end{cases}$$

$$\underline{\phantom{xxxxxxx}}$$

$$2y = 5$$

$$y = 2.5$$

Pratap can row 6.5 mph in still water.

The rate of the current is 2.5 mph.

**25.**

| $d$ | = | $r$ | $\cdot$ | $t$ |
|---|---|---|---|---|
| With the wind | 780 | | $x + y$ | $1\frac{1}{2}$ |
| Into the wind | 780 | | $x - y$ | 2 |

$$\begin{cases} \dfrac{3}{2}(x + y) = 780 \\ 2(x - y) = 780 \end{cases}$$

Multiply the first equation by $\dfrac{2}{3}$ and

the second equation by $\dfrac{1}{2}$.

$$\begin{cases} x + y = 520 \\ x - y = 390 \end{cases} \quad \text{Simplified system}$$

$$2x \quad = 910$$

$$x \quad = 455$$

Multiply the second equation of the

simplified system by $-1$.

$$\begin{cases} x + y = 520 \\ -x + y = -390 \end{cases}$$

$$\underline{\phantom{xxxxxxx}}$$

$$2y = 130$$

$$y = 65$$

The plane can fly 455 mph in still air.

The speed of the wind is 65 mph.

**27.** Let $x$ = ounces of 4% solution and $y$ = ounces of 12% solution.

| | Concentration Rate | Ounces of Solution | Ounces of Pure Acid |
|---|---|---|---|
| First solution | 0.04 | $x$ | $0.04x$ |
| Second solution | 0.12 | $y$ | $0.12y$ |
| Mixture | 0.09 | 12 | $0.09(12)$ |

$$\begin{cases} x + y = 12 \\ 0.04x + 0.12y = 0.09(12) \end{cases}$$

Multiply the first equation by $-4$ and the second equation by 100.

$$-4x - 4y = -48$$
$$\underline{4x + 12y = 108}$$
$$8y = 60$$
$$y = 7.5$$

Let $y = 7.5$ in the first equation.

$$x + 7.5 = 12$$
$$x = 4.5$$

$4\dfrac{1}{2}$ ounces of 4% solution and

$7\dfrac{1}{2}$ ounces of 12% solution.

**29.** Let $x$ = pounds of $4.95 per pound beans and $y$ = pounds of $2.65 per pound beans.

| | Cost Rate | Pounds of Beans | Dollars Cost |
|---|---|---|---|
| High Quality | 4.95 | $x$ | $4.95x$ |
| Low Quality | 2.65 | $y$ | $2.65y$ |
| Mixture | 3.95 | 200 | $3.95(200)$ |

$$\begin{cases} x + y = 200 \\ 4.95x + 2.65y = 3.95(200) \end{cases}$$

Solve the first equation for $y$.

$$y = 200 - x$$

Substitute $200 - x$ for $y$ in the second equation.

$$4.95x + 2.65(200 - x) = 3.95(200)$$
$$4.95x + 530 - 2.65x = 790$$
$$2.30x = 260$$
$$x = 113.04$$

Let $x = 113.04$ in the first equation.

$$113.04 + y = 200$$
$$y = 86.96$$

He needs 113 pounds of $4.95 per pound beans and 87 pounds of $2.65 per pound beans.

**31.** Let $x$ = the first angle and $y$ = the second angle.

$$\begin{cases} x + y = 90 \\ x = 2y \end{cases}$$

Substitute $2y$ for $x$ in the first equation.

$$2y + y = 90$$
$$3y = 90$$
$$y = 30$$

Let $y = 30$ in the second equation.

$$x = 2(30)$$
$$x = 60$$

The angles are 60° and 30°.

**33.** Let $x$ = the first angle and
$y$ = the second angle.

$$\begin{cases} x + y = 90 \\ x = 3y + 10 \end{cases}$$

Substitute $3y + 10$ for $x$ in the
first equation.

$$3y + 10 + y = 90$$
$$4y = 80$$
$$y = 20$$

Let $y = 20$ in the second equation.

$$x = 3(20) + 10$$
$$x = 70$$

The angles are $70°$ and $20°$.

**35.** Let $x$ = liters of 20% solution and
$y$ = liters of 70% solution.

|  | Concentration Rate | Liters of Solution | Liters of Alcohol |
|---|---|---|---|
| First solution | 0.20 | $x$ | $0.20x$ |
| Second solution | 0.70 | $y$ | $0.70y$ |
| Mixture | 0.60 | 50 | $0.60(50)$ |

$$\begin{cases} x + y = 50 \\ 0.20x + 0.70y = 0.60(50) \end{cases}$$

Multiply the first equation by $-20$
and the second equation by 100.

$$-20x - 20y = -1000$$
$$\underline{20x + 70y = 3000}$$
$$50y = 2000$$
$$y = 40$$

Let $y = 40$ in the first equation.

$$x + 40 = 50$$
$$x = 10$$

10 liters of 20% solution and
40 liters of 70% solution.

**37.** Let $x$ = the number sold at \$9.50 and
$y$ = the number sold at \$7.50.

$$\begin{cases} x + y = 90 \\ 9.5x + 7.5y = 721 \end{cases}$$

Solve the first equation for $y$.

$$y = 90 - x$$

Substitute $90 - x$ for $y$ in the
second equation.

$$9.5x + 7.5(90 - x) = 721$$
$$9.5x + 675 - 7.5x = 721$$
$$2x = 46$$
$$x = 23$$

Let $x = 23$ in $y = 90 - x$.

$$y = 90 - 23$$
$$y = 67$$

They sold 23 at \$9.50 and 67 at \$7.50.

**39.** Let $x$ = the width and $y$ = the length.

$$\begin{cases} 2x + 2y = 144 \\ y = x + 12 \end{cases}$$

Substitute $x + 12$ for $y$ in the
first equation.

$$2x + 2(x + 12) = 144$$
$$2x + 2x + 24 = 144$$
$$4x = 120$$
$$x = 30$$

Let $x = 30$ in the second equation.

$$y = 30 + 12$$
$$y = 42$$

Width = 30 inches, length = 42 inches.

**41.** Let $x$ = the time spent walking and $y$ = the time spent on the bicycle.

| | $r$ | $\cdot$ | $t$ | $=$ | $d$ |
|---|---|---|---|---|---|
| Walking | 4 | | $x$ | | $4x$ |
| Biking | 40 | | $y$ | | $40y$ |

$$\begin{cases} x + y = 6 \\ 4x + 40y = 186 \end{cases}$$

Multiply the first equation by $-4$.

$$-4x - 4y = -24$$
$$\underline{4x + 40y = 186}$$
$$36y = 162$$
$$y = 4.5$$

He spent $4\dfrac{1}{2}$ hours on the bicycle.

**43.** Let $x$ = the speed of the eastbound train and $y$ = the speed of the westbound.

| | $r$ | $\cdot$ | $t$ | $=$ | $d$ |
|---|---|---|---|---|---|
| eastbound | $x$ | | 1.25 | | $1.25x$ |
| westbound | $y$ | | 1.25 | | $1.25y$ |

$$\begin{cases} 1.25x + 1.25y = 150 \\ y = 2x \end{cases}$$

Multiply the first equation by 4.

$$\begin{cases} 5x + 5y = 600 \\ y = 2x \end{cases} \quad \text{Simplified system}$$

Substitute $2x$ for $y$ in the first equation.

$$5x + 5(2x) = 600$$
$$15x = 600$$
$$x = 40$$

Let $x = 40$ in the second equation.

$$y = 2(40) = 80$$

Eastbound: 40 mph. Westbound: 80mph.

**45.** Let $x$ = the rate of the faster group and $y$ = the rate of the slower group.

| | $r$ | $\cdot$ | $t$ | $=$ | $d$ |
|---|---|---|---|---|---|
| Slower group | $x$ | | 240 | | $240x$ |
| Faster group | $y$ | | 240 | | $240y$ |

$$\begin{cases} x = y - \dfrac{1}{2} \\ 240x + 240y = 1200 \end{cases}$$

Substitute $y - \dfrac{1}{2}$ for $x$ in the second equation.

$$240\left(y - \dfrac{1}{2}\right) + 240y = 1200$$
$$240y - 120 + 240y = 1200$$
$$480y = 1320$$
$$y = \dfrac{1320}{480} = 2\dfrac{3}{4}$$

Let $y = 2\dfrac{3}{4}$ in the first equation.

$$x = 2\dfrac{3}{4} - \dfrac{1}{2} = 2\dfrac{1}{4}$$

The rate of the faster group is $2\dfrac{3}{4}$ mph.

The rate of the slower group is $2\dfrac{1}{4}$ mph.

**47.** Let $x$ = gallons of 30% solution and $y$ = gallons of 60% solution.

| | Concentration Rate | Gallons of Solution | Gallons of Pure Fertilizer |
|---|---|---|---|
| First solution | 0.30 | $x$ | $0.30x$ |
| Second solution | 0.60 | $y$ | $0.60y$ |
| Mixture | 0.50 | 150 | $0.50(150)$ |

$$\begin{cases} x + y = 150 \\ 0.30x + 0.60y = 0.50(150) \end{cases}$$

Multiply the first equation by $-3$ and the second equation by 10.

$$\begin{array}{r} -3x - 3y = -450 \\ 3x + 6y = \phantom{0}750 \\ \hline 3y = 300 \\ y = 100 \end{array}$$

Let $y = 100$ in the first equation.

$$x + 100 = 150$$
$$x = 50$$

50 gallons of 30% solution and 100 gallons of 60% solution.

**49.** $-3x < -9$

$$\frac{-3x}{-3} > \frac{-9}{-3}$$

$$x > 3, \ (3, \infty)$$

**51.** $4(2x - 1) \geq 0$

$$8x - 4 \geq 0$$
$$8x \geq 4$$
$$x \geq \frac{1}{2}, \ \left[\frac{1}{2}, \infty\right)$$

**53.** Let $x = $ the width and $y = $ the length.

$$\begin{cases} 2x + y = 33 \\ y = 2x - 3 \end{cases}$$

Substitute $2x - 3$ for $y$ in the first equation.

$$2x + 2x - 3 = 33$$
$$4x = 36$$
$$x = 9$$

Let $x = 9$ in the second equation.

$$y = 2(9) - 3$$
$$y = 15$$

Width $= 9$ feet, length $= 15$ feet.

**55.a.** $\begin{cases} y = 3.1x + 21.8 \\ y = -2.1x + 49.3 \end{cases}$

Substitute $3.1x + 21.8$ for $y$ in the second equation.

$$3.1x + 21.8 = -2.1x + 49.3$$

$$5.2x = 27.5$$
$$x = 5.29$$

Let $x = 5.29$ in $y = 3.1x + 21.8$

$$y = 3.1(5.29) + 21.8$$
$$y = 38.20$$

The solution is $(5.3, 38.2)$.

**b.** In 5.3 years after 1995, cable viewers equaled network viewers which numbered 38.2 million.

**c.** Cable: increasing
Networks: decreasing.

**Mental Math 4.5**

**1.** Yes

**2.** No

**3.** Yes

**4.** No

**5.** $x + y > -5, \quad (0,0)$

$0 + 0 \overset{?}{>} -5$

$0 \overset{?}{>} -5$

Yes

**6.** $2x + 3y < 10, \quad (0,0)$

$2(0) + 3(0) \overset{?}{<} 10$

$0 \overset{?}{<} 10$

Yes

**7.** $x - y \le -1, \quad (0,0)$

$0 - 0 \overset{?}{\le} -1$

$0 \overset{?}{\le} -1$

No

**8.** $\dfrac{2}{3}x + \dfrac{5}{6}y > 4, \quad (0,0)$

$\dfrac{2}{3}(0) + \dfrac{5}{6}(0) \overset{?}{>} 4$

$0 \overset{?}{>} 4$

No

**Exercise Set 4.5**

**1.** $x - y > 3$

$(2,-1), \quad 2 - (-1) \overset{?}{>} 3$

$2 + 1 \overset{?}{>} 3$

$3 \overset{?}{>} 3, \text{ False}$

$(2,-1)$ is not a solution

$(5,1), \quad 5 - 1 \overset{?}{>} 3$

$4 \overset{?}{>} 3, \text{ True}$

$(5,1)$ is a solution

**3.** $3x - 5y \le -4$

$(-1,-1), \quad 3(-1) - 5(-1) \overset{?}{\le} -4$

$-3 + 5 \overset{?}{\le} -4$

$2 \overset{?}{\le} -4, \text{ False}$

$(-1,-1)$ is not a solution

$(4,0), \quad 3(4) - 5(0) \overset{?}{\le} -4$

$12 - 0 \overset{?}{\le} -4$

$12 \overset{?}{\le} -4, \text{ False}$

$(4,0)$ is not a solution

**5.** $x < -y$

$(0,2), \quad 0 \overset{?}{<} -2, \text{ False}$

$(0,2)$ is not a solution

$(-5,1), \quad -5 \overset{?}{<} -1, \text{ True}$

$(-5,1)$ is a solution

**7.** $x + y \le 1$

Test $(0,0)$

$0 + 0 \overset{?}{\le} 1, \text{ True}$

Shade below.

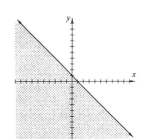

**9.** $2x + y > -4$

Test $(0,0)$

$2(0) + 0 \overset{?}{>} -4$

True

Shade above.

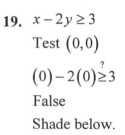

**11.** $x + 6y \le -6$

Test $(0,0)$

$0 + 6(0) \overset{?}{\le} -6$

False

Shade below.

**13.** $2x + 5y > -10$

Test $(0,0)$

$2(0) + 5(0) \overset{?}{>} -10$

True

Shade above.

**15.** $x + 2y \le 3$

Test $(0,0)$

$0 + 2(0) \overset{?}{\le} 3$

True

Shade below.

**17.** $2x + 7y > 5$

Test $(0,0)$

$2(0) + 7(0) \overset{?}{>} 5$

False

Shade above.

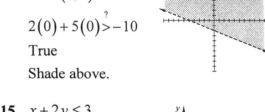

**19.** $x - 2y \ge 3$

Test $(0,0)$

$(0) - 2(0) \overset{?}{\ge} 3$

False

Shade below.

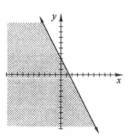

**21.** $5x + y < 3$

Test $(0,0)$

$5(0) + 0 \overset{?}{<} 3$

True

Shade below.

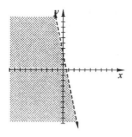

**23.** $4x + y < 8$

Test $(0,0)$

$4(0) + 0 \overset{?}{<} 8$

True

Shade below.

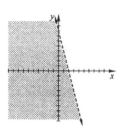

**25.** $y \ge 2x$

Test $(1,0)$

$0 \overset{?}{\ge} 2(1)$

False

Shade above.

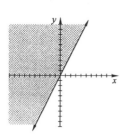

**27.** $x \ge 0$

Test $(1,0)$

$1 \overset{?}{\ge} 0$

True

Shade right.

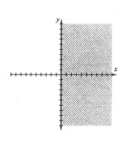

**29.** $y \leq -3$
Shade below.

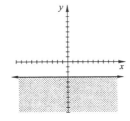

**31.** $2x - 7y > 0$
Test $(1, 0)$

$2(1) - 7(0) \overset{?}{>} 0$
True
Shade below.

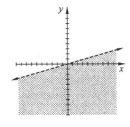

**33.** $3x - 7y \geq 0$
Test $(1, 0)$

$3(1) - 7(0) \overset{?}{\geq} 0$
True
Shade below.

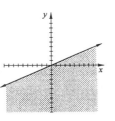

**35.** $x > y$
Test $(0, 1)$

$0 \overset{?}{>} 1$
False
Shade below.

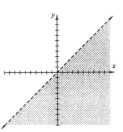

**37.** $x - y \leq 6$
Test $(0, 0)$

$0 - 0 \overset{?}{\leq} 6$
True
Shade above.

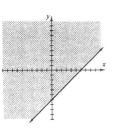

**39.** $-\dfrac{1}{4}y + \dfrac{1}{3}x > 1$
Test $(0, 0)$

$-\dfrac{1}{4}(0) + \dfrac{1}{3}(0) \overset{?}{>} 1$
False
Shade below.

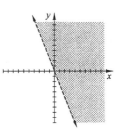

**41.** $-x < 0.4y$
Test $(1, 0)$

$-(1) \overset{?}{<} 0$
True
Shade above.

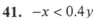

**43. e**

**45. c**

**47. f**

**49.** $2^3 = 2 \cdot 2 \cdot 2 = 8$

**51.** $(-2)^5 = (-2)(-2)(-2)(-2)(-2) = -32$

**53.** $3 \cdot 4^2 = 3 \cdot 4 \cdot 4 = 48$

**55.** Let $x = -5$.
$x^2 = (-5)(-5) = 25$

**57.** Let $x = -1$.
$2x^3 = 2(-1)(-1)(-1) = -2$

**59.** $x + y \geq 13$

Test $(0,0)$

$0 + 0 \overset{?}{\geq} 13$

False

Shade above.

**Exercise Set 4.6**

**1.** $y \geq x + 1$             $y \geq 3 - x$

Test $(0,0)$        Test $(0,0)$

$0 \overset{?}{\geq} 0 + 1$       $0 \overset{?}{\geq} 3 - 0$

False              False

Shade above.       Shade above.

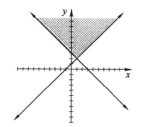

**61.** Answers may vary

**63.** $2.5x + 0.25y \leq 20$

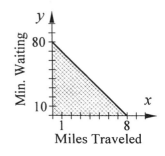

**3.** $y < 3x - 4$             $y \leq x + 2$

Test $(0,0)$        Test $(0,0)$

$0 \overset{?}{<} 3(0) - 4$      $0 \overset{?}{\leq} 0 + 2$

False              True

Shade below.       Shade below.

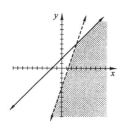

**65.** Answers may vary

**67.a.** $30x + 0.15y \leq 500$

**b.**

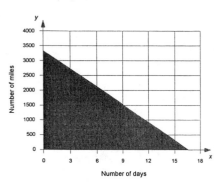

**c.** Answers may vary.

**5.** $y \leq -2x - 2$            $y \geq x + 4$

Test $(0,0)$        Test $(0,0)$

$0 \overset{?}{\leq} -2(0) - 2$    $0 \overset{?}{\geq} 0 + 4$

False              False

Shade below.       Shade above.

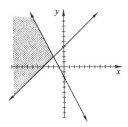

**7.** $y \geq -x + 2$      $y \leq 2x + 5$

Test $(0,0)$      Test $(0,0)$

$0 \overset{?}{\geq} -0 + 2$      $0 \overset{?}{\leq} 2(0) + 5$

False      True

Shade above.      Shade below.

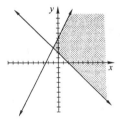

**9.** $x \geq 3y$      $x + 3y \leq 6$

Test $(1,0)$      Test $(0,0)$

$1 \overset{?}{\geq} 3(0)$      $0 + 3(0) \overset{?}{\leq} 6$

True      True

Shade below.      Shade below.

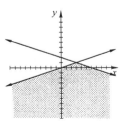

**11.** $y + 2x \geq 0$      $5x - 3y \leq 12$

Test $(1,0)$      Test $(0,0)$

$0 + 2(1) \overset{?}{\geq} 0$      $5(0) - 3(0) \overset{?}{\leq} 12$

True      True

Shade above.      Shade above.

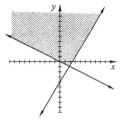

**13.** $3x - 4y \geq -6$      $2x + y \leq 7$

Test $(0,0)$      Test $(0,0)$

$3(0) - 4(0) \overset{?}{\geq} -6$      $2(0) + (0) \overset{?}{\leq} 7$

True      True

Shade below.      Shade below.

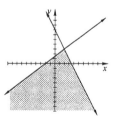

**15.** $x \leq 2$      $y \geq -3$

Shade left.      Shade above.

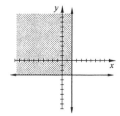

**17.** $y \geq 1$          $x < -3$

Shade above.          Shade left.

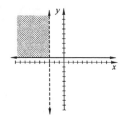

**19.** $2x + 3y \leq -8$          $x \geq -4$

Test $(0,0)$          Shade right.

$2(0) + 3(0) \overset{?}{\leq} -8$

False

Shade below.

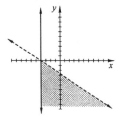

**21.** $2x - 5y \leq 9$          $y \leq -3$

Test $(0,0)$          Shade below.

$2(0) - 5(0) \overset{?}{\leq} 9$

True

Shade above.

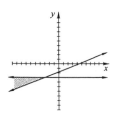

**23.** $y \geq \dfrac{1}{2}x + 2$          $y \leq \dfrac{1}{2}x - 3$

Test $(0,0)$          Test $(0,0)$

$0 \overset{?}{\geq} \dfrac{1}{2}(0) + 2$          $0 \overset{?}{\leq} \dfrac{1}{2}(0) - 3$

False          False

Shade above.          Shade below.

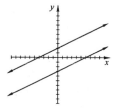

**25.** $(4)^2 = (4)(4) = 16$

**27.** $(6x)^2 = (6x)(6x) = 36x^2$

**29.** $(10y^3)^2 = (10y^3)(10y^3) = 100y^6$

**31.** C

**33.** D

**35.** Answers may vary.

**37.** $\begin{cases} 2x - y \leq 6 \\ x \geq 3 \\ y > 2 \end{cases}$

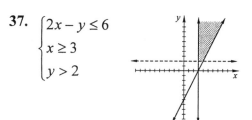

**Chapter 4 Review**

**1. a.** Let $x = 12$ and $y = 4$.

$$2x - 3y = 12 \qquad\qquad 3x + 4y = 1$$

$$2(12) - 3(4) \overset{?}{=} 12 \qquad 3(12) + 4(4) \overset{?}{=} 1$$

$$24 - 12 \overset{?}{=} 12 \qquad\qquad 36 + 16 \overset{?}{=} 1$$

$$12 = 12 \qquad\qquad\qquad 52 = 1$$

$$\text{True} \qquad\qquad\qquad \text{False}$$

$(12, 4)$ is not a solution of the system.

**b.** Let $x = 3$ and $y = -2$.

$$2x - 3y = 12 \qquad\qquad 3x + 4y = 1$$

$$2(3) - 3(-2) \overset{?}{=} 12 \qquad 3(3) + 4(-2) \overset{?}{=} 1$$

$$6 + 6 \overset{?}{=} 12 \qquad\qquad 9 - 8 \overset{?}{=} 1$$

$$2 = 12 \qquad\qquad\qquad 1 = 1$$

$$\text{True} \qquad\qquad\qquad \text{True}$$

$(3, -2)$ is a solution of the system

**c.** Let $x = -3$ and $y = 6$.

$$2x - 3y = 12 \qquad\qquad 3x + 4y = 1$$

$$2(-3) - 3(6) \overset{?}{=} 12 \qquad 3(-3) + 4(6) \overset{?}{=} 1$$

$$-6 - 18 \overset{?}{=} 12 \qquad\qquad -9 + 24 \overset{?}{=} 1$$

$$-24 = 12 \qquad\qquad\qquad 15 = 1$$

$$\text{False} \qquad\qquad\qquad \text{False}$$

$(-3, 6)$ is not a solution of the system

**2. a.** Let $x = \dfrac{3}{4}$ and $y = -3$.

$$4x + y = 0 \qquad\qquad -8x - 5y = 9$$

$$4\left(\dfrac{3}{4}\right) - 3 \overset{?}{=} 0 \qquad -8\left(\dfrac{3}{4}\right) - 5(-3) \overset{?}{=} 9$$

$$3 - 3 \overset{?}{=} 0 \qquad\qquad -6 + 15 \overset{?}{=} 9$$

$$0 = 0 \qquad\qquad\qquad 9 = 9$$

$$\text{True} \qquad\qquad\qquad \text{True}$$

$\left(\dfrac{3}{4}, -3\right)$ is a solution of the system.

**b.** Let $x = -2$ and $y = 8$.

$$4x + y = 0 \qquad\qquad -8x - 5y = 9$$

$$4(-2) + 8 \overset{?}{=} 0 \qquad -8(-2) - 5(8) \overset{?}{=} 9$$

$$-8 + 8 \overset{?}{=} 0 \qquad\qquad 16 - 40 \overset{?}{=} 9$$

$$0 = 0 \qquad\qquad\qquad -24 = 9$$

$$\text{True} \qquad\qquad\qquad \text{False}$$

$(-2, 8)$ is not a solution of the system

**c.** Let $x = \dfrac{1}{2}$ and $y = -2$.

$$4x + y = 0 \qquad\qquad -8x - 5y = 9$$

$$4\left(\dfrac{1}{2}\right) - 2 \overset{?}{=} 0 \qquad -8\left(\dfrac{1}{2}\right) - 5(-2) \overset{?}{=} 9$$

$$2 - 2 \overset{?}{=} 0 \qquad\qquad -4 + 10 \overset{?}{=} 9$$

$$0 = 0 \qquad\qquad\qquad 6 = 9$$

$$\text{True} \qquad\qquad\qquad \text{False}$$

$\left(\dfrac{1}{2}, -2\right)$ is not a solution of the system.

**3. a.** Let $x = -6$ and $y = -8$.

$$5x - 6y = 18 \qquad 2y - x = -4$$

$$5(-6) - 6(-8) \overset{?}{=} 18 \qquad 2(-8) - (-6) \overset{?}{=} -4$$

$$-30 + 48 \overset{?}{=} 18 \qquad -16 + 6 \overset{?}{=} -4$$

$$18 = 18 \qquad\qquad -10 = -4$$

True           False

$(-6, -8)$ is not a solution of the system.

**b.** Let $x = 3$ and $y = \dfrac{5}{2}$.

$$5x - 6y = 18 \qquad 2y - x = -4$$

$$5(3) - 6\left(\dfrac{5}{2}\right) \overset{?}{=} 18 \qquad 2\left(\dfrac{5}{2}\right) - 3 \overset{?}{=} -4$$

$$15 - 15 \overset{?}{=} 18 \qquad 5 - 3 \overset{?}{=} -4$$

$$0 = 18 \qquad\qquad 2 = -4$$

False          False

$\left(3, \dfrac{5}{2}\right)$ is not a solution of the system

**c.** Let $x = 3$ and $y = -\dfrac{1}{2}$.

$$5x - 6y = 18 \qquad 2y - x = -4$$

$$5(3) - 6\left(-\dfrac{1}{2}\right) \overset{?}{=} 18 \qquad 2\left(-\dfrac{1}{2}\right) - 3 \overset{?}{=} -4$$

$$15 + 3 \overset{?}{=} 18 \qquad -1 - 3 \overset{?}{=} -4$$

$$18 = 18 \qquad\qquad -4 = -4$$

True          True

$\left(3, -\dfrac{1}{2}\right)$ is a solution of the system

**4. a.** Let $x = 2$ and $y = 2$.

$$2x + 3y = 1 \qquad 3y - x = 4$$

$$2(2) + 3(2) \overset{?}{=} 1 \qquad 3(2) - (2) \overset{?}{=} 4$$

$$4 + 6 \overset{?}{=} 1 \qquad 6 - 2 \overset{?}{=} 4$$

$$10 = 1 \qquad\qquad 4 = 4$$

False          True

$(2, 2)$ is not a solution of the system.

**b.** Let $x = -1$ and $y = 1$.

$$2x + 3y = 1 \qquad 3y - x = 4$$

$$2(-1) + 3(1) \overset{?}{=} 1 \qquad 3(1) - (-1) \overset{?}{=} 4$$

$$-2 + 3 \overset{?}{=} 1 \qquad 3 + 1 \overset{?}{=} 4$$

$$1 = 1 \qquad\qquad 4 = 4$$

True          True

$(-1, 1)$ is a solution of the system

**c.** Let $x = 2$ and $y = -1$.

$$2x + 3y = 1 \qquad 3y - x = 4$$

$$2(2) + 3(-1) \overset{?}{=} 1 \qquad 3(2) - (-1) \overset{?}{=} 4$$

$$4 - 3 \overset{?}{=} 1 \qquad 6 + 1 \overset{?}{=} 4$$

$$1 = 1 \qquad\qquad 5 = 4$$

True          False

$(2, -1)$ is not a solution of the system.

**5.** $\begin{cases} 2x + y = 5 \\ 3y = -x \end{cases}$

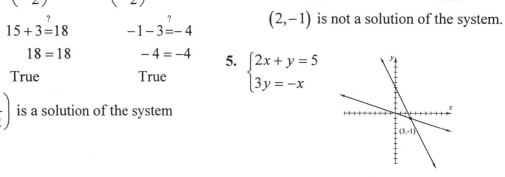

The solution of the system is $(3, -1)$.

**6.** $\begin{cases} 3x + y = -2 \\ 2x - y = -3 \end{cases}$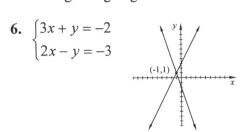

The solution of the system is $(-1,1)$.

**7.** $\begin{cases} y - 2x = 4 \\ x + y = -5 \end{cases}$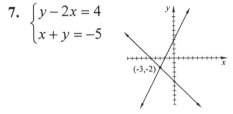

The solution of the system is $(-3, -2)$.

**8.** $\begin{cases} y - 3x = 0 \\ 2y - 3 = 6x \end{cases}$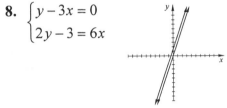

There is no solution of the system.

**9.** $\begin{cases} 3x + y = 2 \\ 3x - 6 = -9y \end{cases}$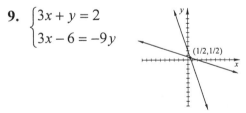

The solution of the system is $\left(\dfrac{1}{2}, \dfrac{1}{2}\right)$.

**10.** $\begin{cases} 2y + x = 2 \\ x - y = 5 \end{cases}$

The solution of the system is $(4, -1)$.

**11.** intersecting, one solution

**12.** parallel no solution

**13.** identical,

There is an infinite number of solutions.

**14.** intersecting, one solution

**15.** $\begin{cases} y = 2x + 6 \\ 3x - 2y = -11 \end{cases}$

Substitute $2x + 6$ for $y$ in the second equation.

$$3x - 2(2x + 6) = -11$$
$$3x - 4x - 12 = -11$$
$$-x = 1$$
$$x = -1$$

Let $x = -1$ in the first equation.

$$y = 2(-1) + 6$$
$$y = 4$$

The solution is $(-1, 4)$.

**16.** $\begin{cases} y = 3x - 7 \\ 2x - 3y = 7 \end{cases}$

Substitute $3x - 7$ for $y$ in the second equation.

$$2x - 3(3x - 7) = 7$$

$$2x - 9x + 21 = 7$$
$$-7x = -14$$
$$x = 2$$

Let $x = 2$ in the first equation.

$$y = 3(2) - 7$$
$$y = -1$$

The solution is $(2, -1)$.

**17.** $\begin{cases} x + 3y = -3 \\ 2x + y = 4 \end{cases}$

Solve the first equation for $x$.

$$x = -3y - 3$$

Substitute $-3y - 3$ for $x$ in the second equation.

$$2(-3y - 3) + y = 4$$
$$-6y - 6 + y = 4$$
$$-5y = 10$$
$$y = -2$$

Let $y = -2$ in $x = -3y - 3$.

$$x = -3(-2) - 3$$
$$x = 3$$

The solution is $(3, -2)$.

**18.** $\begin{cases} 3x + y = 11 \\ x + 2y = 12 \end{cases}$

Solve the first equation for $y$.

$$y = 11 - 3x$$

Substitute $11 - 3x$ for $y$ in the second equation.

$$x + 2(11 - 3x) = 12$$
$$x + 22 - 6x = 12$$

$$-5x = -10$$
$$x = 2$$

Let $x = 2$ in $y = 11 - 3x$.

$$y = 11 - 3(2)$$
$$y = 5$$

The solution is $(2, 5)$.

**19.** $\begin{cases} 4y = 2x - 3 \\ x - 2y = 4 \end{cases}$

Solve the second equation for $x$.

$$x = 4 + 2y$$

Substitute $4 + 2y$ for $x$ in the first equation.

$$4y = 2(4 + 2y) - 3$$
$$4y = 8 + 4y - 3$$
$$0 = 5$$

The system has no solution.

**20.** $\begin{cases} 2x = 3y - 18 \\ x + 4y = 2 \end{cases}$

Solve the second equation for $x$.

$$x = 2 - 4y$$

Substitute $2 - 4y$ for $x$ in the first equation.

$$2(2 - 4y) = 3y - 18$$
$$4 - 8y = 3y - 18$$
$$-11y = -22$$
$$y = 2$$

Let $y = 2$ in $x = 2 - 4y$.

$$x = 2 - 4(2)$$
$$x = -6$$

The solution is $(-6, 2)$.

**21.**
$$\begin{cases} 2(3x - y) = 7x - 5 \\ 6x - 2y = 7x - 5 \\ 5 - 2y = x \\ \\ 3(x - y) = 4x - 6 \\ 3x - 3y = 4x - 6 \\ 6 - 3y = x \end{cases}$$

Substitute $5 - 2y$ for $x$ in the second equation.

$6 - 3y = 5 - 2y$

$\quad\quad -y = -1$

$\quad\quad\quad y = 1$

Let $y = 1$ in $5 - 2y = x$.

$x = 5 - 2(1)$

$x = 3$

The solution is $(3,1)$.

**22.**
$$\begin{cases} 4(x - 3y) = 3x - 1 \\ 4x - 12y = 3x - 1 \\ x = 12y - 1 \\ \\ 3(4y - 3x) = 1 - 8x \\ 12y - 9x = 1 - 8x \\ 12y - 1 = x \end{cases}$$

Substitute $12y - 1$ for $x$ in the second equation.

$12y - 1 = 12y - 1$

$\quad\quad\quad 0 = 0$

The system is dependent.

There are an infinite number of solutions.

**23.**
$$\begin{cases} \dfrac{3}{4}x + \dfrac{2}{3}y = 2 \\ 3x + y = 18 \end{cases}$$

Multiply the first equation by 12.

$$\begin{cases} 9x + 8y = 24 \\ 3x + y = 18 \end{cases}$$

Solve the second equation of the simplified system for $y$

$y = 18 - 3x$

Substitute $18 - 3x$ for $y$ in the first equation..

$9x + 8(18 - 3x) = 24$

$9x + 144 - 24x = 24$

$\quad\quad\quad -15x = -120$

$\quad\quad\quad\quad\quad x = 8$

Substitute 8 for $x$ in $y = 18 - 3x$.

$y = 18 - 3(8) = -6$

The solution of the system is $(8, -6)$

**24.**
$$\begin{cases} \dfrac{2}{5}x + \dfrac{3}{4}y = 1 \\ x + 3y = -2 \end{cases}$$

Multiply the first equation by 20.

$$\begin{cases} 8x + 15y = 20 \\ x + 3y = -2 \end{cases}$$

Solve the second equation of the simplified system for $x$

$x = -3y - 2$

Substitute $-3y - 2$ for $x$ in the first equation..

$8(-3y - 2) + 15y = 20$

$-24y - 16 + 15y = 20$

$$-9y = 36$$
$$y = -4$$

Substitute $-4$ for $y$ in $x = -3y - 2$.

$$x = -3(-4) - 2 = 10$$

The solution of the system is $(10, -4)$

**25.** $\begin{cases} 2x + 3y = -6 \\ \phantom{2}x - 3y = -12 \end{cases}$

$$3x \phantom{+3y} = -18$$
$$x \phantom{+3y} = -6$$

Let $x = -6$ in the first equation.

$$2(-6) + 3y = -6$$
$$-12 + 3y = -6$$
$$3y = 6$$
$$y = 2$$

The solution of the system is $(-6, 2)$

**26.** $\begin{cases} \phantom{-}4x + \phantom{3}y = \phantom{-1}15 \\ -4x + 3y = -19 \end{cases}$

$$4y = -4$$
$$y = -1$$

Let $y = -1$ in the first equation.

$$4x + (-1) = 15$$
$$4x - 1 = 15$$
$$4x = 16$$
$$x = 4$$

The solution of the system is $(4, -1)$.

**27.** $\begin{cases} 2x - 3y = -15 \\ x + 4y = 31 \end{cases}$

Multiply the second equation by $-2$.

$$2x - 3y = -15$$
$$\underline{-2x - 8y = -62}$$
$$-11y = -77$$
$$y = 7$$

Let $y = 7$ in the second equation.

$$x + 4(7) = 31$$
$$x + 28 = 31$$
$$x = 3$$

The solution of the system is $(3, 7)$.

**28.** $\begin{cases} x - 5y = -22 \\ 4x + 3y = 4 \end{cases}$

Multiply the first equation by $-4$.

$$-4x + 20y = 88$$
$$\underline{4x + \phantom{2}3y = 4}$$
$$23y = 92$$
$$y = 4$$

Let $y = 4$ in the first equation.

$$x - 5(4) = -22$$
$$x - 20 = -22$$
$$x = -2$$

The solution of the system is $(-2, 4)$.

**29.** $\begin{cases} 2x = 6y - 1 \\ \dfrac{1}{3}x - y = \dfrac{-1}{6} \end{cases}$

Rearrange the first equation.

Multiply the second equation by $-6$..

$$2x - 6y = -1$$
$$\underline{-2x + 6y = \phantom{-}1}$$
$$0 = 0$$

The system is dependent.

There are an infinite number of solutions.

**30.** $\begin{cases} 8x = 3y - 2 \\ \dfrac{4}{7}x - y = \dfrac{-5}{2} \end{cases}$

Rearrange the first equation.

Multiply the second equation by $-6$.

$$8x - \phantom{1}3y = -2$$
$$\underline{-8x + 14y = 35}$$
$$11y = 33$$
$$y = 3$$

Let $y = 3$ in the first equation.

$$8x = 3(3) - 2$$

$$8x = 7$$

$$x = \frac{7}{8}$$

The solution of the system is $\left( \dfrac{7}{8}, 3 \right)$

**31.** $\begin{cases} 5x = 6y + 25 \\ -2y = 7x - 9 \end{cases}$

Rearrange both equations.

$\begin{cases} \phantom{-}5x - 6y = 25 \\ -7x - 2y = -9 \end{cases}$

Multiply the second equation by $-3$.

$$5x - 6y = 25$$
$$\underline{21x + 6y = 27}$$
$$26x \phantom{+6y} = 52$$
$$x = 2$$

Let $x = 2$ in the second equation.

$$-2y = 7(2) - 9$$

$$-2y = 5$$

$$y = -\frac{5}{2}$$

The solution of the system is $\left( 2, -2\tfrac{1}{2} \right)$

**32.** $\begin{cases} -4x = 8 + 6y \\ -3y = 2x - 3 \end{cases}$

Rearrange both equations.

$\begin{cases} -4x - 6y = 8 \\ -2x - 3y = -3 \end{cases}$

Multiply the second equation by $-2$.

$$-4x - 6y = \phantom{-}8$$
$$\underline{4x + 6y = -6}$$
$$0 = 2$$

The system is inconsistent

and has no solution.

**33.** $\begin{cases} 3(x - 4) = -2y \\ 3x - 12 = -2y \\ 3x + 2y = 12 \\ \\ 2x = 3(y - 19) \\ 2x = 3y - 57 \\ 2x - 3y = -57 \end{cases}$

Multiply the first equation by 3 and

the second equation by 2.

$$9x + 6y = 36$$
$$\underline{4x - 6y = -114}$$
$$13x \phantom{+6y} = -78$$
$$x \phantom{+6y} = -6$$

Let $x = -6$ in the first equation.

$3(-6) + 2y = 12$

$-18 + 2y = 12$

$2y = 30$

$y = 15$

The solution of the system is $(-6, 15)$.

**34.** $\begin{cases} 4(x+5) = -3y \\ 4x + 20 = -3y \\ 4x + 3y = -20 \\ \\ 3x - 2(y+18) = 0 \\ 3x - 2y - 36 = 0 \\ 3x - 2y = 36 \end{cases}$

Multiply the first equation by 2 and the second equation by 3.

$8x + 6y = -40$

$\underline{9x - 6y = 108}$

$17x \qquad = 68$

$x \qquad = 4$

Let $x = 4$ in the first equation.

$4(4) + 3y = -20$

$16 + 3y = -20$

$3y = -36$

$y = -12$

The solution of the system is $(4, -12)$.

**35.** $\begin{cases} \dfrac{2x+9}{3} = \dfrac{y+1}{2} \\ \dfrac{x}{3} = \dfrac{y-7}{6} \end{cases}$

Multiply both equations by 6.

$\begin{cases} 2(2x+9) = 3(y+1) \\ 4x + 18 = 3y + 3 \\ 4x - 3y = -15 \\ \\ 2x = y - 7 \\ 2x - y = -7 \end{cases}$

Multiply the second equation by $-3$.

$4x - 3y = -15$

$\underline{-6x + 3y = \phantom{0}21}$

$-2x \qquad = 6$

$x \qquad = -3$

Let $x = -3$ in the second equation.

$2(-3) - y = -7$

$-6 - y = -7$

$-y = -1$

$y = 1$

The solution of the system is $(-3, 1)$.

**36.** $\begin{cases} \dfrac{2-5x}{4} = \dfrac{2y-4}{2} \\ \dfrac{x+5}{3} = \dfrac{y}{5} \end{cases}$

Multiply the first equation by 4 and the second equation by 15

$$\begin{cases} 2 - 5x = 2(2y - 4) \\ 2 - 5x = 4y - 8 \\ 5x + 4y = 10 \\ \\ 5(x + 5) = 3y \\ 5x + 25 = 3y \\ 5x - 3y = -25 \end{cases}$$

Multiply the second equation by $-1$.

$$5x + 4y = 10$$
$$\underline{-5x + 3y = 25}$$
$$7y = 35$$
$$y = 5$$

Let $y = 5$ in the second equation.

$$5x - 3(5) = -25$$
$$5x - 15 = -25$$
$$5x = -10$$
$$x = -2$$

The solution of the system is $(-2, 5)$.

**37.** Let $x$ = the larger number and $y$ = the smaller number.

$$\begin{cases} x + y = 16 \\ \underline{3x - y = 72} \\ 4x \quad = 88 \\ x = 22 \end{cases}$$

Let $x = 22$ in the first equation.

$$22 + y = 16$$
$$y = -6$$

The numbers are $-6$ and $22$.

**38.** Let $x$ = the number of orchestra seats and $y$ = the number of balcony seats.

$$\begin{cases} x + y = 360 \\ 45x + 35y = 15{,}150 \end{cases}$$

Solve the first equation for $x$.

$$x = 360 - y$$

Substitute $360 - y$ for $x$ in the second equation.

$$45(360 - y) + 35y = 15{,}150$$
$$16{,}200 - 45y + 35y = 15{,}150$$
$$-10y = -1050$$
$$y = 105$$

Let $y = 105$ in $x = 360 - y$.

$$x = 360 - 105$$
$$x = 255$$

There were 255 orchestra seats.

**39.** Let $x$ = the riverboat's speed in still water and $y$ = the rate of the current.

| | $d$ | = $r$ | $\cdot$ $t$ |
|---|---|---|---|
| Downriver | 340 | $x + y$ | 14 |
| Upriver | 340 | $x - y$ | 19 |

$$\begin{cases} 14(x + y) = 340 \\ 19(x - y) = 340 \end{cases}$$

Multiply the first equation by $\dfrac{1}{14}$ and the second equation by $\dfrac{1}{19}$.

$$\begin{cases} x + y = \dfrac{340}{14} \approx 24.29 \\ x - y = \dfrac{340}{19} \approx 17.89 \end{cases}$$

$$\begin{aligned} 2x \quad\quad &\approx 42.18 \\ x \quad\quad &\approx 21.09 \end{aligned}$$

Multiply the second equation of the simplified system by $-1$.

$$\begin{cases} x + y \approx 24.29 \\ -x + y \approx -17.89 \end{cases}$$

$$\begin{aligned} 2y &\approx 6.4 \\ y &\approx 3.2 \end{aligned}$$

The riverboat's speed in still water is 21.1 mph. The rate of the current is 3.2 mph.

**40.** Let $x =$ amount invested at 6% and $y =$ amount invested at 10%.

$$\begin{cases} x + y = 9000 \\ 0.06x + 0.10y = 652.80 \end{cases}$$

Multiply the first equation by $-6$ and the second equation by 100.

$$\begin{aligned} -6x - 6y &= -54,000 \\ \underline{6x + 10y} &= \underline{65,280} \\ 4y &= 11,280 \\ y &= 2820 \end{aligned}$$

Let $y = 2820$ in the first equation.

$$\begin{aligned} x + 2820 &= 9000 \\ x &= 6180 \end{aligned}$$

$6180 invested at 6% and $2820 invested at 10%.

**41.** Let $x =$ the width and $y =$ the length.

$$\begin{cases} 2x + 2y = 6 \\ y = 1.6x \end{cases}$$

Substitute $1.6x$ for $y$ in the first equation.

$$\begin{aligned} 2x + 2(1.6x) &= 6 \\ 2x + 3.2x &= 6 \\ 5.2x &= 6 \\ x &\approx 1.154 \end{aligned}$$

Let $x = 1.154$ in the second equation.

$$y = 1.6(1.154)$$

$$y \approx 1.846$$

Width $= 1.15$ feet, length $= 1.85$ feet.

**42.** Let $x =$ liters of 6% solution and $y =$ liters of 14% solution.

| | Concentration Rate | Ounces of Solution | Ounces of Pure Acid |
|---|---|---|---|
| First solution | 0.06 | $x$ | $0.06x$ |
| Second solution | 0.14 | $y$ | $0.14y$ |
| Mixture | 0.12 | 50 | $0.12(50)$ |

$$\begin{cases} x + y = 50 \\ 0.06x + 0.14y = 0.12(50) \end{cases}$$

Multiply the first equation by $-6$ and the second equation by 100.

$$\begin{aligned} -6x - 6y &= -300 \\ \underline{6x + 14y} &= \underline{600} \\ 8y &= 300 \\ y &= 37.5 \end{aligned}$$

Let $y = 37.5$ in the first equation.

$x + 37.5 = 50$

$\qquad x = 12.5$

$12\dfrac{1}{2}$ cc of 6% solution and

$37\dfrac{1}{2}$ cc of 14% solution.

**43.** Let $x =$ the cost of an egg and

$y =$ the cost of a strip of bacon.

$$\begin{cases} 3x + 4y = 3.80 \\ 2x + 3y = 2.75 \end{cases}$$

Multiply the first equation by $-2$ and the second equation by 3.

$-6x - 8y = -7.60$

$\underline{6x + 9y = \phantom{0}8.25}$

$\qquad y = \phantom{00}0.65$

Let $y = 0.65$ in the first equation.

$3x + 4(0.65) = 3.80$

$\qquad 3x + 2.60 = 3.80$

$\qquad\qquad 3x = 1.20$

$\qquad\qquad x = 0.40$

An egg costs 40¢ and a strip of bacon costs 65¢.

**44.** Let $x =$ the time spent walking and

$y =$ the time spent jogging.

| | $r$ | $\cdot$ | $t$ | $=$ | $d$ |
|---|---|---|---|---|---|
| Walking | 4 | | $x$ | | $4x$ |
| Jogging | 7.5 | | $y$ | | $7.5y$ |

$$\begin{cases} x + y = 3 \\ 4x + 7.5y = 15 \end{cases}$$

Multiply the first equation by $-4$.

$-4x - \phantom{0}4y = -12$

$\underline{4x + 7.5y = \phantom{0}15}$

$\qquad 3.5y = 3$

$\qquad\qquad y \approx 0.857$

Let $y = 0.857$ in the first equation.

$x + 0.857 = 3$

$\qquad x = 2.143$

He spent 2.14 hours walking and 0.86 hours jogging.

**45.** $3x - 4y \le 0$

Test $(1, 0)$

$3(1) - 4(0) \overset{?}{\le} 0$

False

Shade above.

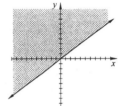

**46.** $3x - 4y \ge 0$

Test $(1, 0)$

$3(1) - 4(0) \overset{?}{\ge} 0$

True

Shade below.

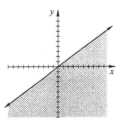

**47.** $x + 6y < 6$

Test $(0, 0)$

$0 + 6(0) \overset{?}{<} 6$

True

Shade below.

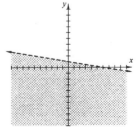

175

**48.** $x + y > -2$

Test $(0,0)$

$0 + 0 \overset{?}{>} -2$

True

Shade above.

**49.** $y \geq -7$

Test $(0,0)$

$0 \overset{?}{\geq} -7$

True

Shade above.

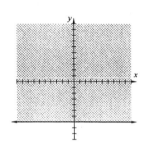

**50.** $y \leq -4$

Test $(0,0)$

$0 \overset{?}{\leq} -4$

False

Shade below.

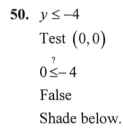

**51.** $-x \leq y$

Test $(1,0)$

$-1 \overset{?}{\leq} 0$

True

Shade above.

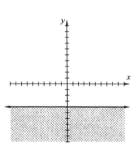

**52.** $x \geq -y$

Test $(1,0)$

$1 \overset{?}{\geq} 0$

True

Shade above.

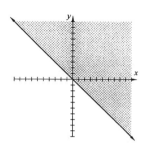

**53.** $y \geq 2x - 3$    $y \leq -2x + 1$

Test $(0,0)$    Test $(0,0)$

$0 \overset{?}{\leq} 2(0) - 3$    $0 \overset{?}{\leq} -2(0) + 1$

True    True

Shade above.    Shade below.

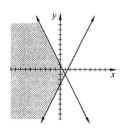

**54.** $y \leq -3x - 3$    $y \leq 2x + 7$

Test $(0,0)$    Test $(0,0)$

$0 \overset{?}{\leq} -3(0) - 3$    $0 \overset{?}{\leq} 2(0) + 7$

False    True

Shade below.    Shade below.

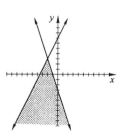

**55.** $x + 2y > 0$      $x - y \leq 6$

   Test $(1, 0)$      Test $(0, 0)$

   $1 + 2(0) \overset{?}{>} 0$      $(0) - (0) \overset{?}{\leq} 6$

   True             True

   Shade above.    Shade above.

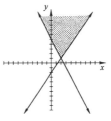

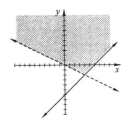

**56.** $x - 2y \geq 7$      $x + y \leq -5$

   Test $(0, 0)$      Test $(0, 0)$

   $0 - 2(0) \overset{?}{\geq} 7$      $(0) + (0) \overset{?}{\leq} -5$

   False           False

   Shade below.    Shade below.

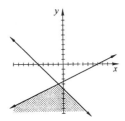

**57.** $3x - 2y \leq 4$      $2x + y \geq 5$

   Test $(0, 0)$      Test $(0, 0)$

   $3(0) - 2(0) \overset{?}{\leq} 4$      $2(0) + (0) \overset{?}{\geq} 5$

   True           False

   Shade above.    Shade above.

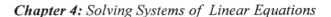

**58.** $4x - y \leq 0$      $3x - 2y \geq -5$

   Test $(1, 0)$      Test $(0, 0)$

   $4(1) - (0) \overset{?}{\leq} 4$      $3(0) - 2(0) \overset{?}{\geq} -5$

   True           True

   Shade above.    Shade below.

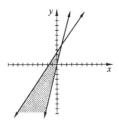

**59.** $-3x + 2y > -1$      $y < -2$

   Test $(0, 0)$      Shade below.

   $-3(0) + 2(0) \overset{?}{>} -1$

   True

   Shade above.

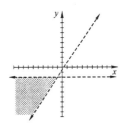

**60.** $-2x + 3y > -7$

Test $(0,0)$

$$-2(0) + 3(0) \overset{?}{>} -7$$

True

Shade above.

$x \geq -2$

Shade right.

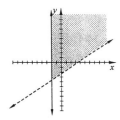

**Chapter 4 Test**

**1.** False

**2.** False

**3.** True

**4.** False

**5.** Let $x = 1$ and $y = -1$.

| | |
|---|---|
| $2x - 3y = 5$ | $6x + y = 1$ |
| $2(1) - 3(-1) \overset{?}{=} 5$ | $6(1) + (-1) \overset{?}{=} 1$ |
| $2 + 3 \overset{?}{=} 5$ | $6 - 1 \overset{?}{=} 1$ |
| $5 = 5$ | $5 = 1$ |
| True | False |

$(1,-1)$ is not a solution of the system.

**6.** Let $x = 3$ and $y = -4$.

| | |
|---|---|
| $4x - 3y = 24$ | $4x + 5y = -8$ |
| $4(3) - 3(-4) \overset{?}{=} 24$ | $4(3) + 5(-4) \overset{?}{=} -8$ |
| $12 + 12 \overset{?}{=} 24$ | $12 - 20 \overset{?}{=} -8$ |
| $24 = 24$ | $-8 = -8$ |
| True | True |

$(3,-4)$ is a solution of the system

**7.** $\begin{cases} y - x = 6 \\ y + 2x = -6 \end{cases}$

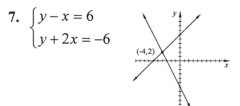

$(-4,2)$ is the solution of the system

**8.** $\begin{cases} 3x - 2y = -14 \\ x + 3y = -1 \end{cases}$

Solve the second equation for $x$.

$x = -3y - 1$

Substitute $-3y - 1$ for $x$ in the first equation.

$$3(-3y - 1) - 2y = -14$$
$$-9y - 3 - 2y = -14$$
$$-11y = -11$$
$$y = 1$$

Let $y = 1$ in $x = -3y - 1$.

$$x = -3(1) - 1$$
$$x = -4$$

The solution is $(-4,1)$.

**9.** $\begin{cases} \dfrac{1}{2}x + 2y = -\dfrac{15}{4} \\ 4x = -y \end{cases}$

Multiply the first equation by 4 and the second equation by $-1$.

$\begin{cases} 2x + 8y = -15 \\ -4x = y \end{cases}$

Substitute $-4x$ for $y$ in the first equation.

$2x + 8(-4x) = -15$

$2x - 32x = -15$

$-30x = -15$

$x = \dfrac{1}{2}$

Let $x = \dfrac{1}{2}$ in the second equation.

$-4\left(\dfrac{1}{2}\right) = y$

$y = -2$

The solution is $\left(\dfrac{1}{2}, -2\right)$.

**10.** $\begin{cases} 3x + 5y = 2 \\ 2x - 3y = 14 \end{cases}$

Multiply the first equation by 2 and the second equation by $-3$.

$6x + 10y = 4$

$\dfrac{-6x + 9y = -42}{19y = -38}$

$y = -2$

Let $y = -2$ in the first equation.

$3x + 5(-2) = 2$

$3x - 10 = 2$

$3x = 12$

$x = 4$

The solution of the system is $(4, -2)$.

**11.** $\begin{cases} 4x - 6y = 7 \\ -2x + 3y = 0 \end{cases}$

Multiply the second equation by 2.

$4x - 6y = 7$

$\dfrac{-4x + 6y = 0}{0 = 7}$

The system is inconsistent. There is no solution.

**12.** $\begin{cases} 3x + y = 7 \\ 4x + 3y = 1 \end{cases}$

Solve the first equation for $y$.

$y = 7 - 3x$

Substitute $7 - 3x$ for $y$ in the second equation.

$4x + 3(7 - 3x) = 1$

$4x + 21 - 9x = 1$

$-5x = -20$

$x = 4$

Let $x = 4$ in $y = 7 - 3x$.

$y = 7 - 3(4)$

$y = -5$

The solution is $(4, -5)$.

**13.**
$$\begin{cases} 3(2x+y) = 4x+20 \\ 6x+3y = 4x+20 \\ 2x+3y = 20 \\ \\ x-2y = 3 \end{cases}$$

Multiply the second equation by $-2$.
$$2x+3y = 20$$
$$\underline{-2x+4y = -6}$$
$$7y = 14$$
$$y = 2$$
Let $y = 2$ in the second equation.
$$x-2(2) = 3$$
$$x-4 = 3$$
$$x = 7$$

The solution of the system is $(7,2)$.

**14.**
$$\begin{cases} \dfrac{x-3}{2} = \dfrac{2-y}{4} \\ \dfrac{7-2x}{3} = \dfrac{y}{2} \end{cases}$$

Multiply the first equation by 4
and the second equation by 6
$$\begin{cases} 2(x-3) = 2-y \\ 2x-6 = 2-y \\ 2x+y = 8 \\ \\ 2(7-2x) = 3y \\ 14-4x = 3y \\ 4x+3y = 14 \end{cases}$$

Multiply the first equation by $-3$.

$$-6x-3y = -24$$
$$\underline{4x+3y = \phantom{0}14}$$
$$-2x \phantom{+3y} = -10$$
$$x = 5$$
Let $x = 5$ in the first equation.
$$2(5)+y = 8$$
$$10+y = 8$$
$$y = -2$$
The solution of the system is $(5,-2)$.

**15.** Let $x = $ cc's of 12% solution and
$y = $ cc's of 16% solution.

|  | Concentration Rate | cc's of Solution | cc's of Salt |
|---|---|---|---|
| First solution | 12% | $x$ | $0.12x$ |
| Second solution | 22% | 80 | $0.22(80)$ |
| Mixture | 16% | $y$ | $0.16y$ |

$$\begin{cases} x+80 = y \\ 0.12x+0.22(80) = 0.16y \end{cases}$$
Multiply the first equation by $-16$
and the second equation by 100.
$$-16x-1280 = -16y$$
$$\underline{12x+1760 = \phantom{0}16y}$$
$$-4x \phantom{0} +480 = \phantom{00} 0$$
$$-4x = -480$$
$$x = 120$$
Should add 120 cc's of 12% solution

**16.** Let $x = $ amount invested at 5% and
$y = $ amount invested at 9%.
$$\begin{cases} x+y = 4000 \\ 0.05x+0.09y = 311 \end{cases}$$
Multiply the first equation by $-5$
and the second equation by 100.

$$-5x - 5y = -20,000$$
$$\underline{5x + 9y = \phantom{0}31,100}$$
$$4y = \phantom{0}11,100$$
$$y = 2775$$

Let $y = 2775$ in the first equation.

$$x + 2775 = 4000$$
$$x = 1225$$

$1225 invested at 5% and
$2775 invested at 9%.

**17.** Let $x =$ the number of thousands of farms in Texas and $y =$ the number thousands of farms in Missouri.

$$\begin{cases} x + y = 336 \\ x - y = 116 \end{cases}$$
$$2x \phantom{xxx} = 452$$
$$x \phantom{xxxx} = 226$$

Let $x = 226$ in the first equation.

$$226 + y = 336$$
$$y = 110$$

There are 226,000 farms in Texas and 110,000 farms in Missouri.

**18.** $y \geq -4x$

Test $(1, 0)$

$$0 \overset{?}{\geq} -4(1)$$

True

Shade above.

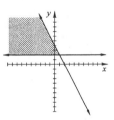

**19.** $2x - 3y > -6$

Test $(0, 0)$

$$2(0) - 3(0) \overset{?}{>} -6$$

True

Shade below.

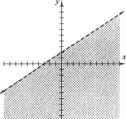

**20.** $y + 2x \leq 4$      $y \geq 2$

Test $(0, 0)$      Shade above.

$$0 + 2(0) \overset{?}{\leq} 4$$

True

Shade below.

**21.** $2y - x \geq 1$      $x + y \geq -4$

Test $(0, 0)$      Test $(0, 0)$

$$2(0) - 0 \overset{?}{\geq} 1 \qquad (0) + (0) \overset{?}{\geq} -4$$

False            True

Shade above.    Shade above.

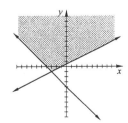

**Cumulative Review Chapter 4**

**1. a.** $-1 < 0$   **b.** $7 = \dfrac{14}{2}$   **c.** $-5 > -6$

**2. a.** $5^2 = 5 \cdot 5 = 25$   **b.** $2^5 = 2 \cdot 2 \cdot 2 \cdot 2 \cdot 2 = 32$

**3. a.** commutative property of multiplication
   **b.** associative property of addition
   **c.** identity element for addition
   **d.** commutative property of multiplication
   **e.** multiplicative inverse property
   **f.** additive inverse property
   **g.** commutative and associative properties of multiplication

**4.** Let $x = 8$, $y = 5$
$$y^2 - 3x = 5^2 - 3(8) = 25 - 24 = 1$$

**5.** $(2x - 3) - (4x - 2) = 2x - 3 - 4x + 2$
$$= -2x - 1$$

**6.** $7 - 12 + (-5) - 2 + (-2)$
$$= 7 + (-12) + (-5) + (-2) + (-2)$$
$$= 7 + (-21)$$
$$= -14$$

**7.** $5t - 5 = 6t + 2$
$$-t - 5 = 2$$
$$-t = 7$$
$$t = -7$$

**8.** Let $x = -7$, $y = -3$
$$2y^2 - x^2 = 2(-3)^2 - (-7)^2$$
$$= 2(9) - 49$$
$$= 18 - 49$$
$$= -31$$

**9.** $\dfrac{5}{2}x = 15$
$$5x = 30$$
$$x = 6$$

**10.** $0.4y - 6.7 + y - 0.3 - 2.6y$
$$= 0.4y + y + (-2.6y) + (-6.7) + (-0.3)$$
$$= -1.2y - 7.0$$

**11.** $\dfrac{x}{2} - 1 = \dfrac{2}{3}x - 3$
$$6\left(\dfrac{x}{2} - 1\right) = 6\left(\dfrac{2}{3}x - 3\right)$$
$$3x - 6 = 4x - 18$$
$$-x - 6 = -18$$
$$-x = -12$$
$$x = 12$$

**12.** $7(x - 2) - 6(x + 1) = 20$
$$7x - 14 - 6x - 6 = 20$$
$$x - 20 = 20$$
$$x = 40$$

**13.**   Let $x =$ the number
$$2(x + 4) = 4x - 12$$
$$2x + 8 = 4x - 12$$
$$-2x + 8 = -12$$
$$-2x = -20$$
$$x = 10$$
The number is 10.

**14.** $5(y-5)=5y+10$

$5y-25=5y+10$

$-25=10$

False statement.  There is no solution.

**15.** $\qquad y=mx+b$

$y-b=mx+b-b$

$y-b=mx$

$\dfrac{y-b}{m}=\dfrac{mx}{m}$

$\dfrac{y-b}{m}=x$

**16.** $\qquad$ Let $x=$ the number

$5(x-1)=6x$

$5x-5=6x$

$-x-5=$

$-x=5$

$x=-5$

The number is $-5$.

**17.** $-2x\le-4$

$\dfrac{-2x}{-2}\ge\dfrac{-4}{-2}$

$x\ge2,\ [2,\infty)$

**18.** $\qquad P=a+b+c$

$P-a-c=a+b+c-a-c$

$P-a-c=b$

**19.** $x=-2y$

| $x$ | $y$ |
|---|---|
| 0 | 0 |
| $-4$ | 2 |

**20.** $3x+7\ge x-9$

$2x+7\ge-9$

$2x\ge-16$

$x\ge-8,\ [-8,\infty)$

**21.** $(-1,5)$ and $(2,-3)$

$m=\dfrac{y_2-y_1}{x_2-x_1}=\dfrac{-3-5}{2-(-1)}=\dfrac{-8}{3}=-\dfrac{8}{3}$

**22.** $x-3y=3$

| $x$ | $y$ |
|---|---|
| 0 | $-1$ |
| 3 | 0 |
| 9 | 2 |

**23.** $y=\dfrac{3}{4}x+6$

$y=mx+b$

$m=\dfrac{3}{4}$

**24.** $(-1,3)$ and $(2,-8)$

$m=\dfrac{y_2-y_1}{x_2-x_1}=\dfrac{-8-3}{2-(-1)}=-\dfrac{11}{3}$

Parallel line has the same slope.

Slope is $-\dfrac{11}{3}$

**25.** $3x-4y=4$

$-4y=-3x+4$

$y=\dfrac{-3x}{-4}+\dfrac{4}{-4}$

$y=\dfrac{3}{4}x-1$

$y = mx + b$

$m = \dfrac{3}{4}, b = -1$

Slope is $\dfrac{3}{4}$, $y$-intercept is $(0, -1)$

**26.** $y = 7x + 0$

$y = mx + b$

$m = 7, \ b = 0$

Slope is 7. $y$-intercept is $(0, 0)$.

**27.** $m = -2$, with point $(-1, 5)$

$y - y_1 = m(x - x_1)$

$y - 5 = -2[x - (-1)]$

$y - 5 = -2x - 2$

$2x + y = 3$

**28.** Line 1: $y = 4x - 5 \Rightarrow m_1 = 4$

Line 2: $-4x + y = 7 \Rightarrow y = 4x + 7$

$\Rightarrow m_2 = 4$

$m_2 = m_1$  The lines are parallel.

**29.** A vertical line has an equation $x = c$.

Point, $(-1, 5)$

$x = -1$

**30.** $m = -5$, with point $(-2, 3)$

$y - y_1 = m(x - x_1)$

$y - 3 = -5[x - (-2)]$

$y - 3 = -5x - 10$

$5x + y = -7$

**31.** Domain is $\{-1, 0, 3\}$

Range is $\{-2, 0, 2, 3\}$

**32.** $f(x) = 5x^2 - 6$

$f(0) = 5(0)^2 - 6 = -6$

$f(-2) = 5(-2)^2 - 6 = 5(4) - 6 = 14$

**33. a.** function

**b.** not a function

**34. a.** not a function

**b.** function

**c.** not a function

**35.** $\begin{cases} 3x - y = 4 \\ y = 3x - 4, \ m = 3 \\ \\ x + 2y = 8 \\ y = -\dfrac{1}{2}x + 4, \ m = -\dfrac{1}{2} \end{cases}$

Because they have different slopes, there is only one solution.

**36. a.** Let $x = 1$ and $y = -4$.

$2x - y = 6 \qquad\qquad 3x + 2y = -5$

$2(1) - (-4) \overset{?}{=} 6 \quad 3(1) + 2(-4) \overset{?}{=} -5$

$2 + 4 \overset{?}{=} 6 \qquad\qquad 3 - 8 \overset{?}{=} -5$

$6 = 6 \qquad\qquad\quad -5 = -5$

True                    True

$(1, -4)$ is a solution of the system.

**b.** Let $x = 0$ and $y = 6$.

$$2x - y = 6 \qquad 3x + 2y = -5$$

$$2(0) - (6) \overset{?}{=} 6 \qquad \text{Test not needed}$$

$$-6 \overset{?}{=} 6$$

$$-6 = 6$$

False

$(0, 6)$ is not a solution of the system

**c.** Let $x = 3$ and $y = 0$.

$$2x - y = 6 \qquad 3x + 2y = -5$$

$$2(3) - (0) \overset{?}{=} 6 \qquad 3(3) + 2(0) \overset{?}{=} -5$$

$$6 - 0 \overset{?}{=} 6 \qquad 9 + 0 \overset{?}{=} -5$$

$$6 = 6 \qquad 9 = -5$$

True          False

$(3, 0)$ is not a solution of the system

**37.** $\begin{cases} x + 2y = 7 \\ 2x + 2y = 13 \end{cases}$

Solve the first equation for $x$.

$$x = 7 - 2y$$

Substitute $7 - 2y$ for $x$
in the second equation.

$$2(7 - 2y) + 2y = 13$$

$$14 - 4y + 2y = 13$$

$$-2y = -1$$

$$y = \frac{1}{2}$$

Let $y = \frac{1}{2}$ in $x = 7 - 2y$.

$$x = 7 - 2\left(\frac{1}{2}\right)$$

$$x = 6$$

The solution is $\left(6, \frac{1}{2}\right)$

**38.** $\begin{cases} 3x - 4y = 10 \\ y = 2x \end{cases}$

Substitute $2x$ for $y$ in the first equation.

$$3x - 4(2x) = 10$$

$$3x - 8x = 10$$

$$-5x = 10$$

$$x = -2$$

Let $x = -2$ in the second equation.

$$y = 2(-2) = -4$$

The solution is $(-2, -4)$.

**39.** $\begin{cases} x + y = 7 \\ x - y = 5 \end{cases}$

$$\begin{array}{rl} 2x & = 12 \\ x & = 6 \end{array}$$

Let $x = 6$ in the first equation.

$$6 + y = 7$$

$$y = 1$$

The solution to the system is $(6, 1)$

**40.** $\begin{cases} x = 5y - 3 \\ x = 8y + 4 \end{cases}$

Substitute $8y + 4$ for $x$
in the first equation.

$8y + 4 = 5y - 3$

$3y + 4 = -3$

$3y = -7$

$y = -\dfrac{7}{3}$

Let $y = -\dfrac{7}{3}$ in the second equation.

$x = 8\left(-\dfrac{7}{3}\right) + 4$

$x = -\dfrac{56}{3} + \dfrac{12}{3}$

$x = -\dfrac{44}{3}$

The solution is $\left(-\dfrac{44}{3}, -\dfrac{7}{3}\right)$.

**41.** Let $x$ = the first number and
$y$ = the second number.

$\begin{cases} x + y = 37 \\ x - y = 21 \end{cases}$

$2x \quad\ = 58$

$\ x \quad\ = 29$

Let $x = 29$ in the first equation.

$29 + y = 37$

$y = 8$

The numbers are 29 and 8.

**42.** $x > 1$
Shade right

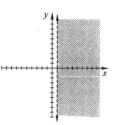

**43.** $2x - y \geq 3$
Test $(0, 0)$

$2(0) - 0 \overset{?}{\geq} 3$

False

Shade below

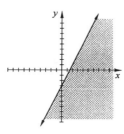

**44.** $2x + 3y < 6$     $y < 2$
Test $(0, 0)$     Shade left

$2(0) + 3(0) \overset{?}{<} 6$

True

Shade below

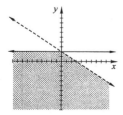

# Chapter 5

## Mental Math 5.1

**1.** $3^2$

base: 3

exponent: 2

**2.** $5^4$

base: 5

exponent: 4

**3.** $(-3)^6$

base: $-3$

exponent: 6

**4.** $-3^7$

base: 3

exponent: 7

**5.** $-4^2$

base: 4

exponent: 2

**6.** $(-4)^3$

base: $-4$

exponent: 3

**7.** $5 \cdot 3^4$

base: 5; exponent: 1

base: 3; exponent: 4

**8.** $9 \cdot 7^6$

base: 9; exponent: 1

base: 7; exponent: 6

**9.** $5x^2$

base: 5; exponent: 1

base: $x$; exponent: 2

**10.** $(5x)^2$

base: $5x$

exponent: 2

## Exercise Set 5.1

**1.** $7^2 = 7 \cdot 7 = 49$

**3.** $(-5)^1 = -5$

**5.** $-2^4 = -2 \cdot 2 \cdot 2 \cdot 2 = -16$

**7.** $(-2)^4 = (-2)(-2)(-2)(-2) = 16$

**9.** $(0.1)^5 = (0.1)(0.1)(0.1)(0.1)(0.1)$
$= 0.00001$

**11.** $\left(\frac{1}{3}\right)^4 = \left(\frac{1}{3}\right)\left(\frac{1}{3}\right)\left(\frac{1}{3}\right)\left(\frac{1}{3}\right) = \frac{1}{81}$

**13.** $7 \cdot 2^5 = 7 \cdot 2 \cdot 2 \cdot 2 \cdot 2 \cdot 2 = 224$

**15.** $-2 \cdot 5^3 = -2 \cdot 5 \cdot 5 \cdot 5 = -250$

**17.** Answers may vary.

**19.** $x^2 = (-2)^2 = (-2)(-2) = 4$

**21.** $5x^3 = 5(3)^3 = 5 \cdot 3 \cdot 3 \cdot 3 = 135$

**23.** $2xy^2 = 2(3)(5)^2 = 2(3)(5)(5) = 150$

**25.** $\dfrac{2z^4}{5} = \dfrac{2(-2)^4}{5} = \dfrac{2(-2)(-2)(-2)(-2)}{5} = \dfrac{32}{5}$

**27.** $V = x^3 = 7^3 = 7 \cdot 7 \cdot 7 = 343$

The volume is 343 cubic meters.

**29.** We use the volume formula

**31.** $x^2 \cdot x^8 = x^{2+8} = x^{10}$

**33.** $(-3)^3 \cdot (-3)^9 = (-3)^{3+9} = (-3)^{12}$

**35.** $\left(5y^4\right)(3y) = 5(3)y^{4+1} = 15y^5$

**37.** $\left(4z^{10}\right)\left(-6z^7\right)\left(z^3\right) = 4(-6)z^{10+7+3} = -24z^{20}$

**39.** $(pq)^7 = p^7 q^7$

**41.** $\left(\dfrac{m}{n}\right)^9 = \dfrac{m^9}{n^9}$

**43.** $\left(x^2 y^3\right)^5 = x^{2\cdot5} y^{3\cdot5} = x^{10} y^{15}$

**45.** $\left(\dfrac{-2xz}{y^5}\right)^2 = \dfrac{(-2)^2 x^2 z^2}{y^{5\cdot2}} = \dfrac{4x^2 z^2}{y^{10}}$

**47.** $\dfrac{x^3}{x} = \dfrac{x^3}{x^1} = x^{3-1} = x^2$

**49.** $\dfrac{(-2)^5}{(-2)^3} = (-2)^{5-3} = (-2)^2 = 4$

**51.** $\dfrac{p^7 q^{20}}{pq^{15}} = p^{7-1} q^{20-15} = p^6 q^5$

**53.** $\dfrac{7x^2 y^6}{14x^2 y^3} = \dfrac{7}{14} x^{2-2} y^{6-3} = \dfrac{1}{2} x^0 y^3 = \dfrac{y^3}{2}$

**55.** $(2x)^0 = 1$

**57.** $-2x^0 = -2(1) = -2$

**59.** $5^0 + y^0 = 1 + 1 = 2$

**61.** $\left(\dfrac{-3a^2}{b^3}\right)^3 = \dfrac{(-3)^3 a^{2\cdot3}}{b^{3\cdot3}} = -\dfrac{27a^6}{b^9}$

**63.** $\dfrac{\left(x^5\right)^7 \cdot x^8}{x^4} = \dfrac{x^{5\cdot7} \cdot x^8}{x^4}$

$= \dfrac{x^{35} x^8}{x^4}$

$= x^{35+8-4}$

$= x^{39}$

**65.** $\dfrac{\left(z^3\right)^6}{(5z)^4} = \dfrac{z^{3\cdot6}}{5^4 z^4} = \dfrac{z^{18}}{625z^4} = \dfrac{z^{18-4}}{625} = \dfrac{z^{14}}{625}$

**67.** $\dfrac{(6mn)^5}{mn^2} = \dfrac{6^5 \cdot m^5 \cdot n^5}{mn^2}$

$= 7776m^{5-1} n^{5-2}$

$= 7776m^4 n^3$

**69.** $-5^2 = -5 \cdot 5 = -25$

**71.** $\left(\dfrac{1}{4}\right)^3 = \dfrac{1^3}{4^3} = \dfrac{1}{64}$

**73.** $(9xy)^2 = 9^2 x^2 y^2 = 81x^2 y^2$

**75.** $(6b)^0 = 1$

**77.** $2^3 + 2^5 = 8 + 32 = 40$

**79.** $b^4 b^2 = b^{4+2} = b^6$

**81.** $a^2a^3a^4 = a^{2+3+4} = a^9$

**83.** $\left(2x^3\right)\left(-8x^4\right) = 2(-8)x^{3+4} = -16x^7$

**85.** $(4a)^3 = 4^3a^3 = 64a^3$

**87.** $\left(-6xyz^3\right)^2 = (-6)^2x^2y^2z^{3\cdot2} = 36x^2y^2z^6$

**89.** $\left(\dfrac{3y^5}{6x^4}\right)^3 = \dfrac{3^3y^{5\cdot3}}{6^3x^{4\cdot3}} = \dfrac{27y^{15}}{216x^{12}} = \dfrac{y^{15}}{8x^{12}}$

**91.** $\dfrac{x^5}{x^4} = x^{5-4} = x$

**93.** $\dfrac{2x^3y^2z}{xyz} = 2x^{3-1}y^{2-1}z^{1-1} = 2x^2y$

**95.** $\dfrac{\left(3x^2y^5\right)^5}{x^3y} = \dfrac{3^5x^{2\cdot5}y^{5\cdot5}}{x^3y}$

$$= \dfrac{243x^{10}y^{25}}{x^3y}$$

$$= 243x^{10-3}y^{25-1}$$

$$= 243x^7y^{24}$$

**97.** Answers may vary.

**99.** $y - 10 + y = 2y - 10$

**101.** $7x + 2 - 8x - 6 = -x - 4$

**103.** $2(x - 5) + 3(5 - x) = 2x - 10 + 15 - 3x$

$$= -x + 5$$

**105.** $\left(4x^2\right)\left(5x^3\right) = 4(5)x^{2+3} = 20x^5$ sq ft

**107.** $\pi(5y)^2 = \pi(5)^2y^2 = 25y^2\pi$ sq. cm

**109.** $\left(3y^4\right)^3 = 3^3y^{4\cdot3} = 27y^{11}$

The volume is $27y^{12}$ cubic feet.

**111.** $x^{5a}x^{4a} = x^{5a+4a} = x^{9a}$

**113.** $\dfrac{x^{9a}}{x^{4a}} = x^{9a-4a} = x^{5a}$

**115.** $\left(x^ay^bz^c\right)^{5a} = x^{5a\cdot a}y^{5a\cdot b}z^{5a\cdot c} = x^{5a^2}y^{5ab}z^{5ac}$

**117.** $A = P\left(1 + \dfrac{r}{12}\right)^6$

$$A = 1000\left(1 + \dfrac{0.09}{12}\right)^6$$

$$A = 1000(1.0075)^6$$

$$A = 1045.85$$

You need $1045.85 to pay off the loan.

**Mental Math 5.2**

**1.** $-9y - 5y = -14y$

**2.** $6m^5 + 7m^5 = 13m^5$

**3.** $4y^3 + 3y^3 = 7y^3$

**4.** $21y^5 - 19y^5 = 2y^5$

**5.** $x + 6x = 7x$

**6.** $7z - z = 6z$

**7.** $5m^2 + 2m = 5m^2 + 2m$

189

**8.** $8p^3 + 3p^2 = 8p^3 + 3p^2$

## Exercise Set 5.2

**1.** $x + 2$

The degree is 1 since $x$ is $x^1$. It is a binomial because it has two terms.

**3.** $9m^3 - 5m^2 + 4m - 8$

The degree is 3, the greatest degree of any of its terms. It is none of these because it has more than three terms.

**5.** $12x^4y - x^2y^2 - 12x^2y^4$

The degree is 6, the greatest degree of any of its terms. It is a trinomial because it has three terms.

**7.** $3zx - 5x^2$

The degree is 2 because2 is the degree of the term with the highest degree . It is a binomial because it has two terms.

**9.** Degree 3

**11.** Degree 2

**13.** Answers may vary.

**15.** Answers may vary.

**17.** a. $x + 6 = 0 + 6 = 6$
   b. $x + 6 = -1 + 6 = 5$

**19.** a. $x^2 - 5x - 2 = 0^2 - 5(0) - 2 = -2$
   b. $x^2 - 5x - 2 = (-1)^2 - 5(-1) - 2$
      $= 1 + 5 - 2 = 4$

**21.** a. $x^3 - 15 = 0^3 - 15 = -15$
   b. $x^3 - 15 = (-1)^3 - 15 = -1 - 15 = -16$

**23.** $-16t^2 + 1150 = -16(9)^2 + 1150$
   $= -1296 + 1150 = -146$ feet
   The object has reached the ground.

**25.** $14x^2 + 9x^2 = (14 + 9)x^2 = 23x^2$

**27.** $15x^2 - 3x^2 - y$
   $(15 - 3)x^2 - y = 12x^2 - y$

**29.** $8s - 5s + 4s = (8 - 5 + 4)s = 7s$

**31.** $0.1y^2 - 1.2y^2 + 6.7 - 1.9$
   $= (0.1 - 1.2)y^2 + (6.7 - 1.9)$
   $= -1.1y^2 + 4.8$

**33.** $\dfrac{2}{5}x^2 - \dfrac{1}{3}x^3 + x^2 - \dfrac{1}{4}x^3 + 6$
   $= -\dfrac{7}{12}x^3 + \dfrac{7}{5}x^2 + 6$

**35.** $6a^2y - 4ab + 7b^2 - a^2 - 5ab + 9b^2$
   $= 5a^2 - 9ab + 16b^2$

**37.** $(3x + 7) + (9x + 5) = 3x + 7 + 9x + 5$
   $= (3x + 9x) + (7 + 5) = 12x + 12$

**39.** $(-7x + 5) + (-3x^2 + 7x + 5)$
   $= -7x + 5 + (-3x^2) + 7x + 5$
   $= -3x^2 + (-7x + 7x) + (5 + 5)$
   $= -3x^2 + 10$

**41.** $(2x^2 + 5) - (3x^2 - 9) = 2x^2 + 5 - 3x^2 + 9$
$$= -x^2 + 14$$

**43.** $3x - (5x - 9) = 3x + (-5x + 9)$
$$= 3x + (-5x) + 9 = -2x + 9$$

**45.** $(2x^2 + 3x - 9) - (-4x + 7)$
$$= (2x^2 + 3x - 9) + (4x - 7)$$
$$= 2x^2 + 3x - 9 + 4x - 7$$
$$= 2x^2 + (3x + 4x) + (-9 - 7)$$
$$= 2x^2 + 7x - 16$$

**47.** $(-x^2 + 3x) + (2x^2 + 5) + (4x - 1)$
$$= -x^2 + 3x + 2x^2 + 5 + 4x - 1$$
$$= (x^2 + 7x + 4) \text{ feet}$$

**49.** $(4y^2 + 4y + 1) - (y^2 - 10)$
$$= 4y^2 + 4y + 1 - y^2 + 10$$
$$= (3y^2 + 4y + 11) \text{ meters}$$

**51.**
$$\begin{array}{r} 3t^2 + 4 \\ +5t^2 - 8 \\ \hline 8t^2 - 4 \end{array}$$

**53.**
$$\begin{array}{r} 4z^2 - 8z + 3 \\ -(6z^2 + 8z - 3) \\ \hline \end{array}$$

$$\begin{array}{r} 4z^2 - 8z + 3 \\ +(-6z^2 - 8z + 3) \\ \hline -2z^2 - 16z + 6 \end{array}$$

**55.**
$$\begin{array}{r} 5x^3 - 4x^2 + 6x - 2 \\ -(3x^3 + 2x^2 - x - 4) \\ \hline \end{array}$$

$$\begin{array}{r} 5x^3 - 4x^2 + 6x - 2 \\ +(-3x^3 + 2x^2 + x + 4) \\ \hline 2x^3 - 2x^2 + 7x + 2 \end{array}$$

**57.** $81x^2 + 10 - (19x^2 + 5) = 81x^2 + 10 - 19x^2 - 5$
$$= 62x^2 + 5$$

**59.** $[(8x + 1) + (6x + 3)] - (2x + 2)$
$$= 8x + 1 + 6x + 3 - 2x - 2$$
$$= 8x + 6x - 2x + 1 + 3 - 2$$
$$= 12x + 2$$

**61.** $-15x - (-4x) = -15x + 4x = -11x$

**63.** $2x - 5 + 5x - 8 = 7x - 13$

**65.** $(-3y^2 - 4y) + (2y^2 + y - 1)$
$$= -3y^2 - 4y + 2y^2 + y - 1$$
$$= (-3y^2 + 2y^2) + (-4y + y) - 1$$
$$= -y^2 - 3y - 1$$

**67.** $(5x + 8) - (-2x^2 - 6x + 8)$
$$= (5x + 8) + (2x^2 + 6x - 8)$$
$$= 5x + 8 + 2x^2 + 6x - 8$$
$$= 2x^2 + (5x + 6x) + (8 - 8)$$
$$= 2x^2 + 11x$$

**69.** $\left(-8x^4 + 7x\right) + \left(-8x^4 + x + 9\right)$

$= -8x^4 + 7x - 8x^4 + x + 9$

$= -16x^4 + 8x + 9$

**71.** $\left(3x^2 + 5x - 8\right) + \left(5x^2 + 9x + 12\right) - \left(x^2 - 14\right)$

$= 3x^2 + 5x - 8 + 5x^2 + 9x + 12 - x^2 + 14$

$= 7x^2 + 14x + 18$

**73.** $(7x - 3) - 4x = 7x - 3 - 4x = 3x - 3$

**75.** $\left(7x^2 + 3x + 9\right) - \left(5x + 7\right)$

$= 7x^2 + 3x + 9 - 5x - 7$

$= 7x^2 - 2x + 2$

**77.** $\left[\left(8y^2 + 7\right) + \left(6y + 9\right)\right] - \left(4y^2 - 6y - 3\right)$

$= 8y^2 + 7 + 6y + 9 - 4y^2 + 6y + 3$

$= 4y^2 + 12y + 19$

**79.** $\left[\left(-x^2 - 2x\right) + \left(5x^2 + x + 9\right)\right]$

$\qquad\qquad\qquad - \left(-2x^2 + 4x - 12\right)$

$= -x - 2x + 5x^2 + x + 9 + 2x^2 - 4x + 12$

$= 6x^2 - 5x + 21$

**81.** $\left(x^3 + x^2 + 1\right) + \left(5x^3 - 2x^2 + 9\right)$

$\qquad\qquad\qquad - \left(3x^3 - x + 4\right)$

$= x^3 + x^2 + 1 + 5x^3 - 2x^2 + 9$

$\qquad\qquad\qquad - 3x^3 + x - 4$

$= 3x^3 - x^2 + x + 6$

**83.** $3x(2x) = 3 \cdot 2 \cdot x \cdot x = 6x^2$

**85.** $\left(12x^3\right)\left(-x^5\right) = \left(12x^3\right)\left(-1x^5\right)$

$= (12)(-1)\left(x^3\right)\left(x^5\right) = -12x^8$

**87.** $10x^2\left(20xy^2\right) = 10 \cdot 20x^2 \cdot x \cdot y^2 = 200x^3 y^2$

**89.** $(2x)^2 + 7x + x^2 + 5x = 4x^2 + x^2 + 7x + 5x$

$= 5x^2 + 12x$

**91.** $9x + 10 + 3x + 12 + 4x + 15 + 2x + 7$

$= \left(9x + 3x + 4x + 2x\right) + \left(10 + 12 + 15 + 7\right)$

$= 18x + 44$

**93.** $x^2 + x^2 + xy + xy + xy + xy = 2x^2 + 4xy$

**95.** $\left(9a + 6b - 5\right) + \left(-11a - 7b + 6\right)$

$= 9a + 6b - 5 - 11a - 7b + 6$

$= -2a - b + 1$

**97.** $\left(4x^2 + y^2 + 3\right) - \left(x^2 + y^2 - 2\right)$

$= 4x^2 + y^2 + 3 - x^2 - y^2 + 2$

$= 3x^2 + 5$

**99.** $\left(x^2 + 2xy - y^2\right) + \left(5x^2 - 4xy + 20y^2\right)$

$= x^2 + 2xy - y^2 + 5x^2 - 4xy + 20y^2$

$= 6x^2 - 2xy + 19y^2$

**101.** $\left(11r^r s + 16rs - 3 - 2r^2 s^2\right)$

$\qquad\qquad\qquad - \left(3sr^2 + 5 - 9r^2 s^2\right)$

$= 11r^2 s + 16rs - 3 - 2r^2 s^2$

$\qquad\qquad\qquad - 3sr^2 - 5 + 9r^2 s^2$

$= 8r^2 s + 16rs + 7r^2 s^2 - 8$

**103.** $7.75x + 9.16x^2 - 1.27 - 14.58x^2 - 18.34$

$= (9.16 - 14.58)x^2 + 7.75x$

$\qquad\qquad + (-1.27 - 18.34)$

$= -5.42x^2 + 7.75x - 19.61$

**105.** $\left[ (7.9y^4 - 6.8y^3 + 3.3y) + (6.1y^3 - 5) \right]$

$\qquad\qquad - (4.2y^4 + 1.1y - 1)$

$= 7.9y^4 - 6.8y^3 + 3.3y + 6.1y^3$

$\qquad\qquad - 5 - 4.2y^4 - 1.1y + 1$

$= 3.7y^4 - 0.7y^3 + 2.2y - 4$

**107.** $-16t^2 + 200t$

a. $t = 1: -16(1)^2 + 200(1)$

$\qquad = -16 + 200 = 184$ feet

b. $t = 5: -16(5)^2 + 200(5)$

$\qquad = -400 + 1000 = 600$ feet

c. $t = 7.6: -16(7.6)^2 + 200(7.6)$

$\qquad = -924.16 + 1520 = 595.84$ feet

d. $t = 10.3: -16(10.3)^2 + 200(10.3)$

$\qquad = -1697.44 + 2060 = 362.56$ feet

**109.** Let $x = 8$

$270.9x^2 - 3240.2x + 10084$

$= 270.9(8)^2 - 3240.2(8) + 10084$

$= 1500$

Expect there to be 1500 boating deaths in 2005.

**111.** $\left( 4.45x^2 - 15.45x + 72.6 \right)$

$\qquad\qquad - \left( 4.4x^2 - 8.8x + 23 \right)$

$= 4.45x^2 - 15.45x + 72.6$

$\qquad\qquad - 4.4x^2 + 8.8x - 23$

$= 0.05x^2 - 6.65x + 49.6$

**113.** Let $x = 13$

$0.08x^3 - 1.19x^2 + 6.45x + 69.93$

$= 0.08(13)^3 - 1.19(13)^2 + 6.45(13)$

$\qquad\qquad\qquad + 69.93$

$= 175.76 - 201.11 + 83.85 + 69.93$

$= 128.43$

Expect the per capita consumption to be 128.43 pounds in 2003.

**Mental Math 5.3**

**1.** $10xy$

**2.** $28ab$

**3.** $x^7$

**4.** $z^5$

**5.** $18x^3$

**6.** $15a^4$

**7.** $-27x^5$

**8.** $32x^8$

**9.** $a^7$

**10.** $a^{10}$

**11.** Cannot simplify

**12.** $a^3$

**Exercise Set 5.3**

**1.** $2a(2a-4) = 2a(2a) - 2a(4) = 4a^2 - 8a$

**3.** $7x(x^2 + 2x - 1)$
$= 7x(x^2) + 7x(2x) + 7x(-1)$
$= 7x^3 + 14x^2 - 7x$

**5.** $3x^2(2x^2 - x) = 3x^2(2x^2) + 3x^2(-x)$
$\qquad\qquad\quad = 6x^4 - 3x^3$

**7.** $x(x+3) = x^2 + 3x$

**9.** $(a+7)(a-2) = a(a)a(-2) + 7(a) + 7(-2)$
$\qquad\qquad\quad = a^2 - 2a + 7a - 14$
$\qquad\qquad\quad = a^2 + 5a - 14$

**11.** $(2y-4)^2 = (2y-4)(2y-4)$
$= 2y(2y) + 2y(-4) - 4(2y) - 4(-4)$
$= 4y^2 - 8y - 8y + 16$
$= 4y^2 - 16y + 16$

**13.** $(5x - 9y)(6x - 5y)$
$= 5x(6x) + 5x(-5y) - 9y(6x) - 9y(-5y)$
$= 30x^2 - 25xy - 54xy + 45y^2$
$= 30x^2 - 79xy + 45y^2$

**15.** $(2x^2 - 5)^2 = (2x^2 - 5)(2x^2 - 5)$
$= 2x^2(2x^2) + 2x^2(-5) - 5(2x^2) - 5(-5)$
$= 4x^4 - 10x^2 - 10x^2 + 25$
$= 4x^4 - 20x^2 + 25$

**17.** $x \cdot x + 3 \cdot x + 2 \cdot x + 2 \cdot 3 = x^2 + 5x + 6$

**19.** $(x-2)(x^2 - 3x + 7)$
$= x(x^2) + x(-3x) + x(7)$
$\qquad -2(x^2) - 2(-3x) - 2(7)$
$= x^3 - 3x^2 + 7x - 2x^2 + 6x - 14$
$= x^3 - 5x^2 + 13x - 14$

**21.** $(x+5)(x^3 - 3x + 4) = x(x^3) + x(-3x)$
$\qquad + x(4) + 5(x^3) + 5(-3x) + 5(4)$
$= x^4 - 3x^2 + 4x + 5x^3 - 15x + 20$
$= x^4 + 5x^3 - 3x^2 - 11x + 20$

**23.** $(2a-3)(5a^2 - 6a + 4) = 2a(5a^2) + 2a(-6a)$
$\qquad + 2a(4) - 3(5a^2) - 3(-6a) - 3(4)$
$= 10a^3 - 12a^2 + 8a - 15a^2 + 18a - 12$
$= 10a^3 - 27a^2 + 26a - 12$

**25.** $(x+2)^3 = (x+2)(x+2)(x+2)$
$\qquad = (x^2 + 2x + 2x + 4)(x+2)$
$\qquad = (x^2 + 4x + 4)(x+2)$
$\qquad = (x^2 + 4x + 4)x + (x^2 + 4x + 4)2$
$\qquad = x^3 + 4x^2 + 4x + 2x^2 + 8x + 8$
$\qquad = x^3 + 6x^2 + 12x + 8$

**27.** $(2y-3)^3 = (2y-3)(2y-3)(2y-3)$

$\quad = (4y^2 - 6y - 6y + 9)(2y - 3)$

$\quad = (4y^2 - 12y + 9)(2y - 3)$

$\quad = (4y^2 - 12y + 9)2y + (4y^2 - 12y + 9)(-3)$

$\quad = 8y^3 - 24y^2 + 18y - 12y^2 + 36y - 27$

$\quad = 8y^3 - 36y^2 + 54y - 27$

**29.**

$$\begin{array}{r} 2x^2 + 4x - 1 \\ x + 3 \\ \hline 6x^2 + 12x - 3 \\ 2x^3 + 4x^2 - x \\ \hline 2x^3 + 10x^2 + 11x - 3 \end{array}$$

**31.**

$$\begin{array}{r} x^3 + 5x - 7 \\ \times \quad x^2 - 9 \\ \hline -9x^3 \quad\quad - 45x + 63 \\ x^5 + 5x^3 - 7x^2 \\ \hline x^5 - 4x^3 - 7x^2 - 45x + 63 \end{array}$$

**33.** a. $(2+3)^2 = 5^2 = 25$

$\quad\quad 2^2 + 3^2 = 4 + 9 = 13$

$\quad$ b. $(8+10)^2 = (18)^2 = 324$

$\quad\quad 8^2 + 10^2 = 64 + 100 = 164$

$\quad$ c. No; Answers may vary.

**35.** $2a(a+4) = 2a(a) + 2a(4) = 2a^2 + 8a$

**37.** $3x(2x^2 - 3x + 4)$

$\quad = 3x(2x^2) + 3x(-3x) + 3x(4)$

$\quad = 6x^3 - 9x^2 + 12x$

**39.** $(5x+9y)(3x+2y)$

$\quad = 5x(3x) + 5x(2y) + 9y(3x) + 9y(2y)$

$\quad = 15x^2 + 10xy + 27xy + 18y^2$

$\quad = 15x^2 + 37xy + 18y^2$

**41.** $(x+2)(x^2 + 5x + 6)$

$\quad = x(x^2) + x(5x) + x(6) + 2(x^2) + 2(5x) + 2(6)$

$\quad = x^3 + 5x^2 + 6x + 2x^2 + 10x + 12$

$\quad = x^3 + 7x^2 + 16x + 12$

**43.** $(7x+4)^2 = (7x+4)(7x+4)$

$\quad\quad = 7x(7x) + 7x(4) + 4(7x) + 4(4)$

$\quad\quad = 49x^2 + 28x + 28x + 16$

$\quad\quad = 49x^2 + 56x + 16$

**45.** $-2a^2(3a^2 - 2a + 3)$

$\quad = -2a^2(3a^2) - 2a^2(-2a) - 2a^2(3)$

$\quad = -6a^4 + 4a^3 - 6a^2$

**47.** $(x+3)(x^2 + 7x + 12)$

$\quad = x(x^2) + x(7x) + x(12) + 3(x^2) + 3(7x) + 3(12)$

$\quad = x^3 + 7x^2 + 12x + 3x^2 + 21x + 36$

$\quad = x^3 + 10^2 + 33x + 36$

**49.** $(a+1)^3 = (a+1)(a+1)(a+1)$

$\quad\quad = (a^2 + a + a + 1)(a+1)$

$\quad\quad = (a^2 + 2a + 1)(a+1)$

$\quad\quad = (a^2 + 2a + 1)a + (a^2 + 2a + 1)1$

$\quad\quad = a^3 + 2a^2 + a + a^2 + 2a + 1$

$\quad\quad = a^3 + 3a^2 + 3a + 1$

**51.** $(x+y)(x+y) = x(x)+x(y)+y(x)+y(y)$
$$= x^2 + xy + xy + y^2$$
$$= x^2 + 2xy + y^2$$

**53.** $(x-7)(x-6) = x(x)+x(-6)-7(x)-7(-6)$
$$= x^2 - 6x - 7x + 42$$
$$= x^2 - 13x + 42$$

**55.** $3a(a^2+2) = 3a(a^2)+3a(2) = 3a^3 + 6a$

**57.** $-4y(y^2+3y-11)$
$$= -4y(y^2)-4y(3y)-4y(-11)$$
$$= -4y^3 - 12y^2 + 44y$$

**59.** $(5x+1)(5x-1)$
$$= 5x(5x)+5x(-1)+1(5x)+1(-1)$$
$$= 25x^2 - 5x + 5x - 1$$
$$= 25x^2 - 1$$

**61.** $(5x+4)(x^2-x+4)$
$$= 5x(x^2)+5x(-x)+5x(4)+4(x^2)+4(-x)+4(4)$$
$$= 5x^3 - 5x^2 + 20x + 4x^2 - 4x + 16$$
$$= 5x^3 - x^2 + 16x + 16$$

**63.** $(2x-5)^3 = (2x-5)(2x-5)(2x-5)$
$$= (4x^2-10x+25)(2x-5)$$
$$= (4x^2-20x+25)(2x-5)$$
$$= (4x^2-20x+25)2x+(4x^2-20x+25)(-5)$$
$$= 8x^3 - 40x^2 + 50x - 20x^2 + 100x - 125$$
$$= 8x^3 - 60x^2 + 150x - 125$$

**65.** $(4x+5)(8x^2+2x-4)$
$$= 4x(8x^2)+4x(2x)+4x(-4)+5(8x^2)+5(2x)+5(-4)$$
$$= 32x^3 + 8x^2 - 16x + 40x^2 + 10x - 20$$
$$= 32x^3 + 48x^2 - 6x - 20$$

**67.** $(7xy-y)^2 = (7xy-y)(7xy-y)$
$$= 7xy(7xy)+7xy(-y)-y(7xy)-y(-y)$$
$$= 49x^2y^2 - 7xy^2 - 7xy^2 + y^2$$
$$= 49x^2y^2 - 14xy^2 + y^2$$

**69.**
$$\begin{array}{r} 5y^2 - y + 3 \\ \times \quad y^2 - 3y - 2 \\ \hline -10y^2 + 2y - 6 \\ -15y^3 + 3y^2 - 9y \\ 5y^4 - \quad y^3 + 3y^2 \\ \hline 5y^4 - 16y^3 - 4y^2 - 7y - 6 \end{array}$$

**71.**
$$\begin{array}{r} 3x^2 + 2x - 4 \\ \times \quad 2x^2 - 4x + 3 \\ \hline 9x^2 + 6x - 12 \\ -12x^3 - 8x^2 + 16x \\ 6x^4 + 4x^3 - 8x^2 \\ \hline 6x^4 - 8x^3 - 7x^2 + 22 - 12 \end{array}$$

**73.** $(5x)^2 = 5^2 x^2 = 25x^2$

**75.** $(-3y^3)^2 = (-3)^2 y^{3\cdot 2} = 9y^6$

**77.** At $t = 0$, value = $7000

**79.** At $t = 0$, value $= \$7000$
$\qquad$ At $t = 1$, value $= \$6500$
$\qquad$ $\$7000 - \$6500 = \$500$

**81.** Answers may vary.

**83.** $(2x-5)(2x+5)$

$= 2x(2x) + 2x(5) - 5(2x) - 5(5)$

$= 4x^2 + 10x - 10x - 25 = 4x^2 - 25$

$(4x^2 - 25)$ square yards

**85.** $\dfrac{1}{2}(3x-2)(4x) = 2x(3x-2)$

$= 2x(3x) + 2x(-2) = 6x^2 - 4x$

$(6x^2 - 4x)$ square inches

**87.** $(x+3)(x+3) - 2 \cdot 2$

$= x^2 + 3x + 3x + 9 - 4$

$= (x^2 + 6x + 5)$ square units

**89.** a. $(a+b)(a-b) = a^2 - ab + ab - b^2$

$= a^2 - b^2$

b. $(2x+3y)(2x-3y)$

$= (2x)^2 - 6xy + 6xy - (3y)^2$

$= 4x^2 - 9y^2$

c. $(4x+7)(4x-7)$

$= (4x)^2 - 28x + 28x - 7^2$

$= 16x^2 - 49$

d. Answers may vary.

**Mental Math 5.4**

**1.** False

**2.** True

**3.** False

**4.** False

**Exercise Set 5.4**

**1.** $(x+3)(x+4) = x^2 + 4x + 3x + 12$

$= x^2 + 7x + 12$

**3.** $(x-5)(x+10) = x^2 + 10x - 5x - 50$

$= x^2 + 5x - 50$

**5.** $(5x-6)(x+2) = 5x^2 + 10x - 6x - 12$

$= 5x^2 + 4x - 12$

**7.** $(y-6)(4y-1) = 4y^2 - 1y - 24y + 6$

$= 4y^2 - 25y + 6$

**9.** $(2x+5)(3x-1) = 6x^2 - 2x + 15x - 5$

$= 6x^2 + 13x - 5$

**11.** $(x-2)^2 = x^2 - 2(x)(2) + 2^2$

$= x^2 - 4x + 4$

**13.** $(2x-1)^2 = (2x)^2 - 2(2x)(1) + (1)^2$

$= 4x^2 - 4x + 1$

**15.** $(3a-5)^2 = (3a)^2 - 2(3a)(5) + 5^2$

$= 9a^2 - 30a + 25$

**17.** $(5x+9)^2 = (5x)^2 + 2(5x)(9) + 9^2$

$= 25x^2 + 90x + 81$

**19.** Answers may vary.

**21.** $(a-7)(a+7) = a^2 - 7^2 = a^2 - 49$

**23.** $(3x-1)(3x-1) = (3x)^2 - 1^2 = 9x^2 - 1$

**25.** $\left(3x - \dfrac{1}{2}\right)\left(3x + \dfrac{1}{2}\right) = (3x)^2 - \left(\dfrac{1}{2}\right)^2$

$$= 9x^2 - \dfrac{1}{4}$$

**27.** $(9x + y)(9x - y) = (9x)^2 - y^2 = 81x^2 - y^2$

**29.** $(2x + 0.1)(2x - 0.1) = (2x)^2 - (0.1)^2$

$$= 4x^2 - 0.01$$

**31.** $(a + 5)(a + 4) = a^2 + 4a + 5a + 20$

$$= a^2 + 9a + 20$$

**33.** $(a + 7)^2 = a^2 + 2(a)(7) + 7^2$

$$= a^2 + 14a + 49$$

**35.** $(4a + 1)(3a - 1) = 12a^2 - 4a + 3a - 1$

$$= 12a^2 - a - 1$$

**37.** $(x + 2)(x - 2) = x^2 - 2^2 = x^2 - 4$

**39.** $(3a + 1)^2 = (3a)^2 + 2(3a)(1) + 1^2$

$$= 9a^2 + 6a + 1$$

**41.** $\left(x^2 + y\right)\left(4x - y^4\right)$

$$= x^2(4x) - x^2 y^4 + y(4x) - y \cdot y^4$$

$$= 4x^3 - x^2 y^4 + 4xy - y^5$$

**43.** $(x + 3)\left(x^2 - 6x + 1\right)$

$$= x\left(x^2\right) + x(-6x) + x(1) + 3\left(x^2\right) + 3(-6x) + 3(1)$$

$$= x^3 - 6x^2 + x + 3x^2 - 18x + 3$$

$$= x^3 - 3x^2 - 17x + 3$$

**45.** $(2a - 3)^2 = (2a)^2 - 2(2a)(3) + (3)^2$

$$= 4a^2 - 12a + 9$$

**47.** $(5x - 6z)(5x + 6z) = (5x)^2 - (6z)^2$

$$= 25x^2 - 36z^2$$

**49.** $\left(x^5 - 3\right)\left(x^5 - 5\right) = x^{10} - 5x^5 - 3x^5 + 15$

$$= x^{10} - 8x^5 + 15$$

**51.** $\left(x - \dfrac{1}{3}\right)\left(x + \dfrac{1}{3}\right) = (x)^2 - \left(\dfrac{1}{3}\right)^2 = x^2 - \dfrac{1}{9}$

**53.** $\left(a^3 + 11\right)\left(a^4 - 3\right) = a^7 - 3a^3 + 11a^4 - 33$

**55.** $3(x - 2)^2 = \left[(x)^2 - 2(x)(2) + (2)^2\right]$

$$= 3\left(x^2 - 4x + 4\right)$$

$$= 3x^2 - 12x + 12$$

**57.** $(3b + 7)(2b - 5) = 6b^2 - 15b + 14b - 35$

$$= 6b^2 - b - 35$$

**59.** $(7p - 8)(7p + 8) = (7p)^2 - (8)^2$

$$= 49p^2 - 64$$

**61.** $\left(\dfrac{1}{3}a^2 - 7\right)\left(\dfrac{1}{3}a^2 + 7\right) = \left(\dfrac{1}{3}a^2\right) - (7)^2$

$$= \dfrac{1}{9}a^4 - 49$$

**63.** $5x^3\left(3x^2 - x + 2\right)$

$$= 5x^2\left(3x^2\right) + 5x^2(-x) + 5x^2(2)$$

$$= 15x^4 - 5x^3 + 10x^2$$

**65.** $(2r-3s)(2r+3s) = (2r)^2 - (3s)^2$
$$= 4r^2 - 9s^2$$

**67.** $(3x-7y)^2 = (3x)^2 - 2(3x)(7y) + (7y)^2$
$$= 9x^2 - 42xy + 49y^2$$

**69.** $(4x+5)(4x-5) = (4x)^2 - 5^2$
$$= 16x^2 - 25$$

**71.** $(8x+4)^2 = (8x)^2 + 2(8x)(4) + (4)^2$
$$= 64x^2 + 64x + 16$$

**73.** $\left(a - \dfrac{1}{2}y\right)\left(a + \dfrac{1}{2}y\right) = a^2 - \left(\dfrac{1}{2}y\right)^2$
$$= a^2 - \dfrac{1}{4}y^2$$

**75.** $\left(\dfrac{1}{5}x - y\right)\left(\dfrac{1}{5}x + y\right) = \left(\dfrac{1}{5}x\right)^2 - (y)^2$
$$= \dfrac{1}{25}x^2 - y^2$$

**77.** $(a+1)(3a^2 - a + 1)$
$$= a(3a^2) + a(-a) + a(1) + 1(3a^2) + 1(-a) + 1(1)$$
$$= 3a^3 - a^2 + a + 3a^2 - a + 1$$
$$= 3a^3 + 2a^2 + 1$$

**79.** $\dfrac{50b^{10}}{70b^5} = \dfrac{5 \cdot 10 b^{10-5}}{7 \cdot 10} = \dfrac{5b^5}{7}$

**81.** $\dfrac{8a^{17}b^{15}}{-4a^7b^{10}} = \dfrac{4 \cdot 2a^{17-7}b^{15-10}}{-4}$
$$= -\dfrac{2a^{10}b^{+5}}{1}$$
$$= -2a^{10}b^5$$

**83.** $\dfrac{2x^4 y^{12}}{3x^4 y^4} = \dfrac{2x^{4-4} y^{12-4}}{3} = \dfrac{2x^0 y^8}{3} = \dfrac{2y^8}{3}$

**85.** $(-1,1)$ and $(2,2)$
$$m = \dfrac{y_2 - y_1}{x_2 - x_1} = \dfrac{2-1`}{2-(-1)} = \dfrac{1}{3}$$

**87.** $(-1,-2)$ and $(1,0)$
$$m = \dfrac{y_2 - y_1}{x_2 - x_1} = \dfrac{0-(-2)}{1-(-1)} = \dfrac{2}{2} = 1$$

**89.** $(2x+1)^2 = (2x)^2 + 2(2x)(1) + 1^2$
$$= 4x^2 + 4x + 1$$
$$(4x^2 + 4x + 1) \text{ square feet}$$

**91.** $\dfrac{1}{2}(5a+b)(5a-b)$
$$= \dfrac{1}{2}\left(25a^2 - b^2\right)$$
$$= \left(\dfrac{25a^2}{2} - \dfrac{b^2}{2}\right) \text{square units}$$

**93.** $(5x-3)^2 - (x+1)^2$
$$= \left[(5x)^2 - 2(5x)(3) + 3^2\right]$$
$$- \left[x^2 + 2(x)(1) + 1^2\right]$$
$$= (25x^2 - 30x + 9) - (x^2 + 2x + 1)$$
$$= 25x^2 - 30x + 9 - x^2 - 2x - 1$$
$$= (224x^2 - 32x + 8) \text{ square meters}$$

**95.** $x \cdot x + x(5) + x(5) + 5 \cdot 5$
$$= x^2 + 5x + 5x + 25$$
$$= (x^2 + 10x + 25) \text{ square units}$$

**97.** $(x+2)(x+2) = x^2 + 2(x)(2) + (2)^2$
$$= x^2 + 4x + 4$$

$(x+2)(x-2) = (x)^2 - (2)^2 = x^2 - 4$

Answers may vary.

**99.** $\left[(a+c) - 5\right]\left[(a+c) + 5\right]$
$$= (a+c)^2 - 5^2$$
$$= a^2 + 2ac + c^2 - 25$$

**101.** $\left[(x-2) + y\right]\left[(x-2) - y\right]$
$$= (x-2)^2 - y^2$$
$$= x^2 - 4x + 4 - y^2$$

**Integrated Review-Exponents and Operations on Polynomials**

**1.** $(5x^2)(7x^3) = (5 \cdot 7)(x^2 \cdot x^3)$
$$= 35x^5$$

**2.** $(4y^2)(8y^7) = (4 \cdot 8)(y^2 \cdot y^7)$
$$= 32y^9$$

**3.** $-4^2 = -(4 \cdot 4) = -16$

**4.** $(-4)^2 = (-4)(-4) = 16$

**5.** $(x-5)(2x+1) = 2x^2 + x - 10x - 5$
$$= 2x^2 - 9x - 5$$

**6.** $(3x-2)(x+5) = 3x^2 + 15x - 2x - 10$
$$= 3x^2 + 13x - 10$$

**7.** $(x-5) + (2x+1) = x - 5 + 2x + 1$
$$= 3x - 4$$

**8.** $(3x-2) + (x+5) = 3x - 2 + x + 5$
$$= 4x + 3$$

**9.** $\dfrac{7x^9 y^{12}}{x^3 y^{10}} = 7x^{9-3} y^{12-10}$
$$= 7x^6 y^2$$

**10.** $\dfrac{20a^2 b^8}{14a^2 b^2} = \dfrac{10a^{2-2} b^{8-2}}{7}$
$$= \dfrac{10b^6}{7}$$

**11.** $(12m^7 n^6)^2 = 12^2 m^{7 \cdot 2} n^{6 \cdot 2}$
$$= 144 m^{14} n^{12}$$

**12.** $(4y^9 z^{10})^3 = 4^3 y^{9 \cdot 3} z^{10 \cdot 3}$
$$= 64 y^{27} z^{30}$$

**13.** $3(4y-3)(4y+3) = 3\left[(4y)^2 - 3^2\right]$
$$= 3(16y^2 - 9)$$
$$= 48y^2 - 27$$

**14.** $2(7x-1)(7x+1) = 2\left[(7x)^2 - 1^2\right]$
$$= 2(49x^2 - 1)$$
$$= 98x^2 - 2$$

**15.** $(x^7 y^5)^9 = x^{63} y^{45}$

**16.** $(3^1 x^9)^3 = 3^3 x^{27}$
$$= 27 x^{27}$$

**17.** $\left(7x^2 - 2x + 3\right) - \left(5x^2 + 9\right)$

$\quad = 7x^2 - 2x + 3 - 5x^2 - 9$

$\quad = 2x^2 - 2x - 6$

**18.** $\left(10x^2 + 7x - 9\right) - \left(4x^2 - 6x + 2\right)$

$\quad = 10x^2 + 7x - 9 - 4x^2 + 6x - 2$

$\quad = 6x^2 + 13x - 11$

**19.** $0.7y^2 - 1.2 + 1.8y^2 - 6y + 1$

$\quad = 2.5y^2 - 6y - 0.2$

**20.** $7.8x^2 - 6.8x + 3.3 + 0.6x^2 - 9$

$\quad = 8.4x^2 - 6.8x - 5.7$

**21.** $\left(x + 4y\right)^2 = \left(x + 4y\right)\left(x + 4y\right)$

$\quad\quad\quad\quad = x^2 + 2(x)(4y) + (4y)^2$

$\quad\quad\quad\quad = x^2 + 8xy + 16y^2$

**22.** $\left(y - 9z\right)^2 = \left(y - 9z\right)\left(y - 9z\right)$

$\quad\quad\quad\quad = y^2 - 2(y)(9z) + (9z)^2$

$\quad\quad\quad\quad = y^2 - 18yz + 81z^2$

**23.** $\left(x + 4y\right) + \left(x + 4y\right) = x + 4y + x + 4y$

$\quad\quad\quad\quad\quad\quad\quad\quad = 2x + 8y$

**24.** $\left(y - 9z\right) + \left(y - 9z\right) = y - 9z + y - 9z$

$\quad\quad\quad\quad\quad\quad\quad\quad = 2y - 18z$

**25.** $7x^2 - 6xy + 4\left(y^2 - xy\right)$

$\quad = 7x^2 - 6xy + 4y^2 - 4xy$

$\quad = 7x^2 - 10xy + 4y^2$

**26.** $5a^2 - 3ab + 6\left(b^2 - a^2\right)$

$\quad = 5a^2 - 3ab + 6b^2 - 6a^2$

$\quad = -a^2 - 3ab + 6b^2$

**27.** $\left(x - 3\right)\left(x^2 + 5x - 1\right)$

$\quad = x\left(x^2\right) + x(5x) + x(-1) - 3\left(x^2\right)$

$\quad\quad - 3(5x) - 3(-1)$

$\quad = x^3 + 5x^2 - x - 3x^2 - 15x + 3$

$\quad = x^3 + 2x^2 - 16x + 3$

**28.** $\left(x + 1\right)\left(x^2 - 3x - 2\right)$

$\quad = x\left(x^2\right) + x(-3x) + x(-2) + 1\left(x^2\right)$

$\quad\quad + 1(-3x) + 1(-2)$

$\quad = x^3 - 3x^2 - 2x + x^2 - 3x - 2$

$\quad = x^3 - 2x^2 - 5x - 2$

**29.**

$\quad \left(2x^3 - 7\right)\left(3x^2 + 10\right)$

$\quad = 2x^3\left(3x^2\right) + 2x^3(10) - 7\left(3x^2\right) - 7(10)$

$\quad = 6x^5 + 20x^3 - 21x^2 - 70$

**30.** $\left(5x^3 - 1\right)\left(4x^4 + 5\right)$

$\quad = 5x^3\left(4x^4\right) + 5x^3(5) - 1\left(4x^4\right) - 1(5)$

$\quad = 20x^7 + 25x^3 - 4x^4 - 5$

**31.** $\left(2x - 7\right)\left(x^2 - 6x + 1\right)$

$\quad = 2x\left(x^2\right) - 2x(6x) + 2x(1) - 7\left(x^2\right)$

$\quad\quad - 7(-6x) - 7(1)$

$\quad = 2x^3 - 12x^2 + 2x - 7x^2 + 42x - 7$

$\quad = 2x^3 - 19x^2 + 44x - 7$

**32.** $(5x-1)(x^2+2x-3)$

$= 5x(x^2)+5x(2x)+5x(-3)-1(x^2)$
$\qquad -1(2x)-1(-3)$

$= 5x^3+10x^2-15x-x^2-2x+3$

$= 5x^3+9x^2-17x+3$

**33.** Cannot simplify

**34.** $(5x^3)(5y^3)=25x^3y^3$

**35.** $(5x^3)^3 = 5^3x^{3\cdot3} = 125x^9$

**36.** $\dfrac{5x^3}{5y^3}=\dfrac{x^3}{y^3}$

**37.** $x+x=2x$

**38.** $x\cdot x = x^2$

**Calculator Explorations 5.5**

**1.** $5.31\times10^3 = 5.31\text{ EE }3$

**3.** $6.6\times10^{-9} = 6.6\text{ EE }-9$

**5.** $3,000,000\times5,000,000 = 1.5\times10^{13}$

**7.** $(3.26\times10^6)(2.5\times10^{13}) = 8.15\times10^{19}$

**Mental Math 5.5**

**1.** $5x^{-2}=\dfrac{5}{x^2}$

**2.** $3x^{-3}=\dfrac{3}{x^3}$

**3.** $\dfrac{1}{y^{-6}}=y^6$

**4.** $\dfrac{1}{x^{-3}}=x^3$

**5.** $\dfrac{4}{y^{-3}}=4y^3$

**6.** $\dfrac{16}{y^{-7}}=16y^7$

**Exercise Set 5.5**

**1.** $4^{-3}=\dfrac{1}{4^3}=\dfrac{1}{64}$

**3.** $(-2)^{-4}=\dfrac{1}{(-2)^4}=\dfrac{1}{16}$

**5.** $7x^{-3}=7\cdot\dfrac{1}{x^3}=\dfrac{7}{x^3}$

**7.** $\left(\dfrac{1}{2}\right)^{-5}=\dfrac{1^{-5}}{2^{-5}}=\dfrac{2^5}{1^5}=32$

**9.** $\left(-\dfrac{1}{4}\right)^{-3}=\dfrac{(-1)^{-3}}{(4)^{-3}}=\dfrac{4^3}{(-1)^3}=\dfrac{64}{-1}=-64$

**11.** $3^{-1}+2^{-1}=\dfrac{1}{3}+\dfrac{1}{2}=\dfrac{2}{6}+\dfrac{3}{6}=\dfrac{5}{6}$

**13.** $\dfrac{1}{p^{-3}}=p^3$

**15.** $\dfrac{p^{-5}}{q^{-4}}=\dfrac{q^4}{p^5}$

**17.** $\dfrac{x^{-2}}{x} = x^{-2-1} = x^{-3} = \dfrac{1}{x^3}$

**19.** $2^0 + 3^{-1} = 1 + \dfrac{1}{3} = \dfrac{3}{3} + \dfrac{1}{3} = \dfrac{4}{3}$

**21.** $\dfrac{-1}{p^{-4}} = 1\left(p^4\right) = -p^4$

**23.** $-2^0 - 3^0 = -1(1) - 1 = -2$

**25.** $\dfrac{x^2 x^5}{x^3} = x^{2+5-3} = x^4$

**27.** $\dfrac{p^2 p}{p^{-1}} = p^{2+1-(-1)} = p^{2+1+1} = p^4$

**29.** $\dfrac{\left(m^5\right)^4 m}{m^{10}} = m^{5(4)+1-10} = m^{20+1-10} = m^{11}$

**31.** $\dfrac{r}{r^{-3} r^{-2}} = r^{1-(-3)-(-2)} = r^{1+3+2} = r^6$

**33.** $\left(x^5 y^3\right)^{-3} = x^{5(-3)} y^{3(-3)} = x^{-15} y^{-9} = \dfrac{1}{x^{15} y^9}$

**35.** $\dfrac{\left(x^2\right)^3}{x^{10}} = \dfrac{x^6}{x^{10}} = x^{6-10} = x^{-4} = \dfrac{1}{x^4}$

**37.** $\dfrac{\left(a^5\right)^2}{\left(a^3\right)^4} = \dfrac{a^{10}}{a^{12}} = a^{10-12} = a^{-2} = \dfrac{1}{a^2}$

**39.** $\dfrac{8k^4}{2k} = \dfrac{8}{2} \cdot k^{4-1} = 4k^3$

**41.** $\dfrac{-6m^4}{-2m^3} = \dfrac{-6}{-2} \cdot m^{4-3} = 3m$

**43.** $\dfrac{-24a^6 b}{6ab^2} = \dfrac{-24}{6} \cdot a^{6-1} b^{1-2} = -4a^5 b^{-1}$

$\quad = -\dfrac{4a^5}{b}$

**45.** $\left(-2x^3 y^{-4}\right)\left(3x^{-1} y\right) = -2(3) x^3 x^{-1} y^{-4} y$

$\quad = -6x^2 y^{-3}$

$\quad = -\dfrac{6x^2}{y^3}$

**47.** $\left(a^{-5} b^2\right)^{-6} = a^{-5(-6)} b^{2(-6)} = a^{30} b^{-12} = \dfrac{a^{30}}{b^{12}}$

**49.** $\left(\dfrac{x^{-2} y^4}{x^3 y^7}\right)^2 = \dfrac{x^{-2(2)} y^{4(2)}}{x^{3(2)} y^{7(2)}} = \dfrac{x^{-4} y^8}{x^6 y^{14}}$

$\quad = x^{-4-6} y^{8-14} = x^{-10} y^{-6} = \dfrac{1}{x^{10} y^6}$

**51.** $\dfrac{4^2 z^{-3}}{4^3 z^{-5}} = 4^{2-3} z^{-3-(-5)} = 4^{-1} z^2 = \dfrac{z^2}{4}$

**53.** $\dfrac{2^{-3} x^{-4}}{2^2 x} = 2^{-3-2} x^{-4-1} = 2^{-5} x^{-5} = \dfrac{1}{2^5 x^5} = \dfrac{1}{32x^5}$

**55.** $\dfrac{7ab^{-4}}{7^{-1} a^{-3} b^2} = 7^{1-(-1)} a^{1-(-3)} b^{-4-2}$

$\quad = 7^2 a^4 b^{-6}$

$\quad = \dfrac{49a^4}{b^6}$

**57.** $\left(\dfrac{a^{-5} b}{ab^3}\right)^{-4} = \dfrac{a^{-5(-4)} b^{-4}}{a^{-4} b^{3(-4)}} = \dfrac{a^{20} b^{-4}}{a^{-4} b^{-12}}$

$\quad = a^{20-(-4)} b^{-4-(-12)}$

$\quad = a^{24} b^8$

**59.** $\dfrac{\left(xy^3\right)^5}{\left(xy\right)^{-4}} = \dfrac{x^5 y^{3(5)}}{x^{-4}y^{-4}} = \dfrac{x^5 y^{15}}{x^{-4}y^{-4}}$

$\qquad = x^{5-(-4)}y^{15-(-4)}$

$\qquad = x^9 y^{19}$

**61.** $\dfrac{\left(-2xy^{-3}\right)^{-3}}{\left(xy^{-1}\right)^{-1}} = \dfrac{\left(-2\right)^{-3}x^{-3}y^9}{x^{-1}y^1}$

$\qquad = \left(-2\right)^{-3}x^{-3-(-1)}y^{9-1}$

$\qquad = -\dfrac{y^8}{8x^2}$

**63.** $\dfrac{6x^2 y^3}{-7xy^5} = -\dfrac{6}{7}x^{2-1}y^{3-5} = -\dfrac{6}{7}x^1 y^{-2}$

$\qquad = -\dfrac{6x}{7y^2}$

**65.** $78,000 = 7.8 \times 10^4$

**67.** $0.00000167 = 1.67 \times 10^{-6}$

**69.** $0.00635 = 6.35 \times 10^{-3}$

**71.** $1,160,000 = 1.16 \times 10^6$

**73.** $20,000,000 = 2.0 \times 10^7$

**75.** $15,600,000 = 1.56 \times 10^7$

**77.** $13,600 = 1.36 \times 10^4$

**79.** $292,000,000 = 2.92 \times 10^8$

**81.** $8.673 \times 10^{-10} = 0.0000000008673$

**83.** $3.3 \times 10^{-2} = 0.033$

**85.** $2.032 \times 10^4 = 20,320$

**87.** $6.25 \times 10^{18} = 6,250,000,000,000,000,000$

**89.** $9.460 \times 10^{12} = 9,460,000,000,000$

**91.** $\left(1.2 \times 10^{-3}\right)\left(3 \times 10^{-2}\right) = 1.2 \cdot 3 \cdot 10^{-3} \cdot 10^{-2}$

$\qquad = 3.6 \times 10^{-5}$

$\qquad = 0.000036$

**93.** $\left(4 \times 10^{-10}\right)\left(7 \times 10^{-9}\right) = 4 \cdot 7 \cdot 10^{-10} \cdot 10^{-9}$

$\qquad = 28 \times 10^{-19}$

$\qquad = 0.0000000000000000028$

**95.** $\dfrac{8 \times 10^{-1}}{16 \times 10^5} = \dfrac{8}{16} \times 10^{-1-5}$

$\qquad = 0.5 \times 10^{-6}$

$\qquad = 0.0000005$

**97.** $\dfrac{1.4 \times 10^{-2}}{7 \times 10^{-8}} = \dfrac{1.4}{7} \times 10^{-2-(-8)}$

$\qquad = 0.2 \times 10^6$

$\qquad = 200,000$

**99.** $\dfrac{5x^7}{3x^4} = \dfrac{5x^{7-4}}{3} = \dfrac{5x^3}{3}$

**101.** $\dfrac{15z^4 y^3}{21zy} = \dfrac{3 \cdot 5}{3 \cdot 7}z^{4-1}y^{3-1} = \dfrac{5z^3 y^2}{7}$

**103.** $\dfrac{1}{y}\left(5y^2 - 6y + 5\right)$

$\qquad = \dfrac{1}{y}\left(5y^2\right) + \dfrac{1}{y}\left(-6y\right) + \dfrac{1}{y}\left(5\right)$

$\qquad = 5y - 6 + \dfrac{5}{y}$

**105.** $\left(\dfrac{3x^{-2}}{z}\right)^3 = \dfrac{3^3 x^{-6}}{z^3} = \dfrac{27}{x^6 z^3}$

The volume is $\dfrac{27}{x^6 z^3}$ cubic inches.

**107.** $\left(2.63 \times 10^{12}\right)\left(-1.5 \times 10^{-10}\right)$

$= 2.63 \cdot (-1.5) \cdot 10^{12} \cdot 10^{-10}$

$= -3.945 \times 10^2 = -394.5$

**109.** $d = r \cdot t$

$238{,}857 = \left(1.86 \times 10^5\right)t$

$t = \dfrac{238{,}857}{1.86 \times 10^5}$

$t = \dfrac{2.38857}{1.86} \times 10^{5-5}$

$t = 1.3$ seconds

**111.** Answers may vary.

**113.** $7.5 \times 10^5 \dfrac{\text{gallons}}{\text{second}} \left(\dfrac{3600 \text{ seconds}}{1 \text{ hour}}\right)$

$= 27{,}000 \times 10^5 = 2.7 \times 10^4 \times 10^5$

$= 2.7 \times 10^9$

$2.7 \times 10^9$ gallons flows over Niagra Falls in one hour.

**115.** Answers may vary.

**117.** $a^{-4m} \cdot a^{5m} = a^{-4m+5m} = a^m$

**119.** $\dfrac{y^{4a}}{y^{-a}} = y^{4a-(-a)} = y^{5a}$

**121.** $\left(z^{3a+2}\right)^{-2} = z^{-2(3a+2)} = \dfrac{1}{z^{6a+4}}$

**Mental Math 5.6**

**1.** $\dfrac{a^6}{a^4} = a^2$

**2.** $\dfrac{y^2}{y} = y$

**3.** $\dfrac{a^3}{a} = a^2$

**4.** $\dfrac{p^8}{p^3} = p^5$

**5.** $\dfrac{k^5}{k^2} = k^3$

**6.** $\dfrac{k^7}{k^5} = k^2$

**7.** $\dfrac{p^8}{p^3} = p^5$

**8.** $\dfrac{k^5}{k^2} = k^3$

**9.** $\dfrac{k^7}{k^5} = k^2$

**Exercise Set 5.6**

**1.** $\dfrac{15p^3 + 18p^2}{3p} = \dfrac{15p^3}{3p} + \dfrac{18p^2}{3p} = 5p^2 + 6p$

**3.** $\dfrac{-9x^4 + 18x^5}{6x^5} = \dfrac{-9x^4}{6x^5} + \dfrac{18x^5}{6x^5} = -\dfrac{3}{2x} + 3$

**5.** $\dfrac{-9x^5+3x^4-12}{3x^3}=\dfrac{-9x^5}{3x^3}+\dfrac{3x^4}{3x^3}-\dfrac{12}{3x^3}$

$\qquad\qquad\qquad = -3x^2+x-\dfrac{4}{x^3}$

**7.** $\dfrac{4x^4-6x^3+7}{-4x^4}=\dfrac{4x^4}{-4x^4}-\dfrac{6x^3}{-4x^4}+\dfrac{7}{-4x^4}$

$\qquad\qquad\qquad = -1+\dfrac{3}{2x}-\dfrac{7}{4x^4}$

**9.** $\dfrac{25x^5-15x^3+5}{5x^2}=\dfrac{25x^5}{5x^2}-\dfrac{15x^3}{5x^2}+\dfrac{5}{5x^2}$

$\qquad\qquad\qquad = 5x^3-3x+\dfrac{1}{x^2}$

**11.** $\dfrac{12x^3+4x-16}{4}=\dfrac{12x^3}{4}+\dfrac{4x}{x}-\dfrac{16}{4}$

$= 3x^3+x-4$

Each side is $\left(3x^3+x-4\right)$ feet.

**13.** $x+3\overline{)x^2+4x+3}$ with quotient $x+1$

$\qquad\quad \underline{x^2+3x}$

$\qquad\qquad\quad x+3$

$\qquad\qquad\quad \underline{x+3}$

$\qquad\qquad\qquad 0$

$\dfrac{x^2+4x+3}{x+3}=x+1$

**15.** $x+5\overline{)2x^2+13x+15}$ with quotient $2x+3$

$\qquad\quad \underline{2x^2+10x}$

$\qquad\qquad\quad 3x+15$

$\qquad\qquad\quad \underline{3x+15}$

$\qquad\qquad\qquad 0$

$\dfrac{2x^2+13x+15}{x+5}=2x+3$

**17.** $x-4\overline{)2x^2-7x+3}$ with quotient $2x+1$

$\qquad\quad \underline{2x^2-8x}$

$\qquad\qquad\quad x+3$

$\qquad\qquad\quad \underline{x-4}$

$\qquad\qquad\qquad 7$

$\dfrac{2x^2-7x+3}{x-4}=2x+1+\dfrac{7}{x-4}$

**19.** $2x-3\overline{)8x^2+6x-27}$ with quotient $4x+9$

$\qquad\quad \underline{8x^2-12x}$

$\qquad\qquad\quad 18x-27$

$\qquad\qquad\quad \underline{18x-27}$

$\qquad\qquad\qquad 0$

$\dfrac{8x^2+6x-27}{2x-3}=4x+9$

**21.** $3a+2\overline{)9a^3-3a^2-3a+4}$ with quotient $3a^2-3a+1$

$\qquad\quad \underline{9a^3+6a^2}$

$\qquad\qquad\quad -9a^2-3a$

$\qquad\qquad\quad \underline{-9a^2-6a}$

$\qquad\qquad\qquad 3a+4$

$\qquad\qquad\qquad \underline{3a+2}$

$\qquad\qquad\qquad\quad 2$

$\dfrac{9a^3-3a^2-3a+4}{3a+2}=3a^2-3a+1+\dfrac{2}{3a+2}$

**23.**

$$\begin{array}{r} 2b^2 + b + 2 \\ b + 4 \overline{) 2b^3 + 9b^2 + 6b - 4} \\ \underline{2b^3 + 8b^2} \\ b^2 + 6b \\ \underline{b^2 + 4b} \\ 2b - 4 \\ \underline{2b + 8} \\ -12 \end{array}$$

$$\frac{2b^3 + 9b^2 + 6b - 4}{b + 4} = 2b^2 + b + 2 - \frac{12}{b + 4}$$

**25.** Answers may vary.

**27.**

$$\begin{array}{r} 2x + 5 \\ 5x + 3 \overline{) 10x^2 + 31x + 15} \\ \underline{10x^2 + 6x} \\ 25 + 15 \\ \underline{25 + 15} \\ 0 \end{array}$$

The height is $(2x + 5)$ meters.

**29.** $\dfrac{20x^2 + 5x + 9}{5x^3} = \dfrac{20x^2}{5x^3} + \dfrac{5x}{5x^3} + \dfrac{9}{5x^3}$

$$= \frac{4}{x} + \frac{1}{x^2} + \frac{9}{5x^3}$$

**31.**

$$\begin{array}{r} 5x - 2 \\ x + 6 \overline{) 5x^2 + 28x - 10} \\ \underline{5x^2 + 30x} \\ -2x - 10 \\ \underline{-2x - 12} \\ 2 \end{array}$$

$$5x - 2 + \frac{2}{x + 6}$$

**33.** $\dfrac{10x^3 - 24x^2 - 10x}{10x} = \dfrac{10x^3}{10x} - \dfrac{24x^2}{10x} - \dfrac{10x}{10x}$

$$= x^2 - \frac{12x}{5} - 1$$

**35.**

$$\begin{array}{r} 6x - 1 \\ x + 3 \overline{) 6x^2 + 17x - 4} \\ \underline{6x^2 + 18x} \\ -\ \ x - 4 \\ \underline{-\ \ x - 3} \\ -1 \end{array}$$

$$6x - 1 - \frac{1}{x + 3}$$

**37.** $\dfrac{12x^4 + 3x^2}{3x^2} = \dfrac{12x^4}{3x^2} + \dfrac{3x^2}{3x^2} = 4x^2 + 1$

**39.**

$$\begin{array}{r} 2x^2 + 6x - 5 \\ x - 2 \overline{) 2x^3 + 2x^2 - 17x + 8} \\ \underline{2x^3 - 4x^2} \\ 6x^2 - 17x \\ \underline{6x^2 - 12x} \\ -5x + 8 \\ \underline{-5x + 10} \\ -2 \end{array}$$

$$2x^2 + 6x - 5 - \frac{2}{x - 2}$$

**41.**

$$
\begin{array}{r}
6x-1 \\
5x-2\overline{\smash{)}30x^2-17x+2} \\
\underline{30x^2-12x} \\
-5x+2 \\
\underline{-5x+2} \\
0
\end{array}
$$

$$6x-1$$

**43.** $\dfrac{3x^4-9x^3+12}{-3x}=\dfrac{3x^4}{-3x}+\dfrac{-9x^3}{-3x}+\dfrac{12}{-3x}$

$$=-x^3+3x^2-\dfrac{4}{x}$$

**45.**

$$
\begin{array}{r}
4x+3 \\
2x+1\overline{\smash{)}8x^2+10x+1} \\
\underline{8x^2+4x} \\
6x+1 \\
\underline{6x+3} \\
-2
\end{array}
$$

$$\dfrac{8x^2+10x+1}{2x+1}=4x+3-\dfrac{2}{2x+1}$$

**47.**

$$
\begin{array}{r}
2x+9 \\
2x-9\overline{\smash{)}4x^2+0x-81} \\
\underline{4x^2-18x} \\
18x-81 \\
\underline{18x-81} \\
0
\end{array}
$$

$$2x+9$$

**49.**

$$
\begin{array}{r}
2x^2+3x-4 \\
2x+3\overline{\smash{)}4x^3+12x^2+x-12} \\
\underline{4x^3+6x^2} \\
6x^2+x \\
\underline{6x^2+9x} \\
-8x-12 \\
\underline{-8x-12} \\
0
\end{array}
$$

$$2x^2+3x-4$$

**51.**

$$
\begin{array}{r}
x^2+3x+9 \\
x-3\overline{\smash{)}x^3+0x^2+0x-27} \\
\underline{x^3-3x^2} \\
3x^2+0x \\
\underline{3x^2-9x} \\
9x-27 \\
\underline{9x-27} \\
0
\end{array}
$$

$$\dfrac{x^3-27}{x-3}=x^2+3x+9$$

**53.**

$$
\begin{array}{r}
x^2-x+1 \\
x+1\overline{\smash{)}x^3+0x^2+0x+1} \\
\underline{x^3+x^2} \\
-x^2+0x \\
\underline{-x^2-x} \\
x+1 \\
\underline{x+1} \\
0
\end{array}
$$

$$\dfrac{x^3+1}{x+1}=x^2-x+1$$

**55.**

$$x+2\overline{)-3x^2+0x+1}$$

$$\begin{array}{r} -3x+6 \\ \underline{-3x^2-6x} \\ 6x+\ 1 \\ \underline{6x+12} \\ -11 \end{array}$$

$$\frac{1-3x^2}{x+2}=-3x+6-\frac{11}{x+2}$$

**57.**

$$2b-1\overline{)4b^2-4b-5}$$

$$\begin{array}{r} 2b-1 \\ \underline{4b^2-2b} \\ -2b-5 \\ \underline{-2b+1} \\ -6 \end{array}$$

$$\frac{-4b+4b^2-5}{2b-1}=2b-1-\frac{6}{2b-1}$$

**59.** $2a\left(a^2+1\right)=2a\left(a^2\right)+2a\left(1\right)$

$$=2a^3+2a$$

**61.** $2x\left(x^2+7x-5\right)$

$$=2x\left(x^2\right)+2x\left(7x\right)+2x\left(-5\right)$$

$$=2x^3+14x^2-10x$$

**63.** $-3xy\left(xy^2+7x^2y+8\right)$

$$=-3xy\left(xy^2\right)-3xy\left(7x^2y\right)-3xy\left(8\right)$$

$$=-3x^2y^3-21x^3y^2-24xy$$

**65.** $9ab\left(ab^2c+4bc-8\right)$

$$=9ab\left(ab^2c\right)+9ab\left(4bc\right)+9ab\left(-8\right)$$

$$=9a^2b^3c+36ab^2c-72ab$$

**67.** The Rolling Stones (1994)

**69.** $110 million

**71.**

$$x^2+x\overline{)x^5+0x^4+0x^3+x^2}$$

$$\begin{array}{r} x^3-x^2+x \\ \underline{x^5+\ x^4} \\ -x^4+0x^3 \\ \underline{-x^4-\ x^3} \\ x^3+x^2 \\ \underline{x^3+x^2} \\ 0 \end{array}$$

$$\frac{x^5+x^2}{x^2+x}=x^3-x^2+x$$

**73.** $\dfrac{18x^{10a}-12x^{8a}+14x^{5a}-2x^{3a}}{2x^{3a}}$

$$=\frac{18x^{10a}}{2x^{3a}}-\frac{12x^{8a}}{2x^{3a}}+\frac{14x^{5a}}{2x^{3a}}-\frac{2x^{3a}}{2x^{3a}}$$

$$=9x^{7a}-6x^{5a}+7x^{2a}-1$$

**75.** Answers may vary.

**Chapter 5 Review**

**1.** $3^2$

base: 3

exponent: 2

**2.** $(-5)^4$

base: $-5$

exponent: 4

**3.** $-5^4$

base: 5

exponent: 4

**4.** $8^3=8\cdot8\cdot8=512$

**5.** $(-6)^2 = (-6)(-6) = 36$

**6.** $-6^2 = -6 \cdot 6 = -36$

**7.** $-4^3 - 4^0 = -4 \cdot 4 \cdot 4 - 1 = -65$

**8.** $(3b)^0 = 1$

**9.** $\dfrac{8b}{8b} = 1$

**10.** $5b^3 b^5 a^6 = 5a^6 b^8$

**11.** $2^3 \cdot x^0 = 8 \cdot 1 = 8$

**12.** $\left[(-3)^2\right]^3 = (9)^3 = 9 \cdot 9 \cdot 9 = 729$

**13.** $(2x^3)(-5x^2) = (2)(-5)(x^3)(x^2) = -10x^5$

**14.** $\left(\dfrac{mn}{q}\right)^2 \cdot \left(\dfrac{mn}{q}\right) = \dfrac{m^2 n^2}{q^2} \cdot \dfrac{mn}{q} = \dfrac{m^3 n^3}{q^3}$

**15.** $\left(\dfrac{3ab^2}{6ab}\right)^4 = \left(\dfrac{b}{2}\right)^4 = \dfrac{b^4}{2^4} = \dfrac{b^4}{16}$

**16.** $\dfrac{x^9}{x^4} = x^{9-4} = x^5$

**17.** $\dfrac{2x^7 y^8}{8xy^2} = \dfrac{2}{8} \cdot x^{7-1} y^{8-2}$

$= \dfrac{x^6 y^6}{4}$

**18.** $\dfrac{3x^4 y^{10}}{12xy^6} = \dfrac{3}{12} \cdot x^{4-1} \cdot y^{10-6} = \dfrac{x^3 y^4}{4}$

**19.** $5a^7 \left(2a^4\right)^3$

$= 5a^7 \left(2^3 a^{4 \cdot 3}\right)$

$= 5a^7 \left(8a^{12}\right)$

$= 5 \cdot 8 a^{7+12}$

$= 40a^{19}$

**20.** $(2x)^2 (9x)$

$= \left(2^2 \cdot x^2\right)(9x)$

$= 4x^2 \cdot 9x$

$= 4 \cdot 9 \cdot x^{2+1}$

$= 36x^3$

**21.** $\dfrac{(-4)^2 (3^3)}{(4)^5 (3^2)} = 4^{2-5} 3^{3-2} = 4^{-3} \cdot 3 = \dfrac{3}{4^3} = \dfrac{3}{64}$

**22.** $\dfrac{(-7)^2 (3^5)}{(-7)^3 (3^4)} = \dfrac{3}{-7} = -\dfrac{3}{7}$

**23.** $\dfrac{(2x)^0 (-4)^2}{16x} = \dfrac{1 \cdot 16}{16x} = \dfrac{1}{x}$

**24.** $\dfrac{(8xy)(3xy)}{18x^2 y^2} = \dfrac{24x^2 y^2}{18x^2 y^2} = \dfrac{4}{3}$

**25.** $m^0 + p^0 + 3q^0 = 1 + 1 + 3(1)$

$= 1 + 1 + 3$

$= 5$

**26.** $(-5a)^0 + 7^0 + 8^0 = 1 + 1 + 1 = 3$

**27.** $(3xy^2 + 8x + 9)^0 = 1$

**28.** $8x^0 + 9^0 = 8(1) + 1 = 9$

**29.** $6(a^2b^3)^3 = 6(a^6b^9) = 6a^6b^9$

**30.** $\dfrac{(x^3z)^a}{x^2z^2} = \dfrac{x^{3a}z^a}{x^2z^2} = x^{3a-2}z^{a-2}$

**31.** $-5x^4y^3$

The degree is $4 + 3 = 7$.

**32.** $10x^3y^2z$

The degree is $3 + 2 + 1 = 6$.

**33.** $35a^5bc^2$

The degree is $5 + 1 + 2 = 8$.

**34.** $95xyz$

The degree is $1 + 1 + 1 = 3$.

**35.** The degree is 5 because $y^5$ is the term with the highest degree.

**36.** The degree is 2 because $9y^2$ is the term with the highest degree.

**37.** The degree is 5 because $-28x^2y^3$ is the term with the highest degree.

**38.** The degree is 6 because $6x^2y^2z^2$ is the term with the highest degree.

**39. a**. $3a^2b - 2a^2 + ab - b^2 - 6$

| Term | Numerical Coefficient | Degree of Term |
|------|------|------|
| $3a^2b$ | 3 | 3 |
| $-2a^2$ | $-2$ | 2 |
| $ab$ | 1 | 2 |
| $-b^2$ | $-1$ | 2 |
| $-6$ | $-6$ | 0 |

**b**. Degree 3

**40. a**. $x^2y^2 + 5x^2 - 7y^2 + 11xy - 1$

| Term | Numerical Coefficient | Degree of Term |
|------|------|------|
| $x^2y^2$ | 1 | 4 |
| $5x^2$ | 5 | 2 |
| $-7y^2$ | $-7$ | 2 |
| $11xy$ | 11 | 2 |
| $-1$ | $-1$ | 0 |

**b**. Degree 4

**41.** $2x^2 + 20x:$

$x = 1: \ 2(1)^2 + 20(1) = 22$

$x = 3: \ 2(3)^2 + 20(3) = 78$

$x = 5.1: \ 2(5.1)^2 + 20(5.1) = 154.02$

$x = 10: \ 2(10)^2 + 20(10) = 400$

**42.** $6a^2b^2 + 4ab + 9a^2b^2 = (6+9)a^2b^2 + 4ab$
$$= 15a^2b^2 + 4ab$$

**43.** $21x^2y^3 + 3xy + x^2y^3 + 6$
$$= (21+1)x^2y^3 + 3xy + 6$$
$$= 22x^2y^3 + 3xy + 6$$

**44.** $4a^2b - 3b^2 - 8q^2 - 10a^2b + 7q^2$

$= \left(4a^2b - 10a^2b\right) - 3b^2 + \left(-8q^2 + 7q^2\right)$

$= -6a^2b - 3b^2 - q^2$

**45.** $2s^{14} + 3s^{13} + 12s^{12} - s^{10}$

Cannot be combined.

**46.** $\left(3k^2 + 2k + 6\right) + \left(5k^2 + k\right)$

$= 3k^2 + 2k + 6 + 5k^2 + k$

$= 8k^2 + 3k + 6$

**47.** $\left(2s^5 + 3s^4 + 4s^3 + 5s^2\right) - \left(4s^2 + 7s + 6\right)$

$= 2s^5 + 3s^4 + 4s^3 + 5s^2 - 4s^2 - 7s - 6$

$= 2s^5 + 3s^4 + 4s^3 + s^2 - 7s - 6$

**48.** $\left(2m^7 + 3x^4 + 7m^6\right) - \left(8m^7 + 4m^2 + 6x^4\right)$

$= 2m^7 + 3x^4 + 7m^6 - 8m^7 - 4m^2 - 6x^4$

$= -6m^7 - 3x^4 + 7m^6 - 4m^2$

**49.** $\left(11r^2 + 16rs - 2s^2\right) - \left(3r^2 + 5rs - 9s^2\right)$

$= 11r^2 + 16rs - 2s^2 - 3r^2 - 5rs + 9s^2$

$= 8r^2 + 11rs + 7s^2$

**50.** $\left(3x^2 - 6xy + y^2\right) - \left(11x^2 - xy + 5y^2\right)$

$= 3x^2 - 6xy + y^2 - 11x^2 + xy - 5y^2$

$= -8x^2 - 5xy - 4y^2$

**51.** $\left(7x - 14y\right) - \left(3x - y\right)$

$= 7x - 14y - 3x + y$

$= 4x - 13y$

**52.** $\left[\left(x^2 + 7x + 9\right) + \left(x^2 + 4\right)\right] - \left(4x^2 + 8x - 7\right)$

$= x^2 + 7x + 9 + x^2 + 4 - 4x^2 - 8x + 7$

$= -2x^2 - x + 20$

**53.** Let $x = 20$.

$72.5x^2 - 17.5x + 120$

$= 72.5\left(20\right)^2 - 17.5\left(20\right) + 120 = 28,770$

Expect 28,770,000 trademark

registrations in 2010.

**54.** $9x\left(x^2y\right) = 9x^3y$

**55.** $-7\left(8xz^2\right) = -56xz^2$

**56.** $\left(6xa^2\right)\left(xya^3\right) = 6x^2ya^5$

**57.** $\left(4xy\right)\left(-3xa^2y^3\right) = -12x^2a^2y^4$

**58.** $6\left(x + 5\right) = 6x + 6\left(5\right)$

$= 6x + 30$

**59.** $9\left(x - 7\right) = 9x - 9\left(7\right)$

$= 9x - 63$

**60.** $4\left(2a + 7\right) = 4\left(2a\right) + 4\left(7\right)$

$= 8a + 28$

**61.** $9\left(6a - 3\right) = 9\left(6a\right) - 9\left(3\right)$

$= 54a - 27$

**62.** $-7x\left(x^2 + 5\right) = -7\left(x^2\right) - 7x\left(5\right)$

$= -7x^3 - 35x$

**63.** $-8y\left(4y^2-6\right)=-8y\left(4y^2\right)-8y\left(-6\right)$
$$=-32y^3+48y$$

**64.** $-2\left(x^3-9x^2+x\right)=-2\left(x^3\right)-2\left(-9x^2\right)-2\left(x\right)$
$$=-2x^3+18x^2-2x$$

**65.** $-3a\left(a^2b+ab+b^2\right)$
$$=-3a\left(a^2b\right)-3a\left(ab\right)-3a\left(b^2\right)$$
$$=-3a^3b-3a^2b-3ab^2$$

**66.** $\left(3a^3-4a+1\right)\left(-2a\right)$
$$=3a^3\left(-2a\right)-4a\left(-2a\right)+1\left(-2a\right)$$
$$=-6a^4+8a^2-2a$$

**67.** $\left(6b^3-4b+2\right)\left(7b\right)$
$$=6b^3\left(7b\right)-4b\left(7b\right)+2\left(7b\right)$$
$$=42b^4-28b^2+14b$$

**68.** $\left(2x+5\right)\left(x-7\right)$
$$=2x\left(x\right)+2x\left(-7\right)+5\left(x\right)+5\left(-7\right)$$
$$=2x^2-9x-35$$

**69.** $\left(2x-5\right)\left(3x+2\right)$
$$=2x\left(3x\right)+2x\left(2\right)-5\left(3x\right)-5\left(2\right)$$
$$=6x^2+4x-15x-10$$
$$=6x^2-11x-10$$

**70.** $\left(4a-1\right)\left(a+7\right)=4a^2+28a-a-7$
$$=4a^2+27a-7$$

**71.** $\left(6a-1\right)\left(7a+3\right)=42a^2+18a-7a-3$
$$=42a^2+11a-3$$

**72.** $\left(x+7\right)\left(x^3+4x-5\right)$
$$=x^4+4x^2-5x+7x^3+28x-35$$
$$=x^4+7x^3+4x^2+23x-35$$

**73.** $\left(x+2\right)\left(x^5+x+1\right)$
$$=x^6+x^2+x+2x^5+2x+2$$
$$=x^6+2x^5+x^2+3x+2$$

**74.** $\left(x^2+2x+4\right)\left(x^2+2x-4\right)$
$$=x^4+2x^3-4x^2+2x^3+4x^2-8x$$
$$\phantom{=}+4x^2+8x-16$$
$$=x^4+4x^3+4x^2-16$$

**75.** $\left(x^3+4x+4\right)\left(x^3+4x-4\right)$
$$=x^6+4x^4-4x^3+4x^4+16x^2-16x$$
$$\phantom{=}+4x^3+16x-16$$
$$=x^6+8x^4+16x^2-16$$

**76.** $\left(x+7\right)^3$
$$=\left(x+7\right)\left(x+7\right)\left(x+7\right)$$
$$=\left(x^2+7x+7x+49\right)\left(x+7\right)$$
$$=\left(x^2+14x+49\right)\left(x+7\right)$$
$$=x^3+7x^2+14x^2+98x+49x+343$$
$$=x^3+21x^2+147x+343$$

**77.** $\left(2x-5\right)^3$
$$=\left(2x-5\right)\left(2x-5\right)\left(2x-5\right)$$
$$=\left(4x^2-10x-10x+25\right)\left(2x-5\right)$$
$$=\left(4x^2-20x+25\right)\left(2x-5\right)$$
$$=8x^3-20x^2-40x^2+100x+50x-125$$
$$=8x^3-60x^2+150x-125$$

**78.** $2x(3x^2 - 7x + 1)$

$= 2x(3x^2) - 2x(7x) + 2x(1)$

$= 6x^3 - 14x^2 + 2x$

**79.** $3y(5y^2 - y + 2)$

$= 3y(5y^2) - 3y(y) + 3y(2)$

$= 15y^3 - 3y^2 + 6y$

**80.** $(6x^5 - 1)(4x^2 + 3)$

$= 6x^5(4x^2) + 6x^5(3) - 1(4x^2) - 1(3)$

$= 24x^7 + 18x^5 - 4x^2 - 3$

**81.** $(4a^3 - 1)(3a^2 + 7)$

$= 4a^3(3a^2) + 4a^3(7) - 1(3a^2) - 1(7)$

$= 12a^5 + 28a^3 - 3a^2 - 7$

**82.** $(x^2 + 7y)^2 = (x^2)^2 + 2(x^2)(7y) + (7y)^2$

$= x^4 + 14x^2y + 49y^2$

**83.** $(x^3 - 5y)^2 = (x^3)^2 - 2(x^3)(5y) + (5y)^2$

$= x^6 - 10x^3y + 25y^2$

**84.** $(3x - 7)^2 = (3x)^2 - 2(3x)(7) + 7^2$

$= 9x^2 - 42x + 49$

**85.** $(4x + 2)^2 = (4x)^2 + 2(4x)(2) + 2^2$

$= 16x^2 + 16x + 4$

**86.** $(y + 1)(y^2 - 6y - 5)$

$= y(y^2) - y(6y) - y(5) + y^2 - 6y - 5$

$= y^3 - 6y^2 - 5y + y^2 - 6y - 5$

$= y^3 - 5y^2 - 11y - 5$

**87.** $(x - 2)(x^2 - x - 2)$

$= x(x^2) - x(x) - 2x - 2x^2 + 2x - 2(-2)$

$= x^3 - x^2 - 2x - 2x^2 + 2x + 4$

$= x^3 - 3x^2 + 4$

**88.** $(5x - 9)^2 = (5x)^2 - 2(5x)(9) + 9^2$

$= 25x^2 - 90x + 81$

**89.** $(5x + 1)(5x - 1) = (5x)^2 - 1^2$

$= 25x^2 - 1$

**90.** $(7x + 4)(7x - 4) = (7x)^2 - 4^2$

$= 49x^2 - 16$

**91.** $(a + 2b)(a - 2b) = a^2 - (2b)^2$

$= a^2 - 4b^2$

**92.** $(2x - 6)(2x + 6) = (2x)^2 - 6^2$

$= 4x^2 - 36$

**93.** $(4a^2 - 2b)(4a^2 + 2b) = (4a^2)^2 - (2b)^2$

$= 16a^4 - 4b^2$

**94.** $7^{-2} = \dfrac{1}{7^2} = \dfrac{1}{49}$

**95.** $-7^{-2} = -\dfrac{1}{7^2} = -\dfrac{1}{49}$

**96.** $2x^{-4} = \dfrac{2}{x^4}$

**97.** $(2x)^{-4} = \dfrac{1}{(2x)^4} = \dfrac{1}{16x^4}$

**98.** $\left(\dfrac{1}{5}\right)^{-3} = \dfrac{1^{-3}}{5^{-3}} = \dfrac{5^3}{1^3} = 125$

**99.** $\left(\dfrac{-2}{3}\right)^{-2} = \dfrac{(-2)^{-2}}{3^{-2}} = \dfrac{3}{(-2)^2} = \dfrac{9}{4}$

**100.** $2^0 + 2^{-4} = 1 + \dfrac{1}{2^4} = \dfrac{16}{16} + \dfrac{1}{16} = \dfrac{17}{16}$

**101.** $6^{-1} - 7^{-1} = \dfrac{1}{6} - \dfrac{1}{7} = \dfrac{7}{42} - \dfrac{6}{42} = \dfrac{1}{42}$

**102.** $\dfrac{1}{(2q)^{-3}} = 1 \cdot (2q)^3 = 1 \cdot 2^3 \cdot q^3 = 1 \cdot 8 \cdot q^3 = 8q^3$

**103.** $\dfrac{-1}{(qr)^{-3}} = -1 \cdot (qr)^3 = -1 \cdot q^3 \cdot r^3 = -q^3 r^3$

**104.** $\dfrac{r^{-3}}{s^{-4}} = \dfrac{s^4}{r^3}$

**105.** $\dfrac{rs^{-3}}{r^{-4}} = r^{1-(-4)}s^{-3} = \dfrac{r^5}{s^3}$

**106.** $(-x^{-3}y^5)(5xy^{-2}) = -5x^{-3+1}y^{5-2}$

$$= -5x^{-2}y^3$$

$$= -\dfrac{5y^3}{x^2}$$

**107.** $(-3x^5y^{-2})(-4x^{-5}y) = 12x^{5-5}y^{-2+1}$

$$= 12x^0 y^{-1}$$

$$= \dfrac{12}{y}$$

**108.** $(2x^{-5})^{-3} = 2^{-3}x^{15} = \dfrac{x^{15}}{2^3} = \dfrac{x^{15}}{8}$

**109.** $(3y^{-6})^{-1} = 3^{-1}y^6 = \dfrac{y^6}{3}$

**110.** $(3a^{-1}b^{-1}c^{-2})^{-2} = 3^{-2}a^2b^2c^4$

$$= \dfrac{a^2b^2c^4}{3^2}$$

$$= \dfrac{a^2b^2c^4}{9}$$

**111.** $(4x^{-2}y^{-3}z)^{-3} = 4^{-3}x^6y^9z^{-3} = \dfrac{x^6y^9}{4^3z^3} = \dfrac{x^6y^9}{64z^3}$

**112.** $\dfrac{5^{-2}x^8}{5^{-3}x^{11}} = 5^{-2-(-3)}x^{8-11}$

$$= 5^{-2+3}x^{8-11}$$

$$= 5^1 x^{-3}$$

$$= \dfrac{5}{x^3}$$

**113.** $\dfrac{7^5 y^{-2}}{7^7 y^{-10}} = 7^{5-7} \cdot y^{-2-(-10)} = 7^{-2} \cdot y^8 = \dfrac{y^8}{7^2} = \dfrac{y^8}{49}$

**114.** $\left(\dfrac{bc^{-2}}{bc^{-3}}\right)^4 = \dfrac{b^4c^{-8}}{b^4c^{-12}} = b^{4-4}c^{-8-(-12)} = c^4$

**115.** $\left(\dfrac{x^{-3}y^{-4}}{x^{-2}y^{-5}}\right)^{-3} = \dfrac{x^9y^{12}}{x^6y^{15}} = x^{9-6}y^{12-15} = \dfrac{x^3}{y^3}$

**116.** $\dfrac{x^{-4}y^{-6}z^3}{x^2y^7z^3} = x^{-4-2}y^{-6-7}z^{3-3}$

$\qquad = x^{-6}y^{-13}z^0$

$\qquad = \dfrac{1}{x^6 y^{13}}$

**117.** $\dfrac{a^5 b^{-5} c^4}{a^{-5} b^5 c^4} = a^{5-(-5)}b^{-5-5}c^{4-4}$

$\qquad = a^{10}b^{-10}c^0$

$\qquad = \dfrac{a^{10}}{b^{10}}$

**118.** $-2^0 + 2^{-4} = -1 \cdot 2^0 + \dfrac{1}{2^4}$

$\qquad = -1 \cdot 1 + \dfrac{1}{16}$

$\qquad = -1 + \dfrac{1}{16}$

$\qquad = -\dfrac{16}{16} + \dfrac{1}{16}$

$\qquad = -\dfrac{15}{16}$

**119.** $-3^{-2} - 3^{-3} = -\dfrac{1}{3^2} - \dfrac{1}{3^3}$

$\qquad = -\dfrac{1}{9} - \dfrac{1}{27}$

$\qquad = -\dfrac{3}{27} - \dfrac{1}{27}$

$\qquad = -\dfrac{4}{27}$

**120.** $a^{6m}a^{5m} = a^{6m+5m} = a^{11m}$

**121.** $\dfrac{\left(x^{5+h}\right)^3}{x^5} = \dfrac{x^{3(5+h)}}{x^5}$

$\qquad = \dfrac{x^{15+3h}}{x^5}$

$\qquad = x^{15+3h-5}$

$\qquad = x^{10+3h}$

**122.** $\left(3xy^{2z}\right)^3 = 3^3 x^3 y^{2z(3)} = 27x^3 y^{6z}$

**123.** $a^{m+2} \cdot a^{m+3} = a^{m+2m+3} = a^{2m+5}$

**124.** $0.00027 = 2.7 \times 10^{-4}$

**125.** $0.8868 = 8.868 \times 10^{-1}$

**126.** $80,800,000 = 8.08 \times 10^7$

**127.** $868,000 = 8.68 \times 10^5$

**128.** $109,379,000 = 1.09379 \times 10^8$ kg

**129.** $150,000 = 1.5 \times 10^5$ light years

**130.** $8.67 \times 10^5 = 867,000$

**131.** $3.86 \times 10^{-3} = 0.00386$

**132.** $8.6 \times 10^{-4} = 0.00086$

**133.** $8.936 \times 10^5 = 893,600$

**134.** $1.43128 \times 10^{15}$
$\qquad = 1,431,280,000,000,000$ cu km

**135.** $1 \times 10^{-10} = 0.0000000001$ m

216

**136.** $\left(8\times10^4\right)\left(2\times10^{-7}\right)$

$= \left(8\times2\right)\times\left(10^4\times10^{-7}\right)$

$= 16\times10^{-3}$

$= 0.016$

**137.** $\dfrac{8\times10^4}{2\times10^{-7}}$

$= \dfrac{8}{2}\times\left(10^{4-(-7)}\right)$

$= 4\times10^{11}$

$= 400,000,000,000$

**138.** $\dfrac{x^2+21x+49}{7x^2} = \dfrac{x^2}{7x^2}+\dfrac{21x}{7x^2}+\dfrac{49}{7x^2}$

$= \dfrac{1}{7}+\dfrac{3}{x}+\dfrac{7}{x^2}$

**139.** $\dfrac{5a^3b-15ab^2+20ab}{-5ab}$

$= \dfrac{5a^3b}{-5ab}-\dfrac{15ab^2}{-5ab}+\dfrac{20ab}{-5ab}$

$= -a^2+3b-4$

**140.**
$$\begin{array}{r} a+1 \\ a-2{\overline{\smash{\big)}\,a^2-a+4}} \\ \underline{a^2-2a} \\ a+4 \\ \underline{a-2} \\ 6 \end{array}$$

$\left(a^2-a+4\right)\div\left(a-2\right) = a+1+\dfrac{6}{a-2}$

**141.**
$$\begin{array}{r} 4x \\ x+5{\overline{\smash{\big)}\,4x^2+20x+7}} \\ \underline{4x^2+20x} \\ 7 \end{array}$$

$\left(4x^2+20x+7\right)\div\left(x+5\right) = 4x+\dfrac{7}{x+5}$

**142.**
$$\begin{array}{r} a^2+3a+8 \\ a-2{\overline{\smash{\big)}\,a^3+\ a^2+2a+6}} \\ \underline{a^3-2a^2} \\ 3a^2+2a \\ \underline{3a^2-6a} \\ 8a+\ 6 \\ \underline{8a-16} \\ 22 \end{array}$$

$\dfrac{a^3+a^2+2a+6}{a-2} = a^2+3a+8+\dfrac{22}{a-2}$

**143.**
$$\begin{array}{r} 3b^2-4b \\ 3b-2{\overline{\smash{\big)}\,9b^3-18b^2+8b-1}} \\ \underline{9b^3-6b^2} \\ -12b^2+8b \\ \underline{-12b^2+8b} \\ -1 \end{array}$$

$\dfrac{9b^3-18b^2+8b-1}{3b-2} = 3b^2-4b-\dfrac{1}{3b-2}$

**144.**
$$\begin{array}{r} 2x^3-x^2+2 \\ 2x-1{\overline{\smash{\big)}\,4x^4-4x^3+x^2+4x-3}} \\ \underline{4x^4-2x^3} \\ -2x^3+x^2 \\ \underline{-2x^2+x^2} \\ 4x-3 \\ \underline{4x-2} \\ -1 \end{array}$$

$$\frac{4x^4 - 4x^3 + x^2 + 4x - 3}{2x - 1}$$

$$= 2x^3 - x^2 + 2 - \frac{1}{2x - 1}$$

**145.**

$$\begin{array}{r} -x^2 - 16x - 117 \\ x - 6 \overline{\smash{\big)}\, -x^3 - 10x^2 - 21x + 18} \\ \underline{-x^3 + 10x^2} \\ -16x^2 - 21x \\ \underline{-16x^2 + 96x} \\ -117x + 18 \\ \underline{-117x + 702} \\ -684 \end{array}$$

$$\frac{-10x^2 - x^3 - 21x + 18}{x - 6}$$

$$= -x^2 - 16x - 117 - \frac{684}{x - 6}$$

## Chapter 5 Test

**1.** $2^5 = 2 \cdot 2 \cdot 2 \cdot 2 \cdot 2 = 32$

**2.** $(-3)^4 = (-3)(-3)(-3)(-3) = 81$

**3.** $-3^4 = -3 \cdot 3 \cdot 3 \cdot 3 = -81$

**4.** $4^{-3} = \frac{1}{4^3} = \frac{1}{64}$

**5.** $(3x^2)(-5x^9) = (3)(-5)(x^2 \cdot x^9)$
$$= -15x^{11}$$

**6.** $\dfrac{y^7}{y^2} = y^{7-2} = y^5$

**7.** $\dfrac{r^{-8}}{r^{-3}} = r^{-8-(-3)} = r^{-5} = \dfrac{1}{r^5}$

**8.** $\left(\dfrac{x^2 y^3}{x^3 y^{-4}}\right)^2 = \dfrac{x^4 y^6}{x^6 y^{-8}}$
$$= x^{4-6} y^{6-(-8)}$$
$$= x^{-2} y^{14}$$
$$= \dfrac{y^{14}}{x^2}$$

**9.** $\left(\dfrac{6^2 x^{-4} y^{-1}}{6^3 x^{-3} y^7}\right) = 6^{2-3} x^{-4-(-3)} y^{-1-7}$
$$= 6^{2-3} x^{-4-(-3)} y^{-1-7}$$
$$= 6^{-1} x^{-1} y^{-8}$$
$$= \dfrac{1}{6xy^8}$$

**10.** $563,000 = 5.63 \times 10^5$

**11.** $0.0000863 = 8.63 \times 10^{-5}$

**12.** $1.5 \times 10^{-3} = 0.0015$

**13.** $6.23 \times 10^4 = 62,300$

**14.** $(1.2 \times 10^5)(3 \times 10^{-7})$
$$= (1.2)(3) \times 10^{5-7}$$
$$= 3.6 \times 10^{-2}$$
$$= 0.036$$

**15.** **a.** $4xy^2 + 7xyz + x^3 y - 2$

| Term | Numerical Coefficient | Degree of Term |
|------|------|------|
| $4xy^2$ | 4 | 3 |
| $7xyz$ | 7 | 3 |
| $x^3 y$ | 1 | 4 |
| $-2$ | $-2$ | 0 |

**b.** Degree 4

**16.** $5x^2 + 4xy - 7x^2 + 11 + 8xy$

$= \left(5x^2 - 7x^2\right) + \left(4xy + 8xy\right) + 11$

$= -2x^2 + 12xy + 11$

**17.** $\left(8x^3 + 7x^2 + 4x - 7\right) + \left(8x^3 - 7x - 6\right)$

$= 8x^3 + 7x^2 + 4x - 7 + 8x^3 - 7x - 6$

$= 16x^3 + 7x^2 - 3x - 13$

**18.** $\quad 5x^3 + x^2 + 5x - 2$

$\quad \underline{-\left(8x^3 - 4x^2 + x - 7\right)}$

$\quad 5x^3 + x^2 + 5x - 2$

$\quad \underline{-8x^3 + 4x^2 - x + 7}$

$\quad -3x^3 + 5x^2 + 4x + 5$

**19.** $\left[\left(8x^2 + 7x + 5\right) + \left(x^3 - 8\right)\right] - \left(4x + 2\right)$

$= 8x^2 + 7x + 5 + x^3 - 8 - 4x - 2$

$= x^3 + 8x^2 + 3x - 5$

**20.** $\left(3x + 7\right)\left(x^2 + 5x + 2\right)$

$= 3x^3 + 15x^2 + 6x + 7x^2 + 35x + 14$

$= 3x^3 + 22x^2 + 41x + 14$

**21.** $3x^2\left(2x^2 - 3x + 7\right)$

$= 3x^2\left(2x^2\right) - 3x^2\left(3x\right) + 3x^2\left(7\right)$

$= 6x^4 - 9x^3 + 21x^2$

**22.** $\left(x + 7\right)\left(3x - 5\right) = 3x^2 - 5x + 21x - 35$

$= 3x^2 + 16x - 35$

**23.** $\left(4x - 2\right)^2 = \left(4x\right)^2 - 2\left(4x\right)\left(2\right) + 2^2$

$= 16x^2 - 16x + 4$

**24.** $\left(x^2 - 9b\right)\left(x^2 + 9b\right) = \left(x^2\right)^2 - \left(9b\right)^2$

$= x^4 - 81b^2$

**25.** $-16t^2 + 1001$

$t = 0: \ -16\left(0\right)^2 + 1001 = 1001 \text{ ft}$

$t = 1: \ -16\left(1\right)^2 + 1001 = 985 \text{ ft}$

$t = 3: \ -16\left(3\right)^2 + 1001 = 857 \text{ ft}$

$t = 5: \ -16\left(5\right)^2 + 1001 = 601 \text{ ft}$

**26.** $\dfrac{4x^2 + 24xy - 7x}{8xy} = \dfrac{4x^2}{8xy} + \dfrac{24xy}{8xy} - \dfrac{7x}{8xy}$

$= \dfrac{x}{2y} + 3 - \dfrac{7}{8y}$

**27.** 
$$
\begin{array}{r}
x + 2 \phantom{0000} \\
x+5\overline{\smash{\big)}\,x^2 + 7x + 10} \\
\underline{x^2 + 5x}\phantom{0000} \\
2x + 10 \\
\underline{2x + 10} \\
0
\end{array}
$$

$\dfrac{x^2 + 7x + 10}{x + 5} = x + 2$

**28.** 
$$
\begin{array}{r}
9x^2 - 6x + 4 \phantom{00} \\
3x+2\overline{\smash{\big)}\,27x^3 + 0x^2 + 0x - 8} \\
\underline{27x^3 + 18x^2}\phantom{0000000} \\
-18x^2 + \ 0x \phantom{000} \\
\underline{-18x^2 - 12x}\phantom{000} \\
12x - 8 \\
\underline{12x + 8} \\
-16
\end{array}
$$

$\dfrac{27x^3 - 8}{3x + 2} = 9x^2 - 6x + 4 - \dfrac{16}{3x + 2}$

**Cumulative Review Chapter 5**

**1.** a. True

   b. True

   c. False

   d. True

**2.** a. $|-7.2| = 7.2$

   b. $|0| = 0$

   c. $\left|-\dfrac{1}{2}\right| = \dfrac{1}{2}$

**3.** a. $\dfrac{4}{5} \div \dfrac{5}{16} = \dfrac{4}{5} \cdot \dfrac{16}{5} = \dfrac{64}{25}$

   b. $\dfrac{7}{10} \div 14 = \dfrac{7}{10} \cdot \dfrac{1}{14} = \dfrac{7}{10 \cdot 7 \cdot 2} = \dfrac{1}{20}$

   c. $\dfrac{3}{8} \div \dfrac{3}{10} = \dfrac{3}{8} \cdot \dfrac{10}{3} = \dfrac{3 \cdot 2 \cdot 5}{3 \cdot 2 \cdot 4} = \dfrac{5}{4}$

**4.** a. $\dfrac{3}{4} \cdot \dfrac{7}{21} = \dfrac{3 \cdot 7}{4 \cdot 3 \cdot 7} = \dfrac{1}{4}$

   b. $\dfrac{1}{2} \cdot 4\dfrac{5}{6} = \dfrac{1}{2} \cdot \dfrac{29}{6} = \dfrac{29}{12} = 2\dfrac{5}{12}$

**5.** a. $3^2 = 3 \cdot 3 = 9$

   b. $5^3 = 5 \cdot 5 \cdot 5 = 125$

   c. $2^4 = 2 \cdot 2 \cdot 2 \cdot 2 = 16$

   d. $7^1 = 7$

   e. $\left(\dfrac{3}{7}\right)^2 = \left(\dfrac{3}{7}\right)\left(\dfrac{3}{7}\right) = \dfrac{9}{49}$

**6.** Let $x = 5$ and $y = 1$.

   $\dfrac{|2x| - |7y|}{x^2} = \dfrac{|2(5)| - |7(1)|}{5^2} = \dfrac{|10| - |7|}{25}$

   $= \dfrac{10 - 7}{25} = \dfrac{3}{25}$

**7.** a. $-3 + (-7) = -10$

   b. $-1 + (-20) = -21$

   c. $-2 + (-10) = -12$

**8.** $8 + 3(2 \cdot 6 - 1) = 8 + 3(12 - 1) = 8 + 3(11)$

   $= 8 + 33 = 41$

**9.** $-4 - (8) = -4 + (-8) = -12$

**10.** $x = 1$

   $5x^2 + 2 = x - 8$

   $5(1)^2 + 2 \overset{?}{=} 1 - 8$

   $5 + 2 \overset{?}{=} -7$

   $7 \overset{?}{=} -7$   False

   $x = 1$ is not a solution.

**11.** a. $\dfrac{1}{22}$   b. $\dfrac{16}{3}$   c. $-\dfrac{1}{10}$   d. $-\dfrac{13}{9}$

**12.** a. $7 - 40 = 7 + (-40) = -33$

   b. $-5 - (-10) = -5 + 10 = 5$

**13.** a. $5 + (4 + 6) = (5 + 4) + 6$

   b. $(-1 \cdot 2) \cdot 5 = -1 \cdot (2 \cdot 5)$

**14.** $\dfrac{4(-3) + (-8)}{5 + (-5)} = \dfrac{-12 + (-8)}{0}$ is undefined

**15. a.** Alaska Village Electric

  **b.** American Electric Power

  **c.** American Electric Power:

      5 ¢ per kilowatt-hour

      Green Mountain Power:

      11 ¢ per kilowatt-hour

      Montana PowerCo.:

      6 ¢ per kilowatt-hour

      Alaska Village Electric:

      42 ¢ per kilowatt-hour

  **d.** $42\text{¢} - 5\text{¢} = 37\text{¢}$ per kilowatt-hour

**16.** $-2(x+3y-z)$

$= -2(x)+(-2)(3y)-(-2)(z)$

$= -2x-6y+2z$

**17. a.** $5(x+2) = 5x+5(2) = 5x+10$

  **b.** $-2(y+0.3z-1)$

  $= -2(y)+(-2)(0.3z)-(-2)(1)$

  $= -2y-0.6z+2$

  **c.** $-(x+y-2z+6) = -1(x+y-2z+6)$

  $= -1(x)+(-1)(y)-(-1)(2z)+(-1)(6)$

  $= -x-y+2z-6$

**18.** $2(6x-1)-(x-7) = 12x-2-x+7$

$= 11x+5$

**19.** $x-7 = 10$

$x-7+7 = 10+7$

$x = 17$

**20.** Let $x =$ a number.

$(x+7)-2x$

**21.** $\frac{5}{2}x = 15$

$\frac{2}{5}\cdot\frac{5}{2}x = \frac{2}{5}\cdot 15$

$x = 6$

**22.** $2x+\frac{1}{8} = x-\frac{3}{8}$

$x+\frac{1}{8} = -\frac{3}{8}$

$x = -\frac{4}{8}$

$x = -\frac{1}{2}$

**23.** Let $x =$ a number

$7+2x = x-3$

$7+x = -3$

$x = -10$

The number is $-10$

**24.** $10 = 5j-2$

$12 = 5j$

$\frac{12}{5} = j$

**25.** Let $x =$ a number

$2(x+4) = 4x-12$

$2x+8 = 4x-12$

$-2x+8 = -12$

$-2x = -20$

$x = 10$

The number is 10.

**26.**   $\dfrac{7x+5}{3} = x+3$

$3\left(\dfrac{7x+5}{3}\right) = 3(x+3)$

$7x+5 = 3x+9$

$4x+5 = 9$

$4x = 4$

$x = 1$

**27.** Let $x$ = the width and $3x-2$ = the length.

$2L+2W = P$

$2(3x-2)+2x = 28$

$6x-4+2x = 28$

$8x-4 = 28$

$8x = 32$

$x = 4$

$3x-2 = 3(4)-2 = 10$

The width is 4 feet and
the length is 10 feet.

**28.** $x < 5, \ (-\infty, 5)$

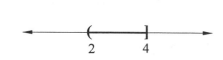

**29.**   $F = \dfrac{9}{5}C+32$

$F-32 = \dfrac{9}{5}C$

$\dfrac{5}{9}(F-32) = C$

$\dfrac{5F-160}{9} = C$

**30.**  a.  $x = -1$ is a vertical line
and the slope is undefined.

b.  $y = 7$ is a horizontal line
and the slope is zero.

**31.** $2 < x \le 4$

**32.** $m = \dfrac{y_2-y_1}{x_2-x_1} = \dfrac{2}{20} = \dfrac{1}{10} \cdot 100\% = 10\%$

**33.** $3x+y = 12$

a.  $(0, \ ): 3(0)+y = 12$

$y = 12, \qquad (0,12)$

b.  $( \ , 6): 3x+(6) = 12$

$3x = 6$

$x = 2, \qquad (2,6)$

c.  $(-1, \ ): 3(-1)+y = 12$

$-3+y = 12$

$y = 15, \qquad (-1,15)$

**34.** $\begin{cases} 3x+2y = -8 \\ 2x-6y = -9 \end{cases}$

Multiply the first equation by 3.

$9x+6y = -24$

$\underline{2x-6y = \ -9}$

$11x \qquad = -33$

$x = -3$

Replace $x$ in the first equation with $-3$

$3(-3) + 2y = -8$

$-9 + 2y = -8$

$2y = 1$

$y = \dfrac{1}{2}$

The solution to the system is $\left(-3, \dfrac{1}{2}\right)$

**35.** $2x + y = 5$

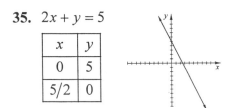

| $x$ | $y$ |
|-----|-----|
| 0   | 5   |
| 5/2 | 0   |

**36.** $\begin{cases} x = -3y + 3 \\ 2x + 9y = 5 \end{cases}$

Replace $x$ in the second
equation by $-3y + 3$.

$2(-3y + 3) + 9y = 5$

$-6y + 6 + 9y = 5$

$3y + 6 = 5$

$3y = -1$

$y = -\dfrac{1}{3}$

Replace $y$ in the first equation with $-\dfrac{1}{3}$.

$x = -3\left(-\dfrac{1}{3}\right) + 3$

$x = 4$

The solution to the system is $\left(4, -\dfrac{1}{3}\right)$

**37.** $x = 2$ for
all values of $y$

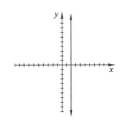

**38. a.** $(-5)^2 = (-5)(-5) = 25$

  **b.** $-5^2 = -(5)(5) = -25$

  **c.** $2 \cdot 5^2 = 2 \cdot 5 \cdot 5 = 50$

**39.** $x = 5$ is a vertical line
and the slope is undefined.

**40.** $\dfrac{(z^2)^3 \cdot z^7}{z^9} = \dfrac{z^6 \cdot z^7}{z^9} = z^{6+7-9} = z^4$

**41.** $x + y < 7$
Test $(0, 0)$

$0 + 0 \overset{?}{<} 7$

True
Shade below

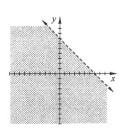

**42.** $(5y^2 - 6) - (y^2 + 2) = 5y^2 - 6 - y^2 - 2$
$= 4y^2 - 8$

**43.** $(2x^2)(-3x^5) = -6x^{2+5} = -6x^7$

**44.** $-x^2$

  **a.** $-(2)^2 = -4$        **b.** $-(-2)^2 = -4$

**45.** $(-2x^2 + 5x - 1) + (-2x^2 + x + 3)$
$= -2x^2 + 5x - 1 - 2x^2 + x + 3$
$= -4x^2 + 6x + 2$

**46.** $\left(10x^2 - 3\right)\left(10x^2 + 3\right) = \left(10x^2\right)^2 - 3^2$

$$= 100x^4 - 9$$

**47.** $\left(2x - y\right)^2 = \left(2x\right)^2 - 2\left(2x\right)\left(y\right) + \left(y\right)^2$

$$= 4x^2 - 4xy + y^2$$

**48.** $\left(10x^2 + 3\right)^2 = \left(10x^2\right)^2 + 2\left(10x^2\right)\left(3\right) + 3^2$

$$= 100x^4 + 60x^2 + 9$$

**49.** $\dfrac{6m^2 + 2m}{2m} = \dfrac{6m^2}{2m} + \dfrac{2m}{2m} = 3m + 1$

**50.** a. $5^{-1} = \dfrac{1}{5}$      b. $7^{-2} = \dfrac{1}{7^2} = \dfrac{1}{49}$

# Chapter 6

**Mental Math 6.1**

**1.** $14 = 2 \cdot 7$

**2.** $15 = 3 \cdot 5$

**3.** $10 = 2 \cdot 5$

**4.** $70 = 2 \cdot 5 \cdot 7$

**5.** $6 = 2 \cdot 3$
$15 = 3 \cdot 5$
$GCF = 3$

**6.** $20 = 2 \cdot 2 \cdot 5$
$15 = 3 \cdot 5$
$GCF = 5$

**7.** $18 = 2 \cdot 3 \cdot 3$
$3 = 3$
$GCF = 3$

**8** .$14 = 2 \cdot 7$
$35 = 5 \cdot 7$
$GCF = 7$

**Exercise Set 6.1**

**1.** $32 = 2 \cdot 2 \cdot 2 \cdot 2 \cdot 2$
$36 = 2 \cdot 2 \cdot 3 \cdot 3$
$GCF = 2 \cdot 2 = 4$

**3.** $12 = 2 \cdot 2 \cdot 3 = 2^2 \cdot 3$
$18 = 2 \cdot 3 \cdot 3 = 2 \cdot 3^2$
$36 = 3 \cdot 2 \cdot 3 \cdot 3 = 2^2 \cdot 3^2$
$GCF = 2 \cdot 3 = 6$

**5.** $y^2$

**7.** $xy^2$

**9.** $8x = 2 \cdot 2 \cdot 2 \cdot x$
$4 = 2 \cdot 2$
$GCF = 2 \cdot 2 = 4$

**11.** $12y^4 = 2 \cdot 2 \cdot 3 \cdot y^4$
$20y^3 = 2 \cdot 2 \cdot 5 \cdot y^3$
$GCF = 2 \cdot 2 \cdot y^3 = 4y^3$

**13.** $12x^3 = 2 \cdot 2 \cdot 3 \cdot x^3$
$6x^4 = 2 \cdot 3 \cdot x^4$
$3x^5 = 3 \cdot x^5$
$GCF = 3 \cdot x^3 = 3x^3$

**15.** $18x^2 y = -2 \cdot 3 \cdot 3 \cdot x^2 \cdot y$
$9x^3 y^3 = 3 \cdot 3 \cdot x^3 \cdot y^3$
$36x^3 y = 2 \cdot 2 \cdot 3 \cdot 3 \cdot x^3 \cdot y$
$GCF = 3 \cdot 3 \cdot x^2 \cdot y = 9x^2 y$

**17.** $30x - 15 = 15(2x - 1)$

**19.** $24cd^3 - 18c^2 d; \ GCF = 6cd$
$6cd \cdot 4d^2 - 6cd \cdot 3c = 6cd(4d^2 - 3c)$

**21.** $-24a^4 x + 18a^3 x; \ GCF = -6a^3 x$
$-6a^3 x(4a) - 6a^3 x(-3) = -6a^3 x(4a - 3)$

**23.** $12x^3 + 16x^2 - 8x = 4x(3x^2 + 4x - 2)$

**25.** $5x^3y - 15x^2y + 10xy = 5xy(x^2 - 3x + 2)$

**27.** Answers may vary.

**29.** $y(x+2) + 3(x+2) = (x+2)(y+3)$

**31.** $x(y-3) - 4(y-3)$; GCF $= (y-3)$
$(y-3)(x-4)$

**33.** $2x(x+y) - (x+y)$; GCF $= (x+y)$
$2x(x+y) + (x+y)(-1) = (x+y)(2x-1)$

**35.** $5x + 15 + xy + 3y = 5(x+3) + y(x+3)$
$= (x+3)(5+y)$

**37.** $2y - 8 + xy - 4x = 2(y-4) + x(y-4)$
$= (y-4)(2+x)$

**39.** $3xy - 6x + 8y - 16 = 3x(y-2) + 8(y-2)$
$= (y-2)(3x+8)$

**41.** $y^3 + 3y^2 + y + 3 = y^2(y+3) + 1(y+3)$
$= (y+3)(y^2+1)$

**43.** Subtract the area of the innter rectangle from the area of the outer rectangle.
Outer rectangle: $A = l \cdot w$
$$A = 12x \cdot x^2 = 12x^3$$
Inner rectangle: $A = l \cdot w$
$$A = 2 \cdot x = 2x$$
The area of the shaded region is given by the expression $12x^3 - 2x = 2x(6x^2 - 1)$.

**45.** $20x(10) + \pi \cdot 5^2 = 200x + 25\pi = 25(8x+\pi)$

**47.** $3x - 6$; GCF $= 3$
$3 \cdot x - 3 \cdot 2 = 3(x-2)$

**49.** $32xy - 18x^2$; GCF $= 2x$
$2x(16y) - 2x(9x) = 2x(16y - 9x)$

**51.** $4x - 8y + 4 = 4(x - 2y + 1)$

**53.** $8(x+2) - y(x+2) = (x+2)(8-y)$

**55.** $-40x^8y^6 - 16x^9y^5$; GCF $= -8x^8y^5$
$-8x^8y^5 \cdot 5y - 8x^8y^5 \cdot 2x = -8x^8y^5(5y + 2x)$

**57.** $-3x + 12$; GCF $= -3$
$= -3 \cdot x - 3(-4) = -3(x-4)$

**59.** $18x^3y^3 - 12x^3y^2 + 6x^5y^2$; GCF $= 6x^3y^2$
$6x^3y^2 \cdot 3y + 6x^3y^2(-2) + 6x^3y^2 \cdot x^2$
$= 6x^3y^2(3y - 2 + x^2)$

**61.** $y^2(x-2) + (x-2)$; GCF $= (x-2)$
$y^2(x-2) + 1(x-2) = (x-2)(y^2+1)$

**63.** $5xy + 15x + 6y + 18 = 5x(y+3) + 6(y+3)$
$= (y+3)(5x+6)$

**65.** $4x^2 - 8xy - 3x + 6y = 4x(x-26) - 3(x=2y)$
$= (x-2y)(4x-3)$

**67.** $126x^3yz + 210y^4z^3$; GCF $= 42yz$
$42yz \cdot 3x^3 + 42yz \cdot 5y^3z^2$
$= 42yz(3x^3 + 5y^3z^2)$

**69.** $3y - 5x + 15 - xy$

$= 3y + 15 - 5x - xy$

$= 3(y + 5) - x(5 + y)$

$= 3(y + 5) - x(y + 5)$

$= (y + 5)(3 - x)$ or $(3 - x)(y + 5)$

**71.** $12x^2 y - 42x^2 - 4y + 14;\ \text{GCF} = 2$

$2\left[6x^2 y - 21x^2 - 2y + 7\right]$

$= 2\left[3x^2(2y - 7) - 1(2y - 7)\right]$

$= 2\left[(2y - 7)(3x^2 - 1)\right]$

$= 2(2y - 7)(3x^2 - 1)$

|       | Two Numbers | Their Product | Their Sum |
| ----- | ----------- | ------------- | --------- |
| **73.** | 2, 6       | 12            | 8         |
| **75.** | −1, −8     | 8             | −9        |
| **77.** | −2, 5      | −10           | 3         |
| **79.** | −8, 3      | −24           | −5        |

**81.** $(x + 2)(x + 5) = x^2 + 2x + 5x + 10$

$= x^2 + 7x + 10$

**83.** $(a - 7)(a - 8) = a^2 - 8a - 7a + 56$

$= a^2 - 15a + 56$

**85.** Answers may vary.

**87.** $(a + 6)(b - 2)$ is factored.

**89.** $5(2y + z) - b(2y + z)$ is not factored.

**91.** $\dfrac{4n^4 - 24n}{4n} = \dfrac{4n^4}{4n} - \dfrac{25n}{4n} = \left(n^3 - 6\right)$ units

**93.** $x^{2n} + 2x^n + 3x^n + 6$

$= x^n\left(x^n + 2\right) + 3\left(x^n + 2\right)$

$= \left(x^n + 2\right)\left(x^n + 3\right)$

**95.** $3x^{2n} + 21x^n - 5x^n - 35$

$= 3x^n\left(x^n + 7\right) - 5\left(x^n + 7\right)$

$= \left(x^n + 7\right)\left(3x^n - 5\right)$

**97.** $8x^2 - 56x + 124$

a. Let $x = 2$

$8(2)^2 - 56(2) + 124$

$= 44$ million

b. Let $x = 7$

$8(7)^2 - 56(7) + 124$

$= 124$ million

c. $8x^2 - 56x + 124 = 4\left(2x^2 - 14x + 31\right)$

**Mental Math 6.2**

**1.** $x^2 + 9x + 20 = (x + 4)(x + 5)$

**2.** $x^2 + 12x + 35 = (x + 5)(x + 7)$

**3.** $x^2 - 7x + 12 = (x - 4)(x - 3)$

**4.** $x^2 - 13x + 22 = (x - 2)(x - 11)$

**5.** $x^2 + 4x + 4 = (x + 2)(x + 2)$

**6.** $x^2 + 10x + 24 = (x + 6)(x + 4)$

**Exercise Set 6.2**

**1.** $x^2 + 7x + 6 = (x + 6)(x + 1)$

**3.** $x^2 + x - 20 = (x + 5)(x - 4)$

**5.** $x^2 - 8x + 15 = (x - 3)(x - 5)$

**7.** $x^2 - 10x + 9 = (x - 9)(x - 1)$

**9.** $x^2 - 15x + 5$ is a prime polynomial.

**11.** $x^2 - 3x - 18 = (x - 6)(x + 3)$

**13.** $x^2 + 5x + 2$ is a prime polynomial.

**15.** $x^2 + 8xy + 15y^2 = (x + 5y)(x + 3y)$

**17.** $x^2 - 2xy + y^2 = (x - y)(x - y)$

**19.** $x^2 - 3xy - 4y^2 = (x - 4y)(x + y)$

**21.** $2z^2 + 20z + 32 = (z^2 + 10z + 16)$
$\qquad = 2(z + 8)(z + 2)$

**23.** $2x^3 - 18x^2 + 40x = 2x(x^2 - 9x + 20)$
$\qquad\qquad = 2x(x - 5)(x - 4)$

**25.** $7x^2 + 14xy - 21y^2 = 7(x^2 + 2xy - 3y^2)$
$\qquad\qquad = 7(x - y)(x + 3y)$

**27.** $6 \cdot 7 = 42$
$\quad 6 + 7 = 13$
The numbers are 6 and 7.

**29.** $x^2 + 15x + 36 = (x + 12)(x + 3)$

**31.** $x^2 - x - 2 = (x - 2)(x + 1)$

**33.** $r^2 - 16r + 48 = (r - 12)(r - 4)$

**35.** $x^2 - 4x - 21 = (x - 7)(x + 3)$

**37.** $x^2 + 7xy + 10y^2 = (x + 5y)(x + 2y)$

**39.** $r^2 - 3r + 6$ is a prime polynomial.

**41.** $2t^2 + 24t + 64$
$\quad = 2(t^2 + 12t + 32)$
$\quad = 2(t + 8)(t + 4)$

**43.** $x^3 - 2x^2 - 24x = x(x^2 - 2x - 24)$
$\qquad\qquad = x(x - 6)(x + 4)$

**45.** $x^2 - 16x + 63 = (x - 9)(x - 7)$

**47.** $x^2 + xy - 2y^2 = (x + 2y)(x - y)$

**49.** $3x^2 - 60x + 108 = 3(x^2 - 20x + 36)$
$\qquad\qquad = 3(x - 18)(x - 2)$

**51.** $x^2 - 18x - 144 = (x - 24)(x + 6)$

**53.** $6x^3 + 54x^2 + 120x = 6x(x^2 + 9x + 20)$
$\qquad\qquad = 6x(x + 5)(x + 4)$

**55.** $2t^5 - 14t^4 + 24t^3 = 2t^3(t^2 - 7t + 12)$
$\qquad\qquad = 2t^3(t - 4)(t - 3)$

**57.** $5x^3y - 25x^2y^2 - 120xy^3$

$\quad = 5xy\left(x^2 - 5xy - 24y^2\right)$

$\quad = 5xy\left(x - 8y\right)\left(x + 3y\right)$

**59.** $4x^2y + 4xy - 12y = 4y\left(x^2 + x - 3\right)$

**61.** $2a^2b - 20ab^2 + 42b^3$

$\quad = 2b\left(a^2 - 10b + 21b^2\right)$

$\quad = 2b\left(a - 3b\right)\left(a - 7b\right)$

**63.** $\left(2x + 1\right)\left(x + 5\right) = 2x^2 + x + 10x + 5$

$\qquad\qquad\qquad\quad = 2x^2 + 11x + 5$

**65.** $\left(5y - 4\right)\left(3y - 1\right) = 15y^2 - 12y - 5y + 4$

$\qquad\qquad\qquad\quad = 15y^2 - 17y + 4$

**67.** $\left(a + 3\right)\left(9a - 4\right) = 9a^2 + 27a - 4a - 12$

$\qquad\qquad\qquad\quad = 9a^2 + 23a - 12$

**69.** $x^2 + bx + 15$ is factorable when b is 8 or 16.

**71.** $m^2 + bm - 27$ is factorable when b is 6 or 26.

**73.** $x^2 + 6x + c$ if factorable when $c$ is 5,8 or 9.

**75.** $y^2 - 4y + c$ if factorable when $c$ is 3 or 4.

**77.** $P = 2l + 2w$

$\quad l = x^2 + 10x$ and $w = 4x + 33$, so

$\quad P = 2\left(x^2 + 10x\right) + 2\left(4x + 33\right)$

$\qquad = 2x^2 + 20x + 8x + 66$

$\qquad = 2x^2 + 28x + 66 = \left(x^2 + 14x + 33\right)$

$\qquad = 2\left(x + 11\right)\left(x + 3\right)$

The perimeter of the rectangle is given by the polynomial $2x^2 + 28x + 66$ which factors as $2\left(x + 11\right)\left(x + 3\right)$.

**79.** Answers may vary.

**81.** $2x^2y + 30xy + 100y$; GCF $= 2y$

$\quad 2y\left(x^2 + 15x + 50\right) = 2y\left(x + 5\right)\left(x + 10\right)$

**83.** $-12x^2y^3 - 24xy^3 - 36y^3$; GCF $= -12y^3$

$\quad = -12y^3\left(x^2 + 2x + 3\right)$

**85.** $y^2\left(x + 1\right) - 2y\left(x + 1\right) - 15\left(x + 1\right)$;

$\quad$GCF $= \left(x + 1\right)$

$\quad \left(x + 1\right)\left(y^2 - 2y - 15\right) = \left(x + 1\right)\left(y - 5\right)\left(y + 3\right)$

**87.** $x^{2n} + 5x^n + 6 = \left(x^n + 2\right)\left(x^n + 3\right)$

**Mental Math 6.3**

**1.** Yes

**2.** Yes

**3.** No

**4.** No

**5.** Yes

**6.** Yes

**Exercise Set 6.3**

**1.** $2x^2 + 13x + 15 = \left(2x + 3\right)\left(x + 5\right)$

**3.** $2x^2 - 9x - 5 = (2x+1)(x-5)$

**5.** $2y^2 - y - 6 = (2y+3)(y-2)$

**7.** $16a^2 - 24a + 9 = (4a-3)(4a-3)$
$$= (4a-3)^2$$

**9.** $36r^2 - 5r - 24 = (9r-8)(4r+3)$

**11.** $10x^2 + 17x + 3 = (5x+1)(2x+3)$

**13.** $21x^2 - 48x - 45 = 3(7x^2 - 16x - 15)$
$$= 3(7x+5)(x-3)$$

**15.** $12x^2 - 14x - 6;\ \text{GCF} = 2$
$$2(6x^2 - 7x - 3) = 2(2x-3)(3x+1)$$

**17.** $4x^3 - 9x^2 - 9x = x(4x^2 - 9x - 9)$
$$= x(4x+3)(x-3)$$

**19.** $x^2 + 22x + 121 = x^2 + 2 \cdot x \cdot 11 + 11^2$
$$= (x+11)^2$$

**21.** $x^2 - 16x + 64 = x^2 - 2 \cdot x \cdot 8 + 8^2$
$$= (x-8)^2$$

**23.** $16y^2 - 40y + 25 = (4y)^2 - 2 \cdot 4y \cdot 5 + 5^2$
$$= (4y-5)^2$$

**25.** $x^2 y^2 - 10xy + 25 = (xy)^2 - 2 \cdot xy \cdot 5 + 5^2$
$$= (xy-5)^2$$

**27.** Answers may vary.

**29.** $2x^2 - 7x - 99 = (2x+11)(x-9)$

**31.** $4x^2 - 8x - 21 = (2x-7)(2x+3)$

**33.** $30x^2 - 53x + 21 = (6x-7)(5x-3)$

**35.** $24x^2 - 58x + 9 = (4x-9)(6x-1)$

**37.** $9x^2 - 24xy + 16y^2$
$$= (3x)^2 - 2 \cdot 3x \cdot 4y + (4y)^2$$
$$= (3x-4y)^2$$

**39.** $x^2 - 14xy + 49y^2 = x^2 - 2 \cdot x \cdot 7y + (7y)^2$
$$= (x-7y)^2$$

**41.** $2x^2 + 7x + 5 = 2x^2 + 5x + 2x + 5$
$$= x(2x+5) + 1(2x+5) = (2x+5)(x+1)$$

**43.** $3x^2 - 5x + 1$
not factorable, prime

**45.** $-2y^2 + y + 10 = 10 + y - 2y^2$
$$= (5-2y)(2+y)$$

**47.** $16x^2 + 24xy + 9y^2$
$$= (4x+3y)(4x+3y)$$
$$= (4x+3y)^2$$

**49.** $8x^2 y + 34xy - 84y = 2y(4x^2 + 17x - 42)$
$$= 2y(4x-7)(x+6)$$

**51.** $3x^2 + x - 2 = (3x - 2)(x + 1)$

**53.** $x^2 y^2 + 4xy + 4 = (xy + 2)(xy + 2)$

$$= (xy + 2)^2$$

**55.** $49y^2 + 42xy + 9x^2 = (7y + 3x)(7y + 3x)$

$$= (7y + 3x)^2$$

**57.** $3x^2 - 42x + 63 = 3(x^2 - 14x + 21)$

**59.** $42a^2 - 43a + 6 = (7a - 6)(6a - 1)$

**61.** $18x^2 - 9x - 14 = (6x - 7)(3x + 2)$

**63.** $25p^2 - 70pq + 49q^2 = (5p - 7q)(5p - 7q)$

$$= (5p - 7q)^2$$

**65.** $15x^2 - 16x - 15 = (5x + 3)(3x - 5)$

**67.** $-27t + 7t^2 - 4 = 7t^2 - 27t - 4$

$$= (7t + 1)(t - 4)$$

**69.** $a^2 + ab + ab + b^2 = a^2 + 2ab + b^2$

**71.** $(x - 2)(x + 2) = x^2 - 2x + 2x - 4 = x^2 - 4$

**73.** $(a + 3)(a^2 - 3a + 9)$

$$= a^3 - 3a^2 + 9a + 3a^2 - 9a + 27$$

$$= a^3 + 27$$

**75.** $(y - 5)(y^2 + 5y + 25)$

$$= y^3 + 5y^2 + 25y - 5y^2 - 25y - 125$$

$$= y^3 - 125$$

**77.** \$75,000 and above.

**79.** Answers may vary

**81.** $3x^2 + bx - 5$ is factorable when $b$ is 2 or 14.

**83.** $2z^2 + bz - 7$ is factorable when $b$ is 5 or 13.

**85.** $5x^2 + 7x + c$ is factorable when $c$ is 2.

**87.** $3x^2 - 8x + c$ is factorable when c is 4 or 5.

**89.** $-12x^3 y^2 + 3x^2 y^2 + 15xy^2$

$$= -3xy^2 (4x^2 - x - 5)$$

$$= -3xy^2 (4x - 5)(x + 1)$$

**91.** $-30p^3 q + 88p^2 q^2 + 6pq^3$; GCF $= -2pq$

$$-2pq(15p^2 - 44pq - 3q^2)$$

$$= -2pq(15p + q)(p - 3q)$$

**93.** $4x^2 (y - 1)^2 + 10x(y - 1)^2 + 25(y - 1)^2$

$$= (y - 1)^2 (4x^2 + 10x + 25)$$

**95.** $3x^{2n} + 17x^n + 10 = (3x^n + 2)(x^n + 5)$

**Graphing Calculator Explorations 6.4**

| $x$ | $x^2 - 2x + 1$ | $x^2 - 2x - 1$ | $(x - 1)^2$ |
|---|---|---|---|
| 5 | 16 | 14 | 16 |
| $-3$ | 16 | 14 | 16 |
| 2.7 | 2.89 | 0.89 | 2.89 |
| $-12.1$ | 171.61 | 169.61 | 171.61 |
| 0 | 1 | $-1$ | 1 |

**Mental Math 6.4**

**1.** $1 = 1^2$

**2.** $25 = 5^2$

**3.** $81 = 9^2$

**4.** $64 = 8^2$

**5.** $9 = 3^2$

**6.** $100 = 10^2$

**7.** $1 = 1^3$

**8.** $64 = 4^3$

**9.** $8 = 2^3$

**10.** $27 = 3^3$

**Exercise Set 6.4**

**1.** $x^2 - 4 = x^2 - 2^2 = (x+2)(x-2)$

**3.** $y^2 - 49 = y^2 - 7^2 = (y+7)(y-7)$

**5.** $25y^2 - 9 = (5y)^2 - 3^2 = (5y+3)(5y-3)$

**7.** $121 - 100x^2 = 11^2 - (10x)^2$
$$= (11+10x)(11-10x)$$

**9.** $12x^2 - 27 = 3(4x^2 - 9) = 3\left((2x)^2 - 3^2\right)$
$$= 3(2x+3)(2x-3)$$

**11.** $169a^2 - 49b^2 = (13a)^2 - (7b)^2$
$$= (13a+7b)(13a-7b)$$

**13.** $x^2y^2 - 1 = (xy)^2 - 1^2$
$$= (xy+1)(xy-1)$$

**15.** $x^4 - 9 = (x^2)^2 - 3^2 = (x^2+3)(x^2-3)$

**17.** $49a^4 - 16 = (7a^2)^2 - 4^2 = (7a^2+4)(7a^2-4)$

**19.** $x^4 - y^{10} = (x^2)^2 - (y^5)^2 = (x^2+y^5)(x^2-y^5)$

**21.** $x+6$ since
$$(x+6)(x-6) = x^2 + 6x - 6x - 36$$
$$= x^2 - 36 = x^2 - 6^2$$

**23.** $a^3 + 27 = a^3 + 3^3$
$$= (a+3)(a^2 - 3a + 9)$$

**25.** $8a^3 + 1 = (2a)^3 + 1^3$
$$= (2a+1)(4a^2 - 2a + 1)$$

**27.** $5k^3 + 40$; GCF $= 5$
$$5(k^3 + 8) = 5(k^3 + 2^3)$$
$$= 5(k+2)(k^2 - 2k + 4)$$

**29.** $x^3y^3 - 64 = (xy)^3 - 4^3$
$$= (xy-4)(x^2y^2 + 4xy + 16)$$

**31.** $x^3 + 125 = x^3 + 5^3$
$$= (x+5)(x^2 - 5x + 25)$$

**33.** $24x^4 - 81xy^3$; GCF $= 3x$

$3x(8x^3 - 27y^3)$

$= 3x\left[(2x)^3 - (3y)^3\right]$

$= 3x(2x - 3y)(4x^2 + 6xy + 9y^2)$

**35.** $2x + y$, since

$(2x + y)(4x^2 - 2xy + y^2)$

$= (2x + y)\left((2x)^2 - 2x \cdot y + (y)^2\right)$

$= (2x)^3 + (y)^3$

**37.** $x^2 - 121 = (x)^2 - 11^2 = (x + 11)(x - 11)$

**39.** $81 - p^2 = 9^2 - p^2 = (9 + p)(9 - p)$

**41.** $4r^2 - 1 = (2r)^2 - 1^2 = (2r + 1)(2r - 1)$

**43.** $9x^2 - 16 = (3x)^2 - 4^2 = (3x + 4)(3x - 4)$

**45.** $16r^2 + 1$ is the sum of two squares,

$(4r)^2 + 1^2$, not the difference of two

squares. $16r^2 + 1$ is a prime polynomial.

**47.** $27 - t^3 = 3^3 - t^3$

$= (3 - t)(9 + 3t + t^2)$

**49.** $8r^3 - 64$; GCF $= 8$

$8(r^3 - 8) = 8(r^3 - 2^3)$

$= 8(r - 2)(r^2 + 2r + 4)$

**51.** $t^3 - 343 = t^3 - 7^3$

$= (t - 7)(t^2 + 7t + 49)$

**53.** $x^2 - 169y^2 = x^2 - (13y)^2$

$= (x + 13y)(x - 13y)$

**55.** $x^2y^2 - z^3 = (xy)^2 - z^2$

$= (xy - z)(xy + z)$

**57.** $x^3y^3 + 1 = (xy)^3 + 1^3$

$= (xy + 1)(x^2y^2 - xy + 1)$

**59.** $s^3 - 64t^3 = s^3 - (4t)^3$

$= (s - 4t)(s^2 + 4st + 16t^2)$

**61.** $18r^2 - 8 = 2(9r^2 - 4)$

$= 2\left((3r)^2 - 2^2\right)$

$= 2(3r + 2)(3r - 2)$

**63.** $9xy^2 - 4x = x(9y^2 - 4)$

$= x\left((3y)^2 - 2^2\right)$

$= x(3y + 2)(3y - 2)$

**65.** $25y^4 - 100y^2 = 25y^2(y^2 - 4)$

$= 25y^2(y^2 - 2^2)$

$= 25y^2(y + 2)(y - 2)$

**67.** $x^3y - 4xy^3 = xy(x^2 - 4y^2)$

$= xy(x^2 - (2y)^2)$

$= xy(x + 2y)(x - 2y)$

**69.** $8s^6t^3 + 100s^3t^6$; GCF $= 4s^3t^3$

$4s^3t^3(2s^3 + 25t^3)$

**71.** $27x^2 y^3 - xy^2$; GCF $= xy^2$

$xy^2 (27xy - 1)$

**73.** $\dfrac{8x^4 + 4x^3 - 2x + 6}{2x}$

$= \dfrac{8x^4}{2x} + \dfrac{4x^3}{2x} - \dfrac{2x}{2x} + \dfrac{6}{2x}$

$= 4x^3 + 2x^2 - 1 + \dfrac{3}{x}$

**75.** $\begin{array}{r} 2x + 1 \\ x - 2 \overline{) 2x^2 - 3x - 2} \end{array}$

$\underline{2x^2 - 4x}$

$x - 2$

$\underline{x - 2}$

$0$

$2x + 1$

**77.** $\begin{array}{r} 3x + 4 \\ x + 3 \overline{) 3x^2 + 13x + 10} \end{array}$

$\underline{3x^2 + 9x}$

$4x + 10$

$\underline{4x + 12}$

$-2$

$3x + 4 - \dfrac{2}{x + 3}$

**79.** $(x + 2)^2 - y^2 = (x + 2 + y)(x + 2 - y)$

**81.** $a^2 (b - 4) - 16(b - 4) = (b - 4)(a^2 - 16)$

$= (b - 4)(a^2 - 4^2) = (b - 4)(a + 4)(a - 4)$

**83.** $(x^2 + 6x + 9) - 4y^2$

$= [(x + 3)(x + 3)] - 4y^2 = (x + 3)^2 - (2y)^2$

$= (x + 3 + 2y)(x + 3 - 2y)$

**85.** $x^{2n} - 100 = (x^n)^2 - 10^2$

$= (x^n + 10)(x^n - 10)$

**87.** a. Let $t = 2$.

$841 - 16t^2 = 841 - 16(2)^2$

$= 841 - 16(4) = 841 - 64 = 777$

After 2 seconds, the height of the object is 777 feet.

b. Let $t = 5$.

$841 - 16t^2 = 841 - 16(5)^2$

$= 841 - 16(25)$

$841 - 400 = 441$

After 5 seconds the height of the object is 441 feet.

c. When the object hits the ground, its height is zero feet. Thus, to find the time, $t$, when the object's height is zero feet above the ground, we set the expression $841 - 16t^2$ equal to 0 and solve for $t$.

$841 - 16t^2 = 0$

$841 - 16t^2 + 16t^2 = 0 + 16t^2$

$841 = 16t^2$

$\dfrac{841}{16} = \dfrac{16t^2}{16}$

$52.5625 = t^2$

$\sqrt{52.5625} = \sqrt{t^2}$

$7.25 = t$

Thus, the object will hit the ground after approximately 7 seconds.

   d. $841 - 16t^2 = 29^2 - (4t)^2$

       $= (29 + 4t)(29 - 4t)$

**89.**  a.  Let $t = 3$

       $1444 - 16t^2 = 1444 - 16(3)^2 = 1300$

       After 3 seconds the height is 1300 feet.

   b.  Let $t = 7$

       $1444 - 16t^2 = 1444 - 16(7)^2$

       $= 660$

       After 7 seconds the height is 660 feet.

   c.  When it hits the ground, the height is 0.

       Let $0 = 1444 - 16t^2$

       $16t^2 = 1444$

       $t^2 = 90.25$

       $t = \sqrt{90.25}$

       $t = 9.5$

       Thus, it will hit the ground after about 10 seconds.

   d.  $1444 - 16t^2 = 4(361 - 4t^2)$

       $= 4\left[(19)^2 - (2t)^2\right]$

       $= 4(19 + 2t)(19 - 2t)$

**91.** Answers may vary.

**Integrated Review -Choosing a Factoring Strategy**

**1.** $a^2 + 2ab + b^2 = (a + b)(a + b) = (a + b)^2$

**2.** $a^2 - 2ab + b^2 = (a - b)(a - b) = (a - b)^2$

**3.** $a^2 + a - 12 = (a - 3)(a + 4)$

**4.** $a^2 - 7a + 10 = (a - 5)(a - 2)$

**5.** $a^2 - a - 6 = (a - 3)(a + 2)$

**6.** $a^2 + 2a + 1 = (a + 1)(a + 1) = (a + 1)^2$

**7.** $x^2 + 2x + 1 = (x + 1)(x + 1) = (x + 1)^2$

**8.** $x^2 + x - 2 = (x + 2)(x - 1)$

**9.** $x^2 + 4x + 3 = (x + 3)(x + 1)$

**10.** $x^2 + x - 6 = (x + 3)(x - 2)$

**11.** $x^2 + 7x + 12 = (x + 4)(x + 3)$

**12.** $x^2 + x - 12 = (x + 4)(x - 3)$

**13.** $x^2 + 3x - 4 = (x + 4)(x - 1)$

**14.** $x^2 - 7x + 10 = (x - 5)(x - 2)$

**15.** $x^2 + 2x - 15 = (x + 5)(x - 3)$

**16.** $x^2 + 11x + 30 = (x + 6)(x + 5)$

**17.** $x^2 - x - 30 = (x - 6)(x + 5)$

**18.** $x^2 + 11x + 24 = (x + 8)(x + 3)$

**19.** $2x^2 - 98 = (x^2 - 49)$

$\qquad = 2(x^2 - 7^2)$

$\qquad = 2(x+7)(x-7)$

**20.** $3x^2 - 75 = 3(x^2 - 25)$

$\qquad = 3(x^2 - 5^2)$

$\qquad = 3(x+5)(x-5)$

**21.** $x^2 + 3x + xy + 3y = x(x+3) + y(x+3)$

$\qquad = (x+3)(x+y)$

**22.** $3y - 21 + xy - 7x = 3(y-7) + x(y-7)$

$\qquad = (y-7)(3+x)$

**23.** $x^2 + 6x - 16 = (x+8)(x-2)$

**24.** $x^2 - 3x - 28 = (x-7)(x+4)$

**25.** $4x^3 + 20x^2 - 56x = 4x(x^2 + 5x - 14)$

$\qquad = 4x(x+7)(x-2)$

**26.** $6x^3 - 6x^2 - 120x = 6x(x^2 - x - 20)$

$\qquad = 6x(x-5)(x+4)$

**27.** $12x^2 + 34x + 24 = 2(6x^2 + 17x + 12)$

$\qquad = 2(6x^2 + 9x + 8x + 12)$

$\qquad = 2[3x(2x+3) + 4(2x+3)]$

$\qquad = 2(2x+3)(3x+4)$

**28.** $8a^2 + 6ab - 5b^2 = 8a^2 + 10ab - 4ab - 5b^2$

$\qquad = 2a(4a+5b) - b(4a+5b)$

$\qquad = (4a+5b)(2a-b)$

**29.** $4a^2 - b^2 = (2a)^2 - b^2 = (2a+b)(2a-b)$

**30.** $28 - 13x - 6x^2 = 28 - 21x + 8x - 6x^2$

$\qquad = 7(4-3x) + 2x(4-3x) = (4-3x)(7+2x)$

**31.** $20 - 3x - 2x^2 = 20 - 8x + 5x - 2x^2$

$\qquad = 4(5-2x) + x(5-2x)$

$\qquad = (5-2x)(4+x)$

**32.** $x^2 - 2x + 4$ is a prime polynomial.

**33.** $a^2 + a - 3$ is a prime polynomial.

**34.** $6y^2 + y - 15 = 6y^2 + 10y - 9y - 15$

$\qquad = 2y(3y+5) - 3(3y+5)$

$\qquad = (3y+5)(2y-3)$

**35.** $4x^2 - x - 5 = 4x^2 - 5x + 4x - 5$

$\qquad = x(4x-5) + 1(4x-5)$

$\qquad = (4x-5)(x+1)$

**36.** $x^2y - y^3 = y(x^2 - y^2) = y(x-y)(x+y)$

**37.** $4t^2 + 36;\ \text{GCF} = 4$

$\qquad 4(t^2 + 9)$

**38.** $x^2 + x + xy + y = x(x+1) + y(x+1)$

$\qquad = (x+1)(x+y)$

**39.** $ax + 2x + a + 2 = x(a+1) + 1(a+2)$

$\qquad = (a+2)(x+1)$

**40.** $18x^3 - 63x^2 + 9x = 9x(2x^2 - 7x + 1)$

**41.** $12a^3 - 24a^2 + 4a = 4a(3a^2 - 6a + 1)$

**42.** $x^2 + 14x - 32 = (x + 16)(x - 2)$

**43.** $x^2 - 14x - 48$ is prime

**44.** $16a^2 - 56ab + 49b^2$
$$= (4a)^2 - 2(4a)(7b) + (7b)^2$$
$$= (4a - 7b)^2$$

**45.** $25p^2 - 70pq + 49q^2$
$$= (5p)^2 - 2(5p)(7q) + (7q)^2$$
$$= (5p - 7q)^2$$

**46.** $7x^2 + 24xy + 9y^2 = 7x^2 + 3xy + 21xy + 9y^2$
$$= x(7x + 3y) + 3y(7x + 3y)$$
$$= (7x + 3y)(x + 3y)$$

**47.** $125 - 8y^3 = 5^3 - (2y)^3$
$$= (5 - 2y)(25 + 10y + 4y^2)$$

**48.** $64x^3 + 27 = (4x)^3 + 3^3$
$$= (4x + 3)(16x^2 - 12x + 9)$$

**49.** $-x^2 - x + 30 = -1(x^2 + x - 30)$
$$= -(x + 6)(x - 5)$$

**50.** $-x^2 + 6x - 8 = -1(x^2 - 6x + 8)$
$$= -(x - 2)(x - 4)$$

**51.** $14 + 5x - x^2 = (7 - x)(2 + x)$

**52.** $3 - 2x - x^2 = (3 + x)(1 - x)$

**53.** $3x^4y + 6x^3y - 72x^2y = 3x^2y(x^2 + 2x - 24)$
$$= 3x^2y(x + 6)(x - 4)$$

**54.** $2x^3y + 8x^2y^2 - 10xy^3 = 2xy(x^2 + 4xy - 5y^2)$
$$= 2xy(x + 5y)(x - y)$$

**55.** $5x^3y^2 - 40x^2y^3 + 35xy^4$; GCF $= 5xy^2$
$$5xy^2(x^2 - 8xy + 7y^2)$$
$$= 5xy^2(x - 7y)(x - y)$$

**56.** $4x^4y - 8x^3y - 60x^2y = 4x^2y(x^2 - 2x - 15)$
$$= 4x^2y(x - 5)(x + 3)$$

**57.** $12x^3y + 243xy = 3xy(4x^2 + 81)$

**58.** $6x^3y^2 + 8xy^2 = 2xy^2(3x^2 + 4)$

**59.** $4 - x^2 = 2^2 - x^2 = (2 + x)(2 - x)$

**60.** $9 - y^2 = 3^2 - y^2 = (3 + y)(3 - y)$

**61.** $3rs - s + 12r - 4 = s(3r - 1) + 4(3r - 1)$
$$= (3r - 1)(s + 4)$$

**62.** $x^3 - 2x^2 + 3x - 6 = x^2(x - 2) + 3(x - 2)$
$$= (x - 2)(x^2 + 3)$$

**63.** $4x^2 - 8xy - 3x + 6y = 4x(x - 2y) - 3(x - 2y)$
$$= (x - 2y)(4x - 3)$$

**64.** $4x^2 - 2xy - 7yz + 14xz$

$= 2x(2x - y) + 7z(-y + 2x)$

$= (2x - y)(2x + 7z)$

**65.** $6x^2 + 18xy + 12y^2 = 6(x^2 + 3xy + 2y^2)$

$= 6(x + 2)(x + y)$

**66.** $12x^2 + 46xy - 8y^2 = 2(6x^2 + 23xy - 4y^2)$

$= 2(6x^2 + 24xy - xy - 4y^2)$

$= 2[6x(x + 4y) - y(x + 4y)]$

$= 2(x + 4y)(6x - y)$

**67.** $xy^2 - 4x + 3y^2 - 12 = x(y^2 - 4) + 3(y^2 - 4)$

$= (y^2 - 4)(x + 3)$

$= (y^2 - 2^2)(x + 3)$

$= (y + 2)(y - 2)(x + 3)$

**68.** $x^2y^2 - 9x^2 + 3y^2 - 27 = x^2(y^2 - 9) + 3(y^2 - 9)$

$= (y^2 - 9)(x^2 + 3)$

$= (y - 3)(y + 3)(x^2 + 3)$

**69.** $5(x + y) + x(x + y) = (x + y)(5 + x)$

**70.** $7(x - y) + y(x - y) = (x - y)(7 + y)$

**71.** $14t^2 - 9t + 1 = 14t^2 - 7t - 2t + 1$

$= 7t(2t - 1) - 1(2t - 1)$

$= (2t - 1)(7t - 1)$

**72.** $3t^2 - 5t + 1$ is a prime polynomial

**73.** $3x^2 + 2x - 5 = 3x^2 + 5x - 3x - 5$

$= x(3x + 5) - 1(3x + 5)$

$= (3x + 5)(x - 1)$

**74.** $7x^2 + 19x - 6 = 7x^2 + 21x - 2x - 6$

$= 7x(x + 3) - 2(x + 3)$

$= (x + 3)(7x - 2)$

**75.** $x^2 + 9xy - 36y^2 = (x + 12y)(x - 3y)$

**76.** $\begin{aligned} & 3x^2 + 10xy - 8y^2 \\ &= 3x^2 - 2xy + 12xy - 8y^2 \\ &= x(3x - 2y) + 4y(3x - 2y) \\ &= (3x - 2y)(x + 4y) \end{aligned}$

**77.** $1 - 8ab - 20a^2b^2$

$= 1 - 10ab + 2ab - 20a^2b^2$

$= 1(1 - 10ab) + 2ab(1 - 10ab)$

$= (1 - 10ab)(1 + 2ab)$

**78.** $1 - 7ab - 60a^2b^2$

$= 1 - 12ab + 5ab - 60a^2b^2$

$= 1(1 - 12ab) + 5ab(1 - 12ab)$

$= (1 - 12ab)(1 + 5ab)$

**79.** $9 - 10x^2 + x^4 = (9 - x^2)(1 - x^2)$

$= (3 + x)(3 - x)(1 + x)(1 - x)$

**80.** $36 - 13x^2 + x^4 = (9 - x^2)(4 - x^2)$

$= (3 + x)(3 - x)(2 + x)(2 - x)$

**81.** $x^4 - 14x^2 - 32 = (x^2 + 2)(x^2 - 16)$
$$= (x^2 + 2)(x + 4)(x - 4)$$

**82.** $x^4 - 22x^2 - 75 = (x^2 + 3)(x^2 - 25)$
$$= (x^2 + 3)(x + 5)(x - 5)$$

**83.** $x^2 - 23x + 120 = (x - 15)(x - 8)$

**84.** $y^2 + 22y + 96 = (y + 16)(y + 6)$

**85.** $6x^3 - 28x^2 + 16x;\ \text{GCF} = 2x$
$$2x(3x^2 - 14x + 8) = 2x(3x - 2)(x - 4)$$

**86.** $6y^3 - 8y^2 - 30y = 2y(3y^2 - 4y - 15)$
$$= 2y(3y + 5)(y - 3)$$

**87.** $27x^3 - 125y^3$
$$= (3x)^3 - (5y)^3$$
$$= (3x - 5y)(9x^2 + 15xy + 25y^2)$$

**88.** $216y^3 - z^3 = (6y)^3 - z^3$
$$= (6y - z)(36y^2 + 6yz + z^2)$$

**89.** $x^3y^3 + 8z^3 = (xy)^3 + (2z)^3$
$$= (xy + 2z)(x^2y^2 - 2xyz + 4z^2)$$

**90.** $27a^3b^3 + 8 = (3ab)^3 + 2^3$
$$= (3ab + 2)(9a^2b^2 - 6ab + 4)$$

**91.** $2xy - 72x^3y = 2xy(1 - 36x^2)$
$$= 2xy(1^2 - (6x)^2) = 2xy(1 + 6x)(1 - 6x)$$

**92.** $2x^3 - 18x = 2x(x^2 - 9)$
$$= 2x(x^2 - 3^2)$$
$$= 2x(x + 3)(x - 3)$$

**93.** $x^3 + 6x^2 - 4x - 24 = x^2(x + 6) - 4(x + 6)$
$$= (x + 6)(x^2 - 4) = (x + 6)(x^2 - 2^2)$$
$$= (x + 6)(x + 2)(x - 2)$$

**94.** $x^3 - 2x^2 - 36x + 72$
$$= x^2(x - 2) - 36(x - 2)$$
$$= (x - 2)(x^2 - 36) = (x - 2)(x^2 - 6^2)$$
$$= (x - 2)(x + 6)(x - 6)$$

**95.** $6a^3 + 10a^2 = 2a^2(3a + 5)$

**96.** $4n^2 - 6n = 2n(2n - 3)$

**97.** $a^2(a + 2) + 2(a + 2) = (a + 2)(a^2 + 2)$

**98.** $a - b + x(a - b) = (a - b)(1 + x)$

**99.** $x^3 - 28 + 7x^2 - 4x = x^3 + 7x^2 - 28 - 4x$
$$= x^2(x + 7) - 4(7 + x)$$
$$= x^2(x + 7) - 4(x + 7)$$
$$= (x + 7)(x^2 - 4)$$
$$= (x + 7)(x + 2)(x - 2)$$

**100.** $a^3 - 45 - 9a + 5a^2 = a^3 + 5a^2 - 9a - 45$
$$= a^2(a + 5) - 9(a + 5)$$
$$= (a + 5)(a^2 - 9)$$
$$= (a + 5)(a + 3)(a - 3)$$

**101.** $(x-y)^2 - z^2 = (x-y+z)(x-y-z)$

**102.** $(x+2y)^2 - 9 = (x+2y)^2 - 3^2$
$$= (x+2y+3)(x+2y-3)$$

**103.**

$81 - (5x+1)^2 = 9^2 - (5x+1)^2$
$$= \left[9+(5x+1)\right]\left[9-(5x+1)\right]$$
$$= (9+5x+1)(9-5x-1)$$

**104.** $b^2 - (4a+c)^2$
$$= \left[b+(4a+c)\right]\left[b-(4a+c)\right]$$
$$= (b+4a+c)(b-4a-c)$$

**105.** Answers may vary.

**106.** Yes. $9\left(x^2 + 9y^2\right)$

**107.** A,C

**Graphing Calculator Explorations 6.5**

**1.** $-0.9$, $2.2$

**3.** no real solutions

**5.** $-1.8$, $2.8$

**Mental Math 6.5**

**1.** $(a-3)(a-7) = 0$
$a-3 = 0$ or $a-7 = 0$
$a = 3$         $a = 7$
The solutions are 3 and 7.

**2.** $(a-5)(a-2) = 0$
$a-5 = 0$ or $a-2 = 0$
$a = 5$         $a = 2$
The solutions are 5 and 2.

**3.** $(x+8)(x+6) = 0$
$x+8 = 0$ or $x+6 = 0$
$x = -8$         $x = -6$
The solutions are $-8$ and $-6$.

**4.** $(x+2)(x+3) = 0$
$x+2 = 0$ or $x+3 = 0$
$x = -2$         $x = -3$
The solutions are $-2$ and $-3$.

**5.** $(x+1)(x-3) = 0$
$x+1 = 0$ or $x-3 = 0$
$x = -1$         $x = 3$
The solutions are $-1$ and 3.

**6.** $(x-1)(x+2) = 0$
$x-1 = 0$ or $x+2 = 0$
$x = 1$         $x = -2$
The solutions are 1 and $-2$.

**Exercise Set 6.5**

**1.** $(x-2)(x+1) = 0$
$x-2 = 0$ or $x+1 = 0$
$x = 2$         $x = -1$
The solutions are 2 and $-1$.

**3.** $x(x+6) = 0$
$x = 0$ or $x+6 = 0$
$x = -6$
The solutions are 0 and $-6$.

**5.** $(2x+3)(4x-5)=0$

$2x+3=0 \quad$ or $\quad 4x-5=0$

$2x=-3 \qquad\qquad 4x=5$

$x=-\dfrac{3}{2} \qquad\qquad x=\dfrac{5}{4}$

The solutions are $-\dfrac{3}{2}$ and $\dfrac{5}{4}$.

**7.** $(2x-7)(7x+2)=0$

$2x-7=0 \quad$ or $\quad 7x+2=0$

$2x=7 \qquad\qquad 7x=-2$

$x=\dfrac{7}{2} \qquad\qquad x=-\dfrac{2}{7}$

The solutions are $\dfrac{7}{2}$ and $-\dfrac{2}{7}$.

**9.** If $x=6$ and $x=-1$ are the solutions, then

$x=6 \quad$ or $\quad x=-1$

$x-6=0 \qquad x+1=0$

$(x-6)(x+1)=0$

**11.** $x^2-13x+36=0$

$(x-9)(x-4)=0$

$x-9=0 \quad$ or $\quad x-4=0$

$x=9 \qquad\qquad x=4$

The solutions are 9 and 4.

**13.** $x^2+2x-8=0$

$(x+4)(x-2)=0$

$x+4=0 \quad$ or $\quad x-2=0$

$x=-4 \qquad\qquad x=2$

The solutions are $-4$ and 2.

**15.** $x^2-4x=32$

$x^2-4x-32=0$

$(x-8)(x+4)=0$

$x-8=0 \quad$ or $\quad x+4=0$

$x=8 \qquad\qquad x=-4$

The solutions are 8 and $-4$.

**17.** $x(3x-1)=14$

$3x^2-x=14$

$3x^2-x-14=0$

$(3x-7)(x+2)=0$

$3x-7=0 \quad$ or $\quad x+2=0$

$3x=7 \qquad\qquad x=-2$

$x=\dfrac{7}{3}$

The solutions are $\dfrac{7}{3}$ and $-2$.

**19.** $3x^2+19x-72=0$

$(3x-8)(x+9)=0$

$3x-8=0 \quad$ or $\quad x+9=0$

$3x=9 \qquad\qquad x=-9$

$x=\dfrac{8}{3}$

The solutions are $\dfrac{8}{3}$ and $-9$.

**21.** If the solutions are $x=5$ and $x=7$,

then, by the zero factor property,

$x=5 \quad$ or $\quad x=7$

$x-5=0 \qquad x-7=0$

$(x-5)(x-7)=0$

$x^2-5x-7x+35=0$

$x^2-12x+35=0$

**23.**  $x^3 - 12x^2 + 32x = 0$

$x(x^2 - 12x + 32) = 0$

$x(x - 8)(x - 4) = 0$

$x = 0$   or   $x - 8 = 0$   or   $x - 4 = 0$

$x = 0$   or   $x = 8$   or   $x = 4$

**25.**  $(4x - 3)(16x^2 - 24x + 9) = 0$

$(4x - 3)(4x - 3)^2 = 0$

$(4x - 3)^3 = 0$

$4x - 3 = 0$

$4x = 3$

$x = \dfrac{3}{4}$

The solution is $\dfrac{3}{4}$.

**27.**          $4x^3 - x = 0$

$x(4x^2 - 1) = 0$

$x(2x + 1)(2x - 1) = 0$

$x = 0$   or   $2x + 1 = 0$   or   $2x - 1 = 0$

$x = 0$   or   $2x = -1$   or   $2x = 1$

$x = 0$   or   $x = -\dfrac{1}{2}$   or   $x = \dfrac{1}{2}$

**29.**      $32x^3 - 4x^2 - 6x = 0$

$2x(16x^2 - 2x - 3) = 0$

$2x(2x - 1)(8x + 3) = 0$

$2x = 0$   or   $2x - 1 = 0$   or   $8x + 3 = 0$

$x = \dfrac{0}{2}$   or   $2x = 1$   or   $8x = -3$

$x = 0$   or   $x = \dfrac{1}{2}$   or   $x = -\dfrac{3}{8}$

**31.**  $x(x + 7) = 0$

$x = 0$   or   $x + 7 = 0$

$x = 0$   or   $x = -7$

**33.**  $(x + 5)(x - 4) = 0$

$x + 5 = 0$   or   $x - 4 = 0$

$x = -5$  or   $x = 4$

**35.**          $x^2 - x = 30$

$x^2 - x - 30 = 0$

$(x - 6)(x + 5) = 0$

$x - 6 = 0$   or   $x + 5 = 0$

$x = 6$   or   $x = -5$

**37.**  $6y^2 - 22y - 40 = 0$

$2(3y^2 - 11y - 20) = 0$

$2(3y + 4)(y - 5) = 0$

$3y + 4 = 0$   or   $y - 5 = 0$

$3y = -4$          $y = 5$

$y = -\dfrac{4}{3}$

The solutions are $-\dfrac{4}{3}$ and 5.

**39.**  $(2x + 3)(2x^2 - 5x - 3) = 0$

$(2x + 3)(2x + 1)(x - 3) = 0$

$2x + 3 = 0$   or   $2x + 1 = 0$   or   $x - 3 = 0$

$2x = -3$          $2x = -1$          $x = 3$

$x = -\dfrac{3}{2}$          $x = -\dfrac{1}{2}$

The solutions are $-\dfrac{3}{2}$, $-\dfrac{1}{2}$, and 3.

**41.** $x^2 - 15 = -2x$

$x^2 + 2x - 15 = 0$

$(x+5)(x-3) = 0$

$x + 5 = 0$    or    $x - 3 = 0$

$x = -5$        $x = 3$

The solutions are $-5$ and $3$.

**43.** $x^2 - 16x = 0$

$x(x-16) = 0$

$x = 0$    or    $x - 16 = 0$

$x = 0$    or      $x = 16$

**45.** $-18y^2 - 33y + 216 = 0$

$-3(6y^2 + 11y - 72) = 0$

$-3(3y - 8)(2y + 9) = 0$

$3y - 8 = 0$    or    $2y + 9 = 0$

$3y = 8$    or      $2y = -9$

$y = \dfrac{8}{3}$    or    $y = -\dfrac{9}{2}$

**47.** $12x^2 - 59x + 55 = 0$

$(4x - 5)(3x - 11) = 0$

$4x - 5 = 0$    or    $3x - 11 = 0$

$4x = 5$    or    $3x = 11$

$x = \dfrac{5}{4}$    or    $x = \dfrac{11}{3}$

**49.** $18x^2 + 9x - 2 = 0$

$(3x + 2)(6x - 1) = 0$

$3x + 2 = 0$    or    $6x - 1 = 0$

$3x = -2$    or    $6x = 1$

$x = -\dfrac{2}{3}$    or    $x = \dfrac{1}{6}$

**51.** $x(6x + 7) = 5$

$6x^2 + 7x = 5$

$6x^2 + 7x - 5 = 0$

$(3x + 5)(2x - 1) = 0$

$3x + 5 = 0$    or    $2x - 1 = 0$

$3x = -5$    or      $2x = 1$

$x = -\dfrac{5}{3}$    or      $x = \dfrac{1}{2}$

**53.** $4(x - 7) = 6$

$4x - 28 = 6$

$4x = 34$

$x = \dfrac{34}{4}$

$x = \dfrac{17}{2}$

The solution is $\dfrac{17}{2}$.

**55.** $5x^2 - 6x - 8 = 0$

$(5x + 4)(x - 2) = 0$

$5x + 4 = 0$    or    $x - 2 = 0$

$5x = -4$    or      $x = 2$

$x = -\dfrac{4}{5}$    or      $x = 2$

**57.** $(y - 2)(y + 3) = 6$

$y^2 - 2y + 3y - 6 = 6$

$y^2 + y - 12 = 0$

$(y + 4)(y - 3) = 0$

$y + 4 = 0$    or    $y - 1 = 0$

$y = -4$        $y = 3$

The solutions are $-4$ and $3$.

**59.** $4y^2 - 1 = 0$

$(2y+1)(2y-1) = 0$

$2y+1 = 0$ or $2y-1 = 0$

$2y = -1$ $\qquad$ $2y = 1$

$y = -\dfrac{1}{2}$ $\qquad$ $y = \dfrac{1}{2}$

The solutions are $-\dfrac{1}{2}$ and $\dfrac{1}{2}$.

**61.** $t^2 + 13t + 22 = 0$

$(t+11)(t-+2) = 0$

$t+11 = 0$ $\quad$ or $\quad$ $t+2 = 0$

$t = -11$ or $\qquad$ $t = -2$

**63.** $5t - 3 = 12$

$5t = 12 + 3$

$5t = 15$

$t = \dfrac{15}{5}$

$t = 3$

**65.** $x^2 + 6x - 17 = -26$

$x^2 + 6x - 17 + 26 = 0$

$x^2 + 6x + 9 = 0$

$(x+3)(x+3) = 0$

$x+3 = 0$

$x = -3$

**67.** $12x^2 + 7x - 12 = 0$

$(4x-3)(3x+4) = 0$

$4x-3 = 0$ $\quad$ or $\quad$ $3x+4 = 0$

$x = \dfrac{3}{4}$ $\qquad$ $x = -\dfrac{4}{3}$

The solutions are $-\dfrac{4}{3}$ and $\dfrac{3}{4}$.

**69.** $10t^3 - 25t - 15t^2 = 0$

$10t^3 - 15t^2 - 25t = 0$

$5t(2t^2 - 3t - 5) = 0$

$5t(2t-5)(t+1) = 0$

$5t = 0$ $\quad$ or $\quad$ $2t-5 = 0$ $\quad$ or $\quad$ $t+1 = 0$

$t = \dfrac{0}{5}$ $\quad$ or $\qquad$ $2t = 5$ $\quad$ or $\qquad$ $t = -1$

$t = 0$ $\quad$ or $\qquad$ $t = \dfrac{5}{2}$ $\quad$ or $\qquad$ $t = -1$

**71.** Let $y = 0$ and solve for $x$.

$y = (3x+4)(x-1)$

$0 = (3x+4)(x-1)$

$3x+4 = 0$ $\quad$ or $\quad$ $x-1 = 0$

$3x = -4$ $\quad$ or $\qquad$ $x = 1$

$x = -\dfrac{4}{3}$

The $x$-intercepts are $\left(-\dfrac{4}{3}, 0\right)$ and $(1, 0)$.

**73.** Let $y = 0$ and solve for $x$.

$y = x^2 - 3x - 10$

$0 = x^2 - 3x - 10$

$0 = (x-5)(x+2)$

$x-5 = 0$ $\quad$ or $\quad$ $x+2 = 0$

$x = 5$ $\quad$ or $\qquad$ $x = -2$

The $x$-intercepts are $(5, 0)$ and $(-2, 0)$.

**75.** Let $y = 0$ and solve for $x$.

$y = 2x^2 + 11x - 6$

$0 = 2x^2 + 11x - 6$

$0 = (2x-1)(x+6)$

$$2x - 1 = 0 \quad \text{or} \quad x + 6 = 0$$
$$2x = 1 \quad \text{or} \quad x = -6$$
$$x = \frac{1}{2}$$

The $x$-intercepts are $\left(\frac{1}{2}, 0\right)$ and $(-6, 0)$.

**77.** E; $x$-intercepts are $(-2, 0), (1, 0)$

**79.** B; $x$-intercepts are $(0, 0), (-3, 0)$

**81.** C; $y = 2x^2 - 8 = 2(x - 2)(x + 2)$

$x$-intercepts are $(2, 0), (-2, 0)$

**83.**
$$\frac{3}{5} + \frac{4}{9} = \frac{3 \cdot 9}{5 \cdot 9} + \frac{4 \cdot 5}{9 \cdot 5}$$
$$= \frac{27}{45} + \frac{20}{45}$$
$$= \frac{27 + 20}{45} = \frac{47}{45}$$

**85.**
$$\frac{7}{10} - \frac{5}{12} = \frac{7 \cdot 6}{10 \cdot 6} - \frac{5 \cdot 5}{12 \cdot 5}$$
$$= \frac{42}{60} - \frac{25}{60}$$
$$= \frac{42 - 25}{60}$$
$$= \frac{17}{60}$$

**87.** $\dfrac{7}{8} \div \dfrac{7}{15} = \dfrac{7}{8} \cdot \dfrac{15}{7} = \dfrac{15}{8}$

**89.** $\dfrac{4}{5} \cdot \dfrac{7}{8} = \dfrac{4 \cdot 7}{5 \cdot 8} = \dfrac{4 \cdot 7}{5 \cdot 2 \cdot 4} = \dfrac{7}{10}$

**91.** $y = -16x^2 + 20x + 300$

a.

| time $x$ | 0 | 1 | 2 | 3 | 4 | 5 | 6 |
|---|---|---|---|---|---|---|---|
| height $y$ | 300 | 304 | 276 | 216 | 124 | 0 | -156 |

b. The compass strikes the ground after 5 seconds, when the height, $y$, is zero feet.

c. The maximum height was approximately 304 feet.

d.

**93.**
$$(x - 3)(3x + 4) = (x + 2)(x - 6)$$
$$3x^2 - 5x - 12 = x^2 - 4x - 12$$
$$2x^2 - x = 0$$
$$x(2x - 1) = 0$$
$$2x - 1 = 0 \quad \text{or} \quad x = 0$$
$$x = \frac{1}{2}$$

The solutions are $\dfrac{1}{2}$ and $0$.

**95.** $(2x-3)(x+8) = (x-6)(x+4)$

$2x^2 + 13x - 24 = x^2 - 2x - 24$

$x^2 + 15x = 0$

$x(x+15) = 0$

$x + 15 = 0$ or $x = 0$

$x = -15$

The solutions are $\dfrac{1}{2}$ and 0.

**Exercise Set 6.6**

**1.** Let $x =$ the width, then $x + 4 =$ the length.

**3.** Let $x =$ the first odd integer, then

$x + 2 =$ the next consecutive odd integer.

**5.** Let $x =$ the base, then $4x + 1 =$ the height.

**7.** Let $x =$ the length of one side.

$A = x^2$

$121 = x^2$

$0 = x^2 - 121$

$0 = x^2 - 11^2$

$0 = (x+11)(x-11)$

$x + 11 = 0$ or $x - 11 = 0$

$x = -11$           $x = 11$

Since the length cannot be negative, the sides are 11 units long.

**9.** The perimeter is the sum of the lengths of the sides.

$120 = (x+5) + (x^2 - 3x) + (3x - 8)(x+3)$

$120 = x + 5 + x^2 - 3x + 3x - 8 + x + 3$

$120 = x^2 + 2x$

$0 = x^2 + 2x - 120$

$x^2 + 2x - 120 = 0$

$(x+12)(x-10) = 0$

$x + 12 = 0$ or $x - 10 = 0$

$x = -12$           $x = 10$

Since the dimensions cannot be negative, the lengths of the sides are:

$10 + 5 = 15$ cm, $10^2 - 3(10) = 70$ cm,

$3(10) - 8 = 22$ cm, and $10 + 3 = 13$ cm.

**11.** $x + 5 =$ the base and $x - 5 =$ the height.

$A = bh$

$96 = (x+5)(x-5)$

$96 = x^2 + 5x - 5x - 25$

$96 = x^2 - 25$

$0 = x^2 - 121$

$x^2 - 121 = 0$

$(x+11)(x-11) = 0$

$x + 11 = 0$ or $x - 11 = 0$

$x = -11$           $x = 11$

Since the dimensions cannot be negative, $x = 11$. The base is $11 + 5 = 16$ miles, and the height is $11 - 5 = 6$ miles.

**13.** Find $t$ when $h = 0$.

$$h = -16t^2 + 64t + 80$$
$$0 = -16t^2 + 64t + 80$$
$$0 = -16\left(t^2 - 4t - 5\right)$$
$$0 = -16(t-5)(t+1)$$
$$t - 5 = 0 \quad \text{or} \quad t + 1 = 0$$
$$t = 5 \qquad\qquad t = -1$$

Since the time $t$ cannot be negative, the object hits the ground after 5 seconds.

**15.** Let $x$ = the width then $2x - 7$ = the length.

$$A = lw$$
$$30 = (2x-7)(x)$$
$$30 = 2x^2 - 7x$$
$$0 = 2x^2 - 7x - 30$$
$$0 = (2x+5)(x-6)$$
$$2x + 5 = 0 \quad \text{or} \quad x - 6 = 0$$

$$x = -\frac{5}{2} \qquad\qquad x = 6$$

Since the dimensions cannot be negative, the width is 6 cm and the length is $2(6) - 7 = 5$ cm.

**17.** Let $n = 12$.

$$D = \frac{1}{2}n(n-3)$$
$$D = \frac{1}{2} \cdot 12(12-3) = 6(9) = 54$$

A polygon with 12 sides has 54 diagonals.

**19.** Let $D = 35$ and solve for $n$.

$$D = \frac{1}{2}n(n-3)$$
$$35 = \frac{1}{2}n(n-3)$$
$$35 = \frac{1}{2}n^2 - \frac{3}{2}n$$
$$0 = \frac{1}{2}n^2 - \frac{3}{2}n - 35$$
$$0 = \frac{1}{2}\left(n^2 - 3n - 70\right)$$
$$0 = \frac{1}{2}(n-10)(n+7)$$
$$n - 10 = 0 \quad \text{or} \quad n + 7 = 0$$
$$n = 10 \qquad\qquad n = -7$$

The polygon has 10 sides.

**21.** Let $x$ = the unknown number.

$$x + x^2 = 132$$
$$x^2 + x - 132 = 0$$
$$(x+12)(x-11) = 0$$

$$x + 12 = 0 \quad \text{or} \quad x - 11 = 0$$
$$x = -12 \qquad\qquad x = 11$$

The two numbers are $-12$ and 11.

**23.** Let $x$ = the rate (in mph) of the slower boat, then $x + 7$ = the rate (in mph) of the faster boat. After one hour, the slower boat has traveled $x$ miles and the faster boat has traveled $x + 7$ miles. By the Pythagorean theorem,

$$x^2 + (x+7)^2 = 17^2$$
$$x^2 + x^2 + 14x + 49 = 289$$
$$2x^2 + 14x + 49 = 289$$

$$2x^2 + 14x - 240 = 0$$
$$2(x^2 + 7x - 120) = 0$$
$$2(x + 15)(x - 8) = 0$$
$$x + 15 = 0 \quad \text{or} \quad x - 8 = 0$$
$$x = -15 \qquad\qquad x = 8$$

Since the rate cannot be negative, the slower boat travels at 8 mph. The faster boat travels at $8 + 7 = 15$ mph.

**25.** Let $x =$ the first number, then $20 - x =$ the other number.

$$x^2 + (20 - x)^2 = 218$$
$$x^2 + 400 - 40x + x^2 = 218$$
$$2x^2 - 40x + 400 = 218$$
$$2x^2 - 40x + 182 = 0$$
$$2(x^2 - 20x + 91) = 0$$
$$2(x - 13)(x - 7) = 0$$

$$x - 13 = 0 \quad \text{or} \quad x - 7 = 0$$
$$x = 13 \qquad\qquad x = 7$$

The numbers are 13 and 7.

**27.** Let $x =$ the length of a side of the original square. Then $x + 3 =$ the length of a side of the larger square.

$$64 = (x + 3)^2$$
$$64 = x^2 + 6x + 9$$
$$0 = x^2 + 6x - 55$$
$$0 = (x + 11)(x - 5)$$
$$x + 11 = 0 \quad \text{or} \quad x - 5 = 0$$
$$x = -11 \qquad\qquad x = -5$$

Since the length cannot be negative, the sides of the original square are 5 inches long.

**29.** Let $x =$ the length of the shorter leg. Then $x + 4 =$ the length of the longer leg and $x + 8 =$ the length of the hypotenuse. By the Pythagorean theorem,

$$x^2 + (x + 4)^2 = (x + 8)^2$$
$$x^2 + x^2 + 8x + 16 = x^2 + 16x + 64$$
$$x^2 - 8x - 48 = 0$$
$$(x - 12)(x + 4) = 0$$
$$x - 12 = 0 \quad \text{or} \quad x + 4 = 0$$
$$x = 12 \qquad\qquad x = -4$$

Since the length cannot be negative, the sides of the triangle are 12 mm, $12 + 4 = 16$ mm, and $12 + 8 = 20$ mm.

**31.** Let $x =$ the height of the triangle, then $2x =$ the base.

$$A = \frac{1}{2}bh$$
$$100 = \frac{1}{2}(2x)(x)$$
$$100 = x^2$$
$$0 = x^2 - 100$$
$$0 = (x + 10)(x - 10)$$
$$x + 10 = 0 \quad \text{or} \quad x - 10 = 0$$
$$x = -10 \qquad\qquad x = 10$$

Since the altitude cannot be negative, the height of the triangle is 10 km.

**33.** Let $x$ = the length of the shorter leg,
then $x + 12$ = the length of the longer leg
and $2x - 12$ = the length of the hypotenuse.
By the Pythagorean theorem,

$$x^2 + (x+12)^2 = (2x-12)^2$$
$$x^2 + x^2 + 24x + 144 = 4x^2 - 48x + 144$$
$$0 = 2x^2 - 72x$$
$$0 = 2x(x - 36)$$
$$2x = 0 \quad \text{or} \quad x - 36 = 0$$
$$x = 0 \qquad\qquad x = 36$$

Since the length cannot be zero feet,
the shorter leg is 36 feet long.

**35.** Find $t$ when $h = 0$.

$$h = -16t^2 + 625$$
$$0 = -16t^2 + 625$$
$$0 = -(4t + 25)(4t - 25)$$

$$4t + 25 = 0 \quad \text{or} \quad 4t - 25 = 0$$
$$4t = -25 \qquad\qquad 4t = 25$$
$$t = -6.25 \qquad\qquad t = 6.25$$

Since the time cannot be negative, the
solution is 6.25 seconds.

**37.** Let $P = 100$ and $A = 144$

$$A = P(1 + r)^2$$
$$144 = 100(1 + r)^2$$
$$1.2 = 1 + r$$
$$0.2 = r$$

The interest rate is 20%.

**39.** Let $x$ = the length and $x - 7$ = the width.

$$A = lw$$
$$120 = (x - 7)(x)$$
$$120 = x^2 - 7x$$
$$0 = x^2 - 7x - 120$$
$$0 = (x + 8)(x - 15)$$
$$x + 8 = 0 \quad \text{or} \quad x - 15 = 0$$
$$x = -8 \qquad\qquad x = 15$$

Since the length cannot be negative,
the length is 15 miles. The width is
$15 - 7 = 8$ miles.

**41.** Let $C = 9500$

$$C = x^2 - 15x + 50$$
$$9500 = x^2 - 15x + 50$$
$$0 = x^2 - 15x - 9450$$
$$0 = (x + 90)(x - 105)$$

$$x + 90 = 0 \quad \text{or} \quad x - 105 = 0$$
$$x = -90 \qquad\qquad x = 105$$

Since the number of units cannot
be negative the solution is 105 units.

**43.** 175 acres

**45.** 6.25 million

**47.** 1966

**49.** Answers may vary

**51.** $\dfrac{24}{32} = \dfrac{2 \cdot 2 \cdot 2 \cdot 3}{2 \cdot 2 \cdot 2 \cdot 2 \cdot 2} = \dfrac{3}{4}$

**53.** $\dfrac{15}{27} = \dfrac{3 \cdot 5}{3 \cdot 3 \cdot 3} = \dfrac{5}{9}$

**55.** $\dfrac{45}{50} = \dfrac{3 \cdot 3 \cdot 5}{2 \cdot 5 \cdot 5} = \dfrac{9}{10}$

**57.** Answers may vary.

**59.** Let $x$ = the width of the walk, then
$24 - 2x$ = the length of the garden,
and $16 - 2x$ = the width of the garden.
$A = lw$
$180 = (24 - 2x)(16 - 2x)$
$180 = 384 - 80x + 4x^2$
$0 = 204 - 80x + 4x^2$

$0 = 4\left(51 - 20x + x^2\right)$
$0 = 4(17 - x)(3 - x)$
$17 - x = 0 \quad$ or $\quad 3 - x = 0$
$\quad\quad 17 = x \quad\quad\quad\quad\quad x = 3$

Since the walk cannot be wider than
8 yards, the width of the walk is 3 yards.

**Chapter 6 Review**

**1.** $6x^2 - 15x = 3x(2x - 5)$

**2.** $2x^3y - 6x^2y^2 - 8xy^3 = 2xy\left(x^2 - 3xy - 4y^2\right)$
$\quad\quad\quad\quad\quad\quad\quad\quad\quad = 2xy(x - 4y)(x + y)$

**3.** $20x^2 + 12x;\ \text{GCF} = 4x$
$\quad 4x(5x + 3)$

**4.** $6x^2y^2 - 3xy^3 = 3xy^2(2x - y)$

**5.** $-8x^3y + 6x^2y^2 = -2x^2y(4x - 3y)$

**6.** $3x(2x + 3) - 5(2x + 3) = (2x + 3)(3x - 5)$

**7.** $5x(x + 1) - (x + 1) = (x + 1)(5x - 1)$

**8.** $3x^2 - 3x + 2x - 2 = 3x(x - 1) + 2(x - 1)$
$\quad\quad\quad\quad\quad\quad\quad\quad = (x - 1)(3x + 2)$

**9.** $6x^2 + 10x - 3x - 5 = 2x(3x + 5) - 1(3x + 5)$
$\quad\quad\quad\quad\quad\quad\quad\quad\quad = (3x + 5)(2x - 1)$

**10.** $3a^2 + 9ab + 3b^2 + ab$
$\quad = 3a(a + 3b) + b(3b + a)$
$\quad = 3a(a + 3b) + b(a + 3b)$
$\quad = (a + 3b)(3a + b)$

**11.** $x^2 + 6x + 8 = (x + 4)(x + 2)$

**12.** $x^2 - 11x + 24 = (x - 8)(x - 3)$

**13.** $x^2 + x + 2$ is a prime polynomial.

**14.** $x^2 - 5x - 6 = (x - 6)(x + 1)$

**15.** $x^2 + 2x - 8 = (x + 4)(x - 2)$

**16.** $x^2 + 4xy - 12y^2 = (x + 6y)(x - 2y)$

**17.** $x^2 + 8xy + 15y^2 = (x + 5y)(x + 3y)$

**18.** $3x^2y + 6xy^2 + 3y^3 = 3y\left(x^2 + 2xy + y^2\right)$
$\quad\quad\quad\quad\quad\quad\quad\quad = 3y(x + y)(x + y)$
$\quad\quad\quad\quad\quad\quad\quad\quad = 3y(x + y)^2$

**19.** $72 - 18x - 2x^2 = 2\left(36 - 9x - x^2\right)$
$\quad\quad\quad\quad\quad\quad = 2(3 - x)(12 + x)$

**20.** $32 + 12x - 4x^2 = 4\left(8 + 3x - x^2\right)$

**21.** $2x^2 + 11x - 6 = \left(2x - 1\right)\left(x + 6\right)$

**22.** $4x^2 - 7x + 4$

not factorable, prime

**23.** $4x^2 + 4x - 3 = 4x^2 + 6x - 2x - 3$
$$= 2x\left(2x + 3\right) - 1\left(2x + 3\right)$$
$$= \left(2x + 3\right)\left(2x - 1\right)$$

**24.** $6x^2 + 5xy - 4y^2 = 6x^2 + 8xy - 3xy - 4y^2$
$$= 2x\left(3x + 4y\right) - y\left(3x + 4y\right)$$
$$= \left(3x + 4y\right)\left(2x - y\right)$$

**25.** $6x^2 - 25xy + 4y^2 = \left(6x - y\right)\left(x - 4y\right)$

**26.** $18x^2 - 60x + 50 = 2\left(9x^2 - 30x + 25\right)$
$$= 2\left(3x - 5\right)\left(3x - 5\right)$$
$$= 2\left(3x - 5\right)^2$$

**27.** $2x^2 - 23xy - 39y^2$
$$= 2x^2 - 26xy + 3xy - 39y^2$$
$$= 2x\left(x - 13y\right) + 3y\left(x - 13y\right)$$
$$= \left(x - 13y\right)\left(2x + 3y\right)$$

**28.** $4x^2 - 28xy + 49y^2 = \left(2x - 7y\right)\left(2x - 7y\right)$
$$= \left(2x - 7y\right)^2$$

**29.** $18x^2 - 9xy - 20y^2$
$$= 18x^2 - 24xy + 15xy - 20y^2$$
$$= 6x\left(3x - 4y\right) + 5y\left(3x - 4y\right)$$
$$= \left(3x - 4y\right)\left(6x + 5y\right)$$

**30.** $36x^3y + 24x^2y^2 - 45xy^3$
$$= 3xy\left(12x^2 + 8xy - 15y^2\right)$$
$$= 3xy\left(12x^2 + 18xy - 10y^2 - 15y^2\right)$$
$$= 3xy\left[6x\left(2x + 3y\right) - 5y\left(2x + 3y\right)\right]$$
$$= 3xy\left(2x + 3y\right)\left(6x - 5y\right)$$

**31.** $4x^2 - 9 = \left(2x\right)^2 - 3^2 = \left(2x + 3\right)\left(2x - 3\right)$

**32.** $9t^2 - 25s^2 = \left(3t\right)^2 - \left(5s\right)^2$
$$= \left(3t + 5s\right)\left(3t - 5s\right)$$

**33.** $16x^2 + y^2$ is a prime polynomial.

**34.** $x^3 - 8y^3 = x^3 - \left(2y\right)^3$
$$= \left(x - 2y\right)\left(x^2 + 2xy + 4y^2\right)$$

**35.** $8x^3 + 27 = \left(2x\right)^3 + 3^3$
$$= \left(2x + 3\right)\left(4x^2 - 6x + 9\right)$$

**36.** $2x^3 + 8x = 2x\left(x^2 + 4\right)$

**37.** $54 - 2x^3y^3;\ \text{GCF} = 2$
$$2\left(27 - x^3y^3\right) = 2\left[3^3 - \left(xy\right)^3\right]$$
$$= 2\left(3 - xy\right)\left(9 + 3xy + x^2y^2\right)$$

**38.** $9x^2 - 4y^2$
$$= \left(3x\right)^2 - \left(2y\right)^2$$
$$= \left(3x - 2y\right)\left(3x + 2y\right)$$

**39.** $16x^4 - 1 = \left(4x^2\right)^2 - 1^2$

$\qquad = \left(4x^2 + 1\right)\left(4x^2 - 1\right)$

$\qquad = \left(4x^2 + 1\right)\left(\left(2x\right)^2 - 1^2\right)$

$\qquad = \left(4x^2 + 1\right)\left(2x + 1\right)\left(2x - 1\right)$

**40.** $x^4 + 16$

not factorable, prime

**41.** $2x^2 + 5x - 12 = \left(2x - 3\right)\left(x + 4\right)$

**42.** $3x^2 - 12 = 3\left(x^2 - 4\right) = 3\left(x - 2\right)\left(x + 2\right)$

**43.** $x\left(x - 1\right) + 3\left(x - 1\right); \text{GCF} = \left(x - 1\right)$

$\left(x - 1\right)\left(x + 3\right)$

**44.** $x^2 + xy - 3x - 3y = x\left(x + y\right) - 3\left(x + y\right)$

$\qquad = \left(x + y\right)\left(x - 3\right)$

**45.** $4x^2 y - 6xy^2; \text{GCF} = 2xy$

$2xy\left(2x - 3y\right)$

**46.** $8x^2 - 15x - x^3 = -x\left(-8x + 15 + x^2\right)$

$\qquad = -x\left(x^2 - 8x + 15\right)$

$\qquad = -x\left(x - 5\right)\left(x - 3\right)$

**47.** $125x^3 + 27 = \left(5x\right)^3 + 3^3$

$\qquad = \left(5x + 3\right)\left(25x^2 - 15x + 9\right)$

**48.** $24x^2 - 3x - 18 = 3\left(8x^2 - x - 6\right)$

**49.** $\left(x + 7\right)^2 - y^2 = \left[\left(x + 7\right) + y\right]\left[\left(x + 7\right) - y\right]$

$\qquad = \left(x + 7 + y\right)\left(x + 7 - y\right)$

**50.** $x^2\left(x + 3\right) - 4\left(x + 3\right) = \left(x + 3\right)\left(x^2 - 4\right)$

$\qquad = \left(x + 3\right)\left(x - 2\right)\left(x + 2\right)$

**51.** $\left(x + 6\right)\left(x - 2\right) = 0$

$x + 6 = 0 \quad \text{or} \quad x - 2 = 0$

$\quad x = -6 \qquad\qquad x = 2$

The solutions are $-6$ and $2$.

**52.** $3x\left(x + 1\right)\left(7x - 2\right) = 0$

$3x = 0 \quad \text{or} \quad x + 1 = 0 \quad \text{or} \quad 7x - 2 = 0$

$x = 0 \qquad\qquad x = -1 \qquad\qquad 7x = 2$

$\qquad\qquad\qquad\qquad\qquad\qquad x = \dfrac{2}{7}$

The solutions are $0$, $-1$, and $\dfrac{2}{7}$.

**53.** $4\left(5x + 1\right)\left(x + 3\right) = 0$

$5x + 1 = 0 \quad \text{or} \quad x + 3 = 0$

$5x = -1 \qquad\qquad x = -3$

$x = -\dfrac{1}{5}$

The solutions are $-\dfrac{1}{5}$ and $-3$.

**54.** $x^2 + 8x + 7 = 0$

$\left(x + 7\right)\left(x + 1\right) = 0$

$x + 7 = 0 \quad \text{or} \quad x + 1 = 0$

$x = -7 \qquad\qquad x = -1$

The solutions are $-7$ and $-1$.

**55.** $x^2 - 2x - 24 = 0$

$\left(x - 6\right)\left(x + 4\right) = 0$

$x - 6 = 0 \quad \text{or} \quad x + 4 = 0$

$x = 6 \qquad\qquad x = -4$

The solutions are $6$ and $-4$.

**56.** $x^2 + 10x = -25$

$x^2 + 10x + 25 = 0$

$(x+5)(x+5) = 0$

$x + 5 = 0$    or    $x + 5 = 0$

     $x = -5$          $x = -5$

The solution is $-5$.

**57.** $x(x-10) = -16$

$x^2 - 10x = -16$

$x^2 - 10x + 16 = 0$

$(x-8)(x-2) = 0$

$x - 8 = 0$    or    $x - 2 = 0$

    $x = 8$          $x = 2$

The solutions are 8 and 2.

**58.** $(3x-1)(9x^2 + 3x + 1) = 0$

$3x - 1 = 0$    or    $9x^2 + 3x + 1 = 0$

$9x^2 + 3x + 1$ is a prime polynomial.

$3x - 1 = 0$

   $3x = 1$

    $x = \dfrac{1}{3}$

**59.** $56x^2 - 5x - 6 = 0$

$56x^2 + 16x - 21x - 6 = 0$

$8x(7x+2)(8x-3) = 0$

$(7x+2)(8x-3) = 0$

$7x + 2 = 0$    or    $8x - 3 = 0$

   $7x = -2$         $8x = 3$

    $x = -\dfrac{2}{7}$       $x = \dfrac{3}{8}$

The solutions are $-\dfrac{2}{7}$ and $\dfrac{3}{8}$.

**60.** $20x^2 - 7x - 6 = 0$

$(4x-3)(5x+2) = 0$

$4x - 3 = 0$    or    $5x + 2 = 0$

   $4x = 3$          $5x = -2$

    $x = \dfrac{3}{4}$    or    $x = -\dfrac{2}{5}$

**61.** $5(3x+2) = 4$

$15x + 10 = 4$

$15x = 4 - 10$

$15x = -6$

    $x = -\dfrac{6}{15} = -\dfrac{2}{5}$

**62.** $6x^2 - 3x + 8 = 0$

no real solution

**63.** $12 - 5t = -3$

$-5t = -3 - 12$

$-5t = -15$

   $t = \dfrac{-15}{-5}$

   $t = 3$

**64.** $5x^3 + 20x^2 + 20x = 0$

$5x(x^2 + 4x + 4) = 0$

$5x(x+2)(x+2) = 0$

$x + 2 = 0$    or    $5x = 0$

$x = -2$    or    $x = 0$

**65.**　$4t^3 - 5t^2 - 21t = 0$

$t\left(4t^2 - 5t - 21\right) = 0$

$t\left(4t + 7\right)\left(t - 3\right) = 0$

$t = 0$　　or　　$4t + 7 = 0$　　or　　$t - 3 = 0$

$t = 0$　　　　　　　$4t = -7$　　　　　$t = 3$

$t = 0$　　　　　　　$t = -\dfrac{7}{4}$　　　　$t = 3$

**66.** Let $x$ = the width of the flag. Then

$2x - 15$ = the length of the flag.

$A = lw$

$500 = \left(2x - 15\right)\left(x\right)$

$500 = 2x^2 - 15x$

$0 = 2x^2 - 15x - 500$

$0 = \left(2x + 25\right)\left(x - 20\right)$

$2x + 25 = 0$　　or　　$x - 20 = 0$

$2x = -25$　　　　　　　$x = 20$

$x = -\dfrac{25}{2}$

Since the dimensions cannot be negative, the width is 20 inches and the length is $2\left(20\right) - 15 = 25$ inches.

**67.** Let $x$ = the height of the sail, then

$4x$ = the base of the sail.

$A = \dfrac{1}{2}bh$

$162 = \dfrac{1}{2}\left(4x\right)\left(x\right)$

$162 = 2x^2$

$0 = 2x^2 - 162$

$0 = 2\left(x^2 - 81\right)$

$0 = 2\left(x + 9\right)\left(x - 9\right)$

$x + 9 = 0$　　or　　$x - 9 = 0$

$x = -9$　　　　　　$x = 9$

Since the dimensions cannot be negative, the height is 9 yards and the base is $4 \cdot 9 = 36$ yards.

**68.** Let $x$ = the first integer. Then

$x + 1$ = the next consecutive integer.

$x\left(x + 1\right)380$

$x^2 + x = 380$

$x^2 + x - 380 = 0$

$\left(x + 20\right)\left(x - 19\right) = 0$

$x + 20 = 0$　　or　　$x - 19 = 0$

$x = -21$　　　　　　$x = 19$

The integers are 19 and 20.

**69.** **A.** Let h = 2800 and solve for $t$.

$h = -16t^2 + 440t$

$2800 = -16t^2 + 440t$

$0 = -16t^2 + 440t - 2800$

$0 = -8\left(2t^2 - 55t + 350\right)$

$0 = -8\left(2t - 35\right)\left(t - 10\right)$

$2t - 35 = 0$　　or　　$t - 10 = 0$

$2t = 35$　　　　　　$t = 10$

$t = \dfrac{35}{2}$

$t = 17.5$

The solutions are 17.5 and 10.

Answers may vary.

b. Find $t$ when $h = 0$.

$$h = -16t^2 + 440t$$

$$0 = 16t^2 + 440t$$

$$0 = -8t(2t - 55)$$

$$-8t = 0 \quad \text{or} \quad 2t - 55 = 0$$

$$t = 0 \qquad\qquad 2t = 55$$

$$t = \frac{55}{2}$$

$$t = 27.5$$

27.5 seconds after being fired, the rocket will reach the ground again.

**70.** Let $x$ = the length of the longer leg, then $x - 8$ = the length of the shorter leg and $x + 8$ = the length of the hypotenuse. By the Pythagorean theorem,

$$x^2 + (x-8)^2 = (x+8)^2$$

$$x^2 + x^2 - 16x + 64 = x^2 + 16x + 64$$

$$x^2 - 32x = 0$$

$$x(x-32) = 0$$

$$x = 0 \quad \text{or} \quad x = 32$$

Since the length cannot be zero cm, the length of the longer leg is 32 cm.

**Chapter 6 Test**

**1.** $y^2 - 8y - 48 = (y-12)(y+4)$

**2.** $x^2 + x - 10$

not factorable, prime

**3.** $9x^3 + 39x^2 + 12x$; GCF $= 3x$

$3x(3x^2 + 13x + 4) = 3x(3x+1)(x+4)$

**4.** $3a^2 + 3ab - 7a - 7b = 3a(a+b) - 7(a+b)$

$$= (a+b)(3a-7)$$

**5.** $3x^2 - 5x + 2 = (3x-2)(x-1)$

**6.** $x^2 + 14xy + 24y^2 = (x+12y)(x+2y)$

**7.** $180 - 5x^2 = 5(36 - x^2)$

$$= 5(6^2 - x^2)$$

$$= 5(6+x)(6-x)$$

**8.** $6t^2 - t - 5 = (6t+5)(t-1)$

**9.** $xy^2 - 7y^2 - 4x + 28$

$$= y^2(x-7) - 4(x-7)$$

$$= (x-7)(y^2 - 4)$$

$$= (x-7)(y^2 - 2^2)$$

$$= (x-7)(y+2)(y-2)$$

**10.** $x - x^5 = x(1 - x^4)$

$$= x\left(1 - \left(x^2\right)^2\right)$$

$$= x(1 + x^2)(1 - x^2)$$

$$= x(1 + x^2)(1 + x)(1 - x)$$

**11.** $-xy^3 - x^3y$; GCF $= -xy$

$$-xy(y^2 + x^2)$$

**12.** $64x^3 - 1 = (4x)^3 - 1^3 = (4x-1)(16x^2 + 4x + 1)$

**13.** $8y^3 - 64 = 8(y^3 - 8) = 8(y^3 - 2^3)$

$= 8(y-2)(y^2 + 2y + 4)$

**14.**    $x^2 + 5x = 14$

$x^2 + 5x - 14 = 0$

$(x+7)(x-2) = 0$

$x + 7 = 0$    or    $x - 2 = 0$

    $x = -7$          $x = 2$

The solutions are $-7$ and 2.

**15.**   $x(x+6) = 7$

$x^2 + 6x = 7$

$x^2 + 6x - 7 = 0$

$(x+7)(x-1) = 0$

$x + 7 = 0$    or    $x - 1 = 0$

    $x = -7$          $x = 1$

The solutions are $-7$ and 1.

**16.**   $3x(2x-3)(3x+4) = 0$

$3x = 0$    or    $2x - 3 = 0$    or    $3x + 4 = 0$

   $x = 0$           $2x = 3$           $3x = -4$

               $x = \dfrac{3}{2}$         $x = -\dfrac{4}{3}$

The solutions are $0$, $\dfrac{3}{2}$, and $-\dfrac{4}{3}$.

**17.**   $5t^3 - 45t = 0$

$5t(t^2 - 0) = 0$

$5t(t+3)(t-3) = 0$

$5t = 0$    or    $t + 3 = 0$    or    $t - 3 = 0$

$t = 0$           $t = -3$           $t = 3$

The solutions are $0$, $-3$, and 3.

**18.** $t^2 - 2t - 15 = 0$

$(t-5)(t+3) = 0$

$x - 5 = 0$    or    $t + 3 = 0$

   $t = 5$           $t = -3$

The solutions are 5 and $-3$.

**19.**   $6x^2 = 15x$

$6x^2 - 15x = 0$

$3x(2x - 5) = 0$

$3x = 0$    or    $2x - 5 = 0$

   $x = 0$           $x = \dfrac{5}{2}$

The solutions are 0 and $\dfrac{5}{2}$.

**20.** Let $x$ = the height of the triangle, then $x + 9$ = the base.

$A = \dfrac{1}{2}bh$

$68 = \dfrac{1}{2}(x+9)(x)$

$68 = \dfrac{1}{2}x^2 + \dfrac{9}{2}x$

$0 = \dfrac{1}{2}x^2 + \dfrac{9}{2}x - 68$

$0 = \dfrac{1}{2}(x^2 + 9x - 136)$

$0 = \dfrac{1}{2}(x+17)(x-8)$

$x + 17 = 0$    or    $x - 8 = 0$

   $x = -17$          $x = 8$

Since the length of the base cannot be negative, the base is $8 + 9 = 17$ feet.

**21.** Let $x$ = the first number, then

$17 - x$ = the other number.

$x^2 + (17 - x)^2 = 145$

$x^2 + 289 - 34x + x^2 = 145$

$2x^2 - 34x + 144 = 0$

$2(x^2 - 17x + 72) = 0$

$2(x - 9)(x - 8) = 0$

$x - 9 = 0$    or    $x - 8 = 0$

    $x = 9$                $x = 8$

The numbers are 8 and 9.

**22.** Find $t$ when $h = 0$.

$h = -16t^2 + 784$

$0 = -16t^2 + 784$

$16t^2 = 784$

$t^2 = 49$

$t = 7$

It reaches the ground after 7 seconds.

**Cumulative Review Chapter 6**

**1.** a. $9 \le 11$

   b. $8 > 1$

   c. $3 \ne 4$

**2.** a. $|-5| > |-3|$

   b. $|0| < |-2|$

**3.** a. $\dfrac{42}{49} = \dfrac{6 \cdot 7}{7 \cdot 7} = \dfrac{6}{7}$

   b. $\dfrac{11}{27} = \dfrac{11}{3 \cdot 3 \cdot 3} = \dfrac{11}{27}$

   c. $\dfrac{88}{20} = \dfrac{4 \cdot 22}{4 \cdot 5} = \dfrac{22}{5}$

**4.** Let $x = 20$ and $y = 10$.

$\dfrac{x}{y} + 5x = \dfrac{20}{10} + 5(20) = 2 + 100 = 102$

**5.** $\dfrac{8 + 2 \cdot 3}{2^2 - 1} = \dfrac{8 + 6}{4 - 1} = \dfrac{14}{3}$

**6.** Let $x = -20$ and $y = 10$.

$\dfrac{x}{y} + 5x = \dfrac{-20}{10} + 5(-20) = -2 - 100 = -102$

**7.** a. $3 + (-7) + (-8) = 3 + (-15) = -12$

   b. $\left[7 + (-10)\right] + \left[-2 + |-4|\right]$

      $= -3 + (-2 + 4)$

      $= -3 + 2$

      $= -1$

**8.** Let $x = -20$ and $y = -10$.

$\dfrac{x}{y} + 5x = \dfrac{-20}{-10} + 5(-20) = 2 - 100 = -98$

**9.** a. $(-6)(4) = -24$

   b. $(2)(-1) = -2$

   c. $(-5)(-10) = 50$

**10.** $5 - 2(3x - 7) = 5 - 6x + 14 = -6x + 19$

**11.** a. $7x - 3x = (7 - 3)x = 4x$

   b. $10y^2 + y^2 = (10 + 1)y^2 = 11y^2$

   c. $8x^2 + 2x - 3x = 8x^2 + (2 - 3)x$

               $= 8x^2 - x$

**12.** $0.8y + 0.2(y - 1) = 1.8$

$0.8y + 0.2y - 0.2 = 1.8$

$1.0y - 0.2 = 1.8$

$y = 2.0$

**13.** $3 - x = 7$

$3 - 3 - x = 7 - 3$

$-x = 4$

$x = -4$

**14.** $\dfrac{x}{-7} = -4$

$-7\left(\dfrac{x}{-7}\right) = -7(-4)$

$x = 28$

**15.** $-3x = 33$

$\dfrac{-3x}{-3} = \dfrac{33}{-3}$

$x = -11$

**16.** $-\dfrac{2}{3}x = -22$

$\left(-\dfrac{3}{2}\right)\left(-\dfrac{2}{3}\right)x = \left(-\dfrac{3}{2}\right)(-22)$

$x = 33$

**17.** $8(2 - t) = -5t$

$16 - 8t = -5t$

$16 - 8t + 5t = -5t + 5t$

$16 - 3t = 0$

$16 - 16 - 3t = -16$

$-3t = -16$

$\dfrac{-3t}{-3} = \dfrac{-16}{-3}$

$t = \dfrac{16}{3}$

**18.** $-z = \dfrac{7z + 3}{5}$

$5(-z) = 5\left(\dfrac{7z + 3}{5}\right)$

$-5z = 7z + 3$

$-5z - 7z = 7z - 7z + 3$

$-12z = 3$

$\dfrac{-12z}{-12} = \dfrac{3}{-12}$

$z = -\dfrac{1}{4}$

**19.** Let $x$ = the length of the shorter piece and $4x$ = the length of the longer piece.

$x + 4x = 10$

$5x = 10$

$x = 2$

$4x = 4(2) = 8$

The pieces are 2 feet and 8 feet in length.

**20.** $3x + 9 \leq 5(x - 1)$

$3x + 9 \leq 5x - 5$

$-2x + 9 \leq -5$

$-2x \leq -14$

$\dfrac{-2x}{-2} \geq \dfrac{-14}{-2}$

$x \geq 7, \ [7, \infty)$

**21.** $y = -\dfrac{1}{3}x$

| $x$ | $y$ |
|-----|-----|
| 0   | 0   |
| 6   | $-2$ |

(0,0)

(6,-2)

**22.** $-7x - 8y = -9$

$$(-1, 2): -7(-1) - 8(2) \overset{?}{=} -9$$

$$7 - 16 \overset{?}{=} -9$$

$$-9 = -9 \quad \text{True}$$

$(-1, 2)$ is a solution to the equation.

**23.** $(-6, 0)$ and $(-2, 3)$

$$m = \frac{y_2 - y_1}{x_2 - x_1} = \frac{3 - 0}{-2 - (-6)} = \frac{3}{4}$$

$(5, 4)$ and $(9, 7)$

$$m = \frac{y_2 - y_1}{x_2 - x_1} = \frac{7 - 4}{9 - 5} = \frac{3}{4}$$

Yes, they are parallel.

**24.** $(5, -6)$ and $(5, 2)$

$$m = \frac{y_2 - y_1}{x_2 - x_1} = \frac{2 - (-6)}{5 - 5} = \frac{8}{0}$$

The slope is undefined.

**25.** a. If $x = 5$, $2x^3 = 2(5)^3 = 2(125) = 250$

    b. If $x = -3$, $\dfrac{9}{x^2} = \dfrac{9}{(-3)^2} = \dfrac{9}{9} = 1$

**26.** $7x - 3y = 2$

$$-3y = -7x + 2$$

$$y = \frac{-7x}{-3} + \frac{2}{-3}$$

$$y = \frac{7}{3}x - \frac{2}{3}$$

$$y = mx + b$$

$$m = \frac{7}{3}, \; b = -\frac{2}{3}, \; \left(0, -\frac{2}{3}\right)$$

**27.** a. $-3x^2$ : Degree 2

    b. $5x^3yz$ : Degree 5

    c. $2$ : Degree 0

**28.** Vertical line has equation $x = c$

Point $(0, 7)$

$x = 0$

**29.** $\left(2x^3 + 8x^2 - 6x\right) - \left(2x^3 - x^2 + 1\right)$

$$= 2x^3 + 8x^2 - 6x - 2x^3 + x^2 - 1$$

$$= 9x^2 - 6x - 1$$

**30.** $m = 4, \; b = \dfrac{1}{2}$

$$y = mx + b$$

$$y = 4x + \frac{1}{2}$$

$$2y = 8x + 1$$

$$8x - 2y = -1$$

**31.** $(3x + 2)(2x - 5)$

$$= 3x(2x) + 3x(-5) + 2(2x) + 2(-5)$$

$$= 6x^2 - 15x + 4x - 10$$

$$= 6x^2 - 11x - 10$$

**32.** $(-4, 0)$ and $(6, -1)$

$$m = \frac{y_2 - y_1}{x_2 - x_1} = \frac{-1 - 0}{6 - (-4)} = -\frac{1}{10}$$

$$m = -\frac{1}{10}, \text{ point } (-4, 0)$$

$$y - y_1 = m(x - x_1)$$

$$y - 0 = -\frac{1}{10}\left[x - (-4)\right]$$

$$y = -\frac{1}{10}x - \frac{4}{10}$$

$$10y = -x - 4$$

$$x + 10y = -4$$

**33.** $(3y + 1)^2 = (3y)^2 + 2(3y)(1) + 1^2$
$$= 9y^2 + 6y + 1$$

**34.** $\begin{cases} -x + 3y = 18 \\ -3x + 2y = 19 \end{cases}$

Multiply the first equation by $-3$

$$3x - 9y = -54$$
$$\underline{-3x + 2y = \phantom{0}19}$$
$$-7y = -35$$
$$y = 5$$

Substitute 5 for $y$ in the first equation.

$$-x + 3(5) = 18$$
$$-x = 3$$
$$x = -3$$

The solution to the system is $(-3, 5)$

**35.**

a. $3^{-2} = \frac{1}{3^2} = \frac{1}{9}$

b. $2x^{-3} = \frac{2}{x^3}$

c. $2^{-1} + 4^{-1} = \frac{1}{2} + \frac{1}{4} = \frac{1 \cdot 2}{2 \cdot 2} + \frac{1}{4}$
$$= \frac{2}{4} + \frac{1}{4} = \frac{2 + 1}{4} = \frac{3}{4}$$

d. $(-2)^{-4} = \frac{1}{(-2)^4} = \frac{1}{16}$

**36.** $\frac{\left(5a^7\right)^2}{a^5} = \frac{25a^{14}}{a^5} = 25a^9$

**37.** a. $367{,}000{,}000 = 3.67 \times 10^8$

b. $0.000003 = 3.0 \times 10^{-6}$

c. $20{,}520{,}000{,}000 = 2.052 \times 10^{10}$

d. $0.00085 = 8.5 \times 10^{-4}$

**38.** $(3x - 7y)^2 = (3x)^2 - 2(3x)(7y) + (7y)^2$
$$= 9x^2 - 42xy + 49y^2$$

**39.**
$$\begin{array}{r} x + 4 \\ x + 3 \overline{)\ x^2 + 7x + 12} \\ \underline{x^2 + 3x} \\ 4x + 12 \\ \underline{4x + 12} \\ 0 \end{array}$$

$x + 4$

**40.** $\frac{(xy)^{-3}}{\left(x^5y^6\right)^3} = \frac{x^{-3}y^{-3}}{x^{15}y^{18}} = x^{-3-15}y^{-3-18} = x^{-18}y^{-21}$

$$= \frac{1}{x^{18}y^{21}}$$

**41.** a. $x^3, x^7, x^5 : \text{GCF} = x^3$

     b. $y, y^4, y^7 : \text{GCF} = y$

**42.** $z^3 + 7z + z^2 + 7 = z\left(z^2 + 7\right) + 1\left(z^2 + 7\right)$

$$= \left(z^2 + 7\right)(z + 1)$$

**43.** $x^2 + 7x + 12 = (x + 4)(x + 3)$

**44.** $2x^3 + 2x^2 - 84x = 2x\left(x^2 + x - 42\right)$

$$= 2x(x + 7)(x - 6)$$

**45.** $8x^2 - 22x + 5 = 8x^2 - 20x - 2x + 5$

$$= 4x(2x - 5) - 1(2x - 5)$$

$$= (2x - 5)(4x - 1)$$

**46.** $-4x^2 - 23x + 6 = -1\left(4x^2 + 23x - 6\right)$

$$= -\left(4x^2 - x + 24x - 6\right)$$

$$= -\left[x(4x - 1) + 6(4x - 1)\right]$$

$$= -(4x - 1)(x + 6)$$

**47.** $25a^2 - 9b^2 = \left(5a\right)^2 - \left(3b\right)^2$

$$= (5a + 3b)(5a - 3b)$$

**48.** $9xy^2 - 16x = x\left(9y^2 - 16\right)$

$$= x\left[\left(3y\right)^2 - 4^2\right]$$

$$= x(3y + 4)(3y - 4)$$

**49.** $(x - 3)(x + 1) = 0$

     $x - 3 = 0$ or $x + 1 = 0$

       $x = 3$ or     $x = -1$

**50.** $\qquad x^2 - 13x = -36$

$$x^2 - 13x + 36 = 0$$

$$(x - 9)(x - 4) = 0$$

$x - 9 = 0$ or $x - 4 = 0$

    $x = 9$ or      $x = 4$

# Chapter 7

**Mental Math 7.1**

**1.** $x = 0$

**2.** $x = 3$

**3.** $x = 0, \ x = 1$

**4.** $x = 5, \ x = 6$

**5.** no

**6.** yes

**7.** yes

**8.** no

**Exercise Set 7.1**

**1.** $\dfrac{x+5}{x+2} = \dfrac{2+5}{2+2} = \dfrac{7}{4}$

**3.** $\dfrac{z-8}{z+2} = \dfrac{-5-8}{-5+2} = \dfrac{-13}{-3} = \dfrac{13}{3}$

**5.** $\dfrac{x^2+8x+2}{x^2-x-6} = \dfrac{2^2+8(2)+2}{2^2-2-6} = \dfrac{4+16+2}{4-8}$

$\quad = \dfrac{22}{-4} = \dfrac{11 \cdot 2}{-2 \cdot 2} = -\dfrac{11}{2}$

**7.** $\dfrac{x+5}{x^2+4x-8} = \dfrac{2+5}{2^2+4(2)-8}$

$\qquad = \dfrac{7}{4+8-8}$

$\qquad = \dfrac{7}{4+0}$

$\qquad = \dfrac{7}{4}$

**9.** $\dfrac{y^3}{y^2-1} = \dfrac{(-2)^3}{(-2)^2-1} = \dfrac{-8}{4-1} = \dfrac{-8}{3} = -\dfrac{8}{3}$

**11.** $3y = 0$

$y = 0$

The expression is undefined when $y = 0$.

**13.** $x + 2 = 0$

$x = -2$

The expression is undefined when $x = -2$.

**15.** $2x - 8 = 0$

$2x = 8$

$x = 4$

The expression is undefined when $x = 4$.

**17.** $15x + 30 = 0$

$15x = -30$

$x = -2$

The expression is undefined when $x = -2$.

**19.** $\qquad x^3 - x = 0$

$\qquad x(x^2 - 1) = 0$

$\quad x(x+1)(x-1) = 0$

$x = 0, \ x = -1, \ \text{or} \ x = 1$

The expression is undefined when $x = 0, -1, 1$.

**21.** The denominator is never zero so there are no values for which

$\dfrac{x^2-5x-2}{x^2+4}$ is undefined.

**23.** Answers may vary.

**25.** $\dfrac{8x^5}{4x^9} = \dfrac{4 \cdot 2x^5}{4x^5 \cdot x^4} = \dfrac{2}{x^4}$

**27.** $\dfrac{5(x-2)}{(x-2)(x+1)} = \dfrac{5}{x+1}$

**29.** $\dfrac{-5a-5b}{a+b} = \dfrac{-5(a+b)}{a+b} = -5$

**31.** $\dfrac{x+5}{x^2-4x-45} = \dfrac{x+5}{(x-9)(x+5)} = \dfrac{1}{x-9}$

**33.** $\dfrac{5x^2+11x+2}{x+2} = \dfrac{(5x+1)(x+2)}{x+2} = 5x+1$

**35.** $\dfrac{x^2+x-12}{2x^2-5x-3} = \dfrac{(x+4)(x-3)}{(2x+1)(x-3)} = \dfrac{x+4}{2x+1}$

**37.** Answers may vary.

**39.** $\dfrac{x-7}{7-x} = \dfrac{x-7}{-1(x-7)} = \dfrac{1}{-1} = -1$

**41.** $\dfrac{y^2-2y}{4-2y} = \dfrac{y(y-2)}{-2(y-2)} = -\dfrac{y}{2}$

**43.** $\dfrac{x^2-4x+4}{4-x^2} = \dfrac{(x-2)(x-2)}{-1(x^2-4)}$

$\qquad = \dfrac{(x-2)(x-2)}{-1(x+2)(x-2)}$

$\qquad = -\dfrac{x-2}{x+2}$

$\qquad = \dfrac{2-x}{x+2}$

**45.** $\dfrac{15x^4y^8}{-5x^8y^3} = \dfrac{3 \cdot 5x^4y^3 \cdot y^5}{-5x^4 \cdot x^4y^3} = -\dfrac{3y^5}{x^4}$

**47.** $\dfrac{(x-2)(x+3)}{5(x+3)} = \dfrac{x-2}{5}$

**49.** $\dfrac{-6a-6b}{a+b} = \dfrac{-6(a+b)}{a+b} = -6$

**51.** $\dfrac{2x^2-8}{4x-8} = \dfrac{2(x^2-4)}{4(x-2)}$

$\qquad = \dfrac{2(x+2)(x-2)}{2 \cdot 2(x-2)}$

$\qquad = \dfrac{x+2}{2}$

**53.** $\dfrac{11x^2-22x^3}{6x-12x^2} = \dfrac{11x^2(1-2x)}{6x(1-2x)}$

$\qquad = \dfrac{11x \cdot x(1-2x)}{6 \cdot x(1-2x)}$

$\qquad = \dfrac{11x}{6}$

**55.** $\dfrac{x+7}{x^2+5x-14} = \dfrac{x+7}{(x-2)(x+7)} = \dfrac{1}{x-2}$

**57.** $\dfrac{2x^2+3x-2}{2x-1} = \dfrac{(2x-1)(x+2)}{2x-1} = x+2$

**59.** $\dfrac{x^2-1}{x^2-2x+1} = \dfrac{(x-1)(x+1)}{(x-1)(x-1)}$

$\qquad = \dfrac{x+1}{x-1}$

**61.** $\dfrac{m^2-6m+9}{m^2-9}=\dfrac{(m-3)(m-3)}{(m+3)(m-3)}$

$\qquad\qquad =\dfrac{m-3}{m+3}$

**63.** $\dfrac{-2a^2+12a-18}{9-a^2}=\dfrac{-2(a^2-6a+9)}{-1(a^2-9)}$

$\qquad\qquad =\dfrac{-2(a-3)(a-3)}{-1(a+3)(a-3)}$

$\qquad\qquad =\dfrac{2(a-3)}{a+3}$ or $\dfrac{2a-6}{a+3}$

**65.** $\dfrac{2-x}{x-2}=\dfrac{-1(x-2)}{x-2}=-1$

**67.** $\dfrac{x^2-1}{1-x}=\dfrac{(x+1)(x-1)}{-1(x-1)}=-(x+1)=-x-1$

**69.** $\dfrac{x^2+7x+10}{x^2-3x-10}=\dfrac{(x+5)(x+2)}{(x-5)(x+2)}=\dfrac{x+5}{x-5}$

**71.** $\dfrac{3x^2+7x+2}{3x^2+13x+4}=\dfrac{(x+2)(3x+1)}{(x+4)(3x+1)}=\dfrac{x+2}{x+4}$

**73.** $\dfrac{1}{3}\cdot\dfrac{9}{11}=\dfrac{1\cdot9}{3\cdot11}=\dfrac{3\cdot3}{3\cdot11}=\dfrac{3}{11}$

**75.** $\dfrac{1}{3}\div\dfrac{1}{4}=\dfrac{1}{3}\cdot\dfrac{4}{1}=\dfrac{4}{3}$

**77.** $\dfrac{5}{6}\cdot\dfrac{10}{11}\cdot\dfrac{2}{3}=\dfrac{5\cdot10\cdot2}{6\cdot11\cdot3}$

$\qquad\qquad =\dfrac{5\cdot2\cdot5\cdot2}{3\cdot2\cdot11\cdot3}$

$\qquad\qquad =\dfrac{5\cdot5\cdot2}{3\cdot11\cdot3}=\dfrac{50}{99}$

**79.** $\dfrac{13}{20}\div\dfrac{2}{9}=\dfrac{13}{20}\cdot\dfrac{9}{2}=\dfrac{13\cdot9}{20\cdot2}=\dfrac{117}{40}$

**81.** $\dfrac{x^2+xy+2x+2y}{x+2}=\dfrac{x(x+y)+2(x+y)}{x+2}$

$\qquad\qquad =\dfrac{(x+y)(x+2)}{x+2}$

$\qquad\qquad =x+y$

**83.** $\dfrac{5x+15-xy-3y}{2x+6}=\dfrac{5(x+3)-y(x+3)}{2(x+3)}$

$\qquad\qquad =\dfrac{(x+3)(5-y)}{2(x+3)}$

$\qquad\qquad =\dfrac{5-y}{2}$

**85.** $\dfrac{x^3+8}{x+2}=\dfrac{(x+2)(x^2-2x+4)}{x+2}=x^2-2x+4$

**87.** $\dfrac{x^3-1}{1-x}=\dfrac{(x-1)(x^2+x+1)}{-1(x-1)}$

$\qquad\qquad =-1(x^2+x+1)$

$\qquad\qquad =-x^2-x-1$

**89.**   a. $R=\dfrac{150x^2}{x^2+3}$

$\qquad\quad =\dfrac{150(1)^2}{1^2+3}$

$\qquad\quad =\dfrac{150}{4}$

$\qquad\quad =\$37.5$ million

b. $R = \dfrac{150x^2}{x^2+3}$

$= \dfrac{150(2)^2}{2^2+3}$

$= \dfrac{600}{7}$

$\approx \$85.7$ million

c. $85.7 = 37.5 = \$48.2$ million

**91.** Let $D = 1000$ and $A = 8$

$C = \dfrac{DA}{A+12} = \dfrac{1000(8)}{8+12} = \dfrac{8000}{20} = 400$

The child should receive 400 mg.

**93.** $B = \dfrac{705w}{h^2};\ w = 148,\ h = 66$

$B = \dfrac{705 \cdot 148}{(66)^2} = \dfrac{104{,}340}{4356} \approx 24.0$

No

**95.** A and c: Answers may vary

**97.** Let $h = 150,\ d = 30,\ t = 1,\ r = 31,\ b = 472$

$S = \dfrac{h+d+2t+3r}{b}$

$S = \dfrac{150+30+2(1)+3(31)}{472}$

$= \dfrac{275}{472}$

$= 0.583$

The slugging percentage is 58.3%.

**99.** $y = \dfrac{x^2-16}{x-4} = \dfrac{(x+4)(x-4)}{x-4}$

$= x+4,\ x \neq 4$

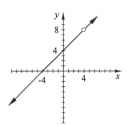

**101.** $y = \dfrac{x^2-6x+8}{x-2}$

$= \dfrac{(x-2)(x-4)}{x-2}$

$= x-4,\ x \neq 2$

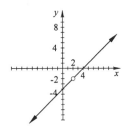

**Mental Math 7.2**

**1.** $\dfrac{2}{y} \cdot \dfrac{x}{3} = \dfrac{2x}{3y}$

**2.** $\dfrac{3x}{4} \cdot \dfrac{1}{y} = \dfrac{3x}{4y}$

**3.** $\dfrac{5}{7} \cdot \dfrac{y^2}{x^2} = \dfrac{5y^2}{7x^2}$

**4.** $\dfrac{x^5}{11} \cdot \dfrac{4}{z^3} = \dfrac{4x^5}{11z^3}$

**5.** $\dfrac{9}{x} \cdot \dfrac{x}{5} = \dfrac{9x}{5x} = \dfrac{9}{5}$

**6.** $\dfrac{y}{7} \cdot \dfrac{3}{y} = \dfrac{3y}{7y} = \dfrac{3}{7}$

**Exercise Set 7.2**

**1.** $\dfrac{8x}{2} \cdot \dfrac{x^5}{4x^2} = \dfrac{8x \cdot x^5}{2 \cdot 4x^2} = \dfrac{2 \cdot 4 \cdot x \cdot x \cdot x^4}{2 \cdot 4 \cdot x \cdot x} = x^4$

**3.** $-\dfrac{5a^2b}{30a^2b^2} \cdot b^3 = \dfrac{5a^2b \cdot b^3}{30a^2b^2}$

$= -\dfrac{5 \cdot a^2 \cdot b \cdot b \cdot b^2}{5 \cdot 6 \cdot a^2 \cdot b^2} = -\dfrac{b \cdot b}{6} = -\dfrac{b^2}{6}$

**5.** $\dfrac{4}{x+2} \cdot \dfrac{x}{7} = \dfrac{4x}{7(x+2)}$

**7.** $\dfrac{x}{2x-14} \cdot \dfrac{x^2-7x}{5} = \dfrac{x \cdot (x^2-7x)}{(2x-14) \cdot 5}$

$= \dfrac{x \cdot x(x-7)}{2(x-7) \cdot 5} = \dfrac{x \cdot x}{2 \cdot 5} = \dfrac{x^2}{10}$

**9.** $\dfrac{6x+6}{5} \cdot \dfrac{10}{36x+36} = \dfrac{(6x+6) \cdot 10}{5 \cdot (36x+36)}$

$= \dfrac{6(x+1) \cdot 2 \cdot 5}{5 \cdot 36(x+1)} = \dfrac{6 \cdot 5 \cdot 2 \cdot (x+1)}{6 \cdot 5 \cdot 2 \cdot 3 \cdot (x+1)}$

$= \dfrac{1}{3}$

**11.** $\dfrac{m^2-n^2}{m+n} \cdot \dfrac{m}{m^2-mn} = \dfrac{(m^2-n^2) \cdot m}{(m+n) \cdot (m^2-mn)}$

$= \dfrac{(m-n)(m+n) \cdot m}{(m+n) \cdot m \cdot (m-n)} = 1$

**13.** $\dfrac{x^2-25}{x^2-3x-10} \cdot \dfrac{x+2}{x} = \dfrac{(x^2-25) \cdot (x+2)}{(x^2-3x-10) \cdot x}$

$= \dfrac{(x-5)(x+5) \cdot (x+2)}{(x-5)(x+2) \cdot x} = \dfrac{x+5}{x}$

**15.** $A = \dfrac{2x}{x^2-25} \cdot \dfrac{x+5}{9x^3} = \dfrac{2x \cdot (x+5)}{(x^2-25) \cdot 9x^3}$

$= \dfrac{2 \cdot x \cdot (x+5)}{9 \cdot x^3 \cdot (x+5)(x-5)} = \dfrac{2}{9x^2(x-5)}$ sq.ft.

**17.** $\dfrac{5x^7}{2x^5} \div \dfrac{10x}{4x^3} = \dfrac{5x^7}{2x^5} \cdot \dfrac{4x^3}{10x}$

$= \dfrac{5 \cdot x^2 \cdot x^5 \cdot 2 \cdot 2x \cdot x^2}{2x^5 \cdot 2 \cdot 5 \cdot x}$

$= x^4$

**19.** $\dfrac{8x^2}{y^3} \div \dfrac{4x^2y^3}{6} = \dfrac{8x^2}{y^3} \cdot \dfrac{6}{4x^2y^3}$

$= \dfrac{2 \cdot 4 \cdot x^2 \cdot 6}{y^3 \cdot 4x^2y^3}$

$= \dfrac{12}{y^6}$

**21.** $\dfrac{(x-6)(x+4)}{4x} \div \dfrac{2x-12}{8x^2}$

$= \dfrac{(x-6)(x+4)}{4x} \cdot \dfrac{8x^2}{2x-12}$

$= \dfrac{(x-6)(x+4) \cdot 2 \cdot 4 \cdot x \cdot x}{4x \cdot 2(x-6)}$

$= x(x+4)$

**23.** $\dfrac{3x^2}{x^2-1} \div \dfrac{x^5}{(x+1)^2} = \dfrac{3x^2}{x^2-1} \cdot \dfrac{(x+1)^2}{x^5}$

$= \dfrac{3x^2 \cdot (x+1)(x+1)}{(x-1)(x+1) \cdot x^2 \cdot x^3}$

$= \dfrac{3(x+1)}{x^3(x-1)}$

**25.** $\dfrac{m^2-n^2}{m+n} \div \dfrac{m}{m^2+nm}$

$= \dfrac{m^2-n^2}{m+n} \cdot \dfrac{m^2+nm}{m}$

$= \dfrac{(m-n)(m+n) \cdot m(m+n)}{(m+n) \cdot m}$

$= (m-n)(m+n) = m^2-n^2$

**27.** $\dfrac{x+2}{7-x} \div \dfrac{x^2-5x+6}{x^2-9x+14} = \dfrac{x+2}{7-x} \cdot \dfrac{x^2-9x+14}{x^2-5x+6}$

$= \dfrac{(x+2) \cdot (x-7)(x-2)}{-1(x-7) \cdot (x-3)(x-2)}$

$= -\dfrac{x+2}{x-3}$

**29.** $\dfrac{x^2+7x+10}{1-x} \div \dfrac{x^2+2x-15}{x-1}$

$= \dfrac{x^2+7x+10}{1-x} \cdot \dfrac{x-1}{x^2+2x-15}$

$= -\dfrac{(x+5)(x+2) \cdot (x-1)}{(x-1) \cdot (x+5)(x-3)} = -\dfrac{x+2}{x-3}$

**31.** Answers may vary.

**33.** $\dfrac{5a^2b}{30a^2b^2} \cdot \dfrac{1}{b^3} = \dfrac{5a^2b}{30a^2b^2 \cdot b^3} = \dfrac{5 \cdot a^2 \cdot b}{5 \cdot 6 \cdot a^2 \cdot b^2}$

$= \dfrac{1}{6b^4}$

**35.** $\dfrac{12x^3y}{8xy^7} \div \dfrac{7x^5y}{6x} = \dfrac{12x^3y}{8xy^7} \cdot \dfrac{6x}{7x^5y}$

$= \dfrac{72x^4y}{56x^6y^8}$

$= \dfrac{9}{7x^2y^7}$

**37.** $\dfrac{5x-10}{12} \div \dfrac{4x-8}{8} = \dfrac{5x-10}{12} \cdot \dfrac{8}{4x-8}$

$= \dfrac{5(x-2) \cdot 2 \cdot 4}{6 \cdot 2 \cdot 4(x-2)} = \dfrac{5}{6}$

**39.** $\dfrac{x^2+5x}{8} \cdot \dfrac{9}{3x+15} = \dfrac{x(x+5) \cdot 3 \cdot 3}{8 \cdot 3(x+5)} \cdot \dfrac{3x}{8}$

**41.** $\dfrac{7}{6p^2+q} \div \dfrac{14}{18p^2+3q}$

$= \dfrac{7}{6p^2+q} \cdot \dfrac{18p^2+3q}{14}$

$= \dfrac{7 \cdot 3(6p^2+q)}{(6p^2+q) \cdot 7 \cdot 2} = \dfrac{3}{2}$

**43.** $\dfrac{3x+4y}{x^2+4xy+4y^2} \cdot \dfrac{x+2y}{2}$

$= \dfrac{(3x+4y) \cdot (x+2y)}{(x+2y)(x+2y) \cdot 2}$

$= \dfrac{3x+4y}{2(x+2y)}$

**45.** $\dfrac{x^2-9}{x^2+8} \div \dfrac{3-x}{2x^2+16}$

$= \dfrac{x^2-9}{x^2+8} \cdot \dfrac{2x^2+16}{3-x}$

$$= -\frac{(x+3)(x-3)}{x^2+8} \cdot \frac{2(x^2+8)}{-1(x-3)}$$

$$= \frac{2 \cdot (x+3) \cdot (x-3) \cdot (x^2+8)}{-1 \cdot (x^2+8) \cdot (x-3)}$$

$$= -2(x+3)$$

**47.** $\dfrac{(x+2)^2}{x-2} \div \dfrac{x^2-4}{2x-4} = \dfrac{(x+2)^2}{x-2} \cdot \dfrac{2x-4}{x^2-4}$

$$= \frac{(x+2)(x+2) \cdot 2(x-2)}{(x-2) \cdot (x+2)(x-2)} = \frac{2(x+2)}{x-2}$$

**49.** $\dfrac{a^2+7a+12}{a^2+5a+6} \cdot \dfrac{a^2+8a+15}{a^2+5a+4}$

$$= \frac{(a+3)(a+4) \cdot (a+5)(a+3)}{(a+3)(a+2) \cdot (a+4)(a+1)}$$

$$= \frac{(a+5)(a+3)}{(a+2)(a+1)}$$

**51.** $\dfrac{1}{-x-4} \div \dfrac{x^2-7x}{x^2-3x-28}$

$$= \frac{1}{-x-4} \cdot \frac{x^2-3x-28}{x^2-7x}$$

$$= \frac{1}{-1(x+4)} \cdot \frac{(x-7)(x+4)}{x(x-7)}$$

$$= \frac{(x-7) \cdot (x+4)}{-1 \cdot x \cdot (x+4) \cdot (x-7)}$$

$$= -\frac{1}{x}$$

**53.** $\dfrac{x^2-5x-24}{2x^2-2x-24} \cdot \dfrac{4x^2+4x-24}{x^2-10x+16}$

$$= \frac{(x-8)(x+3)}{2(x^2-x-12)} \cdot \frac{4(x^2+x-6)}{(x-8)(x-2)}$$

$$= \frac{(x-8)(x+3)}{2(x-4)(x+3)} \cdot \frac{4(x+3)(x-2)}{(x-8)(x-2)}$$

$$= \frac{2 \cdot 2 \cdot (x-8) \cdot (x+3) \cdot (x+3) \cdot (x-2)}{2 \cdot (x-4) \cdot (x+3) \cdot (x-8) \cdot (x-2)}$$

$$= \frac{2(x+3)}{x-4}$$

**55.** $(x-5) \div \dfrac{5-x}{x^2+2} = (x-5) \cdot \dfrac{x^2+2}{5-x}$

$$= (x-5) \cdot \frac{x^2+2}{-1(x-5)}$$

$$= \frac{(x-5) \cdot (x^2+2)}{-1 \cdot (x-5)}$$

$$= -(x^2+2)$$

**57.** $\dfrac{x^2-y^2}{x^2-2xy+y^2} \cdot \dfrac{y-x}{x+y}$

$$= \frac{(x+y)(x-y)}{(x-y)(x-y)} \cdot \frac{-1(x-y)}{x+y}$$

$$= \frac{-1 \cdot (x+y) \cdot (x-y) \cdot (x-y)}{(x-y) \cdot (x-y) \cdot (x+y)}$$

$$= -1$$

**59.** $\dfrac{x^2-9}{2x} \div \dfrac{x+3}{8x^4} = \dfrac{x^2-9}{2x} \cdot \dfrac{8x^4}{x+3}$

$$= \frac{(x+3)(x-3)}{2x} \cdot \frac{8x^4}{x+3}$$

$$= \frac{2x \cdot 4x^3 \cdot (x+3)(x-3)}{2x \cdot (x+3)}$$

$$= 4x^3(x-3)$$

**61.** 1 square foot is 12 inches by 12 inches or 144 square inches.

$$10 \text{ sq ft} \cdot \frac{144 \text{ sq in}}{1 \text{ sq ft}} = 1440 \text{ sq in.}$$

**63.** $3,705,745$ sq ft

$$= 3,705,745 \text{ sq ft} \left( \frac{1 \text{ sq yd}}{9 \text{ sq ft}} \right)$$

$$\approx 411,972 \text{ sq yd}$$

**65.** $\dfrac{50 \text{ miles}}{1 \text{ hour}} \cdot \dfrac{1 \text{ hour}}{3600 \text{ seconds}} = \dfrac{5280 \text{ feet}}{1 \text{ mile}}$

$$= \frac{50 \cdot 5280}{3600} \text{ feet/sec} \approx 73 \text{ feet/sec}$$

**67.** $5023 \text{ ft per sec} \cdot \left( \dfrac{3600 \text{ sec}}{1 \text{ hour}} \right) \left( \dfrac{1 \text{ mile}}{5280 \text{ ft}} \right)$

$$\approx 3424.8 \text{ miles per hour}$$

**69.** $\dfrac{1}{5} + \dfrac{4}{5} = \dfrac{5}{5} = 1$

**71.** $\dfrac{9}{9} - \dfrac{19}{9} = -\dfrac{10}{9}$

**73.** $\dfrac{6}{5} + \left( \dfrac{1}{5} - \dfrac{8}{5} \right) = \dfrac{6}{5} + \left( -\dfrac{7}{5} \right) = -\dfrac{1}{5}$

**75.** $x - 2y = 6$

| $x$ | $y$ |
|----|----|
| 0 | −3 |
| 6 | 0 |

**77.** $\dfrac{a^2 + ac + ba + bc}{a - b} \div \dfrac{a + c}{a + b}$

$$= \frac{a(a+c) + b(a+c)}{a - b} \cdot \frac{a+b}{a+c}$$

$$= \frac{(a+c)(a+b)}{a-b} \cdot \frac{a+b}{a+c}$$

$$= \frac{(a+c) \cdot (a+b) \cdot (a+b)}{(a-b) \cdot (a+c)}$$

$$= \frac{(a+b)^2}{a-b}$$

**79.** $\dfrac{3x^2 + 8x + 5}{x^2 + 8x + 7} \cdot \dfrac{x + 7}{x^2 + 4}$

$$= \frac{(3x+5)(x+1)}{(x+7)(x+1)} \cdot \frac{x+7}{x^2+4}$$

$$= \frac{(3x+5) \cdot (x+1) \cdot (x+7)}{(x+7) \cdot (x+1) \cdot (x^2+4)}$$

$$= \frac{3x+5}{x^2+4}$$

**81.** $\dfrac{x^3 + 8}{x^2 - 2x + 4} \cdot \dfrac{4}{x^2 - 4}$

$$= \frac{(x+2)(x^2 - 2x + 4)}{x^2 - 2x + 4} \cdot \frac{4}{(x+2)(x-2)}$$

$$= \frac{4 \cdot (x+2) \cdot (x^2 - 2x + 4)}{(x+2)(x-2)(x^2 - 2x + 4)}$$

$$= \frac{4}{x-2}$$

**83.** $\dfrac{a^2-ab}{6a^2+6ab} \div \dfrac{a^3-b^3}{a^2-b^2}$

$= \dfrac{a^2-ab}{6a^2+6ab} \cdot \dfrac{a^2-b^2}{a^3-b^3}$

$\dfrac{a(a-b)}{6a(a+b)} \cdot \dfrac{(a-b)(a+b)}{(a-b)\left(a^2+ab+b^2\right)}$

$= \dfrac{a\cdot(a-b)\cdot(a-b)\cdot(a+b)}{6\cdot a\cdot(a+b)\cdot(a-b)\cdot\left(a^2+ab+b^2\right)}$

$= \dfrac{a-b}{6\left(a^2+ab+b^2\right)}$

**85.** $\left(\dfrac{x^2-y^2}{x^2+y^2} \div \dfrac{x^2-y^2}{3x}\right) \cdot \dfrac{x^2+y^2}{6}$

$= \dfrac{x^2-y^2}{x^2+y^2} \cdot \dfrac{3x}{x^2-y^2} \cdot \dfrac{x^2+y^2}{6}$

$= \dfrac{\left(x^2-y^2\right)\cdot 3x\cdot\left(x^2+y^2\right)}{\left(x^2+y^2\right)\cdot\left(x^2-y^2\right)\cdot 2\cdot 3} = \dfrac{x}{2}$

**87.** $\left(\dfrac{2a+b}{b^2} \cdot \dfrac{3a^2-2ab}{ab+2b^2}\right) \div \dfrac{a^2-3ab+2b^2}{5ab-10b^2}$

$= \dfrac{2a+b}{b^2} \cdot \dfrac{3a^2-2ab}{ab+2b^2} \cdot \dfrac{5ab-10b^2}{a^2-3ab+2b^2}$

$= \dfrac{(2a+b)\cdot\left(3a^2-2ab\right)\cdot\left(5ab-10b^2\right)}{b^2\cdot\left(ab+2b^2\right)\cdot\left(a^2-3ab+2b^2\right)}$

$= \dfrac{(2a+b)\cdot a(3a-2b)\cdot 5b(a-2b)}{b^2\cdot b(a+2b)\cdot(a-2b)(a-b)}$

$= \dfrac{5a(2a+b)(3a-2b)}{b^2(a+2b)(a-b)}$

**89.** $\$2000 = \$2000\left(\dfrac{1\text{ euro}}{1.09\text{ dollars}}\right)$

$\approx 1834.86$ euros

**Mental Math 7.3**

**1.** $\dfrac{2}{3}+\dfrac{1}{3}=\dfrac{3}{3}=1$

**2.** $\dfrac{5}{11}+\dfrac{1}{11}=\dfrac{6}{11}$

**3.** $\dfrac{3x}{9}+\dfrac{4x}{9}=\dfrac{7x}{9}$

**4.** $\dfrac{3y}{8}+\dfrac{2y}{8}=\dfrac{5y}{8}$

**5.** $\dfrac{8}{9}-\dfrac{7}{9}=\dfrac{1}{9}$

**6.** $-\dfrac{4}{12}-\dfrac{3}{12}=-\dfrac{7}{12}$

**7.** $\dfrac{7}{5}-\dfrac{10y}{5}=\dfrac{7-10y}{5}$

**8.** $\dfrac{12x}{7}-\dfrac{4x}{7}=\dfrac{8x}{7}$

**Exercise Set 7.3**

**1.** $\dfrac{a}{13}+\dfrac{9}{13}=\dfrac{a+9}{13}$

**3.** $\dfrac{9}{3+y}+\dfrac{y+1}{3+y}=\dfrac{9+y+1}{3+y}=\dfrac{y+10}{3+y}$

**5.** $\dfrac{4m}{3n} + \dfrac{5m}{3n} = \dfrac{4m+5m}{3n} = \dfrac{9m}{3n} = \dfrac{3m}{n}$

**7.** $\dfrac{2x+1}{x-3} + \dfrac{3x+6}{x-3} = \dfrac{2x+1+3x+6}{x-3} = \dfrac{5x+7}{x-3}$

**9.** $\dfrac{7}{8} - \dfrac{3}{8} = \dfrac{4}{8} = \dfrac{1}{2}$

**11.** $\dfrac{4m}{m-6} - \dfrac{24}{m-6} = \dfrac{4m-24}{m-6} = \dfrac{4(m-6)}{m-6} = 4$

**13.** $\dfrac{2x^2}{x-5} - \dfrac{25+x^2}{x-5} = \dfrac{2x^2-(25+x^2)}{x-5}$

$= \dfrac{2x^2-25-x^2}{x-5}$

$= \dfrac{x^2-25}{x-5}$

$= \dfrac{(x+5)(x-5)}{x-5}$

$= x+5$

**15.** $\dfrac{-3x^2-4}{x-4} - \dfrac{12-4x^2}{x-4}$

$= \dfrac{-3x^2-4-(12-4x^2)}{x-4}$

$= \dfrac{-3x^2-4-12+4x^2}{x-4}$

$= \dfrac{x^2-16}{x-4}$

$= \dfrac{(x+4)(x-4)}{x-4}$

$= x+4$

**17.** $\dfrac{2x+3}{x+1} - \dfrac{x+2}{x+1} = \dfrac{2x+3-(x+2)}{x+1}$

$= \dfrac{2x+3-x-2}{x+1}$

$= \dfrac{x+1}{x+1}$

$= 1$

**19.** $\dfrac{3}{x^3} + \dfrac{9}{x^3} = \dfrac{3+9}{x^3} = \dfrac{12}{x^3}$

**21.** $\dfrac{5}{x+4} - \dfrac{10}{x+4} = \dfrac{5-10}{x+4} = -\dfrac{5}{x+4}$

**23.** $\dfrac{x}{x+y} - \dfrac{2}{x+y} = \dfrac{x-2}{x+y}$

**25.** $\dfrac{8x}{2x+5} + \dfrac{20}{2x+5} = \dfrac{8x+20}{2x+5} = \dfrac{4(2x+5)}{2x+5} = 4$

**27.** $\dfrac{5x+4}{x-1} - \dfrac{2x+7}{x-1} = \dfrac{5x+4-(2x+7)}{x-1}$

$= \dfrac{5x+4-2x-7}{x-1}$

$= \dfrac{3x-3}{x-1}$

$= \dfrac{3(x-1)}{x-1}$

$= 3$

**29.** $\dfrac{a}{a^2+2a-15} - \dfrac{3}{a^2+2a-15} = \dfrac{a-3}{a^2+2a-15}$

$= \dfrac{a-3}{(a+5)(a-3)} = \dfrac{1}{a+5}$

**31.** $\dfrac{2x+3}{x^2-x-30} - \dfrac{x-2}{x^2-x-30}$

$= \dfrac{2x+3-(x-2)}{x^2-x-30}$

$= \dfrac{2x+3-x+2}{x^2-x-30} = \dfrac{x+5}{x^2-x-30}$

$= \dfrac{x+5}{(x-6)(x+5)} = \dfrac{1}{x-6}$

**33.** $P = \dfrac{5}{x-2} + \dfrac{5}{x-2} + \dfrac{5}{x-2} + \dfrac{5}{x-2}$

$= \dfrac{5+5+5+5}{x-2} = \dfrac{20}{x-2}$

The perimeter is $\dfrac{20}{x-2}$ meters.

**35.** Answers may vary.

**37.** $3 = 3$

$33 = 3 \cdot 11$

$\text{LCD} = 3 \cdot 11 = 33$

**39.** $2x = 2 \cdot x$

$4x^3 = 2^2 \cdot x^3$

$\text{LCD} = 2^2 \cdot x^3 = 4x^3$

**41.** $8x = 2^3 \cdot x$

$2x + 4 = 2(x+2)$

$\text{LCD} = 2^3 \cdot x \cdot (x+2) = 8x(x+2)$

**43.** $3x + 3 = 3 \cdot (x+1)$

$2x^2 + 4x + 2 = 2(x^2 + 2x + 1) = 2 \cdot (x+1)^2$

$\text{LCD} = 2 \cdot 3(x+1)^2$

$= 6(x+1)^2$

**45.** $x - 8 = x - 8$

$8 - x = -(x-8)$

$\text{LCD} = x - 8 \text{ or } 8 - x$

**47.** $8x^2(x-1)^2 = 2^3 \cdot x^2 \cdot (x-1)^2$

$10x^3(x-1) = 2 \cdot 5 \cdot x^3 \cdot (x-1)$

$\text{LCD} = 2^3 \cdot 5 \cdot x^3 \cdot (x-1)^2 = 40x^3(x-1)^2$

**49.** $2x + 1 = (2x+1)$

$2x - 1 = (2x-1)$

$\text{LCD} = (2x+1)(2x-1)$

**51.** $2x^2 + 7x - 4 = (x+4)(2x-1)$

$2x^2 + 5x - 3 = (x+3)(2x-1)$

$\text{LCD} = (x+4)(x+3)(2x-1)$

**53.** Answers may vary

**55.** $\dfrac{3}{2x} = \dfrac{3(2x)}{2x(2x)} = \dfrac{6x}{4x^2}$

**57.** $\dfrac{6}{3a} = \dfrac{6(4b^2)}{3a(4b^2)} = \dfrac{24b^2}{12ab^2}$

**59.** $\dfrac{9}{x+3} = \dfrac{9(2)}{(x+3)(2)} = \dfrac{18}{2(x+3)}$

**61.** $\dfrac{9a+2}{5a+10} = \dfrac{9a+2}{5(a+2)}$

$= \dfrac{(9a+2)(b)}{5(a+2)(b)}$

$= \dfrac{9ab+2b}{5b(a+2)}$

**63.** $\dfrac{x}{x^2 + 6x + 8} = \dfrac{x}{(x+4)(x+2)}$

$= \dfrac{x(x+1)}{(x+4)(x+2)(x+1)}$

$= \dfrac{x^2 + x}{(x+4)(x+2)(x+1)}$

**65.** $\dfrac{9y-1}{15x^2 - 30} = \dfrac{(9y-1)(2)}{(15x^2 - 30)2} = \dfrac{18y-2}{30x^2 - 60}$

**67.** $\dfrac{5}{2x^2 - 9x - 5} = \dfrac{5}{(2x+1)(x-5)} \cdot \dfrac{3x(x-7)}{3x(x-7)}$

$= \dfrac{15x(x-7)}{3x(2x+1)(x-7)(x-5)}$

**69.** LCD = 21

$\dfrac{2}{3} + \dfrac{5}{7} = \dfrac{2(7)}{3(7)} + \dfrac{5(3)}{7(3)} = \dfrac{14}{21} + \dfrac{15}{21} = \dfrac{29}{21}$

**71.** Since $6 = 2 \cdot 3$ and $4 = 2^2$,

LCD $= 2^2 \cdot 3 = 12$.

$\dfrac{2}{6} - \dfrac{3}{4} = \dfrac{2(2)}{6(2)} - \dfrac{3(3)}{4(3)} = \dfrac{4}{12} - \dfrac{9}{12} = \dfrac{4-9}{12}$

$= -\dfrac{5}{12}$

**73.** $x(x-3) = 0$

$x = 0$    or    $x - 3 = 0$

$x = 3$

**75.**    $x^2 + 6x + 5 = 0$

$(x+5)(x+1) = 0$

$x + 5 = 0$    or    $x + 1 = 0$

$x = -5$   or        $x = -1$

**77.** C

**79.** B

**81.** $\dfrac{5}{2-x} = \dfrac{5(-1)}{(2-x)(-1)} = -\dfrac{5}{x-2}$

**83.** $-\dfrac{7+x}{2-x} = \dfrac{7+x}{(-1)(2-x)} = \dfrac{7+x}{x-2}$

**85.** Since $88 = 2^3 \cdot 11$ and $4332 = 2^2 \cdot 3 \cdot 19^2$

the LCM of 88 and 4332 is

$2^3 \cdot 3 \cdot 11 \cdot 19^2 = 95,304$. They will align

again in 95,304 Earth days.

**87.** Answers may vary.

**Mental Math 7.4**

**1. D**

**2. C**

**3. A**

**4. B**

**Exercise Set 7.4**

**1.** LCD $= 2 \cdot 3 \cdot x = 6x$

$\dfrac{4}{2x} + \dfrac{9}{3x} = \dfrac{4(3)}{2x(3)} + \dfrac{9(2)}{3x(2)} = \dfrac{12}{6x} + \dfrac{18}{6x}$

$= \dfrac{30}{6x} = \dfrac{5(6)}{6x} = \dfrac{5}{x}$

**3.** $\text{LCD} = 5b$

$$\frac{15a}{b} + \frac{6b}{5} = \frac{15a(5)}{b(5)} + \frac{6b(b)}{5(b)} = \frac{75a}{5b} + \frac{6b^2}{5b}$$

$$= \frac{75a + 6b^2}{5b}$$

**5.** $\text{LCD} = 2x^2$

$$\frac{3}{x} + \frac{5}{2x^2} = \frac{3(2x)}{x(2x)} + \frac{5}{2x^2} = \frac{6x}{2x^2} + \frac{5}{2x^2}$$

$$= \frac{6x+5}{2x^2}$$

**7.** $2x + 2 = 2(x+1)$

$\text{LCD} = 2(x+1)$

$$\frac{6}{x+1} + \frac{9}{2x+2} = \frac{6}{x+1} + \frac{9}{2(x+1)}$$

$$= \frac{6(2)}{(x+1)2} + \frac{9}{2(x+1)} = \frac{12}{2(x+1)} + \frac{9}{2(x+1)}$$

$$= \frac{12+9}{2(x+1)} = \frac{21}{2(x+1)}$$

**9.** $2x - 4 = 2(x-2)$

$x^2 - 4 = (x-2)(x+2)$

$\text{LCD} = 2(x-2)(x+2)$

$$\frac{15}{2x-4} + \frac{x}{x^2-4} = \frac{15}{2(x-2)} + \frac{x}{(x-2)(x+2)}$$

$$= \frac{15(x+2)}{2(x-2)(x+2)} + \frac{x(2)}{(x-2)(x+2)(x)}$$

$$= \frac{15x+30}{2(x-2)(x+2)} + \frac{2x}{2(x-2)(x+2)}$$

$$= \frac{15x+30+2x}{2(x-2)(x+2)} = \frac{17x+30}{2(x-2)(x+2)}$$

**11.** $\text{LCD} = 4x(x-2)$

$$\frac{3}{4x} + \frac{8}{x-2} = \frac{3(x-2)}{4x(x-2)} + \frac{8(4x)}{(x-2)(4x)}$$

$$= \frac{3x-6}{4x(x-2)} + \frac{32x}{4x(x-2)} = \frac{3x-6+32x}{4x(x-2)}$$

$$= \frac{35x-6}{4x(x-2)}$$

**13.** $\text{LCD} = y^2(2y+1)$

$$\frac{5}{y^2} - \frac{y}{2y+1} = \frac{5(2y+1)}{y^2(2y+1)} - \frac{yy^2}{(2y+1)y^2}$$

$$= \frac{10y+5}{y^2(2y+1)} - \frac{y^3}{y^2(2y+1)} = \frac{10y+5-y^3}{y^2(2y+1)}$$

**15.** Answers may vary.

**17.** $\dfrac{6}{x-3} + \dfrac{8}{3-x} = \dfrac{6}{x-3} + \dfrac{8}{-(x-3)}$

$$= \frac{6}{x-3} + \frac{-8}{x-3} = \frac{6+(-8)}{x-3} = -\frac{2}{x-3}$$

**19.** $\dfrac{-8}{x^2-1} - \dfrac{7}{1-x^2} = \dfrac{8}{-(x^2-1)} - \dfrac{7}{1-x^2}$

$$= \frac{8}{1-x^2} - \frac{7}{1-x^2} = \frac{8-7}{1-x^2}$$

$$= \frac{1}{1-x^2} \text{ or } -\frac{1}{x^2-1}$$

**21.** $\dfrac{x}{x^2-4} - \dfrac{2}{4-x^2} = \dfrac{x}{x^2-4} - \dfrac{2}{-(x^2-4)}$

$$= \frac{x}{x^2-4} + \frac{2}{x^2-4} = \frac{x+2}{x^2-4}$$

$$= \frac{x+2}{(x+2)(x-2)} = \frac{1}{x-2}$$

**23.** $\dfrac{5}{x} + 2 = \dfrac{5}{x} + \dfrac{2}{1} = \dfrac{5}{x} + \dfrac{2(x)}{1(x)} = \dfrac{5+2x}{x}$

**25.** $\dfrac{5}{x-2} + 6 = \dfrac{5}{x-2} + \dfrac{6}{1} = \dfrac{5}{x-2} + \dfrac{6(x-2)}{1(x-2)}$

$\quad = \dfrac{5}{x-2} + \dfrac{6x-12}{x-2} = \dfrac{5+6x-12}{x-2} = \dfrac{6x-7}{x-2}$

**27.** $\dfrac{y+2}{y+3} - 2 = \dfrac{y+2}{y+3} - \dfrac{2}{1} = \dfrac{y+2}{y+3} - \dfrac{2(y+3)}{y+3}$

$\quad = \dfrac{y+2}{y+3} - \dfrac{2y+6}{y+3} = \dfrac{y+2-(2y+6)}{y+3}$

$\quad = \dfrac{y+2-2y-6}{y+3} = \dfrac{-y-4}{y+3} = \dfrac{-(y+4)}{y+3}$

$\quad = -\dfrac{y+4}{y+3}$

**29.** $90° - \left(\dfrac{40}{x}\right)° = \left(90 - \dfrac{40}{x}\right)°$

$\quad$ LCD $= x$

$\quad \left(90 \cdot \dfrac{x}{x} - \dfrac{40}{x}\right)° = \left(\dfrac{90x}{x} - \dfrac{40}{x}\right)°$

$\quad\quad = \left(\dfrac{90x-40}{x}\right)°$

**31.** $\dfrac{5x}{x+2} - \dfrac{3x-4}{x+2} = \dfrac{5x-(3x-4)}{x+2}$

$\quad = \dfrac{5x-3x+4}{x+2} = \dfrac{2x+4}{x+2} = \dfrac{2(x+2)}{x+2} = 2$

**33.** $\dfrac{3x^4}{x} - \dfrac{4x^2}{x^2} = \dfrac{3x^4(x)}{x(x)} - \dfrac{4x^2}{x^2} = \dfrac{3x^5}{x^2} - \dfrac{4x^2}{x^2}$

$\quad = \dfrac{3x^5 - 4x^2}{x^2} = \dfrac{x^2(3x^3 - 4)}{x^2} = 3x^3 - 4$

**35.** $\dfrac{1}{x+3} - \dfrac{1}{(x+3)^2} = \dfrac{1(x+3)}{(x+3)(x+3)} - \dfrac{1}{(x+3)^2}$

$\quad = \dfrac{x+3}{(x+3)^2} - \dfrac{1}{(x+3)^2} = \dfrac{x+3-1}{(x+3)^2} = \dfrac{x+2}{(x+3)^2}$

**37.** $\dfrac{4}{5b} + \dfrac{1}{b-1} = \dfrac{4(b-1)}{5b(b\text{-}1)} + \dfrac{1(5b)}{(b-1)(5b)}$

$\quad = \dfrac{4b-4}{5b(b\text{-}1)} + \dfrac{5b}{5b(b\text{-}1)} = \dfrac{4b-4+5b}{5b(b\text{-}1)}$

$\quad = \dfrac{9b-4}{5b(b-1)}$

**39.** $\dfrac{2}{m} + 1 = \dfrac{2}{m} + \dfrac{1}{1} = \dfrac{2}{m} + \dfrac{1(m)}{1(m)} = \dfrac{2+m}{m}$

**41.** $\dfrac{6}{1-2x} - \dfrac{4}{2x-1} = \dfrac{6}{1-2x} - \dfrac{4}{-(1-2x)}$

$\quad = \dfrac{6}{1-2x} - \dfrac{-4}{1-2x} = \dfrac{6-(-4)}{1-2x} = \dfrac{10}{1-2x}$

**43.** $\dfrac{7}{(x+1)(x-1)} + \dfrac{8}{(x+1)^2}$

$\quad = \dfrac{7(x+1)}{(x+1)(x-1)(x+1)} + \dfrac{8(x-1)}{(x+1)^2(x-1)}$

$\quad = \dfrac{7x+7}{(x+1)^2(x-1)} + \dfrac{8x-8}{(x+1)^2(x-1)}$

$\quad = \dfrac{7x+7+8x-8}{(x+1)^2(x-1)} = \dfrac{15x-1}{(x+1)^2(x-1)}$

**45.** $\dfrac{x}{x^2-1} - \dfrac{2}{x^2-2x+1}$

$= \dfrac{x}{(x-1)(x+1)} - \dfrac{2}{(x-1)^2}$

$= \dfrac{x(x-1)}{(x-1)(x+1)(x-1)} - \dfrac{2(x+1)}{(x-1)^2(x+1)}$

$= \dfrac{x^2-x}{(x-1)^2(x+1)} - \dfrac{2x+2}{(x-1)^2(x+1)}$

$= \dfrac{x^2-x-(2x+2)}{(x-1)^2(x+1)} = \dfrac{x^2-x-2x-2}{(x-1)^2(x+1)}$

$= \dfrac{x^2-3x-2}{(x-1)^2(x+1)}$

**47.** $\dfrac{3a}{2a+6} - \dfrac{a-1}{a+3} = \dfrac{3a}{2(a+3)} - \dfrac{a-1}{a+3}$

$= \dfrac{3a}{2(a+3)} - \dfrac{(a-1)(2)}{(a+3)(2)}$

$= \dfrac{3a}{2(a+3)} - \dfrac{2a-2}{2(a+3)}$

$= \dfrac{3a-(2a-2)}{2(a+3)} = \dfrac{3a-2a+2}{2(a+3)} = \dfrac{a+2}{2(a+3)}$

**49.** $\dfrac{5}{2-x} + \dfrac{x}{2x-4} = \dfrac{5}{-(x-2)} + \dfrac{x}{2(x-2)}$

$= \dfrac{-5}{x-2} + \dfrac{x}{2(x-2)}$

$= \dfrac{-5(2)}{(x-2)(2)} + \dfrac{x}{2(x-2)}$

$= \dfrac{-10}{2(x-2)} + \dfrac{x}{2(x-2)} = \dfrac{x-10}{2(x-2)}$

**51.** $\dfrac{-7}{y^2-3y+2} - \dfrac{2}{y-1} = \dfrac{-7}{(y-1)(y-2)} - \dfrac{2}{y-1}$

$= \dfrac{-7}{(y-1)(y-2)} - \dfrac{2(y-2)}{(y-1)(y-2)}$

$= \dfrac{-7-(2y-4)}{(y-1)(y-2)} = \dfrac{-7-2y+4}{(y-1)(y-2)}$

$= \dfrac{-3-2y}{(y-2)(y-1)}$

**53.** $\dfrac{13}{x^2-5x+6} - \dfrac{5}{x-3} = \dfrac{13}{(x-3)(x-2)} - \dfrac{5}{x-3}$

$= \dfrac{13}{(x-3)(x-2)} - \dfrac{5(x-2)}{(x-3)(x-2)}$

$= \dfrac{13-(5x-10)}{(x-3)(x-2)} = \dfrac{213-5x+10}{(x-3)(x-2)}$

$= \dfrac{-5x+23}{(x-3)(x-2)}$

**55.** $\dfrac{8}{(x+2)(x-2)} + \dfrac{4}{(x+2)(x-3)}$

$LCD = (x+2)(x-2)(x-3)$

$\dfrac{8}{(x+2)(x-2)} \cdot \dfrac{(x-3)}{(x-3)} + \dfrac{4}{(x+2)(x-3)} \cdot \dfrac{(x-2)}{(x-2)}$

$\dfrac{8(x-3)}{(x+2)(x-2)(x-3)} + \dfrac{4(x-2)}{(x+2)(x-3)(x-2)}$

$= \dfrac{8x-24+4x-8}{(x+2)(x-2)(x-3)}$

$= \dfrac{12x-32}{(x+2)(x-2)(x-3)}$

**57.** $\dfrac{5}{9x^2-4}+\dfrac{2}{3x-2}=\dfrac{5}{(3x+2)(3x-2)}+\dfrac{2}{3x-2}$

$\text{LCD}=(3x+2)(3x-2)$

$=\dfrac{5}{(3x+2)(3x-2)}+\dfrac{2}{(3x-2)}\cdot\dfrac{(3x+2)}{(3x+2)}$

$=\dfrac{5}{(3x+2)(3x-2)}+\dfrac{2(3x+2)}{(3x+-2)(3x+2)}$

$=\dfrac{5+6x+4}{(3x+2)(3x-2)}$

$=\dfrac{6x+9}{(3x+2)(3x-2)}$

**59.** $\dfrac{x+8}{x^2-5x-6}+\dfrac{x+1}{x^2-4x-5}$

$=\dfrac{x+8}{(x-6)(x+1)}+\dfrac{x+1}{(x-5)(x+1)}$

$=\dfrac{(x+8)(x-5)}{(x-6)(x+1)(x-5)}+\dfrac{(x+1)(x-6)}{(x-5)(x+1)(x-6)}$

$=\dfrac{x^2+3x-40+x^2-5x-6}{(x-6)(x+1)(x-5)}$

$=\dfrac{2x^2-2x-46}{(x-6)(x+1)(x-5)}$

**61.** $\quad 3x+5=7$

$3x+5-5=7-5$

$3x=2$

$\dfrac{3x}{3}=\dfrac{2}{3}$

$x=\dfrac{2}{3}$

**63.** $2x^2-x-1=0$

$(2x+1)(x-1)=0$

$2x+1=0\quad$ or $\quad x-1=0$

$2x=-1\qquad\qquad x=1$

$x=-\dfrac{1}{2}$

The solutions are $x=-\dfrac{1}{2}$ and $x=1$.

**65.** $\dfrac{2+x}{x+2}=\dfrac{x+2}{x+2}=1$

**67.** $\dfrac{2-x}{x-2}=\dfrac{-x+2}{x-2}=\dfrac{-(x-2)}{x-2}=-1$

**69.** $P=2\left(\dfrac{3}{y-5}\right)+2\left(\dfrac{2}{y}\right)$

$=\dfrac{6(y)}{(y-5)(y)}+\dfrac{4(y-5)}{y(y-5)}$

$=\dfrac{6y+4(y-5)}{y(y-5)}=\dfrac{6y+4y-20}{y(y-5)}$

$=\dfrac{10y-20}{y(y-5)}$

The perimeter is $\dfrac{10y-20}{y(y-5)}$ ft

$A=\dfrac{3}{y-5}\cdot\dfrac{2}{y}=\dfrac{6}{y(y-5)}$

The area is $\dfrac{6}{y(y-5)}$ sq ft.

**71.** $\dfrac{15x}{x+8}\cdot\dfrac{2x+16}{3x}=\dfrac{15x}{x+8}\cdot\dfrac{2(x+8)}{3x}$

$=\dfrac{2\cdot5\cdot3x\cdot(x+8)}{3x\cdot(x+8)}$

$=10$

**73.** $\dfrac{8x+7}{3x+5} - \dfrac{2x-3}{3x+5} = \dfrac{8x+7(2x-3)}{3x+5}$

$\qquad\qquad = \dfrac{8x+7-2x+3}{3x+5}$

$\qquad\qquad = \dfrac{6x+10}{3x+5}$

$\qquad\qquad = \dfrac{2(3x+5)}{3x+5}$

$\qquad\qquad = 2$

**75.** $\dfrac{5a+10}{18} \div \dfrac{a^2-4}{10a} = \dfrac{5(a+2)}{2\cdot 9} \cdot \dfrac{2\cdot 5a}{(a-2)(a+2)}$

$\qquad\qquad = \dfrac{25a}{9(a-2)}$

**77.** $\dfrac{5}{x^2-3x+2} + \dfrac{1}{x-2} = \dfrac{5}{(x-2)(x-1)} + \dfrac{1}{x-2}$

$\text{LCD} = (x-2)(x-1)$

$\dfrac{5}{(x-2)(x-1)} + \dfrac{1}{(x-2)} \cdot \dfrac{(x-1)}{(x-1)}$

$= \dfrac{5}{(x-2)(x-1)} + \dfrac{x-1}{(x-2)(x-1)}$

$= \dfrac{5+x-1}{(x-2)(x-1)}$

$= \dfrac{x+4}{(x-2)(x-1)}$

**79.** Answers may vary.

**81.** $\dfrac{5}{x^2-4} + \dfrac{2}{x^2-4x+4} - \dfrac{3}{x^2-x-6}$

$= \dfrac{5}{(x-2)(x+2)} + \dfrac{2}{(x-2)^2}$

$\qquad - \dfrac{3}{(x-3)(x+2)}$

$= \dfrac{5(x-2)(x-3)}{(x-2)(x+2)(x-2)(x-3)}$

$\qquad + \dfrac{2(x+2)(x-3)}{(x-2)^2(x+2)(x-3)}$

$\qquad - \dfrac{3(x-2)^2}{(x-3)(x+2)(x-2)^2}$

$= \dfrac{5(x^2-5x+6)}{(x-2)^2(x+2)(x-3)}$

$\qquad + \dfrac{2(x^2-x-6)}{(x-2)^2(x+2)(x-3)}$

$\qquad - \dfrac{3(x^2-4x+4)}{(x-2)^2(x+2)(x-3)}$

$= \dfrac{5x^2-25x+30}{(x-2)^2(x+2)(x-3)}$

$\qquad + \dfrac{2x^2-2x-12}{(x-2)^2(x+2)(x-3)}$

$\qquad - \dfrac{3x^2-12x+12}{(x-2)^2(x+2)(x-3)}$

$= \dfrac{4x^2-15x+6}{(x-2)^2(x+2)(x-3)}$

**83.** $\dfrac{5+x}{x^3-27}+\dfrac{x}{x^3+3x^2+9x}$

$=\dfrac{5+x}{(x-3)(x^2+3x+9)}+\dfrac{x}{x(x^2+3x+9)}$

$\text{LCD}=x(x-3)(x^2+3x+9)$

$\dfrac{(5+x)}{(x-3)(x^2+3x+9)}\cdot\dfrac{x}{x}+\dfrac{x}{x(x^2+3x+9)}\cdot\dfrac{(x-3)}{(x-3)}$

$=\dfrac{(5+x)(x)}{x(x-3)(x^2+3x+9)}+\dfrac{x(x-3)}{x(x-3)(x^2+3x+9)}$

$=\dfrac{5x+x^2+x^2-3x}{x(x-3)(x^2+3x+9)}$

$=\dfrac{2x^2+2x}{x(x-3)(x^2+3x+9)}$

$=\dfrac{2x(x+1)}{x(x-3)(x^2+3x+9)}$

$=\dfrac{2(x+1)}{(x-3)(x^2+3x+9)}$

**85.** $\dfrac{DA}{A+12}-\dfrac{D(A+1)}{24}$

$=\dfrac{24DA}{24(A+12)}-\dfrac{D(A+1)(A+12)}{24(A+12)}$

$=\dfrac{24DA-DA^2-13DA-12D}{24(A+12)}$

$=\dfrac{11DA-DA^2-12D}{24(A+12)}$

**Graphing Calculator Explorations 7.5**

**1.** $y_1=\dfrac{x-4}{2}-\dfrac{x-3}{9}$, $y_2=\dfrac{5}{18}$.

Use INTERSECT

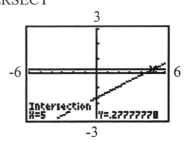

The solution of the equation is 5

**3.** $y_1=3-\dfrac{6}{x}$, $y_2=x+8$.

Use INTERSECT

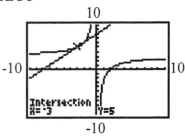

One solution is -3

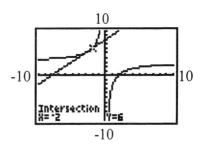

The other solution is -2.

**5.** $y_1 = x + \dfrac{14}{x-2}$, $y_2 = \dfrac{7x}{x-2} + 1$.

Use INTERSECT

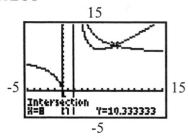

The solution is 8.

**Mental Math 7.5**

**1.** $\dfrac{x}{5} = 2$

$x = 10$

**2.** $\dfrac{x}{8} = 4$

$x = 32$

**3.** $\dfrac{z}{6} = 6$

$z = 36$

**4.** $\dfrac{y}{7} = 8$

$y = 56$

**Exercise Set 7.5**

**1.** $\dfrac{x}{5} + 3 = 9$

$5\left(\dfrac{x}{5} + 3\right) = 5(9)$

$5\left(\dfrac{x}{5}\right) + 5(3) = 5(9)$

$x + 15 = 45$

$x = 30$

Check:

$\dfrac{x}{5} + 3 = 9$

$\dfrac{30}{5} + 3 \overset{?}{=} 9$

$6 + 3 \overset{?}{=} 9$

$9 = 9$    True

The solution is 30.

**3.** $\dfrac{x}{2} + \dfrac{5x}{4} = \dfrac{x}{12}$

$12\left(\dfrac{x}{2} + \dfrac{5x}{4}\right) = 12\left(\dfrac{x}{12}\right)$

$12\left(\dfrac{x}{2}\right) + 12\left(\dfrac{5x}{4}\right) = 12\left(\dfrac{x}{12}\right)$

$6x + 15x = x$

$21x = x$

$20x = 0$

$x = 0$

Check:

$\dfrac{x}{2} + \dfrac{5x}{4} = \dfrac{x}{12}$

$\dfrac{0}{2} + \dfrac{5 \cdot 0}{4} \overset{?}{=} \dfrac{0}{12}$

$0 + \dfrac{0}{4} \overset{?}{=} 0$

$0 = 0$    True

The solution is 0.

**5.**
$$\frac{y}{7} - \frac{3y}{2} = 1$$
$$14\left(\frac{y}{7} - \frac{3y}{2}\right) = 14(1)$$
$$14\left(\frac{y}{7}\right) - 14\left(\frac{3y}{2}\right) = 14$$
$$2y - 21y = 14$$
$$-19y = 14$$
$$y = -\frac{14}{19}$$

Check:
$$\frac{y}{7} - \frac{3y}{2} = 1$$
$$\frac{-(14/19)}{7} - \frac{3(-14/19)}{2} \overset{?}{=} 1$$
$$-\frac{2}{19} + \frac{21}{19} \overset{?}{=} 1$$
$$\frac{19}{19} \overset{?}{=} 1$$
$$1 = 1 \quad \text{True}$$

The solution is $-\dfrac{14}{19}$.

**7.** $2 + \dfrac{10}{x} = x + 5$
$$x\left(2 + \frac{10}{x}\right) = x(x+5)$$
$$x(2) + x\left(\frac{10}{x}\right) = x(x+5)$$
$$2x + 10 = x^2 + 5x$$
$$0 = x^2 + 3x - 10$$
$$0 = (x+5)(x-2)$$
$$x + 5 = 0 \quad \text{or} \quad x - 2 = 0$$
$$x = -5 \qquad\qquad x = 2$$

Check:
$x = -5$:
$$2 + \frac{10}{x} = x + 5$$
$$2 + \frac{10}{-5} \overset{?}{=} -5 + 5$$
$$2 + (-2) \overset{?}{=} -5 + 5$$
$$0 = 0 \quad \text{True}$$
$x = 2$:
$$2 + \frac{10}{x} = x + 5$$
$$2 + \frac{10}{2} \overset{?}{=} 2 + 5$$
$$2 + 5 \overset{?}{=} 2 + 5$$
$$7 = 7 \quad \text{True}$$
Both $-5$ and 2 are solutions.

**9.** $\dfrac{a}{5} = \dfrac{a-3}{2}$
$$10\left(\frac{a}{5}\right) = 10\left(\frac{a-3}{2}\right)$$
$$2a = 5(a-3)$$
$$2a = 5a - 15$$
$$-3a = -15$$
$$a = 5$$

Check:
$$\frac{a}{5} = \frac{a-3}{2}$$
$$\frac{5}{5} \overset{?}{=} \frac{5-3}{2}$$
$$\frac{5}{5} \overset{?}{=} \frac{2}{2}$$
$$1 = 1 \quad \text{True}$$
The solution is 5.

**11.** $\dfrac{x-3}{5}+\dfrac{x-2}{2}=\dfrac{1}{2}$

$10\left(\dfrac{x-3}{5}+\dfrac{x-2}{2}\right)=10\left(\dfrac{1}{2}\right)$

$10\left(\dfrac{x-3}{5}\right)+10\left(\dfrac{x-2}{2}\right)=10\left(\dfrac{1}{2}\right)$

$2(x-3)+5(x-2)=5$

$2x-6+5x-10=5$

$7x-16=5$

$7x=21$

$x=3$

Check:

$\dfrac{x-3}{5}+\dfrac{x-2}{2}=\dfrac{1}{2}$

$\dfrac{3-3}{5}+\dfrac{3-2}{2}\overset{?}{=}\dfrac{1}{2}$

$\dfrac{0}{5}+\dfrac{1}{2}\overset{?}{=}\dfrac{1}{2}$

$0+\dfrac{1}{2}\overset{?}{=}\dfrac{1}{2}$

$\dfrac{1}{2}=\dfrac{1}{2}$    True

The solution is 3.

**13.** $\dfrac{x+1}{3}-\dfrac{x-2}{4}=\dfrac{5}{6}$

$12\left(\dfrac{x+1}{3}-\dfrac{x-2}{4}\right)=12\left(\dfrac{5}{6}\right)$

$12\left(\dfrac{x+1}{3}\right)-12\left(\dfrac{x-2}{4}\right)=10$

$4(x+1)-3(x-2)=10$

$4x+4-3x+6=10$

$x+10=10$

$x=0$

Check:

$\dfrac{x+1}{3}-\dfrac{x-2}{4}=\dfrac{5}{6}$

$\dfrac{0+1}{3}-\dfrac{0-2}{4}\overset{?}{=}\dfrac{5}{6}$

$\dfrac{1}{3}+\dfrac{1}{2}\overset{?}{=}\dfrac{5}{6}$

$\dfrac{5}{6}=\dfrac{5}{6}$    True

The solution is 0.

**15.** $\dfrac{9}{2a-5}=-2$

$(2a-5)\left(\dfrac{9}{2a-5}\right)=(2a-5)(-2)$

$9=-4a+10$

$9-10=-4a$

$-1=-4a$

$\dfrac{1}{4}=a$

**17.** $\dfrac{y}{y+4}+\dfrac{4}{y+4}=3$

$(y+4)\left(\dfrac{y}{y+4}+\dfrac{4}{y+4}\right)=(y+4)(3)$

$(y+4)\left(\dfrac{y}{y+4}\right)+(y+4)\left(\dfrac{4}{y+4}\right)=(y+4)(3)$

$y+4=3y+12$

$y-3y=12-4$

$-2y=8$

$y=\dfrac{8}{-2}=-4$

$-4$ makes a denominator zero. There is no solution.

**19.** $\dfrac{4y}{y-3} - 3 = \dfrac{3y-1}{y+3}$

$(y-3)(y+3)\left(\dfrac{4y}{y-3}\right)$

$\qquad = (y-3)(y+3)\left(\dfrac{3y-1}{y+3}\right)$

$(y-3)(y+3)\left(\dfrac{4y}{y-3}\right) - (y-3)(y+3)(3)$

$\qquad = (y-3)(y+3)\left(\dfrac{3y-1}{y+3}\right)$

$4y(y+3) - 3(y^2 - 9) = (y-3)(3y-1)$

$4y^2 + 12y - 3y^2 + 27 = 3y^2 - 10y + 3$

$y^2 + 12y + 27 = 3y^2 - 10y + 3$

$0 = 2y^2 - 22y - 24$

$0 = (2y+2)(y-12)$

$2y+2 = 0 \quad \text{or} \quad y-12 = 0$

$\qquad 2y = -2$

$\qquad y = -1 \qquad\qquad y = 12$

The solutions are $-1$ and 12.

**21.** $\dfrac{4y}{y-4} + 5 = \dfrac{5y}{y-4}$

$(y-4)\left(\dfrac{4y}{y-4} + 5\right) = (y-4)\left(\dfrac{5y}{y-4}\right)$

$(y-4)\left(\dfrac{4y}{y-4}\right) + (y-4)(5) = (y-4)\left(\dfrac{5y}{y-4}\right)$

$4y + 5y - 20 = 5y$

$9y - 20 = 5y$

$4y - 20 = 0$

$4y = 20$

$y = 5$

The solution is 5.

**23.** $\dfrac{7}{x-2} + 1 = \dfrac{x}{x+2}$

$(x-2)(x+2)\left(\dfrac{7}{x-2} + 1\right)$

$\qquad = (x-2)(x+2)\left(\dfrac{x}{x+2}\right)$

$(x-2)(x+2)\left(\dfrac{7}{x-2}\right) + (x-2)(x+2)$

$\qquad = (x-2)(x+2)\left(\dfrac{x}{x+2}\right)$

$7(x+2) + (x-2)(x+2) = x(x-2)$

$7x + 14 + x^2 - 4 = x^2 - 2x$

$7x + x^2 + 10 = x^2 - 2x$

$9x = -10$

$x = -\dfrac{10}{9}$

The solution is $-\dfrac{10}{9}$

**25.** $\dfrac{x+1}{x+3} = \dfrac{2x^2 - 15x}{x^2 + x - 6} - \dfrac{x-3}{x-2}$

$\dfrac{x+1}{x+3} = \dfrac{2x^2 - 15x}{(x+3)(x-2)} - \dfrac{x-3}{x-2}$

$(x+3)(x-2)\left(\dfrac{x+1}{x+3}\right)$

$\quad = (x+3)(x-2)\left(\dfrac{2x^2 - 15x}{(x+3)(x-2)} - \dfrac{x-3}{x-2}\right)$

$(x+3)(x-2)\left(\dfrac{x+1}{x+3}\right)$

$\quad = (x+3)(x-2)\left(\dfrac{2x^2 - 15x}{(x+3)(x-2)}\right)$

$\qquad - (x+3)(x-2)\left(\dfrac{x-3}{x-2}\right)$

$$(x-2)(x+1) = 2x^2 - 15x - (x+3)(x-3)$$
$$x^2 - x - 2 = 2x^2 - 15x - (x^2 - 9)$$
$$x^2 - x - 2 = 2x^2 - 15x - x^2 + 9$$
$$x^2 - x - 2 = x^2 - 15x + 9$$
$$14x = 11$$
$$x = \frac{11}{14}$$

The solution is $\dfrac{11}{14}$.

**27.** $\dfrac{y}{2y+2} + \dfrac{2y-16}{4y+4} = \dfrac{2y-3}{y+1}$

$$\frac{y}{2(y+1)} + \frac{2y-16}{4(y+1)} = \frac{2y-3}{y+1}$$

$$4(y+1)\left(\frac{y}{2(y+1)} + \frac{2y-16}{4(y+1)}\right)$$
$$= 4(y+1)\left(\frac{2y-3}{y+1}\right)$$

$$4(y+1)\left(\frac{y}{2(y+1)}\right) + 4(y+1)\left(\frac{2y-16}{4(y+1)}\right)$$
$$= 4(y+1)\left(\frac{2y-3}{y+1}\right)$$

$$2y + 2y - 16 = 4(2y-3)$$
$$4y - 16 = 8y - 12$$
$$-4y = 4$$
$$y = -1$$

In the original equation, $-1$ makes a denominator 0.

This equation has no solution.

**29.** $\dfrac{2x}{7} - 5x = 9$

$$7\left(\frac{2x}{7} - 5x\right) = 7(9)$$

$$7\left(\frac{2x}{7}\right) + 7(-5x) = 7(9)$$

$$2x - 35x = 63$$
$$-33x = 63$$
$$x = \frac{63}{33} \quad x = -\frac{21}{11}$$

**31.** $\dfrac{2}{y} + \dfrac{1}{2} = \dfrac{5}{2y}$

$$2y\left(\frac{2}{y} + \frac{1}{2}\right) = 2y\left(\frac{5}{2y}\right)$$

$$2y\left(\frac{2}{y}\right) + 2y\left(\frac{1}{2}\right) = 2y\left(\frac{5}{2y}\right)$$

$$4 + y = 5$$
$$y = 1$$

**33.** $\dfrac{4x+10}{7} = \dfrac{8}{2}$

$$14\left(\frac{4x+10}{7}\right) = 14\left(\frac{8}{2}\right)$$

$$2(4x+10) = 7(8)$$
$$8x = 56 - 20$$
$$8x = 36$$
$$x = \frac{36}{8} = \frac{9}{2}$$

**35.**

$$2 + \frac{3}{a-3} = \frac{a}{a-3}$$

$$(a-3)\left(2 + \frac{3}{a-3}\right) = (a-3)\left(\frac{a}{a-3}\right)$$

$$(a-3)(2) + (a-3)\left(\frac{3}{a-3}\right) = a$$

$$2a - 6 + 3 = a$$

$$2a - 3 = a$$

$$-3 = a - 2a$$

$$-3 = -a$$

$$\frac{-3}{-1} = a$$

$$3 = a$$

3 is an extraneous solution. If $a = 3$, the denominator would equal zero. The equation has no solution.

**37.**

$$\frac{5}{x} + \frac{2}{3} = \frac{7}{2x}$$

$$6x\left(\frac{5}{x} + \frac{2}{3}\right) = 6x\left(\frac{7}{2x}\right)$$

$$6x\left(\frac{5}{x}\right) + 6x\left(\frac{2}{3}\right) = 21$$

$$30 + 4x = 21$$

$$4x = 21 - 30$$

$$4x = -9$$

$$x = -\frac{9}{4}$$

**39.**

$$\frac{2a}{a+4} = \frac{3}{a-1}$$

$$(a+4)(a-1)\left(\frac{2a}{a+4}\right) = (a+4)(a-1)\left(\frac{3}{a-1}\right)$$

$$(a-1)(2a) = (a+4)(3)$$

$$2a^2 - 2a = 3a + 12$$

$$2a^2 - 2a - 3a - 12 = 0$$

$$2a^2 - 5a - 12 = 0$$

$$(2a+3)(a-4) = 0$$

$$2a + 3 = 0 \quad \text{or} \quad a - 4 = 0$$

$$2a = -3 \qquad\qquad a = 4$$

$$a = -\frac{3}{2}$$

**41.**

$$\frac{x+1}{3} - \frac{x-1}{6} = \frac{1}{6}$$

$$6\left(\frac{x+1}{3} - \frac{x-1}{6}\right) = 6\left(\frac{1}{6}\right)$$

$$6\left(\frac{x+1}{3}\right) - 6\left(\frac{x-1}{6}\right) = 6\left(\frac{1}{6}\right)$$

$$2(x+1) - (x-1) = 1$$

$$2x + 2 - x + 1 = 1$$

$$x + 3 = 1$$

$$x = -2$$

The solution is $-2$.

**43.**

$$\frac{4r-1}{r^2 + 5r - 14} + \frac{2}{r+7} = \frac{1}{r-2}$$

$$\frac{4r-1}{(r+7)(r-2)} + \frac{2}{r+7} = \frac{1}{r-2}$$

$$(r+7)(r-2)\left(\frac{4r-1}{(r+7)(r-2)} + \frac{2}{r+7}\right)$$

$$= (r+7)(r-2)\left(\frac{1}{r-2}\right)$$

$$(r+7)(r-2)\left(\frac{4r-1}{(r+7)(r-2)}\right)$$

$$+(r+7)(r-2)\left(\frac{2}{r+7}\right)$$

$$=(r+7)(r-2)\left(\frac{1}{r-2}\right)$$

$$4r-1+2(r-2)=(r+7)(1)$$

$$4r-1+2r-4=r+7$$

$$6r-5=r+7$$

$$5r=12$$

$$r=\frac{12}{5}$$

$$12(x+3)\left(\frac{x}{2(x+3)}\right)$$

$$+12(x+3)\left(\frac{x+1}{3(x+3)}\right)=3(2)$$

$$6x+4(x+1)=6$$

$$6x+4x+4=6$$

$$10x+4=6$$

$$10x=6-4$$

$$10x=2$$

$$x=\frac{2}{10}=\frac{1}{5}$$

**45.**
$$\frac{t}{t-4}=\frac{t+4}{6}$$

$$6(t-4)\left(\frac{t}{t-4}\right)=6(t-4)\left(\frac{t+4}{6}\right)$$

$$6t=(t-4)(t+4)$$

$$6t=t^2-16$$

$$0=t^2-16$$

$$0=(t-8)(t+2)$$

$$t+2=0 \quad \text{or} \quad t-8=0$$

$$t=-2 \qquad t=8$$

**47.**
$$\frac{x}{2x+6}+\frac{x+1}{3x+9}=\frac{2}{4x+12}$$

$$\frac{x}{2(x+3)}+\frac{x+1}{3(x+3)}=\frac{2}{4(x+3)}$$

$$12(x+3)\left(\frac{x}{2(x+3)}+\frac{x+1}{3(x+3)}\right)$$

$$=12(x+3)\left(\frac{2}{4(x+3)}\right)$$

**49.**
$$\frac{D}{R}=T$$

$$R\left(\frac{D}{R}\right)=R(T)$$

$$D=RT$$

$$\frac{D}{T}=R$$

**51.**
$$\frac{3}{x}=\frac{5y}{x+2}$$

$$x(x+2)\left(\frac{3}{x}\right)=x(x+2)\left(\frac{5y}{x+2}\right)$$

$$(x+2)(3)=x(5y)$$

$$3x+6=5xy$$

$$\frac{3x+6}{5x}=y$$

**53.**
$$\frac{3a+2}{3b-2} = -\frac{4}{2a}$$
$$2a(3b-2)\left(\frac{3a+2}{3b-2}\right) = 2a(3b-2)\left(-\frac{4}{2a}\right)$$
$$2a(3a+2) = (3b-2)(-4)$$
$$6a^2 + 4a = -12b + 8$$
$$6a^2 + 4a - 8 = -12b$$
$$\frac{6a^2 + 4 - 8}{2 \cdot 6} = b$$
$$-\frac{2(3a^2 + 2a - 4)}{2 \cdot 6} = b$$
$$-\frac{3a^2 + 2a - 4}{6} = b$$

**55.**
$$\frac{A}{BH} = \frac{1}{2}$$
$$2BH\left(\frac{A}{BH}\right) = 2BH\left(\frac{1}{2}\right)$$
$$2A = BH$$
$$\frac{2A}{H} = B$$

**57.**
$$\frac{C}{\pi r} = 2$$
$$\pi r\left(\frac{C}{\pi r}\right) = \pi r(2)$$
$$C = 2\pi r$$
$$\frac{C}{2\pi} = \frac{2\pi r}{2\pi}$$
$$\frac{C}{2\pi} = r$$

**59.**
$$\frac{1}{a} = \frac{1}{b} + \frac{1}{c}$$
$$abc\left(\frac{1}{a}\right) = abc\left(\frac{1}{b} + \frac{1}{c}\right)$$
$$bc = abc\left(\frac{1}{b}\right) + abc\left(\frac{1}{c}\right)$$
$$bc = ac + ab$$
$$bc = a(c + b)$$
$$\frac{bc}{c+b} = a$$

**61.**
$$\frac{m^2}{6} - \frac{n}{3} = \frac{p}{2}$$
$$6\left(\frac{m^2}{6} - \frac{n}{3}\right) = 6\left(\frac{p}{2}\right)$$
$$6\left(\frac{m^2}{6}\right) - 6\left(\frac{n}{3}\right) = 3p$$
$$m^2 - 2n = 3p$$
$$m^2 - 3p = 2n$$
$$\frac{m^2 - 3p}{2} = n$$

**63.** The graph crosses the *x*-axis at $x = 2$. It crosses the *y*-axis at $y = -2$. The *x*-intercept is $(2, \; 0)$ and the *y*-intercept is $(0, \; -2)$.

**65.** The graph crosses the *x*-axis at $x = -4$, $x = -2$ and $x = 3$. It crosses the *y*-axis at $y = 4$. The *x*-intercept are $(-4, \; 0)$, $(-2, 0)$ and $(3, 0)$, and the *y*-intercept is $(0, 4)$.

**67.** Answers may vary.

**69.** expression

$$\frac{1}{x}+\frac{5}{9}=\frac{1(9)}{x(9)}+\frac{5x}{9x}=\frac{5x+9}{9x}$$

**71.** equation

$$\frac{5}{x-1}-\frac{2}{x}=\frac{5}{x(x-1)}$$

$$x(x-1)\left(\frac{5}{x-1}\right)-x(x-1)\left(\frac{2}{x}\right)$$

$$=x(x-1)\left(\frac{5}{x(x-1)}\right)$$

$$5x-2(x-1)=5$$

$$5x-2x+2=5$$

$$3x=3$$

$$x=1$$

1 makes a denominator zero.

There is no solution

**73.** $\dfrac{20x}{3}+\dfrac{32x}{6}=180$

$$6\left(\frac{180}{3}+\frac{32x}{6}\right)=6(180)$$

$$6\left(\frac{20x}{3}\right)+6\left(\frac{32x}{6}\right)=6(180)$$

$$40x+32x=1080$$

$$72x=1080$$

$$\frac{72x}{72}=\frac{1080}{72}$$

$$x=15$$

$$\frac{20x}{3}=\frac{20(15)}{3}=100$$

$$\frac{32x}{6}=\frac{32(15)}{6}=80$$

The angles are 100° and 80°.

**75.** $\dfrac{150}{x}+\dfrac{450}{x}=90$

$$x\left(\frac{150}{x}+\frac{450}{x}\right)=x(90)$$

$$x\left(\frac{150}{x}\right)+x\left(\frac{450}{x}\right)=x(90)$$

$$150+450=90x$$

$$600=90x$$

$$\frac{600}{90}=x$$

$$\frac{20}{3}=x$$

$$\frac{150}{x}=\frac{150}{\frac{20}{3}}=150\left(\frac{3}{20}\right)=\frac{45}{2}=22.5$$

$$\frac{450}{x}=\frac{450}{\frac{20}{3}}=450\left(\frac{3}{20}\right)=\frac{135}{2}=67.5$$

The angles are 22.5° and 67.5°.

**77.**

$$\frac{5}{a^2+4a+3}+\frac{2}{a^2+a-6}-\frac{3}{a^2-a-2}=0$$

$$\frac{5}{(a+3)(a+1)}+\frac{2}{(a+3)(a-2)}$$

$$-\frac{3}{(a-2)(a+1)}=0$$

$$(a+3)(a+1)(a-2)\left(\begin{array}{c}\dfrac{5}{(a+3)(a+1)}\\[2mm]+\dfrac{2}{(a+3)(a-2)}\\[2mm]-\dfrac{3}{(a-2)(a+1)}\end{array}\right)$$

$$=(a+3)(a+1)(a-2)(0)$$

$$(a+3)(a+1)(a-2)\left(\frac{5}{(a+3)(a+1)}\right)$$

$$+(a+3)(a+1)(a-2)\left(\frac{2}{(a+3)(a-2)}\right)$$

$$-(a+3)(a+1)(a-2)\left(\frac{3}{(a-2)(a+1)}\right)$$

$$=0$$

$$5(a-2)+2(a+1)-3(a+3)=0$$

$$5a-10+2a+2-3a-9=0$$

$$4a-17=0$$

$$4a=17$$

$$a=\frac{17}{4}$$

**Integrated Review-Summary on Rational Expressions**

1. expression

$$\frac{1}{x}+\frac{2}{3}=\frac{1(3)}{x(3)}+\frac{2(x)}{3(x)}=\frac{3}{3x}+\frac{2x}{3x}=\frac{3+2x}{3x}$$

2. expression

$$\frac{3}{a}+\frac{5}{6}=\frac{3(6)}{a(6)}+\frac{5(a)}{6(a)}=\frac{18}{6a}+\frac{5a}{6a}=\frac{18+5a}{6a}$$

3. equation

$$\frac{1}{x}+\frac{2}{3}=\frac{3}{x}$$

$$3x\left(\frac{1}{x}+\frac{2}{3}\right)=3x\left(\frac{3}{x}\right)$$

$$3x\left(\frac{1}{x}\right)+3x\left(\frac{2}{3}\right)=3x\left(\frac{3}{x}\right)$$

$$3+2x=9$$

$$2x=6$$

$$x=3$$

The solution is 3.

4. equation

$$\frac{3}{a}+\frac{5}{6}=1$$

$$6a\left(\frac{3}{a}+\frac{5}{6}\right)=6a(1)$$

$$6a\left(\frac{3}{a}\right)+6a\left(\frac{5}{6}\right)=6a$$

$$18+5a=6a$$

$$18=a$$

The solution is 18.

5. expression

$$\frac{2}{x-1}-\frac{1}{x}=\frac{2(x)}{(x-1)(x)}-\frac{1(x-1)}{x(x-1)}$$

$$=\frac{2x-(x-1)}{x(x-1)}=\frac{x+1}{x(x-1)}$$

**6.** expression

$$\frac{4}{x-3}-\frac{1}{x}=\frac{4(x)}{(x-3)(x)}-\frac{1(x-3)}{x(x-3)}$$

$$=\frac{4x-(x-3)}{x(x-3)}=\frac{4x-x+3}{x(x-3)}=\frac{3x+3}{x(x-3)}$$

$$=\frac{3(x+1)}{x(x-3)}$$

**7.** equation

$$\frac{2}{x+1}-\frac{1}{x}=1$$

$$x(x+1)\left(\frac{2}{x+1}-\frac{1}{x}\right)=x(x+1)(1)$$

$$x(x+1)\left(\frac{2}{x+1}\right)-x(x+1)\left(\frac{1}{x}\right)=x(x+1)$$

$$2x-(x+1)=x(x+1)$$

$$2x-x-1=x^2+x$$

$$x-1=x^2+x$$

$$-1=x^2$$

There is no real number solution.

**8.** equation

$$\frac{4}{x-3}-\frac{1}{x}=\frac{6}{x(x-3)}$$

$$x(x-3)\left(\frac{4}{x-3}-\frac{1}{x}\right)=x(x-3)\left(\frac{6}{x(x-3)}\right)$$

$$x(x-3)\left(\frac{4}{x-3}\right)-x(x-3)\left(\frac{1}{x}\right)=6$$

$$4x-(x-3)=6$$

$$4x-x+3=6$$

$$3x+3=6$$

$$3x=3$$

$$x=1$$

The solution is 1.

**9.** expression

$$\frac{15x}{x+8}\cdot\frac{2x+16}{3x}=\frac{15x\cdot(2x+16)}{(x+8)\cdot3x}$$

$$=\frac{3\cdot5\cdot x\cdot2\cdot(x+8)}{(x+8)\cdot3\cdot x}=5\cdot2=10$$

**10.** expression

$$\frac{9z+5}{15}\cdot\frac{5z}{81z^2-25}=\frac{(9z+5)\cdot5z}{15\cdot(81z^2-25)}$$

$$=\frac{(9z+5)\cdot5\cdot z}{5\cdot3\cdot(9z+5)(9z-5)}=\frac{z}{3(9z-5)}$$

**11.** expression

$$\frac{2x+1}{x-3}+\frac{3x+6}{x-3}=\frac{2x+1+3x+6}{x-3}$$

$$=\frac{5x+7}{x-3}$$

**12.** expression

$$\frac{4p-3}{2p+7}+\frac{3p+8}{2p+7}=\frac{4p-3+3p+8}{2p+7}$$

$$=\frac{7p+5}{2p+7}$$

**13.** equation

$$\frac{x+5}{7} = \frac{8}{2}$$

$$14\left(\frac{x+5}{7}\right) = 14\left(\frac{8}{2}\right)$$

$$2(x+5) = 56$$

$$2x + 10 = 56$$

$$2x = 46$$

$$x = 23$$

The solution is 23.

**14.** equation

$$\frac{1}{2} = \frac{x-1}{8}$$

$$8\left(\frac{1}{2}\right) = 8\left(\frac{x-1}{8}\right)$$

$$4 = x - 1$$

$$5 = x$$

The solution is 5.

**15.** expression

$$\frac{5a+10}{18} \div \frac{a^2-4}{10a} = \frac{5a+10}{18} \cdot \frac{10a}{a^2-4}$$

$$= \frac{5(a+2) \cdot 2 \cdot 5 \cdot a}{2 \cdot 9(a+2)(a-2)}$$

$$= \frac{5 \cdot 5 \cdot a}{9(a-2)} = \frac{25a}{9(a-2)}$$

**16.** expression

$$\frac{9}{x^2-1} + \frac{12}{3x+3}$$

$$= \frac{9(3)}{(x+1)(x-1)(3)} + \frac{12(x-1)}{3(x+1)(x-1)}$$

$$= \frac{27+12x-12}{3(x-1)(x+1)}$$

$$= \frac{15+12x}{3(x+1)(x-1)} = \frac{3(5+4x)}{3(x+1)(x-1)}$$

$$= \frac{4x+5}{(x+1)(x-1)}$$

**17.** expression

$$\frac{x+2}{3x-1} + \frac{5}{(3x-1)^2}$$

$$= \frac{(x+2)(3x-1)}{(3x-1)(3x-1)} + \frac{5}{(3x-1)^2}$$

$$= \frac{3x^2+5x-2+5}{(3x-1)^2}$$

$$= \frac{3x^2+5x+3}{(3x-1)^2}$$

**18.** expression

$$\frac{4}{(2x-5)^2} + \frac{x+1}{2x-5}$$

$$= \frac{4}{(2x-5)^2} + \frac{(x+1)(2x-5)}{(2x-5)(2x-5)}$$

$$= \frac{4+2x^2-3x-5}{(2x-5)^2}$$

$$= \frac{2x^2-3x-1}{(2x-5)^2}$$

**19.** expression

$$\frac{x-7}{x} - \frac{x+2}{5x} = \frac{(x-7)(5)}{x(5)} - \frac{x+2}{5x}$$

$$= \frac{5x-25-x-2}{5x}$$

$$= \frac{4x-37}{5x}$$

**20.** equation

$$\frac{9}{x^2-4}+\frac{2}{x+2}=\frac{-1}{x-2}$$

$$\left(x^2-4\right)\left(\frac{9}{x^2-4}\right)+\left(x^2-4\right)\left(\frac{2}{x+2}\right)$$

$$=\left(x^2-4\right)\left(\frac{-1}{x-2}\right)$$

$$9+(x-2)(2)=(x+2)(-1)$$

$$9+2x-4=-x-2$$

$$2x+5=-x-2$$

$$3x+5=-2$$

$$3x=-7$$

$$x=-\frac{7}{3}$$

The solution is $-\dfrac{7}{3}$.

**21.** equation

$$\frac{3}{x+3}=\frac{5}{x^2-9}-\frac{2}{x-3}$$

$$\left(x^2-9\right)\left(\frac{3}{x+3}\right)$$

$$=\left(x^2-9\right)\left(\frac{5}{x^2-9}\right)-\left(x^2-9\right)\left(\frac{2}{x-3}\right)$$

$$(x-3)(3)=5-(x+3)(2)$$

$$3x-9=5-2x-6$$

$$3x-9=-2x-1$$

$$5x-9=-1$$

$$5x=8$$

$$x=\frac{8}{5}$$

The solution is $\dfrac{8}{5}$.

**22.** expression

$$\frac{10x-9}{x}-\frac{x-4}{3x}=\frac{(10x-9)(3)}{x(3)}-\frac{x-4}{3x}$$

$$=\frac{30x-27-x+4}{3x}$$

$$=\frac{29x-23}{3x}$$

**Exercise Set 7.6**

**1.** $\dfrac{2}{3}=\dfrac{x}{6}$

$$12=3x$$

$$4=x$$

**3.** $\dfrac{x}{10}=\dfrac{5}{9}$

$$9x=50$$

$$x=\frac{50}{9}$$

**5.** $\dfrac{x+1}{2x+3}=\dfrac{2}{3}$

$$3(x+1)=2(2x+3)$$

$$3x+3=4x+6$$

$$3=x+6$$

$$-3=x$$

**7.** $\dfrac{9}{5}=\dfrac{12}{3x+2}$

$$9(3x+2)=5(12)$$

$$27x+18=60$$

$$27x=42$$

$$x=\frac{42}{27}=\frac{14}{9}$$

**9. a**

**11.** Let $x$ = the elephant's weight on Pluto.

$$\frac{100}{3} = \frac{4100}{x}$$

$$100x = 3(4100)$$

$$100x = 12,300$$

$$x = 123$$

The elephant's weight is 123 pounds.

**13.** Let $x$ = the number of calories

in 42.6 grams.

$$\frac{110}{28.4} = \frac{x}{42.6}$$

$$110(42.6) = 28.4x$$

$$4686 = 28.4x$$

$$165 = x.$$

There are 165 calories in 42.6 grams.

**15.** $$\frac{16}{10} = \frac{34}{y}$$

$$16y = 340$$

$$y = 21.25$$

**17.** $$\frac{y}{20} = \frac{8}{28}$$

$$y = \frac{20 \cdot 8}{28}$$

$$y = \frac{40}{7}$$

$$y = 5\frac{5}{7} \text{ ft}$$

**19.** $$3 \cdot \frac{1}{x} = 9 \cdot \frac{1}{6}$$

$$\frac{3}{x} = \frac{9}{6}$$

$$6x\left(\frac{3}{x}\right) = 6x\left(\frac{9}{6}\right)$$

$$18 = 9x$$

$$x = 2$$

The unknown number is 2.

**21.** $$\frac{3+2x}{x+1} = \frac{3}{2}$$

$$2(x+1)\left(\frac{3+2x}{x+1}\right) = 2(x+1)\left(\frac{3}{2}\right)$$

$$2(3+2x) = 3(x+1)$$

$$6+4x = 3x+3$$

$$x = -3$$

The unknown number is $-3$.

**23.**

| | Hours to Complete Total Job | Part of Job Completed in 1 Hour |
|---|---|---|
| Experienced | 4 | 1/4 |
| Apprentice | 5 | 1/5 |
| Together | x | 1/x |

$$\frac{1}{4} + \frac{1}{5} = \frac{1}{x}$$

$$20x\left(\frac{1}{4}\right) + 20x\left(\frac{1}{5}\right) = 20x\left(\frac{1}{x}\right)$$

$$5x + 4x = 20$$

$$9x = 20$$

$$x = \frac{20}{9} \text{ or } 2\frac{2}{9}$$

The experienced surveyor and

apprentice surveyor, working together,

can survey the road in $2\frac{2}{9}$ hours.

**25.**

| | Minutes to Complete Total Job | Part of Job Completed in 1 Minute |
|---|---|---|
| Larger belt | 2 | $1/2$ |
| Smaller belt | 6 | $1/6$ |
| Both belts | $x$ | $1/x$ |

$$\frac{1}{2}+\frac{1}{6}=\frac{1}{x}$$

$$6x\left(\frac{1}{2}\right)+6x\left(\frac{1}{6}\right)=6x\left(\frac{1}{x}\right)$$

$$3x+x=6$$

$$4x=6$$

$$x=\frac{6}{4}=\frac{3}{2}=1\frac{1}{2}$$

Both belts together can move the cans

to the storage area in $1\frac{1}{2}$ minute.

**27.**

| | Distance = | rate · | time |
|---|---|---|---|
| Trip to Park | 12 | $12/x$ | $x$ |
| Return Trip | 18 | $18/(x+1)$ | $x+1$ |

$$\frac{12}{x}=\frac{18}{x+1}$$

$$12(x+1)=18x$$

$$12x+12=18x$$

$$12=6x$$

$$2=x$$

The jogger spends 2 hours on her trip to

the park, so her rate is $\dfrac{12}{2}=6$ miles

per hour.

**29.**

| | Distance = | rate · | time |
|---|---|---|---|
| 1st portion | 20 | $r$ | $20/r$ |
| Cooldown portion | 16 | $r-2$ | $16/r-2$ |

$$\frac{20}{r}=\frac{16}{r-2}$$

$$20(r-2)=16r$$

$$20r-40=16r$$

$$-40=-4r$$

$$r=10 \text{ and } r-2=10-2=8$$

His speed was 10 miles per hour
during the first portion and 8 miles per
hour during the cooldown portion.

**31.** Let $x=$ the minimum floor space
needed by 40 students.

$$\frac{1}{9}=\frac{40}{x}$$

$$1x=9(40)$$

$$x=360$$

40 students need 360 square feet.

**33.**

$$\frac{1}{4}=\frac{x}{8}$$

$$8\left(\frac{1}{4}\right)=8\left(\frac{x}{8}\right)$$

$$2=x$$

The unknown number is 2.

**35.**

| | Hours to Complete Total Job | Part of Job Completed in 1 Hour |
|---|---|---|
| Marcus | 6 | $1/6$ |
| Tony | 4 | $1/4$ |
| Together | $x$ | $1/x$ |

$$\frac{1}{6}+\frac{1}{4}=\frac{1}{x}$$

$$12x\left(\frac{1}{6}\right)+12x\left(\frac{1}{4}\right)=12x\left(\frac{1}{x}\right)$$

$$2x+3x=12$$

$$5x=12$$

$$x=\frac{12}{5}=2\frac{2}{5}$$

$$45\left(\frac{12}{5}\right)=108$$

Together Marcus and Tony work for

$2\frac{2}{5}$ hours at \$45 per hour. The labor

estimate should be \$108.

**37.** Let r = the speed of the car in still air.

|  | Distance = | rate $\cdot$ | time |
|---|---|---|---|
| Into the wind | 10 | $r-3$ | $10/r-3$ |
| With the wind | 11 | $r+3$ | $11/r+3$ |

$$\frac{10}{r-3}=\frac{11}{r+3}$$

$$10(r+3)=11(r-3)$$

$$10r+30=11r-33$$

$$63=r$$

The speed of the car in still air is 63 miles per hour.

**39.** $\dfrac{y}{25\text{ ft}}=\dfrac{3\text{ ft}}{2\text{ ft}}$

$$y\cdot 2\text{ ft}=25\text{ ft}\cdot 3\text{ ft}$$

$$y\cdot 2\text{ ft}=75\text{ sq. ft}$$

$$y=\frac{75\text{ sq. ft}}{2\text{ ft}}$$

$$y=37\frac{1}{2}\text{ ft}$$

**41.** Let $x$ = the number of rushing yards in one game.

$$\frac{x}{1}=\frac{4045}{12}$$

$$12x=1(4045)$$

$$12x=4045$$

$$x\approx 337$$

Ken averaged 337 yards per game.

**43.** $\dfrac{2}{x-3}-\dfrac{4}{x+3}=8\cdot\dfrac{1}{x^2-9}$

$$(x-3)(x+3)\left(\frac{2}{x-3}-\frac{4}{x+3}\right)$$

$$=(x-3)(x+3)\left(\frac{8}{x^2-9}\right)$$

$$(x-3)(x+3)\left(\frac{2}{x-3}\right)$$

$$-(x-3)(x+3)\left(\frac{4}{x+3}\right)=8$$

$$2(x+3)-4(x-3)=8$$

$$2x+6-4x+12=8$$

$$-2x=-10$$

$$x=5$$

The unknown number is 5.

**45.**

|  | Distance = | rate $\cdot$ | time |
|---|---|---|---|
| With wind | 630 | $r+35$ | $\frac{630}{r+35}$ |
| Against wind | 455 | $r-35$ | $\frac{455}{r-35}$ |

$$\frac{630}{r+35} = \frac{455}{r-35}$$
$$630(r-35) = 455(r+35)$$
$$630r - 22,050 = 455r + 15,925$$
$$175r = 37,975$$
$$r = 217$$

The speed in still air is 217 mph.

**47.** Let $x$ = the number of gallons of water needed.

$$\frac{8}{2} = \frac{36}{x}$$
$$8x = 2(36)$$
$$8x = 72$$
$$x = 9$$

Need to mix 9 gallons of water with the entire box.

**49.** Let $w$ = the rate of the wind.

| | Distance = | rate $\cdot$ | time |
|---|---|---|---|
| With the wind | 48 | $16+w$ | $48/16+w$ |
| Into the wind | 16 | $16-w$ | $16/16-w$ |

$$\frac{48}{16+w} = \frac{16}{16-w}$$
$$48(16-w) = 16(16+w)$$
$$768 - 48w = 256 + 16w$$
$$512 = 64w$$
$$w = 8$$

The rate of the wind is 8 miles per hour.

**51.**

| | Hours to Complete Total Job | Part of Job Completed in 1 Hour |
|---|---|---|
| Custodian | 3 | 1/3 |
| 2nd Worker | $x$ | $1/x$ |
| Together | $1\frac{1}{2}$ or $\frac{3}{2}$ | 2/3 |

$$\frac{1}{3} + \frac{1}{x} = \frac{2}{3}$$
$$3x\left(\frac{1}{3}\right) + 3x\left(\frac{1}{x}\right) = 3x\left(\frac{2}{3}\right)$$
$$x + 3 = 2x$$
$$3 = x$$

It takes the second worker 3 hours to do the job alone.

**53.** $\dfrac{x}{8} = \dfrac{20}{6}$

$$x = \frac{160}{6}$$
$$x = \frac{80}{3} = 26\frac{2}{3}$$

The side is $26\frac{2}{3}$ feet long.

**55.** $\dfrac{3}{2} = \dfrac{324}{x}$

$$3 \cdot x = 2 \cdot 324$$
$$3x = 648$$
$$x = \frac{648}{3} = 216 \text{ nuts}$$

**57.**

| | Hours to Complete Total Job | Part of Job Completed in 1 Hour |
|---|---|---|
| 1st Pipe | 20 | 1/20 |
| 2nd Pipe | 15 | 1/15 |
| 3rd Pipe | $x$ | $1/x$ |
| 3 Pipes Together | 6 | 1/6 |

$$\frac{1}{20} + \frac{1}{15} + \frac{1}{x} = \frac{1}{6}$$

$$60x\left(\frac{1}{20}\right) + 60x\left(\frac{1}{15}\right) + 60x\left(\frac{1}{x}\right)$$
$$= 60x\left(\frac{1}{6}\right)$$

$$3x + 4x + 60 = 10x$$

$$7x + 60 = 10x$$

$$60 = 3x$$

$$20 = x$$

It takes the third pipe 20 hours to fill the pond.

**59.**

|  | Time | In one hour |
|---|---|---|
| Andew | 2 | $\frac{1}{2}$ |
| Timothy | 3 | $\frac{1}{3}$ |
| Together | $x$ | $\frac{1}{x}$ |

$$\frac{1}{2} + \frac{1}{3} = \frac{1}{x}$$

$$6x\left(\frac{1}{2} + \frac{1}{3}\right) = 6x\left(\frac{1}{x}\right)$$

$$6x\left(\frac{1}{2}\right) + 6x\left(\frac{1}{3}\right) = 6$$

$$3x + 2x = 6$$

$$5x = 6$$

$$\frac{5x}{5} = \frac{6}{5}$$

$$x = \frac{6}{5} = 1\frac{1}{5}$$

Together it will take them $1\frac{1}{5}$ hours.

**61.**

|  | Time | In one hour |
|---|---|---|
| First cook | 6 | $\frac{1}{6}$ |
| Second cook | 7 | $\frac{1}{7}$ |
| Third cook | $x$ | $\frac{1}{x}$ |
| Together | 2 | $\frac{1}{2}$ |

$$\frac{1}{6} + \frac{1}{7} + \frac{1}{x} = \frac{1}{2}$$

$$42x\left(\frac{1}{6} + \frac{1}{7} + \frac{1}{x}\right) = 42x\left(\frac{1}{2}\right)$$

$$42x\left(\frac{1}{6}\right) + 42x\left(\frac{1}{7}\right) + 42x\left(\frac{1}{x}\right) = 21x$$

$$7x + 6x + 42 = 21x$$

$$13x + 42 = 21x$$

$$42 = 21x - 13x$$

$$42 = 8x$$

$$\frac{42}{8} = x$$

$$\frac{21}{4} = x$$

$$5\frac{1}{4} = x$$

The third cook can prepare the pies in $5\frac{1}{4}$ hours.

**63.**

| | Minutes to Complete Total Job | Part of Job Completed in 1 Minute |
|---|---|---|
| 1st Pump | $3x$ | $1/3x$ |
| 2nd Pump | $x$ | $1/x$ |
| Together | 21 | $1/21$ |

$$\frac{1}{3x}+\frac{1}{x}=\frac{1}{21}$$

$$21x\left(\frac{1}{3x}\right)+21x\left(\frac{1}{x}\right)=21x\left(\frac{1}{21}\right)$$

$$7+21=x$$

$$28=x,\ \ 3x=3(28)=84$$

The 1st pump takes 28 minutes and the 2nd takes 84 minutes.

**65.** $(0,\ 4),\ (2,\ 10)$

$$m=\frac{10-4}{2-0}=\frac{6}{2}=3$$

Since the slope is positive, the lines moves upward.

**67.** $(-2,\ 7),\ (3,\ -2)$

$$m=\frac{-2-7}{3-(-2)}=\frac{-9}{5}=-\frac{9}{5}$$

Since the slope is negative, the lines moves downward.

**69.** $(0,\ -4),\ (2,\ -4)$

$$m=\frac{-4-(-4)}{2-0}=\frac{0}{2}=0$$

The slope is zero.
Since the slope is zero, the line is horizontal.

**71.** The capacity in 2000 is approximately 2650 megawatts. The capacity in 2002 is approximately 4685 megawatts. The increase is approximately
$4685-2650=2035$ megawatts.

**73.** Answers may vary.

**75.**

$$\frac{1}{6}x+\frac{1}{12}x+\frac{1}{7}x+5+\frac{1}{2}x+4=x$$

$$\frac{1}{6}x+\frac{1}{12}x+\frac{1}{7}x+\frac{1}{2}x+9=x$$

$$84\left(\frac{1}{6}x+\frac{1}{12}x+\frac{1}{7}x+\frac{1}{2}x+9\right)=84x$$

$$14x+7x+12x+42x+756=84x$$

$$75x+756=84x$$

$$756=9x$$

$$\frac{756}{9}=\frac{9x}{9}$$

$$84=x$$

He died when he was 84 years old.

**77.**

$$4+\frac{1}{2}x+\frac{1}{6}x+3+\frac{1}{10}x=x$$

$$30\left(7+\frac{1}{2}x+\frac{1}{6}x+\frac{1}{10}x\right)=(30)(x)$$

$$30\cdot 7+30\left(\frac{1}{2}x\right)+30\left(\frac{1}{6}x\right)+30\left(\frac{1}{10}x\right)=30x$$

$$210+15x+5x+3x=30x$$

$$210+23x=30x$$

$$210=30x-23x$$

$$210=7x$$

$$30=x$$

You are 30 years old.

**79.**

| Distance | = | rate | · | time |
|---|---|---|---|---|
| H | $d + 0.5$ | 40 | | $\frac{d+0.5}{40}$ |
| G | $d$ | 32 | | $\frac{d}{32}$ |

$$\frac{d+0.5}{40} = \frac{d}{32}$$

$$32(d+0.5) = 40d$$

$$32d + 16 = 40d$$

$$16 = 8d$$

$$2 = d, \quad \frac{d}{32} = \frac{2}{32} = \frac{1}{16}$$

It will take the hyena $\dfrac{1}{16}$ hour or

3.75 minutes to overtake the giraffe.

## Mental Math 7.7

1. Direct

2. Inverse

3. Inverse

4. Direct

5. Inverse

6. Direct

7. Direct

8. Inverse

## Exercise Set 7.7

**1.** $y = kx$

$$3 = k(6)$$

$$\frac{3}{6} = k$$

$$\frac{1}{2} = k$$

$$y = \frac{1}{2}x$$

**3.** $y = kx$

$$-12 = k(-2)$$

$$6 = k$$

$$y = 6x$$

**5.** $k = \text{slope} = \dfrac{3-0}{1-0} = \dfrac{3}{1} = 3$

$$y = 3x$$

**7.** $k = \text{slope} = \dfrac{2-0}{3-0} = \dfrac{2}{3}$

$$y = \frac{2}{3}x$$

**9.** $y = \dfrac{k}{x}$

$$7 = \frac{k}{1}$$

$$7 = k$$

$$y = \frac{7}{x}$$

**11.**  $y = \dfrac{k}{x}$

$0.05 = \dfrac{k}{10}$

$0.5 = k$

$y = \dfrac{0.5}{x}$

**13.**  $y = kx$

**15.**  $h = \dfrac{k}{t}$

**17.**  $z = kx^2$

**19.**  $y = \dfrac{k}{z^3}$

**21.**  $x = \dfrac{k}{\sqrt{y}}$

**23.**  $y = kx$

$20 = k(5)$

$4 = k$

$y = 4x$

$y = 4(10)$

$y = 40$

**25.**  $y = \dfrac{k}{x}$

$5 = \dfrac{k}{60}$

$300 = k$

$y = \dfrac{300}{x}$

$y = \dfrac{300}{100}$

$y = 3$

**27.**  $z = kx^2$

$96 = k(4)^2$

$96 = 16k$

$6 = k$

$z = 6x^2$

$z = 6(3)^2$

$z = 6(9)$

$z = 54$

**29.**  $a = \dfrac{k}{b^3}$

$\dfrac{3}{2} = \dfrac{k}{2^3}$

$\dfrac{3}{2} = \dfrac{k}{8}$

$2k = 24$

$k = 12$

$a = \dfrac{12}{b^3}$

$a = \dfrac{12}{3^3}$

$a = \dfrac{12}{27}$

$a = \dfrac{4}{9}$

**31.**
$$p = kh$$
$$112.50 = k(18)$$
$$6.25 = k$$

$$p = 6.25h$$
$$p = 6.25(10)$$
$$p = 62.5$$

$62.50 for 10 hours

$277.10 for 34 hours

$$d = \frac{1}{15}w$$
$$d = \frac{1}{15}(80)$$
$$d = \frac{80}{15}$$
$$d = 5\frac{1}{3}$$

$5\frac{1}{3}$ inches with a 80 lb weight.

**33.**
$$c = \frac{k}{n}$$
$$9.00 = \frac{k}{5000}$$
$$45,000 = k$$

$$c = \frac{45,000}{n}$$
$$c = \frac{45,000}{7500}$$
$$c = 6$$

$6.00 to manufacture 7500 headphones

**37.**
$$w = \frac{k}{d^2}$$
$$180 = \frac{k}{4000^2}$$
$$180 = \frac{k}{16,000,000}$$
$$2,880,000,000 = k$$

$$w = \frac{2,880,000,000}{d^2}$$
$$w = \frac{2,880,000,000}{4010^2}$$
$$w = \frac{2,880,000,000}{16,080,100}$$
$$w \approx 179$$

179 pounds,
10 miles above the earth's surface.

**35.**
$$d = kw$$
$$4 = k(60)$$
$$\frac{4}{60} = k$$
$$\frac{1}{15} = k$$

**39.**
$$d = kt^2$$
$$64 = k(2)^2$$
$$64 = 4k$$
$$16 = k$$

$d = 16t^2$

$d = 16(10)^2$

$d = 16(100)$

$d = 1600$

1600 feet in 10 seconds

**41.** $\dfrac{\frac{3}{4}+\frac{1}{4}}{\frac{3}{8}+\frac{13}{8}} = \dfrac{\frac{3+1}{4}}{\frac{3+13}{8}} = \dfrac{\frac{4}{4}}{\frac{16}{8}} = \dfrac{1}{2}$

**43.** $\dfrac{\frac{2}{5}+\frac{1}{5}}{\frac{7}{10}+\frac{7}{10}} = \dfrac{\frac{2+1}{5}}{\frac{7+7}{10}} = \dfrac{\frac{3}{5}}{\frac{14}{10}} = \dfrac{3}{5} \div \dfrac{14}{10} = \dfrac{3}{5} \cdot \dfrac{10}{14}$

$= \dfrac{3 \cdot 2 \cdot 5}{5 \cdot 2 \cdot 7} = \dfrac{3}{7}$

**45.** $y = kx$

If $x$ is tripled, $y$ is also tripled. Answers may vary.

**47.** $p = k\sqrt{l}$

If $l$ is quadrupled (multiplied by 4), since $l$ is square rooted, $\sqrt{4} = 2$. Therefore, $p$ is doubled (multiplied by 2). Answers may vary.

**Mental Math 7.8**

**1.** $\dfrac{y}{5x}$

**2.** $\dfrac{10}{z}$

**3.** $\dfrac{3x}{5}$

**4.** $\dfrac{2a}{b}$

**Exercise Set 7.8**

**1.** $\dfrac{\frac{1}{2}}{\frac{3}{4}} = \dfrac{1}{2} \cdot \dfrac{4}{3} = \dfrac{1 \cdot 2 \cdot 2}{2 \cdot 3} = \dfrac{2}{3}$

**3.** $\dfrac{-\frac{4x}{9}}{-\frac{2x}{3}} = -\dfrac{4x}{9} \cdot -\dfrac{3}{2x} = \dfrac{2 \cdot 2 \cdot 3 \cdot x}{3 \cdot 3 \cdot 2 \cdot x} = \dfrac{2}{3}$

**5.** $\dfrac{\frac{1+x}{6}}{\frac{1+x}{3}} = \dfrac{1+x}{6} \cdot \dfrac{3}{1+x} = \dfrac{3 \cdot (1+x)}{2 \cdot 3 \cdot (1+x)} = \dfrac{1}{2}$

**7.** $\dfrac{\frac{1}{2}+\frac{2}{3}}{\frac{5}{9}-\frac{5}{6}} = \dfrac{\frac{1}{2} \cdot \frac{3}{3}+\frac{2}{3} \cdot \frac{2}{2}}{\frac{5}{9} \cdot \frac{2}{2}-\frac{5}{6} \cdot \frac{3}{3}}$

$= \dfrac{\frac{3}{6}+\frac{4}{6}}{\frac{10}{18}-\frac{15}{18}}$

$= \dfrac{\frac{7}{6}}{-\frac{5}{18}}$

$= \dfrac{7}{6} \cdot -\dfrac{18}{5}$

$= -\dfrac{7 \cdot 3 \cdot 6}{6 \cdot 5}$

$= -\dfrac{21}{5}$

**9.** $\dfrac{2+\frac{7}{10}}{1+\frac{3}{5}} = \dfrac{10\left(2+\frac{7}{10}\right)}{10\left(1+\frac{3}{5}\right)}$

$= \dfrac{10(2)+10\left(\frac{7}{10}\right)}{10(1)+10\left(\frac{3}{5}\right)} = \dfrac{20+7}{10+6} = \dfrac{27}{16}$

**11.** $\dfrac{\frac{1}{3}}{\frac{1}{2}-\frac{1}{4}} = \dfrac{12\left(\frac{1}{3}\right)}{12\left(\frac{1}{2}-\frac{1}{4}\right)} = \dfrac{12\left(\frac{1}{3}\right)}{12\left(\frac{1}{2}\right)-12\left(\frac{1}{4}\right)}$

$= \dfrac{4}{6-3} = \dfrac{4}{3}$

**13.** $\dfrac{-\frac{2}{9}}{-\frac{14}{3}} = -\dfrac{1}{9} \cdot -\dfrac{3}{14} = \dfrac{2 \cdot 3}{3 \cdot 3 \cdot 2 \cdot 7} = \dfrac{1}{21}$

**15.** $\dfrac{-\frac{5}{12x^2}}{\frac{25}{16x^3}} = -\dfrac{5}{12x^2} \cdot \dfrac{16x^3}{25}$

$= -\dfrac{5 \cdot 4 \cdot 4 \cdot x^2 \cdot x}{4 \cdot 3 \cdot x^2 \cdot 5 \cdot 5}$

$= -\dfrac{4x}{15}$

**17.** $\dfrac{\frac{m}{n}-1}{\frac{m}{n}+1} = \dfrac{n\left(\frac{m}{n}-1\right)}{n\left(\frac{m}{n}+1\right)} = \dfrac{n\left(\frac{m}{n}\right)-n(1)}{n\left(\frac{m}{n}\right)+n(1)} = \dfrac{m-n}{m+n}$

**19.** $\dfrac{\frac{1}{5}-\frac{1}{x}}{\frac{7}{10}+\frac{1}{x^2}} = \dfrac{10x^2\left(\frac{1}{5}-\frac{1}{x}\right)}{10x^2\left(\frac{7}{10}+\frac{1}{x^2}\right)}$

$= \dfrac{10x^2\left(\frac{1}{5}\right)-10x^2\left(\frac{1}{x}\right)}{10x^2\left(\frac{7}{10}\right)+10x^2\left(\frac{1}{x^2}\right)} = \dfrac{2x^2-10x}{7x^2+10}$

$= \dfrac{2x(x-5)}{7x^2+10}$

**21.** $\dfrac{1+\frac{1}{y-2}}{y+\frac{1}{y-2}} = \dfrac{(y-2)\left(1+\frac{1}{y-2}\right)}{(y-2)\left(y+\frac{1}{y-2}\right)}$

$= \dfrac{(y-2)(1)+(y-2)\left(\frac{1}{y-2}\right)}{(y-2)(y)+(y-2)\left(\frac{1}{y-2}\right)}$

$= \dfrac{y-2+1}{y^2-2y+1} = \dfrac{y-1}{(y-1)^2} = \dfrac{1}{y-1}$

**23.** $\dfrac{\frac{4y-8}{16}}{\frac{6y-12}{4}} = \dfrac{4y-8}{16} \cdot \dfrac{4}{6y-12} = \dfrac{4(y-2)\cdot 4}{4 \cdot 4 \cdot 6(y-2)}$

$= \dfrac{1}{6}$

**25.** $\dfrac{\frac{x}{y}+1}{\frac{x}{y}-1} = \dfrac{y\left(\frac{x}{y}+1\right)}{n\left(\frac{x}{y}-1\right)} = \dfrac{y\left(\frac{x}{y}\right)+y(1)}{y\left(\frac{x}{y}\right)-y(1)} = \dfrac{x+y}{x-y}$

**27.** $\dfrac{1}{2+\frac{1}{3}} = \dfrac{3(1)}{3\left(2+\frac{1}{3}\right)} = \dfrac{3(1)}{3(2)+3\left(\frac{1}{3}\right)}$

$= \dfrac{3}{6+1} = \dfrac{3}{7}$

**29.** $\dfrac{\frac{ax+ab}{x^2-b^2}}{\frac{x+b}{x-b}} = \dfrac{ax+ab}{x^2-b^2} \cdot \dfrac{x-b}{x+b}$

$= \dfrac{a(x+b)\cdot(x-b)}{(x+b)(x-b)\cdot(x+b)} = \dfrac{a}{x+b}$

**31.** $\dfrac{\frac{-3+y}{4}}{\frac{8+y}{28}} = \dfrac{-3+y}{4} \cdot \dfrac{28}{8+y}$

$= \dfrac{4 \cdot 7 \cdot(-3+y)}{4 \cdot(8+y)}$

$= \dfrac{7(y-3)}{8+y}$

**33.** $\dfrac{3+\frac{12}{x}}{1-\frac{16}{x^2}} = \dfrac{x^2\left(3+\frac{12}{x}\right)}{x^2\left(1-\frac{16}{x^2}\right)}$

$= \dfrac{x^2(3)+x^2\left(\frac{12}{x}\right)}{x^2(1)+x^2\left(-\dfrac{16}{x^2}\right)}$

$= \dfrac{3x^2+12x}{x^2-16}$

$= \dfrac{3x(x+4)}{(x-4)(x+4)}$

$= \dfrac{3x}{x-4}$

**35.** $\dfrac{\frac{8}{x+4}+2}{\frac{12}{x+4}-2} = \dfrac{(x+4)\left(\frac{8}{x+4}+2\right)}{(x+4)\left(\frac{12}{x+4}-2\right)}$

$= \dfrac{(x+4)\left(\frac{8}{x+4}\right)+(x+4)(2)}{(x+4)\left(\frac{12}{x+4}\right)-(x+4)(2)}$

$= \dfrac{8+2x+8}{12-2x-8} = \dfrac{16+2x}{4-2x} = \dfrac{2(8+x)}{2(2-x)}$

$= \dfrac{8+x}{2-x} = -\dfrac{x+8}{x-2}$

**37.** $\dfrac{\frac{s}{r}+\frac{r}{s}}{\frac{s}{r}-\frac{r}{s}} = \dfrac{rs\left(\frac{s}{r}+\frac{r}{s}\right)}{rs\left(\frac{s}{r}-\frac{r}{s}\right)} = \dfrac{rs\left(\frac{s}{r}\right)+rs\left(\frac{r}{s}\right)}{rs\left(\frac{s}{r}\right)-rs\left(\frac{r}{s}\right)}$

$= \dfrac{s^2+r^2}{s^2-r^2}$

**39.** Answers may vary.

**41.** $\sqrt{81} = \sqrt{9^2} = 9$

**43.** $\sqrt{1} = \sqrt{1^2} = 1$

**45.** $\sqrt{\dfrac{1}{25}} = \sqrt{\left(\dfrac{1}{5}\right)^2} = \dfrac{1}{5}$

**47.** $\sqrt{\dfrac{4}{9}} = \sqrt{\left(\dfrac{2}{3}\right)^2} = \dfrac{2}{3}$

**49.** $\dfrac{\frac{1}{3}+\frac{3}{4}}{2} = \dfrac{12\left(\frac{1}{3}+\frac{3}{4}\right)}{12(2)} = \dfrac{12\left(\frac{1}{3}\right)+12\left(\frac{3}{4}\right)}{12(2)}$

$= \dfrac{4+9}{24} = \dfrac{13}{24}$

**51.** $\dfrac{1}{\frac{1}{R_1}+\frac{1}{R_2}} = \dfrac{R_1 R_2 (1)}{R_1 R_2 \left(\frac{1}{R_1}+\frac{1}{R_2}\right)}$

$= \dfrac{R_1 R_2}{R_1 R_2 \left(\frac{1}{R_1}\right)+R_1 R_2 \left(\frac{1}{R_2}\right)} = \dfrac{R_1 R_2}{R_2 + R_1}$

**53.** $t = \dfrac{d}{r}$

$t = \dfrac{\frac{20x}{3}}{\frac{5x}{9}} = \dfrac{20x}{3}\cdot\dfrac{9}{5x} = \dfrac{4\cdot 5x\cdot 3\cdot 3}{3\cdot 5x} = 12$ hours

**55.** $\dfrac{x^{-1}+2^{-1}}{x^{-2}-4^{-1}} = \dfrac{\frac{1}{x}+\frac{1}{2}}{\frac{1}{x^2}-\frac{1}{4}} = \dfrac{4x^2\left(\frac{1}{x}+\frac{1}{2}\right)}{4x^2\left(\frac{1}{x^2}-\frac{1}{4}\right)}$

$= \dfrac{4x+2x^2}{4-x^2} = \dfrac{2x^2+4x}{-\left(x^2-4\right)}$

$= -\dfrac{2x(x+2)}{(x+2)(x-2)} = -\dfrac{2x}{x-2}$

**57.** $\dfrac{x+y^{-1}}{\frac{x}{y}} = \dfrac{x+\frac{1}{y}}{\frac{x}{y}} = \dfrac{y\left(x+\frac{1}{y}\right)}{y\left(\frac{x}{y}\right)} = \dfrac{xy+1}{x}$

**59.** $\dfrac{y^{-2}}{1-y^{-2}} = \dfrac{\frac{1}{y^2}}{1-\frac{1}{y^2}} = \dfrac{y^2\left(\frac{1}{y^2}\right)}{y^2\left(1-\frac{1}{y^2}\right)}$

$= \dfrac{y^2\left(\frac{1}{y^2}\right)}{y^2(1)-y^2\left(\frac{1}{y^2}\right)} = \dfrac{1}{y^2-1}$

## Chapter 7 Review

**1.** The rational expression is undefined when

$x^2-4=0$

$(x-2)(x+2)=0$

$x-2=0 \quad$ or $\quad x+2=0$

$x=2 \qquad\qquad x=-2$

**2.** The rational expression is undefined when

$$4x^2 - 4x - 15 = 0$$

$$(2x+3)(2x-5) = 0$$

$$2x+3 = 0 \quad \text{or} \quad 2x-5 = 0$$

$$2x = -3 \qquad\qquad 2x = 5$$

$$x = -\frac{3}{2} \qquad\qquad x = \frac{5}{2}$$

**3.** $\dfrac{z^2 - z}{z + xy} = \dfrac{(-2)^2 - (-2)}{-2 + 5(7)} = \dfrac{4+2}{-2+35} = \dfrac{6}{33} = \dfrac{2}{11}$

**4.** $\dfrac{x^2 + xy - z^2}{x + y + z} = \dfrac{5^2 + 5 \cdot 7 - (-2)^2}{5 + 7 + (-2)}$

$$= \dfrac{25 + 35 - 4}{10} = \dfrac{56}{10} = \dfrac{28}{5}$$

**5.** $\dfrac{x+2}{x^2 - 3x - 10} = \dfrac{x+2}{(x-5)(x+2)} = \dfrac{1}{x-5}$

**6.** $\dfrac{x+4}{x^2 + 5x + 4} = \dfrac{x+4}{(x+1)(x+4)} = \dfrac{1}{x+1}$

**7.** $\dfrac{x^3 - 4x}{x^2 + 3x + 2} = \dfrac{x(x^2 - 4)}{(x+2)(x+1)}$

$$= \dfrac{x(x-2)(x+2)}{(x+2)(x+1)} = \dfrac{x(x-2)}{x+1}$$

**8.** $\dfrac{5x^2 - 125}{x^2 + 2x - 15} = \dfrac{5(x^2 - 25)}{(x-3)(x+5)}$

$$= \dfrac{5(x-5)(x+5)}{(x-3)(x+5)} = \dfrac{5(x-5)}{x-3}$$

**9.** $\dfrac{x^2 - x - 6}{x^2 - 3x - 10} = \dfrac{(x-3)(x+2)}{(x-5)(x+2)} = \dfrac{x-3}{x-5}$

**10.** $\dfrac{x^2 - 2x}{x^2 + 2x - 8} = \dfrac{x(x-2)}{(x+4)(x-2)} = \dfrac{x}{x+4}$

**11.** $\dfrac{x^2 + 6x + 5}{2x^2 + 11x + 5} = \dfrac{(x+5)(x+1)}{(2x+1)(x+5)} = \dfrac{x+1}{2x+1}$

**12.** $\dfrac{x^2 + xa + xb + ab}{x^2 - xc + bx - bc} = \dfrac{x(x+a) + b(x+a)}{x(x-c) + b(x-c)}$

$$= \dfrac{(x+a)(x+b)}{(x-c)(x+b)} = \dfrac{x+a}{x-c}$$

**13.** $\dfrac{x^2 + 5x - 2x - 10}{x^2 - 3x - 2x + 6} = \dfrac{x(x+5) - 2(x+5)}{x(x-3) - 2(x-3)}$

$$= \dfrac{(x+5)(x-2)}{(x-3)(x-2)} = \dfrac{x+5}{x-3}$$

**14.** $\dfrac{x^2 - 9}{9 - x^2} = \dfrac{(x-3)(x+3)}{(3-x)(3+x)}$

$$= \dfrac{(x-3)(x+3)}{-(x-3)(3+x)}$$

$$= \dfrac{1}{-1}$$

$$= -1$$

**15.** $\dfrac{4-x}{x^3 - 64} = -\dfrac{x-4}{x^3 - 64}$

$$= \dfrac{x-4}{(x-4)(x^2 + 4x + 16)}$$

$$= -\dfrac{1}{x^2 + 4x + 16}$$

**16.** $\dfrac{15x^3y^2}{z} \cdot \dfrac{z}{5xy^3} = \dfrac{15x^3y^2 \cdot z}{z \cdot 5xy^3}$

$= \dfrac{3 \cdot 5 \cdot x^2 \cdot x \cdot y^2 \cdot z}{z \cdot 5 \cdot x \cdot y^2 \cdot y} = \dfrac{3x^2}{y}$

**17.** $\dfrac{-y^3}{8} \cdot \dfrac{9x^2}{y^3} = \dfrac{y^3 \cdot 9x^2}{8 \cdot y^3} = -\dfrac{9x^2}{8}$

**18.** $\dfrac{x^2-9}{x^2-4} \cdot \dfrac{x-2}{x+3} = \dfrac{\left(x^2-9\right) \cdot (x-2)}{\left(x^2-4\right) \cdot (x+3)}$

$= \dfrac{(x-3)(x+3)(x-2)}{(x+2)(x-2)(x+3)} = \dfrac{x-3}{x+2}$

**19.** $\dfrac{2x+5}{x-6} \cdot \dfrac{2x}{-x+6} = \dfrac{2x+5}{x-6} \cdot \dfrac{2x}{-(x-6)}$

$= \dfrac{2x+5}{x-6} \cdot \dfrac{-2x}{x-6} = \dfrac{(2x+5) \cdot (-2x)}{(x-6) \cdot (x-6)}$

$= \dfrac{-2x(2x+5)}{(x-6)^2}$

**20.** $\dfrac{x^2-5x-24}{x^2-x-12} \div \dfrac{x^2-10x+16}{x^2+x-6}$

$= \dfrac{x^2-5x-24}{x^2-x-12} \cdot \dfrac{x^2+x-6}{x^2-10x+16}$

$= \dfrac{(x-8)(x+3) \cdot (x+3)(x-2)}{(x-4)(x+3) \cdot (x-8)(x-2)}$

$= \dfrac{x+3}{x-4}$

**21.** $\dfrac{4x+4y}{xy^2} \div \dfrac{3x+3y}{x^2y} = \dfrac{4x+4y}{xy^2} \cdot \dfrac{x^2y}{3x+3y}$

$= \dfrac{4(x+y) \cdot x \cdot x \cdot y}{x \cdot y \cdot y \cdot 3(x+y)} = \dfrac{4x}{3y}$

**22.** $\dfrac{x^2+x-42}{x-3} \cdot \dfrac{(x-3)^2}{x+7}$

$= \dfrac{(x+7)(x-6) \cdot (x-3)(x-3)}{(x-3) \cdot (x+7)}$

$= (x-6)(x-3)$

**23.** $\dfrac{2a+2b}{3} \cdot \dfrac{a-b}{a^2-b^2} = \dfrac{2(a+b) \cdot (a-b)}{3 \cdot (a+b)(a-b)} = \dfrac{2}{3}$

**24.** $\dfrac{x^2-9x+14}{x^2-5x+6} \cdot \dfrac{x+2}{x^2-5x-14}$

$= \dfrac{(x-7)(x-2) \cdot (x+2)}{(x-3)(x-2) \cdot (x-7)(x+2)} = \dfrac{1}{x-3}$

**25.** $(x-3) \cdot \dfrac{x}{x^2+3x-18}$

$= \dfrac{(x-3) \cdot x}{(x-3)(x+6)} = \dfrac{x}{x+6}$

**26.** $\dfrac{2x^2-9x+9}{8x-12} \div \dfrac{x^2-3x}{2x}$

$= \dfrac{2x^2-9x+9}{9x-12} \cdot \dfrac{2x}{x^2-3x}$

$= \dfrac{(2x-3)(x-3) \cdot 2x}{4(2x-3) \cdot x(x-3)}$

$= \dfrac{2}{4} = \dfrac{1}{2}$

**27.** $\dfrac{x^2-y^2}{x^2+xy} \div \dfrac{3x^2-2xy-y^2}{3x^2+6x}$

$= \dfrac{x^2-y^2}{x^2+xy} \cdot \dfrac{3x^2+6x}{3x^2-2xy-y^2}$

$= \dfrac{(x-y)(x+y)\cdot 3x(x+2)}{x(x+y)\cdot(3x+y)(x-y)}$

$= \dfrac{3(x+2)}{3x+y}$

**28.** $\dfrac{x^2-y^2}{8x^2-16xy+8y^2} \div \dfrac{x+y}{4x-y}$

$= \dfrac{(x-y)(x+y)}{8(x-y)(x-y)} \cdot \dfrac{4x-y}{x+y}$

$= \dfrac{(x-y)(x+y)(4x-y)}{8(x-y)(x-y)(x+y)}$

$= \dfrac{4x-y}{8(x-y)}$

**29.** $\dfrac{x-y}{4} \div \dfrac{y^2-2y-xy+2x}{16x+24}$

$= \dfrac{x-y}{4} \cdot \dfrac{16x+24}{y^2-2y-xy+2x}$

$= \dfrac{x-y}{4} \cdot \dfrac{8(2x+3)}{y(y-2)-x(y-2)}$

$= \dfrac{x-y}{4} \cdot \dfrac{8(2x+3)}{(y-2)(y-x)}$

$= -\dfrac{y-x}{4} \cdot \dfrac{8(2x+3)}{(y-2)(y-x)}$

$= -\dfrac{2\cdot 4(y-x)(2x+3)}{4(y-2)(y-x)}$

$= -\dfrac{2(2x+3)}{y-2}$

**30.** $\dfrac{y-3}{4x+3} \div \dfrac{9-y^2}{4x^2-x-3} = \dfrac{y-3}{4x+3} \cdot \dfrac{4x^2-x-3}{9-y^2}$

$= \dfrac{y-3}{4x+3} \cdot \dfrac{(4x+3)(x-1)}{-1(y-3)(y+3)}$

$= \dfrac{(y-3)(4x+3)(x-1)}{-(4x+3)(y-3)(y+3)}$

$= -\dfrac{x-1}{y+3}$

**31.** $\dfrac{5x-4}{3x-1} + \dfrac{6}{3x-1} = \dfrac{5x-4+6}{3x-1} = \dfrac{5x+2}{3x-1}$

**32.** $\dfrac{4x-5}{3x^2} - \dfrac{2x+5}{3x^2} = \dfrac{4x-5-(2x+5)}{3x^2}$

$= \dfrac{4x-5-2x-5}{3x^2} = \dfrac{2x-10}{3x^2}$

**33.** $\dfrac{9x+7}{6x^2} - \dfrac{3x+4}{6x^2} = \dfrac{9x+7-(3x+4)}{6x^2}$

$= \dfrac{9x+7-3x-4}{6x^2} = \dfrac{6x+3}{6x^2} = \dfrac{3(2x+1)}{3\cdot 2x^2}$

$= \dfrac{2x+1}{2x^2}$

**34.** $2x = 2\cdot x$

$7x = 7\cdot x$

$\text{LCD} = 2\cdot 7\cdot x = 14x$

**35.** $x^2-5x-24 = (x-8)(x+3)$

$x^2+11x+24 = (x+8)(x+3)$

$\text{LCD} = (x-8)(x+3)(x+8)$

**36.** $\dfrac{x+2}{x^2+11x+18} = \dfrac{x+2}{(x+9)(x+2)}$

$= \dfrac{(x+2)(x-5)}{(x+9)(x+2)(x-5)}$

$= \dfrac{x^2-3x-10}{(x+2)(x-5)(x+9)}$

**37.** $\dfrac{3x-5}{x^2+4x+4} = \dfrac{3x-5}{(x+2)^2}$

$= \dfrac{(3x-5)(x+3)}{(x+2)^2(x+3)} = \dfrac{3x^2+4x-15}{(x+2)^2(x+3)}$

$= \dfrac{x^2-3x-10}{(x+2)(x-5)(x+9)}$

**38.** $\dfrac{4}{5x^2} - \dfrac{6}{y} = \dfrac{4(y)}{5x^2(y)} - \dfrac{6(5x^2)}{y(5x^2)} = \dfrac{4y-30x^2}{5x^2 y}$

**39.** $\dfrac{2}{x-3} - \dfrac{4}{x-1}$

$= \dfrac{2(x-1)}{(x-3)(x-1)} - \dfrac{4(x-3)}{(x-1)(x-3)}$

$= \dfrac{2(x-1)-4(x-3)}{(x-3)(x-1)} = \dfrac{2x-2-4x+12}{(x-3)(x-1)}$

$= \dfrac{-2x+10}{(x-3)(x-1)}$

**40.** $\dfrac{x+7}{x+3} - \dfrac{x-3}{x+7}$

$= \dfrac{(x+7)(x+7)}{(x+3)(x+7)} - \dfrac{(x-3)(x+3)}{(x+7)(x+3)}$

$= \dfrac{x^2+14x+49-(x^2-9)}{(x+3)(x+7)}$

$= \dfrac{x^2+14x+49-x^2+9}{(x+3)(x+7)} = \dfrac{14x+58}{(x+3)(x+7)}$

**41.** $\dfrac{4}{x+3} - 2 = \dfrac{4}{x+3} - \dfrac{2(x+3)}{x+3}$

$= \dfrac{4-2(x+3)}{x+3} = \dfrac{4-2x-6}{x+3} = \dfrac{-2x-2}{x+3}$

**42.** $\dfrac{3}{x^2+2x-8} + \dfrac{2}{x^2-3x+2}$

$= \dfrac{3}{(x+4)(x-2)} + \dfrac{2}{(x-1)(x-2)}$

$= \dfrac{3(x-1)}{(x+4)(x-2)(x-1)}$

$\qquad\qquad + \dfrac{2(x+4)}{(x-1)(x-2)(x+4)}$

$= \dfrac{3(x-1)+2(x+4)}{(x+4)(x-2)(x-1)}$

$= \dfrac{3x-3+2x+8}{(x+4)(x-2)(x-1)}$

$= \dfrac{5x+5}{(x+4)(x-2)(x-1)}$

**43.** $\dfrac{2x-5}{6x+9} - \dfrac{4}{2x^2+3x}$

$= \dfrac{2x-5}{3(2x+3)} - \dfrac{4}{x(2x+3)}$

$= \dfrac{(2x-5)(x)}{3(2x+3)(x)} - \dfrac{4(3)}{x(2x+3)(3)}$

$= \dfrac{2x^2-5x-12}{3x(2x+3)} = \dfrac{(2x+3)(x-4)}{3x(2x+3)}$

$= \dfrac{x-4}{3x}$

**44.** $\dfrac{x-1}{x^2-2x+1} - \dfrac{x+1}{x-1} = \dfrac{x-1}{(x-1)^2} - \dfrac{x+1}{x-1}$

$= \dfrac{1}{x-1} = \dfrac{x+1}{x-1} = \dfrac{1-(x+1)}{x-1}$

$= \dfrac{1-x-1}{x-1} = \dfrac{-x}{x-1} = -\dfrac{x}{x-1}$

**45.** $\dfrac{x-1}{x^2+4x+4} + \dfrac{x-1}{x+2}$

$= \dfrac{x-1}{(x+2)^2} + \dfrac{(x-1)(x+2)}{(x+2)(x+2)}$

$= \dfrac{x-1+(x-1)(x+2)}{(x+2)^2}$

$= \dfrac{x-1+x^2+x-2}{(x+2)^2}$

$= \dfrac{x^2+2x-3}{(x+2)^2}$

**46.** $P = 2l + 2w$

$P = 2\left(\dfrac{2}{8}\right) + 2\left(\dfrac{x+2}{4x}\right)$

$= \dfrac{x}{4} + \dfrac{2(x+2)}{4x}$

$= \dfrac{x \cdot x}{4 \cdot x} + \dfrac{2x+4}{4x}$

$= \dfrac{x^2+2x+4}{4x}$

$A = l \cdot w$

$A = \dfrac{x}{8} \cdot \dfrac{x+2}{4x} = \dfrac{x \cdot (x+2)}{8 \cdot 4x} = \dfrac{x+2}{32}$

The perimeter is $\dfrac{x^2+2x+4}{4x}$ units

and the area is $\dfrac{x+2}{32}$ square units.

**47.** $P = \dfrac{3x}{4x-4} + \dfrac{2x}{3x-3} + \dfrac{x}{x-1}$

$= \dfrac{3x}{4(x-1)} + \dfrac{2x}{3(x-1)} + \dfrac{x}{x-1}$

$= \dfrac{3x(3)}{4(x-1)(3)} + \dfrac{2x(4)}{3(x-1)(4)} + \dfrac{x(12)}{(x-1)(12)}$

$= \dfrac{9x+8x+12x}{12(x-1)} = \dfrac{29x}{12(x-1)}$

$A = \dfrac{1}{2} \cdot b \cdot h$

$A = \dfrac{1}{2} \cdot \dfrac{x}{x-1} \cdot \dfrac{6y}{5}$

$= \dfrac{1 \cdot x \cdot 2 \cdot 3y}{2 \cdot (x-1) \cdot 5}$

$= \dfrac{3xy}{5(x-1)}$

The perimeter is $\dfrac{29x}{12(x-1)}$ units and the

area is $\dfrac{3xy}{5(x-1)}$ square units.

**48.**
$$\dfrac{x+4}{9} = \dfrac{5}{9}$$
$$9\left(\dfrac{x+4}{9}\right) = 9\left(\dfrac{5}{9}\right)$$
$$x+4 = 5$$
$$x = 1$$

**49.**
$$\dfrac{n}{10} = 9 - \dfrac{n}{5}$$
$$10\left(\dfrac{n}{10}\right) = 10\left(9 - \dfrac{n}{5}\right)$$
$$10\left(\dfrac{n}{10}\right) = 10(9) - 10\left(\dfrac{n}{5}\right)$$
$$n = 90 - 2n$$

$$3n = 90$$
$$n = 30$$

**50.**
$$\dfrac{5y-3}{7} = \dfrac{15y-2}{28}$$
$$28\left(\dfrac{5y-3}{7}\right) = 28\left(\dfrac{15y-2}{28}\right)$$
$$4(5y-3) = 15y-2$$
$$20y-12 = 15y-2$$
$$5y = 10$$
$$y = 2$$

**51.**
$$\dfrac{2}{x+1} - \dfrac{1}{x-2} = -\dfrac{1}{2}$$
$$2(x+1)(x-2)\left(\dfrac{2}{x+1} - \dfrac{1}{x-2}\right)$$
$$= 2(x+1)(x-2)\left(-\dfrac{1}{2}\right)$$
$$2(x+1)(x-2)\left(\dfrac{2}{x+1}\right)$$
$$-2(x+1)(x-2)\left(\dfrac{1}{x-2}\right)$$
$$= 2(x+1)(x-2)\left(-\dfrac{1}{2}\right)$$
$$4(x-2)-2(x+1) = -(x+1)(x-2)$$
$$4x-8-2x-2 = -(x^2-x-2)$$
$$2x-10 = -x^2+x+2$$
$$x^2+x-12 = 0$$
$$(x+4)(x-3) = 0$$
$$x+4 = 0 \quad \text{or} \quad x-3 = 0$$
$$x = -4 \qquad\qquad x = 3$$

**52.**
$$\dfrac{1}{a+3} + \dfrac{1}{a-3} = -\dfrac{5}{a^2-9}$$
$$(a-3)(a+3)\left(\dfrac{1}{a+3} + \dfrac{1}{a-3}\right)$$
$$= (a-3)(a+3)\left(-\dfrac{5}{(a-3)(a+3)}\right)$$
$$(a-3)(a+3)\left(\dfrac{1}{a+3}\right)$$
$$+(a-3)(a+3)\left(\dfrac{1}{a-3}\right) = -5$$
$$a-3+a+3 = -5$$
$$2a = -5$$
$$a = -\dfrac{5}{2}$$

**53.** $\dfrac{y}{2y+2} + \dfrac{2y-16}{4y+4} = \dfrac{y-3}{y+1}$

$\dfrac{y}{2(y+1)} + \dfrac{2y-16}{4(y+1)} = \dfrac{y-3}{y+1}$

$4(y+1)\left(\dfrac{y}{2(y+1)} + \dfrac{2y-16}{4(y+1)}\right)$

$\qquad = 4(y+1)\left(\dfrac{y-3}{y+1}\right)$

$4(y+1)\left(\dfrac{y}{2(y+1)}\right) + 4(y+1)\left(\dfrac{2y-16}{4(y+1)}\right)$

$\qquad = 4(y+1)\left(\dfrac{y-3}{y+1}\right)$

$2y + 2y - 16 = 4(y-3)$

$4y - 16 = 4y - 12$

$-16 = -12 \quad$ False

This equation has no solution.

**54.** $\dfrac{4}{x+3} + \dfrac{8}{x^2-9} = 0$

$(x-3)(x+3)\left(\dfrac{4}{x+3} + \dfrac{8}{(x-3)(x+3)}\right)$

$\qquad = (x-3)(x+3)(0)$

$(x-3)(x+3)\left(\dfrac{4}{x+3}\right)$

$\qquad + (x-3)(x+3)\left(\dfrac{8}{(x-3)(x+3)}\right) = 0$

$4(x-3) + 8 = 0$

$4x - 12 + 8 = 0$

$4x - 4 = 0$

$4x = 4$

$x = 1$

**55.** $\dfrac{2}{x-3} - \dfrac{4}{x+3} = \dfrac{8}{x^2-9}$

$(x-3)(x+3)\left(\dfrac{2}{x-3} - \dfrac{4}{x+3}\right)$

$\qquad = (x-3)(x+3)\left(\dfrac{8}{(x-3)(x+3)}\right)$

$(x-3)(x+3)\left(\dfrac{2}{x-3}\right)$

$\qquad - (x-3)(x+3)\left(\dfrac{4}{x+3}\right) = 8$

$2(x+3) - 4(x-3) = 8$

$2x + 6 - 4x + 12 = 8$

$-2x + 18 = 8$

$-2x = -10$

$x = 5$

**56.** $\dfrac{x-3}{x+1} - \dfrac{x-6}{x+5} = 0$

$(x+1)(x+5)\left(\dfrac{x-3}{x+1} - \dfrac{x-6}{x+5}\right)$

$\qquad = (x+1)(x+5)(0)$

$(x+1)(x+5)\left(\dfrac{x-3}{x+1}\right)$

$\qquad - (x+1)(x+5)\left(\dfrac{x-6}{x+5}\right) = 0$

$(x+5)(x-3) - (x+1)(x-6) = 0$

$x^2 + 2x - 15 - (x^2 - 5x - 6) = 0$

$x^2 + 2x - 15 - x^2 + 5x + 6 = 0$

$7x - 9 = 0$

$7x = 9$

$x = \dfrac{9}{7}$

**57.** $x + 5 = \dfrac{6}{x}$

$x(x+5) = x\left(\dfrac{6}{x}\right)$

$x^2 + 5x = 6$

$x^2 + 5x - 6 = 0$

$(x+6)(x-1) = 0$

$x + 6 = 0$ or $x - 1 = 0$

$x = -6$      $x = 1$

**58.** $\dfrac{4A}{5b} = x^2$

$4A = 5bx^2$

$\dfrac{4A}{5x^2} = \dfrac{5bx^2}{5x^2}$

$\dfrac{4A}{5x^2} = b$

**59.** $\dfrac{x}{7} + \dfrac{y}{8} = 10$

$56\left(\dfrac{x}{7}\right) + 56\left(\dfrac{y}{8}\right) = 56(10)$

$8x + 7y = 560$

$7y = 560 - 8x$

$y = \dfrac{560 - 8x}{7}$

**60.** $\dfrac{x}{2} = \dfrac{12}{4}$

$4x = 24$

$x = 6$

**61.** $\dfrac{20}{1} = \dfrac{x}{25}$

$500 = x$

**62.** $\dfrac{2}{x-1} = \dfrac{3}{x+3}$

$2(x+3) = 3(x-1)$

$2x + 6 = 3x - 3$

$6 = x - 3$

$9 = x$

**63.** $\dfrac{4}{y-3} = \dfrac{2}{y-3}$

$4(y-3) = 2(y-3)$

$4y - 12 = 2y - 6$

$2y - 12 = -6$

$2y = 6$

$y = 3$

$y = 3$ doesn't check.

No solution

**64.** Let $x$ = the number of parts processed in 45 minutes.

$\dfrac{300}{20} = \dfrac{x}{45}$

$13,500 = 20x$

$675 = x$

675 parts can be processed in 45 minutes.

**65.** Let $x$ = the charge for 3 hours.

$\dfrac{90.00}{8} = \dfrac{x}{3}$

$270.00 = 8x$

$33.75 = x$

He charges $33.75 for 3 hours.

312

**66.** Let $x$ = the number of letters addressed in 55 minutes.

$$\frac{100}{35} = \frac{x}{55}$$

$$5500 = 35x$$

$$157 \approx x$$

He can address 157 letters in 55 minutes.

**67.** $5 \cdot \frac{1}{x} = \frac{3}{2} \cdot \frac{1}{x} + \frac{7}{6}$

$$\frac{5}{x} = \frac{3}{2x} + \frac{7}{6}$$

$$6x\left(\frac{5}{x}\right) = 6x\left(\frac{3}{2x}\right) + 6x\left(\frac{7}{6}\right)$$

$$30 = 9 + 7x$$

$$21 = 7x$$

$$x = 3$$

The unknown number is 3.

**68.** $\frac{1}{x} = \frac{1}{4 - x}$

$$4 - x = x$$

$$4 = 2x$$

$$2 = x$$

The unknown number is 2.

**69.**

| | Distance = | rate | time |
|---|---|---|---|
| 1st car | 90 | $r$ | $90/r$ |
| 2nd car | 60 | $r - 10$ | $60/r - 10$ |

$$\frac{90}{r} = \frac{60}{r - 10}$$

$$90(r - 10) = 60r$$

$$90r - 900 = 60r$$

$$-900 = -30r$$

$30 = r$

$r - 10 = 30 - 10 = 20$

The rate of the first car is 30 miles per hour and the rate of the second car is 20 miles per hour.

**70.**

| | Distance = | rate · | time |
|---|---|---|---|
| Upstream | 48 | $r - 4$ | $48/r - 4$ |
| Downstream | 72 | $r + 4$ | $72/r + 4$ |

$$\frac{48}{r - 4} = \frac{72}{r + 4}$$

$$48(r + 4) = 72(r - 4)$$

$$48r + 192 = 72r - 288$$

$$480 = 24r$$

$$r = 20$$

The speed of the boat in still water is 20 miles per hour.

**71.**

| | Hours to Complete Total Job | Part of Job Completed in 1 Hour |
|---|---|---|
| Mark | 7 | 1/7 |
| Maria | $x$ | $1/x$ |
| Together | 5 | 1/5 |

$$\frac{1}{7} + \frac{1}{x} = \frac{1}{5}$$

$$35x\left(\frac{1}{7}\right) + 35x\left(\frac{1}{x}\right) = 35x\left(\frac{1}{5}\right)$$

$$5x + 35 = 7x$$

$$35 = 2x$$

$$x = \frac{35}{2} \text{ or } 17\frac{1}{2}$$

It takes Maria $17\frac{1}{2}$ hours to complete the job alone.

**72.**

| | Days to Complete Total Job | Part of Job Completed in 1 Day |
|---|---|---|
| Pipe A | 20 | 1/20 |
| Pipe B | 15 | 1/15 |
| Together | $x$ | $1/x$ |

$$\frac{1}{20} + \frac{1}{25} = \frac{1}{x}$$

$$60x\left(\frac{1}{20}\right) + 60x\left(\frac{1}{15}\right) = 60x\left(\frac{1}{x}\right)$$

$$3x + 4x = 60$$

$$7x = 60$$

$$x = \frac{60}{7} = 8\frac{4}{7}$$

Both pipes fill the pond in $8\frac{4}{7}$ days.

**73.** $\dfrac{2}{4} = \dfrac{10}{x}$

$$2x = 40$$

$$x = 20$$

The missing length is 20.

**74.** $\dfrac{12}{4} = \dfrac{18}{x}$

$$12x = 72$$

$$x = 6$$

The missing length is 6.

**75.** $\dfrac{9}{7\frac{1}{5}} = \dfrac{x}{12}$

$$108 = 7\frac{1}{5}x$$

$$108 = \frac{36}{5}x$$

$$540 = 36x$$

$$15 = x$$

The missing length is 15.

**76.** $\dfrac{x}{5} = \dfrac{30}{2.5}$

$$2.5x = 150$$

$$x = 60$$

The missing length is 60.

**77.** $y = kx$

$$40 = k(4)$$

$$10 = k$$

$$y = 10x$$

$$y = 10(11)$$

$$y = 110$$

**78.** $y = \dfrac{k}{x}$

$$4 = \frac{k}{6}$$

$$24 = k$$

$$y = \frac{24}{x}$$

$$y = \frac{24}{48}$$

$$y = \frac{1}{2}$$

**79.** $y = \dfrac{k}{x^3}$

$$12.5 = \frac{k}{2^3}$$

$$12.5 = \frac{k}{8}$$

$$100 = k$$

$$y = \frac{100}{x^3}$$

$$y = \frac{100}{3^3}$$

$$y = \frac{100}{27}$$

**80.**    $y = kx^2$

$$175 = k(5)^2$$

$$175 = 25k$$

$$7 = k$$

$$y = 7x^2$$

$$y = 7(10)^2$$

$$y = 7(100)$$

$$y = 700$$

**81.**        $c = \dfrac{k}{a}$

$$6600 = \frac{k}{3000}$$

$$19{,}800{,}000 = k$$

$$c = \frac{19{,}800{,}000}{a}$$

$$c = \frac{19{,}800{,}000}{5000}$$

$$c = 3960$$

It cost \$3960 to manufacture 5000 *ml* of medicine.

**82.**    $d = kw$

$$8 = k(150)$$

$$\frac{8}{150} = k$$

$$\frac{4}{75} = k$$

$$d = \frac{4}{75}w$$

$$d = \frac{4}{75}(90)$$

$$d = \frac{360}{75}$$

$$d = 4\frac{4}{5}$$

A 90 lb weight would stretch the spring $4\dfrac{4}{5}$ inches.

**83.** $\dfrac{\frac{5x}{27}}{-\frac{10xy}{21}} = \dfrac{5x}{27} \cdot -\dfrac{21}{10xy} = -\dfrac{5x \cdot 3 \cdot 7}{3 \cdot 9 \cdot 5 \cdot 2 \cdot x \cdot y}$

$$= -\frac{7}{18y}$$

**84.** $\dfrac{\frac{8x}{x^2-9}}{\frac{4}{x+3}} = \dfrac{8x}{x^2-9} \cdot \dfrac{x+3}{4}$

$$= \frac{2 \cdot 4 \cdot x(x+3)}{(x-3)(x+3) \cdot 4} = \frac{2x}{x-3}$$

**85.** $\dfrac{\frac{3}{5}+\frac{2}{7}}{\frac{1}{5}+\frac{5}{6}} = \dfrac{\frac{21}{35}+\frac{10}{35}}{\frac{6}{30}+\frac{25}{30}} = \dfrac{\frac{31}{35}}{\frac{31}{30}} = \dfrac{31}{35} \cdot \dfrac{30}{31}$

$$= \frac{31 \cdot 5 \cdot 6}{5 \cdot 7 \cdot 31} = \frac{6}{7}$$

**86.** $\dfrac{\frac{2}{a}+\frac{1}{2a}}{a+\frac{a}{2}} = \dfrac{2a\left(\frac{2}{a}+\frac{1}{2a}\right)}{2a\left(a+\frac{a}{2}\right)} = \dfrac{4+1}{2a^2+a^2} = \dfrac{5}{3a^2}$

**87.** $\dfrac{3-\frac{1}{y}}{2-\frac{1}{y}} = \dfrac{y\left(3-\frac{1}{y}\right)}{y\left(2-\frac{1}{y}\right)} = \dfrac{y(3)-y\left(\frac{1}{y}\right)}{y(2)-y\left(\frac{1}{y}\right)}$

$= \dfrac{3y-1}{2y-1}$

**88.** $\dfrac{2+\frac{1}{x^2}}{\frac{1}{x}+\frac{2}{x^2}} = \dfrac{x^2\left(2+\frac{1}{x^2}\right)}{x^2\left(\frac{1}{x}+\frac{2}{x^2}\right)} = \dfrac{x^2(2)+x^2\left(\frac{1}{x^2}\right)}{x^2\left(\frac{1}{x}\right)+x^2\left(\frac{2}{x^2}\right)}$

$= \dfrac{2x^2+1}{x+2}$

**89.** $\dfrac{\frac{1}{a}+\frac{1}{b}}{\frac{1}{ab}} = \dfrac{ab\left(\frac{1}{a}+\frac{1}{b}\right)}{ab\left(\frac{1}{ab}\right)} = \dfrac{b+a}{1} = b+a$

**90.** $\dfrac{\frac{6}{x+2}+4}{\frac{8}{x+2}-4} = \dfrac{(x+2)\left(\frac{6}{x+2}+4\right)}{(x+2)\left(\frac{8}{x+2}-4\right)}$

$= \dfrac{(x+2)\left(\frac{6}{x+2}\right)+(x+2)(4)}{(x+2)\left(\frac{8}{x+2}\right)-(x+2)(4)}$

$= \dfrac{6+4x+8}{8-4x-8} = \dfrac{4x+14}{-4x} = -\dfrac{2(2x+7)}{2\cdot 2x}$

$= -\dfrac{2x+7}{2x}$

## Chapter 7 Test

**1.** The rational expression is undefined when

$x^2+4x+3=0$

$(x+3)(x+1)=0$

$x+3=0 \quad$ or $\quad x+1=0$

$\quad\quad x=-3 \quad\quad\quad\quad x=-1$

**2. a.** $C = \dfrac{100x+3000}{x}$

$= \dfrac{100(200)+3000}{200}$

$= \dfrac{20,000+3000}{200}$

$= \dfrac{23,000}{200} = 115$

The average cost/desk is \$115.

**b.** $C = \dfrac{100x+3000}{x}$

$= \dfrac{100(1000)+3000}{1000}$

$= \dfrac{100,000+3000}{1000}$

$= \dfrac{103,000}{1000} = 103$

The average cost/desk is \$103.

**3.** $\dfrac{3x-6}{5x-10} = \dfrac{3(x-2)}{5(x-2)} = \dfrac{3}{5}$

**4.** $\dfrac{x+6}{x^2+12x+36} = \dfrac{x+6}{(x+6)^2} = \dfrac{1}{x+6}$

**5.** $\dfrac{x+3}{x^3+27} = \dfrac{x+3}{(x+3)(x^2-3x+9)} = \dfrac{1}{x^2-3x+9}$

**6.** $\dfrac{2m^3-2m^2-12m}{m^2-5m+6} = \dfrac{2m(m^2-m-6)}{(m-3)(m-2)}$

$= \dfrac{2m(m-3)(m+2)}{(m-3)(m-2)} = \dfrac{2m(m+2)}{m-2}$

**8.** $\dfrac{y-x}{x^2-y^2} = \dfrac{-(x-y)}{(x-y)(x+y)} = -\dfrac{1}{x+y}$

**9.** $\dfrac{3}{x-1} \cdot (5x-5) = \dfrac{3}{x-1} \cdot 5(x-1)$

$= \dfrac{3 \cdot 5(x-1)}{x-1} = 15$

**10.** $\dfrac{y^2-5y+6}{2y+4} \cdot \dfrac{y+2}{2y-6}$

$= \dfrac{(y-3)(y-2) \cdot (y+2)}{2(y+2) \cdot 2(y-3)} = \dfrac{y-2}{4}$

**11.** $\dfrac{5}{2x+5} - \dfrac{6}{2x+5} = \dfrac{5-6}{2x+5} = \dfrac{-1}{2x+5}$

**12.** $\dfrac{5a}{a^2-a-6} - \dfrac{2}{a-3}$

$= \dfrac{5a}{(a-3)(a+2)} - \dfrac{2(a+2)}{(a-3)(a+2)}$

$= \dfrac{5a-2(a+2)}{(a-3)(a+2)} = \dfrac{5a-2a-4}{(a-3)(a+2)}$

$= \dfrac{3a-4}{(a-3)(a+2)}$

**13.** $\dfrac{6}{x^2-1} + \dfrac{3}{x+1}$

$= \dfrac{6}{(x+1)(x-1)} + \dfrac{3(x-1)}{(x+1)(x-1)}$

$= \dfrac{6+3x-3}{(x+1)(x-1)} = \dfrac{3x+3}{(x+1)(x-1)}$

$= \dfrac{3(x+1)}{(x+1)(x-1)} = \dfrac{3}{x-1}$

**14.** $\dfrac{x^2-9}{x^2-3x} \div \dfrac{xy+5x+3y+15}{2x+10}$

$= \dfrac{x^2-9}{x^2-3x} \cdot \dfrac{2x+10}{xy+5x+3y+15}$

$= \dfrac{(x-3)(x+3) \cdot 2(x+5)}{x(x-3) \cdot (x+3)(y+5)}$

$= \dfrac{2(x+3)(x+5)}{x(x+3)(y+5)}$

$= \dfrac{2(x+5)}{x(y+5)}$

**15.** $\dfrac{x+2}{x^2+11x+18} + \dfrac{5}{x^2-3x-10}$

$= \dfrac{x+2}{(x+9)(x+2)} + \dfrac{5}{(x-5)(x+2)}$

$= \dfrac{(x+2)(x-5)}{(x+9)(x+2)(x-5)}$

$+ \dfrac{5(x+9)}{(x-5)(x+2)(x+9)}$

$= \dfrac{(x+2)(x-5)+5(x+9)}{(x+9)(x+2)(x-5)}$

$= \dfrac{x^2-3x-10+5x+45}{(x+9)(x+2)(x-5)}$

$= \dfrac{x^2+2x+35}{(x+9)(x+2)(x-5)}$

**16.** $\dfrac{4}{y} - \dfrac{5}{3} = -\dfrac{1}{5}$

$15y\left(\dfrac{4}{y} - \dfrac{5}{3}\right) = 15y\left(-\dfrac{1}{5}\right)$

$15y\left(\dfrac{4}{y}\right) - 15y\left(\dfrac{5}{3}\right) = 15y\left(-\dfrac{1}{5}\right)$

$60 - 25y = -3y$

$60 = 22y$

$\dfrac{60}{22} = y$

$y = \dfrac{30}{11}$

**17.** $\dfrac{5}{y+1} = \dfrac{4}{y+2}$

$5(y+2) = 4(y+1)$

$5y + 10 = 4y + 4$

$y = -6$

**18.** $\dfrac{a}{a-3} = \dfrac{3}{a-3} - \dfrac{3}{2}$

$2(a-3)\left(\dfrac{a}{a-3}\right) = 2(a-3)\left(\dfrac{3}{a-3} - \dfrac{3}{2}\right)$

$2a = 2(a-3)\left(\dfrac{3}{a-3}\right) - 2(a-3)\left(\dfrac{3}{2}\right)$

$2a = 6 - 3(a-3)$

$2a = 6 - 3a + 9$

$2a = 15 - 3a$

$5a = 15$

$a = 3$

In the original equation, 3 makes a denominator 0. This equation has no solution.

**19.**

$x - \dfrac{14}{x-1} = 4 - \dfrac{2x}{x-1}$

$(x-1)\left(x - \dfrac{14}{x-1}\right) = (x-1)\left(4 - \dfrac{2x}{x-1}\right)$

$x(x-1) - 14 = 4(x-1) - 2x$

$x^2 - x - 14 = 4x - 4 - 2x$

$x^2 - x - 14 = 2x - 4$

$x^2 - 3x - 10 = 0$

$(x-5)(x+2) = 0$

$x - 5 = 0 \text{ or } x + 2 = 0$

$x = 5 \text{ or } \quad x = -2$

**20.** $\dfrac{\frac{5x^2}{yz^2}}{\frac{10x}{z^3}} = \dfrac{5x^2}{yz^2} \cdot \dfrac{z^3}{10x} = -\dfrac{5 \cdot x \cdot x \cdot z \cdot z^2}{y \cdot z^2 \cdot 2 \cdot 5 \cdot x}$

$= \dfrac{xz}{2y}$

**21.** $\dfrac{5 - \frac{1}{y^2}}{\frac{1}{y} + \frac{2}{y^2}} = \dfrac{y^2\left(5 - \frac{1}{y^2}\right)}{y^2\left(\frac{1}{y} + \frac{2}{y^2}\right)} = \dfrac{y^2(5) - y^2\left(\frac{1}{y^2}\right)}{y^2\left(\frac{1}{y}\right) + y^2\left(\frac{2}{y^2}\right)}$

$= \dfrac{5y^2 - 1}{y + 2}$

**22.** $y = kx$

$10 = k(15)$

$\dfrac{10}{15} = k$

$\dfrac{2}{3} = k$

$$y = \frac{2}{3}x$$

$$y = \frac{2}{3}(42)$$

$$y = \frac{84}{3}$$

$$y = 28$$

**23.**    $y = \dfrac{k}{x^2}$

$$8 = \frac{k}{5^2}$$

$$8 = \frac{k}{25}$$

$$200 = k$$

$$y = \frac{200}{x^2}$$

$$y = \frac{200}{15^2}$$

$$y = \frac{200}{225}$$

$$y = \frac{8}{9}$$

**24.** Let $x =$ the number of defective bulbs.

$$\frac{85}{3} = \frac{510}{x}$$

$$85x = 1530$$

$$x = 18$$

Expect to find 18 defective bulbs.

**25.**   $x + 5 \cdot \dfrac{1}{x} = 6$

$$x + \frac{5}{x} = 6$$

$$x\left(x + \frac{5}{x}\right) = x(6)$$

$$x(x) + x\left(\frac{5}{x}\right) = x(6)$$

$$x^2 + 5 = 6x$$

$$x^2 - 6x + 5 = 0$$

$$(x - 5)(x - 1) = 0$$

$$x - 5 = 0 \quad \text{or} \quad x - 1 = 0$$

$$x = 5 \qquad\qquad x = 1$$

The unknown number is 5 or 1.

**26.**

| | Distance $=$ | rate $\cdot$ | time |
|---|---|---|---|
| Upstream | 14 | $r - 2$ | $14/r - 2$ |
| Downstream | 16 | $r + 2$ | $16/r + 2$ |

$$\frac{14}{r - 2} = \frac{16}{r + 2}$$

$$14(r + 2) = 16(r - 2)$$

$$14r + 28 = 16r - 32$$

$$60 = 2r$$

$$r = 30$$

The speed of the boat in still water is 30 miles per hour.

**27.**

| | Hours to Complete Total Job | Part of Job Completed in 1 Hour |
|---|---|---|
| 1st pipe | 12 | $1/12$ |
| 2nd pipe | 15 | $1/15$ |
| Together | $x$ | $1/x$ |

$$\frac{1}{12}+\frac{1}{15}=\frac{1}{x}$$

$$60x\left(\frac{1}{12}\right)+60x\left(\frac{1}{15}\right)=60x\left(\frac{1}{x}\right)$$

$$5x+4x=60$$

$$9x=60$$

$$x=\frac{60}{9}=\frac{20}{3}=6\frac{2}{3}$$

Together, the pipes can fill the tank in

$6\frac{2}{3}$ hours.

**28.** $\dfrac{8}{x}=\dfrac{10}{15}$

$$8(15)=10x$$

$$120=10x$$

$$12=x$$

**Cumulative Review Chapter 7**

**1.** a. $\dfrac{15}{x}=4$

b. $12-3=x$

c. $4x+17=21$

**2.** a. $12-x=-45$

b. $12x=-45$

c. $x-10=2x$

**3.** Let $x=$ the amount invested at 9%
for one year.

*Principal · Rate = Interest*

| | | | |
|---|---|---|---|
| 9% | $x$ | .09 | $.09x$ |
| 7% | $20,000-x$ | .07 | $.07(20,000-x)$ |
| Total | 20,000 | | 1550 |

$$.09x+.07(20,000-x)=1550$$
$$.09x+1400-.07x=1550$$
$$.02x+1400=1550$$
$$.02x=150$$
$$x=7500$$
$$20,000-x=20,000-7500=12,500$$

He invested $7500 @ 9% and
$12,500 @ 7%

**4.** Let $x=$ the number of bankruptcies in
1994 and $2x-80,000=$ the number in
2002.

$$x+2x-80,000=2,290,000$$
$$3x-80,000=2,290,000$$
$$3x=2,370,000$$
$$x=790,000$$
$$2x-80,000=2(790,000)-80,000$$
$$=1,500,000$$

There were 790,000 bankruptcies in 1994
and 1,500,000 in 2002.

**5.** $x-3y=6$

| $x$ | $y$ |
|---|---|
| 0 | −2 |
| 6 | 0 |

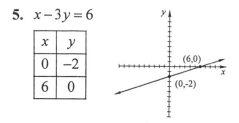

**6.** $7x+2y=9$

$$2y=-7x+9$$

$$y=-\frac{7}{2}x+\frac{9}{2}$$

$$y=mx+b$$

$$m=-\frac{7}{2}$$

**7. a.** $4^2 \cdot 4^5 = 4^{2+5} = 4^7$

   **b.** $x^2 \cdot x^5 = x^{2+5} = x^7$

   **c.** $y^3 \cdot y = y^{3+1} = y^4$

   **d.** $y^3 \cdot y^2 \cdot y^7 = y^{3+2+7} = y^{12}$

   **e.** $(-5)^7 \cdot (-5)^8 = (-5)^{7+8} = (-5)^{15}$

   **f.** $a^2 \cdot b^2 = a^2 b^2$

**8. a.** $\dfrac{x^9}{x^7} = x^{9-7} = x^2$

   **b.** $\dfrac{x^{19} y^5}{xy} = x^{19-1} \cdot y^{5-1} = x^{18} y^4$

   **c.** $\left(x^5 y^2\right)^3 = x^{5\cdot3} y^{2\cdot3} = x^{15} y^6$

   **d.** $\left(-3a^2 b\right)\left(5a^3 b\right) = -15a^{2+3} b^{1+1}$
   $$= -15a^5 b^2$$

**9.** $\left[(8z+11)+(9z-2)\right]-(5z-7)$
   $$= 8z+11+9z-2-5z+7$$
   $$= 12z+16$$

**10.** $(x+1)-\left(9x^2-6x+2\right)$
   $$= x+1-9x^2+6x-2$$
   $$= -9x^2+7x-1$$

**11.** $(3a+b)^3 = (3a+b)(3a+b)^2$
   $$= (3a+b)\left[(3a)^2 + 2(3a)(b)+(b)^2\right]$$
   $$= (3a+b)\left(9a^2 + 6ab + b^2\right)$$
   $$= 27a^3 + 18a^2 b + 3ab^2 + 9a^2 b + 6ab^2 + b^3$$
   $$= 27a^3 + 27a^2 b + 9ab^2 + b^3$$

**12.** $(2x+1)\left(5x^2 - x + 2\right)$
   $$= 2x\left(5x^2 - x + 2\right)+1\left(5x^2 - x + 2\right)$$
   $$= 10x^3 - 2x^2 + 4x + 5x^2 - x + 2$$
   $$= 10x^3 + 3x^2 + 3x + 2$$

**13. a.** $(t+2)^2 = (t)^2 + 2(t)(2)+(2)^2$
   $$= t^2 + 4t + 4$$

   **b.** $(p-q)^2 = (p)^2 + 2(p)(q)+(q)^2$
   $$= p^2 + 2pq + q$$

   **c.** $(2x+5)^2 = (2x)^2 + 2(2x)(5)+(5)^2$
   $$= 4x^2 + 20x + 25$$

   **d.** $\left(x^2 - 7y\right)^2 = \left(x^2\right)^2 - 2\left(x^2\right)(7y)+(7y)^2$
   $$= x^4 - 14x^2 y + 49y^2$$

**14. a.** $(x+9)^2 = (x)^2 + 2(x)(9)+(9)^2$
   $$= x^2 + 18x + 81$$

   **b.** $(2x+1)(2x-1) = (2x)^2 - (1)^2$
   $$= 4x^2 - 1$$

   **c.** $8x\left(x^2+1\right)\left(x^2-1\right) = 8x\left[\left(x^2\right)^2 - (1)^2\right]$
   $$= 8x\left[x^4 - 1\right]$$
   $$= 8x^5 - 8x$$

**15. a.** $\dfrac{1}{x^{-3}} = x^3$

   **b.** $\dfrac{1}{3^{-4}} = 3^4 = 81$

   **c.** $\dfrac{p^{-4}}{q^{-9}} = \dfrac{q^9}{p^4}$

   **d.** $\dfrac{5^{-3}}{2^{-5}} = \dfrac{2^5}{5^3} = \dfrac{32}{125}$

**16.** a. $5^{-3} = \dfrac{1}{5^3} = \dfrac{1}{125}$

b. $\dfrac{9}{x^{-7}} = 9x^7$

c. $\dfrac{11^{-1}}{7^{-2}} = \dfrac{7^2}{11^1} = \dfrac{49}{11}$

**17.**
$$2x+3 \overline{\smash{\big)}\, 8x^3 + 4x^2 + 0x + 7}$$

quotient: $4x^2 - 4x + 6$

$\underline{8x^3 + 12x^2}$

$-8x^2 + 0x$

$\underline{-8x^2 - 12x}$

$12x + 7$

$\underline{12x + 18}$

$-11$

$\dfrac{8x^3 + 4x^2 + 7}{2x+3} = 4x^2 - 4x + 6 - \dfrac{11}{2x+3}$

**18.**
$$x-4 \overline{\smash{\big)}\, 4x^3 + 0x^2 - 9x + 2}$$

quotient: $4x^2 + 16x + 55$

$\underline{4x^3 - 16x^2}$

$16x^2 - 9x$

$\underline{16x^2 - 64x}$

$55x + 2$

$\underline{55x - 220}$

$222$

$\dfrac{4x^3 - 9x + 2}{x-4} = 4x^2 + 16x + 55 + \dfrac{222}{x-4}$

**19.** a. $28 = 2 \cdot 2 \cdot 7$

$40 = 2 \cdot 2 \cdot 2 \cdot 5$

$\text{GCF} = 2^2 = 4$

b. $55 = 5 \cdot 11$

$21 = 3 \cdot 7$

$\text{GCF} = 1$

c. $15 = 3 \cdot 5$

$18 = 2 \cdot 3 \cdot 3$

$66 = 2 \cdot 3 \cdot 11$

$\text{GCF} = 3$

**20.** $9x^2 = 3 \cdot 3 \cdot x^2$

$6x^3 = 2 \cdot 3 \cdot x^3$

$21x^5 = 3 \cdot 7 \cdot x^5$

$\text{GCF} = 3x^2$

**21.** $-9a^5 + 18a^2 - 3a = -3a\left(3a^4 - 6a + 1\right)$

**22.** $7x^6 - 7x^5 + 7x^4 = 7x^4\left(x^2 - x + 1\right)$

**23.** $3m^2 - 24m - 60$

$= 3\left(m^2 - 8m - 20\right)$

$= 3\left(m^2 - 10m + 2m - 20\right)$

$= 3\left[m(m-10) + 2(m-10)\right]$

$= 3(m-10)(m+2)$

**24.** $-2a^2 + 10a + 12 = -2\left(a^2 - 5a - 6\right)$

$= -2(a+1)(a-6)$

**25.** $3x^2 + 11x + 6 = 3x^2 + 2x + 9x + 6$

$= x(3x+2) + 3(3x+2)$

$= (3x+2)(x+3)$

**26.** $10m^2 - 7m + 1$

$= 10m^2 - 2m - 5m + 1$

$= 2m(5m-1) - 1(5m-1)$

$= (2m-1)(5m-1)$

**27.** $x^2 + 12x + 36 = x^2 + 2 \cdot x \cdot 6 + 6^2 = (x+6)^2$

**28.** $4x^2 + 12x + 9 = (2x)^2 + 2(2x)(3) + (3)^2$
$$= (2x+3)^2$$

**29.** $x^2 + 4$ is a prime polynomial

**30.** $x^2 - 4 = (x)^2 - (2)^2 = (x+2)(x-2)$

**31.** $x^3 + 8 = x^3 + 2^3$
$$= (x+2)(x^2 - x \cdot 2 + 2^2)$$
$$= (x+2)(x^2 - 2x + 4)$$

**32.** $27y^3 - 1 = (3y)^3 - (1)^3$
$$= (3y-1)\left[(3y)^2 + 3y(1) + (1)^2\right]$$
$$= (3y-1)(9y^2 + 3y + 1)$$

**33.** $2x^3 + 3x^2 - 2x - 3$
$$= x^2(2x+3) - 1(2x+3)$$
$$= (2x+3)(x^2 - 1)$$
$$= (2x+3)(x^2 - 1^2)$$
$$= (2x+3)(x+1)(x-1)$$

**34.** $3x^3 + 5x^2 - 12x - 20$
$$= x^2(3x+5) - 4(3x+5)$$
$$= (3x+5)(x^2 - 4)$$
$$= (3x+5)(x^2 - 2^2)$$
$$= (3x+5)(x+2)(x-2)$$

**35.** $12m^2 - 3n^2 = 3(4m^2 - n^2)$
$$= 3\left[(2m)^2 - (n)^2\right]$$
$$= 3(2m+n)(2m-n)$$

**36.** $x^5 - x = x(x^4 - 1)$
$$= x\left[(x^2)^2 - 1^2\right]$$
$$= x(x^2+1)(x^2-1)$$
$$= x(x^2+1)(x+1)(x-1)$$

**37.**
$$x(2x-7) = 4$$
$$2x^2 - 7x = 4$$
$$2x^2 - 7x - 4 = 0$$
$$2x^2 - 8x + x - 4 = 0$$
$$2x(x-4) + 1(x-4) = 0$$
$$(x-4)(2x+1) = 0$$
$$2x+1 = 0 \text{ or } x - 4 = 0$$
$$2x = -1 \text{ or } \quad x = 4$$
$$x = -\frac{1}{2}$$

**38.**
$$3x^2 + 5x = 2$$
$$3x^2 + 5x - 2 = 0$$
$$3x^2 + 6x - x - 2 = 0$$
$$3x(x+2) - 1(x+2) = 0$$
$$(x+2)(3x-1) = 0$$
$$3x - 1 = 0 \text{ or } x + 2 = 0$$
$$3x = 1 \text{ or } \quad x = -2$$
$$x = \frac{1}{3}$$

**39.**
$$y = x^2 - 5x + 4$$
$$0 = x^2 - 5x + 4$$
$$0 = (x-4)(x-1)$$
$$x - 1 = 0 \text{ or } x - 4 = 0$$
$$x = 1 \text{ or } \quad x = 4$$
The $x$-intercepts are $(1, 0)$ and $(4, 0)$.

**40.**
$$y = x^2 - x - 6$$
$$0 = x^2 - x - 6$$
$$0 = (x-3)(x+2)$$
$$x + 2 = 0 \text{ or } x - 3 = 0$$
$$x = -2 \text{ or } \quad x = 3$$
The $x$-intercepts are $(-2, 0)$ and $(3, 0)$.

**41.** Let $x =$ the base and $2x - 2 =$ the height.
$$A = \frac{1}{2}bh$$
$$30 = \frac{1}{2}x(2x - 2)$$
$$30 = \frac{1}{2}(2x)(x - 1)$$
$$30 = x(x - 1)$$
$$30 = x^2 - x$$
$$0 = x^2 - x - 30$$
$$0 = (x+5)(x-6)$$
$$x - 6 = 0 \text{ or } x + 5 = 0$$
$$x = 6 \text{ or } \quad x = -5$$
Length cannot be negative, so $x = 6$
$$2x - 2 = 2(6) - 2 = 10$$
The base is 6 m and the height is 10 m.

**42.** Let $x =$ the base and $3x + 5 =$ the height.
$$A = bh$$
$$182 = x(3x + 5)$$
$$182 = 3x^2 + 5x$$
$$0 = 3x^2 + 5x - 182$$
$$0 = 3x^2 + 26x - 21x - 182$$
$$0 = x(3x + 26) - 7(3x + 26)$$
$$0 = (x - 7)(3x + 26)$$
$$x - 7 = 0 \text{ or } 3x + 26 = 0$$
$$x = 7 \text{ or } \quad x = -\frac{26}{3}$$
Length cannot be negative, so $x = 7$
$$3x + 5 = 3(7) + 5 = 26$$
The base is 7 ft and the height is 26 ft.

**43.**
$$\frac{5x - 5}{x^3 - x^2} = \frac{5(x-1)}{x^2(x-1)} = \frac{5}{x^2}$$

**44.**
$$\frac{2x^2 - 50}{4x^4 - 20x^3} = \frac{2(x^2 - 25)}{4x^3(x - 5)}$$
$$= \frac{2(x+5)(x-5)}{4x^3(x-5)}$$
$$= \frac{x+5}{2x^3}$$

**45.**
$$\frac{6x + 2}{x^2 - 1} \div \frac{3x^2 + x}{x - 1}$$
$$= \frac{6x + 2}{x^2 - 1} \cdot \frac{x - 1}{3x^2 + x}$$
$$= \frac{2(3x + 1)}{(x+1)(x-1)} \cdot \frac{x - 1}{x(3x + 1)}$$
$$= \frac{2}{x(x + 1)}$$

**46.** $\dfrac{6x^2-18x}{3x^2-2x} \cdot \dfrac{15x-10}{x^2-9}$

$= \dfrac{6x(x-3)}{x(3x-2)} \cdot \dfrac{5(3x-2)}{(x+3)(x-3)}$

$= \dfrac{30}{x+3}$

**47.** $\dfrac{\dfrac{x+1}{y}}{\dfrac{x}{y}+2}$

$= \dfrac{y\left(\dfrac{x+1}{y}\right)}{y\left(\dfrac{x}{y}+2\right)}$

$= \dfrac{x+1}{y\left(\dfrac{x}{y}\right)+2y}$

$= \dfrac{x+1}{x+2y}$

**48.** $\dfrac{\dfrac{m}{3}+\dfrac{n}{6}}{\dfrac{m+n}{12}} = \dfrac{12}{12} \cdot \dfrac{\dfrac{m}{3}+\dfrac{n}{6}}{\dfrac{m+n}{12}}$

$= \dfrac{12\left(\dfrac{m}{3}\right)+12\left(\dfrac{n}{6}\right)}{12\left(\dfrac{m+n}{12}\right)}$

$= \dfrac{4m+2n}{m+n}$ or $\dfrac{2(2m+n)}{m+n}$

# Chapter 8

**Calculator Explorations 8.1**

1. $\sqrt{7} \approx 2.646$

3. $\sqrt{11} \approx 3.317$

5. $\sqrt{82} \approx 9.055$

7. $\sqrt[3]{40} \approx 3.420$

9. $\sqrt[4]{20} \approx 2.115$

11. $\sqrt[5]{18} \approx 1.783$

**Mental Math 8.1**

1. False   2. True   3. True   4. True
5. True   6. False

**Exercise Set 8.1**

1. $\sqrt{16} = 4$, because $4^2 = 16$ and
   4 is positive.

3. $\sqrt{81} = 9$, because $9^2 = 81$ and
   9 is positive.

5. $\sqrt{\dfrac{1}{25}} = \dfrac{1}{5}$, because $\left(\dfrac{1}{5}\right)^2 = \dfrac{1}{25}$ and
   $\dfrac{1}{5}$ is positive.

7. $-\sqrt{100} = -10$, because $10^2 = 100$ and
   the negative sign indicates the
   negative square root.

9. $\sqrt{-4}$ is not a real number.

11. $-\sqrt{121} = -11$, because $11^2 = 121$ and
    the negative sign indicates the
    negative square root.

13. $\sqrt{\dfrac{9}{25}} = \dfrac{3}{5}$, because $\left(\dfrac{3}{5}\right)^2 = \dfrac{9}{25}$ and
    $\dfrac{3}{5}$ is positive.

15. $\sqrt{144} = 12$, because $12^2 = 144$ and
    12 is positive.

17. $\sqrt{\dfrac{49}{36}} = \dfrac{7}{6}$ because $\left(\dfrac{7}{6}\right)^2 = \dfrac{49}{36}$
    and $\dfrac{7}{6}$ is positive.

19. $-\sqrt{1} = -1$ because the sign in front
    of the radical indicates the negative
    square root of 1.

21. $\sqrt{37} \approx 6.083$

23. $\sqrt{136} \approx 11.662$

25. $\sqrt{2} \approx 1.41$
    $90\sqrt{2} \approx 90 \cdot 1.41 = 126.90$ feet

326

**27.** $\sqrt{z^2} = z$, because $(z)^2 = z^2$.

**29.** $\sqrt{x^4} = x^2$, because $(x^2)^2 = x^4$.

**31.** $\sqrt{9x^8} = 3x^4$, because $(3x^4)^2 = 9x^8$.

**33.** $\sqrt{81x^2} = 9x$, because $(9x)^2 = 81x^2$.

**35.** $\sqrt{\dfrac{x^6}{36}} = \dfrac{x^6}{6}$ because $\left(\dfrac{x^3}{6}\right)^2 = \dfrac{x^6}{36}$.

**37.** $\sqrt{\dfrac{25y^2}{9}} = \dfrac{5y}{3}$ because $\left(\dfrac{5y}{3}\right)^2 = \dfrac{25y^2}{9}$.

**39.** $\sqrt[3]{125} = 5$, because $(5)^3 = 125$.

**41.** $\sqrt[3]{-64} = -4$, because $(-4)^3 = -64$.

**43.** $-\sqrt[3]{125} = -5$, because $\sqrt[3]{125} = 5$.

**45.** $\sqrt[3]{\dfrac{1}{8}} = \dfrac{1}{2}$, because $\left(\dfrac{1}{2}\right)^3 = \dfrac{1}{8}$.

**47.** $\sqrt[3]{-125} = -5$, because $(-5)^3 = -125$.

**49.** $\sqrt[3]{-1000} = -10$ because $(-10)^3 = -1000$.

**51.** Answers may vary.

**53.** $\sqrt[5]{32} = 2$, because $(2)^5 = 32$.

**55.** $\sqrt[4]{-16}$ is not a real number.

**57.** $-\sqrt[4]{625} = -5$, because $\sqrt[4]{625} = 5$.

**59.** $\sqrt[6]{1} = 1$, because $(1)^6 = 1$.

**61.** $\sqrt[5]{-32} = -2$ because $(-2)^5 = -32$.

**63.** $\sqrt[4]{256} = 4$ because $4^4 = 256$.

**65.** $50 = 25 \cdot 2$

**67.** $32 = 16 \cdot 2$  or  $32 = 4 \cdot 8$

**69.** $28 = 4 \cdot 7$

**71.** $27 = 9 \cdot 3$

**73.** Let $A = 49$

The length of a side $= \sqrt{A}$

$\sqrt{A} = \sqrt{49} = 7$

The length of a side $= 7$ miles

**75.** Let $A = 9.61$

The length of a side $= \sqrt{A}$

$\sqrt{A} = \sqrt{9.61} = 3.1$

The length of a side $= 3.1$ inches.

**77.** $\sqrt{\sqrt{81}} = \sqrt{9} = 3$

**79.** $\sqrt{x^2} = \sqrt{(x)^2} = |x|$

**81.** $\sqrt{(x+2)^2} = |x+2|$

**83.** Let $V = 195,112$

The length of a side $= \sqrt[3]{V}$

$\sqrt[3]{V} = \sqrt[3]{195,112} = 58$

The length of a side $= 58$ ft.

**85.** $y = \sqrt[3]{x}$

| $x$ | $-8$ | $-2$ | $-1$ | $0$ | $1$ | $2$ | $8$ |
|-----|------|------|------|-----|-----|-----|-----|
| $y$ | $-2$ | $-1.3$ | $-1$ | $0$ | $1$ | $1.3$ | $2$ |

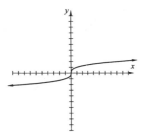

**87.** $\sqrt[3]{x^{21}} = x^7$ because $\left(x^7\right)^3 = x^{21}$.

**89.** $\sqrt[4]{x^{20}} = x^5$ because $\left(x^5\right)^4 = x^{20}$.

**91.** $y = \sqrt{x+3}$

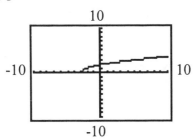

The graph starts at $(-3, 0)$ because
$x + 3 \geq 0$ for $x \geq -3$

**93.** $y = \sqrt{x-5}$

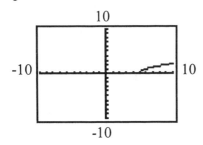

The graph starts at $(5, 0)$ because
$x - 5 \geq 0$ for $x \geq 5$

**Mental Math 8.2**

**1.** $\sqrt{4 \cdot 9} = 6$

**2.** $\sqrt{9 \cdot 36} = 18$

**3.** $\sqrt{x^2} = x$

**4.** $\sqrt{y^4} = y^2$

**5.** $\sqrt{0} = 0$

**6.** $\sqrt{1} = 1$

**7.** $\sqrt{25x^4} = 5x^2$

**8.** $\sqrt{49x^2} = 7x$

**Exercise Set 8.2**

**1.** $\sqrt{20} = \sqrt{4 \cdot 5} = \sqrt{4} \cdot \sqrt{5} = 2\sqrt{5}$

**3.** $\sqrt{18} = \sqrt{9 \cdot 2} = \sqrt{9} \cdot \sqrt{2} = 3\sqrt{2}$

**5.** $\sqrt{50} = \sqrt{25 \cdot 2} = \sqrt{25} \cdot \sqrt{2} = 5\sqrt{2}$

**7.** $\sqrt{33}$ can't be simplified.

**9.** $\sqrt{60} = \sqrt{4 \cdot 15} = \sqrt{4} \cdot \sqrt{15} = 2\sqrt{15}$

**11.** $\sqrt{180} = \sqrt{36 \cdot 5} = \sqrt{36} \cdot \sqrt{5} = 6\sqrt{5}$

**13.** $\sqrt{52} = \sqrt{4 \cdot 13} = \sqrt{4} \cdot \sqrt{13} = 2\sqrt{13}$

**15.** $\sqrt{\dfrac{8}{25}} = \dfrac{\sqrt{8}}{\sqrt{25}} = \dfrac{\sqrt{4}\cdot\sqrt{2}}{5} = \dfrac{2\sqrt{2}}{5}$

**17.** $\sqrt{\dfrac{27}{121}} = \dfrac{\sqrt{27}}{\sqrt{121}} = \dfrac{\sqrt{9}\cdot\sqrt{3}}{11} = \dfrac{3\sqrt{3}}{11}$

**19.** $\sqrt{\dfrac{9}{4}} = \dfrac{\sqrt{9}}{\sqrt{4}} = \dfrac{3}{2}$

**21.** $\sqrt{\dfrac{125}{9}} = \dfrac{\sqrt{125}}{\sqrt{9}} = \dfrac{\sqrt{25}\cdot\sqrt{5}}{3} = \dfrac{5\sqrt{5}}{3}$

**23.** $\sqrt{\dfrac{11}{36}} = \dfrac{\sqrt{11}}{\sqrt{36}} = \dfrac{\sqrt{11}}{6}$

**25.** $-\sqrt{\dfrac{27}{144}} = -\dfrac{\sqrt{27}}{\sqrt{144}}$

$= -\dfrac{\sqrt{9}\cdot\sqrt{3}}{12} = -\dfrac{3\sqrt{3}}{12} = -\dfrac{\sqrt{3}}{4}$

**27.** $\sqrt{x^7} = \sqrt{x^6\cdot x} = \sqrt{x^6}\cdot\sqrt{x} = x^3\sqrt{x}$

**29.** $\sqrt{x^{13}} = \sqrt{x^{12}\cdot x} = \sqrt{x^{12}}\cdot\sqrt{x} = x^6\sqrt{x}$

**31.** $\sqrt{75x^2} = \sqrt{25x^2\cdot 3} = \sqrt{25x^2}\cdot\sqrt{3} = 5x\sqrt{3}$

**33.** $\sqrt{96x^4} = \sqrt{16x^4\cdot 6} = \sqrt{16x^4}\cdot\sqrt{6} = 4x^2\sqrt{6}$

**35.** $\sqrt{\dfrac{12}{y^2}} = \dfrac{\sqrt{12}}{\sqrt{y^2}} = \dfrac{\sqrt{4}\cdot\sqrt{3}}{y} = \dfrac{2\sqrt{3}}{y}$

**37.** $\sqrt{\dfrac{9x}{y^2}} = \dfrac{\sqrt{9x}}{\sqrt{y^2}} = \dfrac{\sqrt{9}\cdot\sqrt{x}}{y} = \dfrac{3\sqrt{x}}{y}$

**39.** $\sqrt{\dfrac{88}{x^4}} = \dfrac{\sqrt{88}}{\sqrt{x^4}} = \dfrac{\sqrt{4}\cdot\sqrt{22}}{x^2} = \dfrac{2\sqrt{22}}{x^2}$

**41.** $\sqrt[3]{24} = \sqrt[3]{8\cdot 3} = \sqrt[3]{8}\cdot\sqrt[3]{3} = 2\sqrt[3]{3}$

**43.** $\sqrt[3]{250} = \sqrt[3]{125\cdot 2} = \sqrt[3]{125}\cdot\sqrt[3]{2} = 5\sqrt[3]{2}$

**45.** $\sqrt[3]{\dfrac{5}{64}} = \dfrac{\sqrt[3]{5}}{\sqrt[3]{64}} = \dfrac{\sqrt[3]{5}}{4}$

**47.** $\sqrt[3]{\dfrac{7}{8}} = \dfrac{\sqrt[3]{7}}{\sqrt[3]{8}} = \dfrac{\sqrt[3]{7}}{2}$

**49.** $\sqrt[3]{\dfrac{15}{64}} = \dfrac{\sqrt[3]{15}}{\sqrt[3]{64}} = \dfrac{\sqrt[3]{15}}{4}$

**51.** $\sqrt[3]{80} = \sqrt[3]{8\cdot 10} = \sqrt[3]{8}\cdot\sqrt[3]{10} = 2\sqrt[3]{10}$

**53.** $\sqrt[4]{48} = \sqrt[4]{16\cdot 3} = \sqrt[4]{16}\cdot\sqrt[4]{3} = 2\sqrt[4]{3}$

**55.** $\sqrt[4]{\dfrac{8}{81}} = \dfrac{\sqrt[4]{8}}{\sqrt[4]{81}} = \dfrac{\sqrt[4]{8}}{3}$

**57.** $\sqrt[5]{96} = \sqrt[5]{32\cdot 3} = \sqrt[5]{32}\cdot\sqrt[5]{3} = 2\sqrt[5]{3}$

**59.** $\sqrt[5]{\dfrac{5}{32}} = \dfrac{\sqrt[5]{5}}{\sqrt[5]{32}} = \dfrac{\sqrt[5]{5}}{2}$

**61.** $\sqrt[3]{80} = \sqrt[3]{8\cdot 10} = \sqrt[3]{8}\cdot\sqrt[3]{10} = 2\sqrt[3]{10}$

The length of each side is $2\sqrt[3]{10}$ inches.

**63.** Answers may vary.

**65.** $6x + 8x = (6+8)x = 14x$

**67.** $(2x+3)(x-5) = 2x^2 - 10x + 3x - 15$
$$= 2x^2 - 7x - 15$$

**69.** $9y^2 - 9y^2 = 0$

**71.** $\dfrac{x}{12} = \dfrac{6}{9}$
$$9x = 12 \cdot 6$$
$$\dfrac{9x}{9} = \dfrac{72}{9}$$
$$x = 8 \text{ cm}$$

**73.** $\sqrt{x^6 y^3} = \sqrt{x^6 y^2 y} = \sqrt{x^6 y^2} \cdot \sqrt{y} = x^3 y \sqrt{y}$

**75.** $\sqrt{x^2 + 4x + 4} = \sqrt{(x+2)^2} = x+2$

**77.** $\sqrt[3]{\dfrac{2}{x^9}} = \dfrac{\sqrt[3]{2}}{\sqrt[3]{x^9}} = \dfrac{\sqrt[3]{2}}{\sqrt[3]{(x^3)^3}} = \dfrac{\sqrt[3]{2}}{x^3}$

**79.** Let $A = 120$. The length of a side $= \sqrt{\dfrac{A}{6}}$
$$\sqrt{\dfrac{A}{6}} = \sqrt{\dfrac{120}{6}} = \sqrt{20} = \sqrt{4} \cdot \sqrt{5} = 2\sqrt{5}$$
The length of a side $= 2\sqrt{5}$ inches

**81.** Let $A = 150$. The length of a side $= \sqrt{\dfrac{A}{6}}$
$$\sqrt{\dfrac{A}{6}} = \sqrt{\dfrac{150}{6}} = \sqrt{25} = 5$$
The length of a side $= 5$ inches.

**83.** Let $n = 1000$
$$C = 100\sqrt[3]{n} + 700$$
$$C = 100\sqrt[3]{1000} + 700$$
$$= 100(10) + 700$$
$$= 1700$$
The cost is \$1700

**85.** Let $h = 169$ and $w = 64$.
$$B = \sqrt{\dfrac{hw}{3600}}$$
$$B = \sqrt{\dfrac{(169)(64)}{3600}} = \sqrt{\dfrac{10,816}{3600}} = \sqrt{\dfrac{676}{225}}$$
$$= \dfrac{26}{15} \approx 1.7$$
The surface area is about 1.7 sq. m.

**Mental Math 8.3**

**1.** $3\sqrt{2} + 5\sqrt{2} = 8\sqrt{2}$

**2.** $3\sqrt{5} + 7\sqrt{5} = 10\sqrt{5}$

**3.** $5\sqrt{x} + 2\sqrt{x} = 7\sqrt{x}$

**4.** $8\sqrt{x} + 3\sqrt{x} = 11\sqrt{x}$

**5.** $5\sqrt{7} - 2\sqrt{7} = 3\sqrt{7}$

**6.** $8\sqrt{6} - 5\sqrt{6} = 3\sqrt{6}$

**Exercise Set 8.3**

**1.** $4\sqrt{3} - 8\sqrt{3} = (4-8)\sqrt{3} = -4\sqrt{3}$

**3.** $3\sqrt{6} + 8\sqrt{6} - 2\sqrt{6} - 5 = (3+8-2)\sqrt{6} - 5$
$$= 9\sqrt{6} - 5$$

**5.** $6\sqrt{5} - 5\sqrt{5} + \sqrt{2} = (6-5)\sqrt{5} + \sqrt{2}$
$$= \sqrt{5} + \sqrt{2}$$

**7.** $2\sqrt[3]{3} + 5\sqrt[3]{3} - \sqrt{3} = (2+5)\sqrt[3]{3} - \sqrt{3}$
$$= 7\sqrt[3]{3} - \sqrt{3}$$

**9.** $2\sqrt[3]{2} - 7\sqrt[3]{2} - 6 = (2-7)\sqrt[3]{2} - 6$
$$= -5\sqrt[3]{2} - 6$$

**11.** Let $l = 3\sqrt{5}$ and $w = \sqrt{5}$
Perimeter $= 2l + 2w$
$$= 2(3\sqrt{5}) + 2(\sqrt{5})$$
$$= 6\sqrt{5} + 2\sqrt{5}$$
$$= 8\sqrt{5} \text{ inches}$$

**13.** Answers may vary.

**15.** $\sqrt{12} + \sqrt{27} = \sqrt{4 \cdot 3} + \sqrt{9 \cdot 3}$
$$= \sqrt{4} \cdot \sqrt{3} + \sqrt{9} \cdot \sqrt{3}$$
$$= 2\sqrt{3} + 3\sqrt{3} = 5\sqrt{3}$$

**17.** $\sqrt{45} + 3\sqrt{20} = \sqrt{9 \cdot 5} + 3\sqrt{4 \cdot 5}$
$$= \sqrt{9} \cdot \sqrt{5} + 3\sqrt{4} \cdot \sqrt{5}$$
$$= 3\sqrt{5} + 3(2)\sqrt{5}$$
$$= 3\sqrt{5} + 6\sqrt{5}$$
$$= 9\sqrt{5}$$

**19.** $2\sqrt{54} - \sqrt{20} + \sqrt{45} - \sqrt{24}$
$$= 2\sqrt{9 \cdot 6} - \sqrt{4 \cdot 5} + \sqrt{9 \cdot 5} - \sqrt{4 \cdot 6}$$
$$= 2\sqrt{9} \cdot \sqrt{6} - \sqrt{4} \cdot \sqrt{5} + \sqrt{9} \cdot \sqrt{5} - \sqrt{4} \cdot \sqrt{6}$$
$$= 2(3)\sqrt{6} - 2\sqrt{5} + 3\sqrt{5} - 2\sqrt{6}$$
$$= 6\sqrt{6} - 2\sqrt{5} + 3\sqrt{5} - 2\sqrt{6}$$
$$= 4\sqrt{6} + \sqrt{5}$$

**21.** $4x - 3\sqrt{x^2} + \sqrt{x} = 4x - 3x + \sqrt{x} = x + \sqrt{x}$

**23.** $\sqrt{25x} + \sqrt{36x} - 11\sqrt{x}$
$$= \sqrt{25} \cdot \sqrt{x} + \sqrt{36} \cdot \sqrt{x} - 11\sqrt{x}$$
$$= 5\sqrt{x} + 6\sqrt{x} - 11\sqrt{x}$$
$$= 0$$

**25.** $\sqrt{16x} - \sqrt{x^3} = \sqrt{16x} - \sqrt{x^2 \cdot x}$
$$= \sqrt{16} \cdot \sqrt{x} - \sqrt{x^2} \cdot \sqrt{x}$$
$$= 4\sqrt{x} - x\sqrt{x}$$
$$= (4-x)\sqrt{x}$$

**27.** $12\sqrt{5} - \sqrt{5} - 4\sqrt{5} = (12-1-4)\sqrt{5} = 7\sqrt{5}$

**29.** $\sqrt{5} + \sqrt[3]{5}$ cannot be simplified.

**31.** $4 + 8\sqrt{2} - 9 = 8\sqrt{2} + 4 - 9 = 8\sqrt{2} - 5$

**33.** $8 - \sqrt{2} - 5\sqrt{2} = 8 + (-1 - 5)\sqrt{2} = 8 - 6\sqrt{2}$

**35.** $5\sqrt{32} - \sqrt{72} = 5\sqrt{16}\sqrt{2} - \sqrt{36}\sqrt{2}$
$$= 5(4)\sqrt{2} - 6\sqrt{2}$$
$$= 20\sqrt{2} - 6\sqrt{2}$$
$$= 14\sqrt{2}$$

**37.** $\sqrt{8} + \sqrt{9} + \sqrt{18} + \sqrt{81}$
$$= \sqrt{4} \cdot \sqrt{2} + 3 + \sqrt{9} \cdot \sqrt{2} + 9$$
$$= 2\sqrt{2} + 3 + 3\sqrt{2} + 9$$
$$= 5\sqrt{2} + 12$$

**39.** $\sqrt{\dfrac{5}{9}} + \sqrt{\dfrac{5}{81}} = \dfrac{\sqrt{5}}{\sqrt{9}} + \dfrac{\sqrt{5}}{\sqrt{81}} = \dfrac{\sqrt{5}}{3} + \dfrac{\sqrt{5}}{9}$
$$= \dfrac{3\sqrt{5}}{9} + \dfrac{\sqrt{5}}{9} = \dfrac{3\sqrt{5} + \sqrt{5}}{9} = \dfrac{4\sqrt{5}}{9}$$

**41.** $\sqrt{\dfrac{3}{4}} - \sqrt{\dfrac{3}{64}} = \dfrac{\sqrt{3}}{\sqrt{4}} - \dfrac{\sqrt{3}}{\sqrt{64}} = \dfrac{\sqrt{3}}{2} - \dfrac{\sqrt{3}}{8}$
$$= \dfrac{4\sqrt{3}}{8} - \dfrac{\sqrt{3}}{8} = \dfrac{4\sqrt{3} - \sqrt{3}}{8} = \dfrac{3\sqrt{3}}{8}$$

**43.** $2\sqrt{45} - 2\sqrt{20} = 2\sqrt{9} \cdot \sqrt{5} - 2\sqrt{4} \cdot \sqrt{5}$
$$= 2(3)\sqrt{5} - 2(2)\sqrt{5}$$
$$= 6\sqrt{5} - 4\sqrt{5}$$
$$= 2\sqrt{5}$$

**45.** $\sqrt{35} - \sqrt{140} = \sqrt{35} - \sqrt{4} \cdot \sqrt{35}$
$$= \sqrt{35} - 2\sqrt{35}$$
$$= -\sqrt{35}$$

**47.** $5\sqrt{2x} + \sqrt{98x} = 5\sqrt{2x} + \sqrt{49 \cdot 2x}$
$$= 5\sqrt{2x} + \sqrt{49} \cdot \sqrt{2x}$$
$$= 5\sqrt{2x} + 7\sqrt{2x}$$
$$= 12\sqrt{2x}$$

**49.** $5\sqrt{x} + 4\sqrt{4x} = 5\sqrt{x} + 4\sqrt{4} \cdot \sqrt{x}$
$$= 5\sqrt{x} + 4(2)\sqrt{x}$$
$$= 5\sqrt{x} + 8\sqrt{x}$$
$$= 13\sqrt{x}$$

**51.** $\sqrt{3x^3} + 3x\sqrt{x} = \sqrt{x^2 \cdot 3x} + 3x\sqrt{x}$
$$= \sqrt{x^2} \cdot \sqrt{3x} + 3x\sqrt{x}$$
$$= x\sqrt{3x} + 3x\sqrt{x}$$

**53.** $\sqrt[3]{81} + \sqrt[3]{24} = \sqrt[3]{27}\sqrt[3]{3} + \sqrt[3]{8}\sqrt[3]{3}$
$$= 3\sqrt[3]{3} + 2\sqrt[3]{3}$$
$$= 5\sqrt[3]{3}$$

**55.** $4\sqrt[3]{9} - \sqrt[3]{243} = 4\sqrt[3]{9} - \sqrt[3]{27}\sqrt[3]{9}$
$$= 4\sqrt[3]{9} - 3\sqrt[3]{9}$$
$$= \sqrt[3]{9}$$

**57.** $2\sqrt[3]{8} + 2\sqrt[3]{16} = 2(2) + 2\sqrt[3]{8 \cdot 2} = 4 + 4\sqrt[3]{2}$

**59.** $\sqrt[3]{8} + \sqrt[3]{54} - 5 = 2\sqrt[3]{27 \cdot 2} - 5 = -3 + 3\sqrt[3]{2}$

**61.** $\sqrt{32x^2} + \sqrt[3]{32} + \sqrt{4x^2}$
$$= \sqrt{16} \cdot \sqrt{x^2} \cdot \sqrt{2} + \sqrt[3]{8} \cdot \sqrt[3]{4} \cdot \sqrt{2} + \sqrt{4} \cdot \sqrt{x^2}$$
$$= 4x\sqrt{2} + 2\sqrt[3]{4} + 2x$$

**63.** 
$$\sqrt{40x} + \sqrt[3]{40} - 2\sqrt{10x} - \sqrt[3]{5}$$
$$= \sqrt{4} \cdot \sqrt{10x} + \sqrt[3]{8} \cdot \sqrt[3]{5} - 2\sqrt{10x} - \sqrt[3]{5}$$
$$= 2\sqrt{10x} + 2\sqrt[3]{5} - 2\sqrt{10x} - \sqrt[3]{5}$$
$$= \sqrt[3]{5}$$

**65.** 
$$(x+6)^2 = x^2 + 2(6)x + 6^2$$
$$= x^2 + 12x + 36$$

**67.** 
$$(2x-1)^2 = (2x)^2 + 2(-1)(2x) + (-1)^2$$
$$= 4x^2 - 4x + 1$$

**69.** 
$$\begin{cases} x = 2y \\ x + 5y = 14 \end{cases}$$
Substitute $2y$ for $x$ in the second equation.
$$2y + 5y = 14$$
$$7y = 14$$
$$y = 2$$

Let $y = 2$ in the first equation.
$$x = 2(2) = 4$$
The solution is $(4, 2)$

**71.** Let $l = 8$ and $w = 3$
Area = area of 2 triangles
            + area of 2 rectangles
$$= 2\left( \frac{3\sqrt{27}}{4} \right) + 2lw$$
$$= \frac{3\sqrt{9} \cdot \sqrt{3}}{2} + 2(8)(3)$$
$$= \frac{9\sqrt{3}}{2} + 48 \text{ square feet}$$

**73.** 
$$\sqrt{\frac{x^3}{16}} - x\sqrt{\frac{9x}{25}} + \frac{\sqrt{81x^3}}{2}$$
$$= \frac{\sqrt{x^2 \cdot x}}{\sqrt{16}} - x\frac{\sqrt{9x}}{\sqrt{25}} + \frac{\sqrt{81x^2 \cdot x}}{2}$$
$$= \frac{x\sqrt{x}}{4} - x\frac{3\sqrt{x}}{5} + \frac{9x\sqrt{x}}{2}$$
$$= \frac{5x\sqrt{x}}{4 \cdot 5} - 4x\frac{3\sqrt{x}}{4 \cdot 5} + \frac{10 \cdot 9x\sqrt{x}}{2 \cdot 10}$$
$$= \frac{5x\sqrt{x} - 12x\sqrt{x} + 90x\sqrt{x}}{20}$$
$$= \frac{83x\sqrt{x}}{20}$$

**Mental Math 8.4**

**1.** $\sqrt{2} \cdot \sqrt{3} = \sqrt{6}$

**2.** $\sqrt{5} \cdot \sqrt{7} = \sqrt{35}$

**3.** $\sqrt{1} \cdot \sqrt{6} = \sqrt{6}$

**4.** $\sqrt{7} \cdot \sqrt{x} = \sqrt{7x}$

**5.** $\sqrt{10} \cdot \sqrt{y} = \sqrt{10y}$

**6.** $\sqrt{x} \cdot \sqrt{y} = \sqrt{xy}$

**Exercise Set 8.4**

**1.** $\sqrt{8} \cdot \sqrt{2} = \sqrt{8 \cdot 2} = \sqrt{16} = 4$

**3.** $\sqrt{10} \cdot \sqrt{5} = \sqrt{10 \cdot 5} = \sqrt{50} = \sqrt{25} \cdot \sqrt{2} = 5\sqrt{2}$

**5.** $\sqrt{10}\left(\sqrt{2} + \sqrt{5}\right) = \sqrt{10} \cdot \sqrt{2} + \sqrt{10} \cdot \sqrt{5}$
$= \sqrt{20} + \sqrt{50} = \sqrt{4} \cdot \sqrt{5} + \sqrt{25} \cdot \sqrt{2}$
$= 2\sqrt{5} + 5\sqrt{2}$

**7.** $\left(3\sqrt{5} - \sqrt{10}\right)\left(\sqrt{5} - 4\sqrt{3}\right)$
$= 3\sqrt{25} - 12\sqrt{15} - \sqrt{50} + 4\sqrt{30}$
$= 3(5) - 12\sqrt{15} - \sqrt{25}\sqrt{2} + 4\sqrt{30}$
$= 15 - 12\sqrt{15} - 5\sqrt{2} + 4\sqrt{30}$

**9.** $\left(\sqrt{x} + 6\right)\left(\sqrt{x} - 6\right) = \left(\sqrt{x}\right)^2 - (6)^2 = x - 36$

**11.** $\left(\sqrt{3} + 8\right)^2 = \left(\sqrt{3}\right)^2 + 2\left(\sqrt{3}\right)(8) + (8)^2$
$= 3 + 16\sqrt{3} + 64$
$= 67 + 16\sqrt{3}$

**13.** Let $l = 13\sqrt{2}$ and $w = 5\sqrt{6}$.
$A = lw$
$= 13\sqrt{2} \cdot 5\sqrt{6} = 65\sqrt{12} = 65\sqrt{4} \cdot \sqrt{3}$
$= 65(2)\sqrt{3} = 130\sqrt{3}$ square meters

**15.** $\dfrac{\sqrt{32}}{\sqrt{2}} = \sqrt{\dfrac{32}{2}} = \sqrt{16} = 4$

**17.** $\dfrac{\sqrt{90}}{\sqrt{5}} = \sqrt{\dfrac{90}{5}} = \sqrt{18} = \sqrt{9} \cdot \sqrt{2} = 3\sqrt{2}$

**19.** $\dfrac{\sqrt{75y^5}}{\sqrt{3y}} = \sqrt{\dfrac{75y^5}{3y}} = \sqrt{25y^4} = 5y^2$

**21.** $\sqrt{\dfrac{3}{5}} = \dfrac{\sqrt{3}}{\sqrt{5}} = \dfrac{\sqrt{3}}{\sqrt{5}} \cdot \dfrac{\sqrt{5}}{\sqrt{5}} = \dfrac{\sqrt{15}}{\sqrt{25}} = \dfrac{\sqrt{15}}{5}$

**23.** $\dfrac{1}{\sqrt{6y}} = \dfrac{1}{\sqrt{6y}} \cdot \dfrac{\sqrt{6y}}{\sqrt{6y}} = \dfrac{\sqrt{6y}}{6y}$

**25.** $\sqrt{\dfrac{5}{18}} = \dfrac{\sqrt{5}}{\sqrt{18}} = \dfrac{\sqrt{5}}{\sqrt{9} \cdot \sqrt{2}} = \dfrac{\sqrt{5}}{3\sqrt{2}}$
$= \dfrac{\sqrt{5}}{3\sqrt{2}} \cdot \dfrac{\sqrt{2}}{\sqrt{2}} = \dfrac{\sqrt{10}}{3(2)} = \dfrac{\sqrt{10}}{6}$

**27.** $\dfrac{3}{\sqrt{2}+1} = \dfrac{3}{\sqrt{2}+1} \cdot \dfrac{\sqrt{2}-1}{\sqrt{2}-1} = \dfrac{3\left(\sqrt{2}-1\right)}{\left(\sqrt{2}\right)^2 - 1^2}$
$= \dfrac{3\left(\sqrt{2}-1\right)}{2-1} = \dfrac{3\left(\sqrt{2}-1\right)}{1} = 3\sqrt{2} - 3$

**29.** $\dfrac{2}{\sqrt{10}-3} = \dfrac{2}{\sqrt{10}-3} \cdot \dfrac{\sqrt{10}+3}{\sqrt{10}+3} = \dfrac{2\left(\sqrt{10}+3\right)}{\left(\sqrt{10}\right)^2 - 3^2}$
$= \dfrac{2\left(\sqrt{10}+3\right)}{10-9} = \dfrac{2\left(\sqrt{10}+3\right)}{1} = 2\sqrt{10} + 6$

**31.** $\dfrac{\sqrt{5}+1}{\sqrt{6}-\sqrt{5}} = \dfrac{\sqrt{5}+1}{\sqrt{6}-\sqrt{5}} \cdot \dfrac{\sqrt{6}+\sqrt{5}}{\sqrt{6}+\sqrt{5}}$
$= \dfrac{\sqrt{30}+5+\sqrt{6}+\sqrt{5}}{\left(\sqrt{6}\right)^2 - \left(\sqrt{5}\right)^2}$
$= \dfrac{\sqrt{30}+5+\sqrt{6}+\sqrt{5}}{6-5}$
$= \sqrt{30} + 5 + \sqrt{6} + \sqrt{5}$

**33.** $\dfrac{6+2\sqrt{3}}{2} = \dfrac{2\left(3+\sqrt{3}\right)}{2} = 3+\sqrt{3}$

**35.** $\dfrac{18-12\sqrt{5}}{6} = \dfrac{6\left(3-2\sqrt{5}\right)}{6} = 3-2\sqrt{5}$

**37.** $\dfrac{15\sqrt{3}+5}{5} = \dfrac{5\left(3\sqrt{3}+1\right)}{5} = 3\sqrt{3}+1$

**39.** $2\sqrt{3}\cdot 4\sqrt{15} = 8\sqrt{45}$
$$= 8\sqrt{9}\sqrt{5}$$
$$= 8(3)\sqrt{5}$$
$$= 24\sqrt{5}$$

**41.** $\left(2\sqrt{5}\right)^2 = 2^2\left(\sqrt{5}\right)^2 = 4(5) = 20$

**43.** $\left(6\sqrt{x}\right)^2 = 6^2\left(\sqrt{x}\right)^2 = 36x$

**45.** $\sqrt{6}\left(\sqrt{5}+\sqrt{7}\right) = \sqrt{6}\cdot\sqrt{5}+\sqrt{6}\cdot\sqrt{7}$
$$= \sqrt{30}+\sqrt{42}$$

**47.** $4\sqrt{5x}\left(\sqrt{x}-3\sqrt{5}\right) = 4\sqrt{5x^2}-12\sqrt{25x}$
$$= 4\sqrt{x^2}\sqrt{5}-12\sqrt{25}\sqrt{x}$$
$$4x\sqrt{5}-12(5)\sqrt{x}$$
$$= 4x\sqrt{5}-60\sqrt{x}$$

**49.** $\left(\sqrt{3}+\sqrt{5}\right)\left(\sqrt{2}-\sqrt{5}\right)$
$$= \sqrt{3}\cdot\sqrt{2}-\sqrt{3}\cdot\sqrt{5}+\sqrt{5}\cdot\sqrt{2}-\left(\sqrt{5}\right)^2$$
$$= \sqrt{6}-\sqrt{15}+\sqrt{10}-5$$

**51.** $\left(\sqrt{7}-2\sqrt{3}\right)\left(\sqrt{7}+2\sqrt{3}\right) = \left(\sqrt{7}\right)^2-\left(2\sqrt{3}\right)^2$
$$= 7-4(3)$$
$$= 7-12$$
$$= -5$$

**53.** $\left(\sqrt{x}-3\right)\left(\sqrt{x}+3\right) = \left(\sqrt{x}\right)^2-(3)^2 = x-9$

**55.** $\left(\sqrt{6}+3\right)^2 = \left(\sqrt{6}\right)^2+2(3)\sqrt{6}+(3)^2$
$$= 6+6\sqrt{6}+9$$
$$= 15+6\sqrt{6}$$

**57.** $\left(3\sqrt{x}-5\right)^2 = \left(3\sqrt{x}\right)^2+2\left(3\sqrt{x}\right)(-5)+(5)^2$
$$= 9x-30\sqrt{x}+25$$

**59.** $\dfrac{\sqrt{150}}{\sqrt{2}} = \sqrt{\dfrac{150}{2}} = \sqrt{75} = \sqrt{25}\cdot\sqrt{3} = 5\sqrt{3}$

**61.** $\dfrac{\sqrt{72y^5}}{\sqrt{3y^3}} = \sqrt{\dfrac{72y^5}{3y^3}} = \sqrt{24y^2}$
$$= \sqrt{4y^2}\cdot\sqrt{6} = 2y\sqrt{6}$$

**63.** $\dfrac{\sqrt{24x^3y^4}}{\sqrt{2xy}} = \sqrt{\dfrac{24x^3y^4}{2xy}} = \sqrt{12x^2y^3}$
$$= \sqrt{4x^2y^2}\cdot\sqrt{3y} = 2xy\sqrt{3y}$$

**65.** $\sqrt{\dfrac{2}{15}} = \dfrac{\sqrt{2}}{\sqrt{15}} = \dfrac{\sqrt{2}}{\sqrt{15}}\cdot\dfrac{\sqrt{15}}{\sqrt{15}} = \dfrac{\sqrt{30}}{15}$

**67.** $\sqrt{\dfrac{3}{20}} = \dfrac{\sqrt{3}}{\sqrt{20}} = \dfrac{\sqrt{3}}{\sqrt{4}\cdot\sqrt{5}} = \dfrac{\sqrt{3}}{2\sqrt{5}}$

$\qquad = \dfrac{\sqrt{3}}{2\sqrt{5}}\cdot\dfrac{\sqrt{5}}{\sqrt{5}} = \dfrac{\sqrt{15}}{2(5)} = \dfrac{\sqrt{15}}{10}$

**69.** $\dfrac{3x}{\sqrt{2x}} = \dfrac{3x}{\sqrt{2x}}\cdot\dfrac{\sqrt{2x}}{\sqrt{2x}} = \dfrac{3x\sqrt{2x}}{2x} = \dfrac{3\sqrt{2x}}{2}$

**71.** $\dfrac{4}{2-\sqrt{5}} = \dfrac{4}{2-\sqrt{5}}\cdot\dfrac{2+\sqrt{5}}{2+\sqrt{5}} = \dfrac{4\left(2+\sqrt{5}\right)}{2^2 - \left(\sqrt{5}\right)^2}$

$\qquad = \dfrac{4\left(2+\sqrt{5}\right)}{4-5} = \dfrac{4\left(2+\sqrt{5}\right)}{-1} = -8 - 4\sqrt{5}$

**73.** $\dfrac{5}{3+\sqrt{10}} = \dfrac{5\left(3-\sqrt{10}\right)}{\left(3+\sqrt{10}\right)\left(3-\sqrt{10}\right)}$

$\qquad = \dfrac{5\left(3-\sqrt{10}\right)}{9-10}$

$\qquad = \dfrac{15-5\sqrt{10}}{-1}$

$\qquad = \dfrac{-1\left(-15+5\sqrt{10}\right)}{-1}$

$\qquad = -15 + 5\sqrt{10}$

**75.** $\dfrac{2\sqrt{3}}{\sqrt{15}+2} = \dfrac{2\sqrt{3}\left(\sqrt{15}-2\right)}{\left(\sqrt{15}+2\right)\left(\sqrt{15}-2\right)}$

$\qquad = \dfrac{2\sqrt{3}\left(\sqrt{15}-2\right)}{15-4}$

$\qquad = \dfrac{2\sqrt{45}-4\sqrt{3}}{11}$

$\qquad = \dfrac{2(3)\sqrt{5}-4\sqrt{3}}{11}$

$\qquad = \dfrac{6\sqrt{5}-4\sqrt{3}}{11}$

**77.** $\dfrac{\sqrt{3}+1}{\sqrt{2}-1} = \dfrac{\sqrt{3}+1}{\sqrt{2}-1}\cdot\dfrac{\sqrt{2}+1}{\sqrt{2}+1}$

$\qquad = \dfrac{\sqrt{6}+\sqrt{3}+\sqrt{2}+1}{2-1}$

$\qquad = \sqrt{6}+\sqrt{3}+\sqrt{2}+1$

**79.** $\sqrt[3]{12}\cdot\sqrt[3]{4} = \sqrt[3]{12\cdot 4}$

$\qquad = \sqrt[3]{48}$

$\qquad = \sqrt[3]{8\cdot 6}$

$\qquad = \sqrt[3]{8}\cdot\sqrt[3]{6}$

$\qquad = 2\sqrt[3]{6}$

**81.** $2\sqrt[3]{5}\cdot 6\sqrt[3]{2} = 12\sqrt[3]{10}$

**83.** $\sqrt[3]{15}\cdot\sqrt[3]{25} = \sqrt[3]{375} = \sqrt[3]{125}\cdot\sqrt[3]{3} = 5\sqrt[3]{3}$

**85.** $\dfrac{\sqrt[3]{54}}{\sqrt[3]{2}} = \sqrt[3]{\dfrac{54}{2}} = \sqrt[3]{27} = 3$

**87.** $\dfrac{\sqrt[3]{120}}{\sqrt[3]{5}} = \sqrt[3]{\dfrac{120}{5}}$

$= \sqrt[3]{24}$

$= \sqrt[3]{8 \cdot 3}$

$= \sqrt[3]{8} \cdot \sqrt[3]{3}$

$= 2\sqrt[3]{3}$

**89.** $\sqrt[3]{\dfrac{5}{4}} = \dfrac{\sqrt[3]{5}}{\sqrt[3]{4}} \cdot \dfrac{\sqrt[3]{2}}{\sqrt[3]{2}} = \dfrac{\sqrt[3]{10}}{\sqrt[3]{8}} = \dfrac{\sqrt[3]{10}}{2}$

**91.** $\dfrac{6}{\sqrt[3]{2}} = \dfrac{6}{\sqrt[3]{2}} \cdot \dfrac{\sqrt[3]{4}}{\sqrt[3]{4}} = \dfrac{6\sqrt[3]{4}}{\sqrt[3]{8}} = \dfrac{6\sqrt[3]{4}}{2} = 3\sqrt[3]{4}$

**93.** $\sqrt[3]{\dfrac{1}{9}} = \dfrac{\sqrt[3]{1}}{\sqrt[3]{9}} = \dfrac{1}{\sqrt[3]{9}} \cdot \dfrac{\sqrt[3]{3}}{\sqrt[3]{3}} = \dfrac{\sqrt[3]{3}}{\sqrt[3]{27}} = \dfrac{\sqrt[3]{3}}{3}$

**95.** $\sqrt[3]{\dfrac{2}{9}} = \dfrac{\sqrt[3]{2}}{\sqrt[3]{9}} = \dfrac{\sqrt[3]{2}}{\sqrt[3]{9}} \cdot \dfrac{\sqrt[3]{3}}{\sqrt[3]{3}} = \dfrac{\sqrt[3]{6}}{3}$

**97.** $\dfrac{3x+12}{3} = \dfrac{3(x+4)}{3} = x+4$

**99.** $\dfrac{6x^2-3x}{3x} = \dfrac{3x(2x-1)}{3x} = 2x-1$

**101.** $x+5 = 7^2$

$x+5 = 49$

$x = 44$

**103.** $4z^2 + 6z - 12 = (2z)^2$

$4z^2 + 6z - 12 = 4z^2$

$6z - 12 = 0$

$6z = 12$

$z = 2$

**105.** $\sqrt{\dfrac{A}{\pi}} = \dfrac{\sqrt{A}}{\sqrt{\pi}} = \dfrac{\sqrt{A}}{\sqrt{\pi}} \cdot \dfrac{\sqrt{\pi}}{\sqrt{\pi}} = \dfrac{\sqrt{A\pi}}{\pi}$

**107.** Answers may vary.

**109.** Answers may vary.

**111.** $\dfrac{\sqrt{2}-2}{2-\sqrt{3}} = \dfrac{\sqrt{2}-2}{2-\sqrt{3}} \cdot \dfrac{\sqrt{2}+2}{\sqrt{2}+2}$

$= \dfrac{2-4}{2\sqrt{2}-4-\sqrt{6}-2\sqrt{3}}$

$= \dfrac{-2}{2\sqrt{2}-4-\sqrt{6}-2\sqrt{3}}$

**Integrated Review-Simplifying Radicals**

**1.** $\sqrt{36} = 6$, because $6^2 = 36$ and 6 is positive.

**2.** $\sqrt{48} = \sqrt{16} \cdot \sqrt{3} = 4\sqrt{3}$

**3.** $\sqrt{x^4} = x^2$, because $\left(x^2\right)^2 = x^4$.

**4.** $\sqrt{y^7} = \sqrt{y^6}\sqrt{y} = y^3\sqrt{y}$

**5.** $\sqrt{16x^2} = 4x$, because $(4x)^2 = 16x^2$.

**6.** $\sqrt{18x^{11}} = \sqrt{9x^{10}}\sqrt{2x} = 3x^5\sqrt{2x}$

**7.** $\sqrt[3]{8} = 2$, because $(2)^3 = 8$.

**8.** $\sqrt[4]{81} = 3$, because $(3)^4 = 81$.

**9.** $\sqrt[3]{-27} = -3$, because $(-3)^3 = -27$.

**10.** $\sqrt{-4}$ is not a real number.

**11.** $\sqrt{\dfrac{11}{9}} = \dfrac{\sqrt{11}}{\sqrt{9}} = \dfrac{\sqrt{11}}{3}$

**12.** $\sqrt[3]{\dfrac{7}{64}} = \dfrac{\sqrt[3]{7}}{\sqrt[3]{64}} = \dfrac{\sqrt[3]{7}}{4}$

**13.** $-\sqrt{16} = -4$

**14.** $-\sqrt{25} = -5$

**15.** $\sqrt{\dfrac{9}{49}} = \dfrac{3}{7}$

**16.** $\sqrt{\dfrac{1}{64}} = \dfrac{1}{8}$

**17.** $\sqrt{a^8 b^2} = a^4 b$

**18.** $\sqrt{x^{10} y^{20}} = x^5 y^{10}$

**19.** $\sqrt{25 m^6} = 5 m^3$

**20.** $\sqrt{9 n^{16}} = 3 n^8$

**21.** $5\sqrt{7} + \sqrt{7} = (5+1)\sqrt{7} = 6\sqrt{7}$

**22.** $\sqrt{50} - \sqrt{8} = \sqrt{25} \cdot \sqrt{2} - \sqrt{4} \cdot \sqrt{2}$
$= 5\sqrt{2} - 2\sqrt{2} = (5-2)\sqrt{2} = 3\sqrt{2}$

**23.** $5\sqrt{2} - 5\sqrt{3}$ cannot be simplified.

**24.** $2\sqrt{x} + \sqrt{25x} - \sqrt{36x} + 3x$
$= 2\sqrt{x} + \sqrt{25} \cdot \sqrt{x} - \sqrt{36} \cdot \sqrt{x} + 3x$
$= 2\sqrt{x} + 5\sqrt{x} - 6\sqrt{x} + 3x$
$= (2 + 5 - 6)\sqrt{x} + 3x$
$= \sqrt{x} + 3x$

**25.** $\sqrt{2} \cdot \sqrt{15} = \sqrt{2 \cdot 15} = \sqrt{30}$

**26.** $\sqrt{3} \cdot \sqrt{3} = \sqrt{3 \cdot 3} = \sqrt{9} = 3$

**27.** $\left(2\sqrt{7}\right)^2 = 2^2 \left(\sqrt{7}\right)^2 = 4(7) = 28$

**28.** $\left(3\sqrt{5}\right)^2 = 3^2 \left(\sqrt{5}\right)^2 = 9(5) = 45$

**29.** $\sqrt{3}\left(\sqrt{11} + 1\right) = \sqrt{3} \cdot \sqrt{11} + \sqrt{3} \cdot 1$
$= \sqrt{33} + \sqrt{3}$

**30.** $\sqrt{6}\left(\sqrt{3} - 2\right) = \sqrt{6} \cdot \sqrt{3} - \sqrt{6} \cdot 2$
$= \sqrt{18} - 2\sqrt{6} = \sqrt{9} \cdot \sqrt{2} - 2\sqrt{6}$
$= 3\sqrt{2} - 2\sqrt{6}$

**31.** $\sqrt{8y} \cdot \sqrt{2y} = \sqrt{8y \cdot 2y} = \sqrt{16 y^2} = 4y$

**32.** $\sqrt{15 x^2} \cdot \sqrt{3 x^2} = \sqrt{15 x^2 \cdot 3 x^2} = \sqrt{45 x^4}$
$= \sqrt{9 x^4} \cdot \sqrt{5} = 3 x^2 \sqrt{5}$

**33.** $\left(\sqrt{x} - 5\right)\left(\sqrt{x} + 2\right) = \sqrt{x^2} + 2\sqrt{x} - 5\sqrt{x} - 10$
$= x - 3\sqrt{x} - 10$

**34.** $\left(3+\sqrt{2}\right)^2 = 3^2 + 2(3)\sqrt{2} + \left(\sqrt{2}\right)^2$

$\qquad\qquad = 9 + 6\sqrt{2} + 2$

$\qquad\qquad = 11 + 6\sqrt{2}$

**35.** $\dfrac{\sqrt{8}}{\sqrt{2}} = \sqrt{\dfrac{8}{2}} = \sqrt{4} = 2$

**36.** $\dfrac{\sqrt{45}}{\sqrt{15}} = \sqrt{\dfrac{45}{15}} = \sqrt{3}$

**37.** $\dfrac{\sqrt{24x^5}}{\sqrt{2x}} = \sqrt{\dfrac{24x^5}{2x}} = \sqrt{12x^4} = \sqrt{4x^4} \cdot \sqrt{3}$

$\qquad\qquad = 2x^2\sqrt{3}$

**38.** $\dfrac{\sqrt{75a^4b^5}}{\sqrt{5ab}} = \sqrt{\dfrac{75a^4b^5}{5ab}} = \sqrt{15a^3b^4}$

$\qquad = \sqrt{a^2b^4} \cdot \sqrt{15a} = ab^2\sqrt{15a}$

**39.** $\sqrt{\dfrac{1}{6}} = \dfrac{\sqrt{1}}{\sqrt{6}} = \dfrac{1}{\sqrt{6}} \cdot \dfrac{\sqrt{6}}{\sqrt{6}} = \dfrac{\sqrt{6}}{6}$

**40.** $\dfrac{x}{\sqrt{20}} = \dfrac{x}{\sqrt{4} \cdot \sqrt{5}} = \dfrac{x}{2\sqrt{5}} = \dfrac{x}{2\sqrt{5}} \cdot \dfrac{\sqrt{5}}{\sqrt{5}}$

$\qquad = \dfrac{x\sqrt{5}}{2(5)} = \dfrac{x\sqrt{5}}{10}$

**41.** $\dfrac{4}{\sqrt{6}+1} = \dfrac{4}{\sqrt{6}+1} \cdot \dfrac{\sqrt{6}-1}{\sqrt{6}-1} = \dfrac{4\left(\sqrt{6}-1\right)}{6-1}$

$\qquad = \dfrac{4\sqrt{6}-4}{5}$

**42.** $\dfrac{\sqrt{2}+1}{\sqrt{x}-5} = \dfrac{\sqrt{2}+1}{\sqrt{x}-5} \cdot \dfrac{\sqrt{x}+5}{\sqrt{x}+5}$

$\qquad = \dfrac{\sqrt{2x}+5\sqrt{2}+\sqrt{x}+5}{x-25}$

**Exercise Set 8.5**

**1.** $\sqrt{x} = 9$

$\qquad \left(\sqrt{x}\right)^2 = 9^2$

$\qquad\qquad x = 81$

**3.** $\sqrt{x+5} = 2$

$\qquad \left(\sqrt{x+5}\right)^2 = 2^2$

$\qquad\qquad x+5 = 4$

$\qquad\qquad x = -1$

**5.** $\sqrt{2x+6} = 4$

$\qquad \left(\sqrt{2x+6}\right)^2 = 4^2$

$\qquad\qquad 2x+6 = 16$

$\qquad\qquad 2x = 10$

$\qquad\qquad x = 5$

**7.** $\sqrt{x} - 2 = 5$

$\qquad \sqrt{x} = 7$

$\qquad \left(\sqrt{x}\right)^2 = 7^2$

$\qquad\qquad x = 49$

**9.** $3\sqrt{x} + 5 = 2$

$\qquad 3\sqrt{x} = -3$

The square root cannot be negative, therefore there is no solution.

**11.** $\sqrt{x+6}+1=3$

$\sqrt{x+6}=2$

$\left(\sqrt{x+6}\right)^2=2^2$

$x+6=4$

$x=-2$

**13.** $\sqrt{2x+1}+3=5$

$\sqrt{2x+1}=2$

$\left(\sqrt{2x+1}\right)^2=2^2$

$2x+1=4$

$2x=3$

$x=\dfrac{3}{2}$

**15.** $\sqrt{x}+3=7$

$\sqrt{x}=4$

$\left(\sqrt{x}\right)^2=4^2$

$x=16$

**17.** $\sqrt{x+6}+5=3$

$\sqrt{x+6}=-2$

The square root cannot be negative, therefore there is no solution.

**19.** $\sqrt{4x-3}=\sqrt{x+3}$

$\left(\sqrt{4x-3}\right)^2=\left(\sqrt{x+3}\right)^2$

$4x-3=x+3$

$3x=6$

$x=2$

**21.** $\sqrt{x}=\sqrt{3x-8}$

$\left(\sqrt{x}\right)^2=\left(\sqrt{3x-8}\right)^2$

$x=3x-8$

$-2x=-8$

$x=4$

**23.** $\sqrt{4x}=\sqrt{2x+6}$

$\left(\sqrt{4x}\right)^2=\left(\sqrt{2x+6}\right)^2$

$4x=2x+6$

$2x=6$

$x=3$

**25.** $\sqrt{9x^2+2x-4}=3x$

$\left(\sqrt{9x^2+2x-4}\right)^2=\left(3x\right)^2$

$9x^2+2x-4=9x^2$

$2x-4=0$

$2x=4$

$x=2$

**27.** $\sqrt{16x^2-3x+6}=4x$

$\left(\sqrt{16x^2-3x+6}\right)^2=\left(4x\right)^2$

$16x^2-3x+6=16x^2$

$-3x+6=0$

$-3x=-6$

$x=2$

**29.** $\sqrt{16x^2+2x+2}=4x$

$\left(\sqrt{16x^2+2x+2}\right)^2=\left(4x\right)^2$

$16x^2+2x+2=16x^2$

$2x+2=0$

$$2x = -2$$
$$x = -1$$

A check shows that $x = -1$ is an extraneous solution. Therefore, there is no solution.

**31.**    $\sqrt{2x^2 + 6x + 9} = 3$

$$\left(\sqrt{2x^2 + 6x + 9}\right)^2 = (3)^2$$
$$2x^2 + 6x + 9 = 9$$
$$2x^2 + 6x = 0$$
$$2x(x + 3) = 0$$
$$2x = 0 \quad \text{or} \quad x + 3 = 0$$
$$x = 0 \qquad\qquad x = -3$$

**33.**    $\sqrt{x + 7} = x + 5$

$$\left(\sqrt{x + 7}\right)^2 = (x + 5)^2$$
$$x + 7 = x^2 + 10x + 25$$
$$0 = x^2 + 9x + 18$$
$$0 = (x + 3)(x + 6)$$
$$x + 3 = 0 \quad \text{or} \quad x + 6 = 0$$
$$x = -3 \qquad\quad x = -6 \,(\text{extraneous})$$

**35.**    $\sqrt{x} = x - 6$

$$\left(\sqrt{x}\right)^2 = (x - 6)^2$$
$$x = x^2 - 12x + 36$$
$$0 = x^2 - 13x + 36$$
$$0 = (x - 9)(x - 4)$$
$$x - 9 = 0 \quad \text{or} \quad x - 4 = 0$$
$$x = 9 \qquad\quad x = 4 \,(\text{extraneous})$$

**37.**    $\sqrt{2x + 1} = x - 7$

$$\left(\sqrt{2x + 1}\right)^2 = (x - 7)^2$$
$$2x + 1 = x^2 - 14x + 49$$
$$0 = x^2 - 16x + 48$$
$$0 = (x - 12)(x - 4)$$
$$x - 12 = 0 \quad \text{or} \quad x - 4 = 0$$
$$x = 12 \qquad\quad x = 4 \,(\text{extraneous})$$

**39.**    $x = \sqrt{2x - 2} + 1$

$$x - 1 = \sqrt{2x - 2}$$
$$(x - 1)^2 = \left(\sqrt{2x - 2}\right)^2$$
$$x^2 - 2x + 1 = 2x - 2$$
$$x^2 - 4x + 3 = 0$$
$$(x - 1)(x - 3) = 0$$
$$x - 1 = 0 \quad \text{or} \quad x - 3 = 0$$
$$x = 1 \qquad\quad x = 3$$

**41.**    $\sqrt{1 - 8x} - x = 4$

$$\sqrt{1 - 8x} = x + 4$$
$$\left(\sqrt{1 - 8x}\right)^2 = (x + 4)^2$$
$$1 - 8x = x^2 + 8x + 16$$
$$0 = x^2 + 16x + 15$$
$$0 = (x + 1)(x + 15)$$
$$x + 1 = 0 \quad \text{or} \quad x + 15 = 0$$
$$x = -1 \qquad\quad x = -15$$
$$(\text{extraneous})$$

**43.**    $\sqrt{2x + 5} - 1 = x$

$$\sqrt{2x + 5} = x + 1$$
$$\left(\sqrt{2x + 5}\right)^2 = (x + 1)^2$$

$$2x + 5 = x^2 + 2x + 1$$
$$0 = x^2 - 4$$
$$0 = (x - 2)(x + 2)$$
$$x - 2 = 0 \quad \text{or} \quad x + 2 = 0$$
$$x = 2 \qquad x = -2$$
$$\text{(extraneous)}$$

**45.**
$$\sqrt{x - 7} = \sqrt{x} - 1$$
$$\left(\sqrt{x - 7}\right)^2 = \left(\sqrt{x} - 1\right)^2$$
$$x - 7 = x - 2\sqrt{x} + 1$$
$$2\sqrt{x} = 8$$
$$\sqrt{x} = 4$$
$$\left(\sqrt{x}\right)^2 = (4)^2$$
$$x = 16$$

**47.**
$$\sqrt{x} + 3 = \sqrt{x + 15}$$
$$\left(\sqrt{x} + 3\right)^2 = \left(\sqrt{x + 15}\right)^2$$
$$x + 6\sqrt{x} + 9 = x + 15$$
$$6\sqrt{x} = 6$$
$$\sqrt{x} = 1$$
$$\left(\sqrt{x}\right)^2 = (1)^2$$
$$x = 1$$

**49.**
$$\sqrt{x + 8} = \sqrt{x} + 2$$
$$\left(\sqrt{x + 8}\right)^2 = \left(\sqrt{x} + 2\right)^2$$
$$x + 8 = x + 4\sqrt{x} + 4$$
$$4 = 4\sqrt{x}$$
$$1 = \sqrt{x}$$
$$1^2 = \left(\sqrt{x}\right)^2$$
$$1 = x$$

**51.**
$$3x - 8 = 19$$
$$3x = 27$$
$$x = 9$$

**53.** Let $x$ = width and $2x$ = length.
$$2(2x + x) = 24$$
$$2(3x) = 24$$
$$6x = 24$$
$$x = 4$$
$$2x = 2(4) = 8$$
The length is 8 inches.

**55.**                Let $x$ = number
$$x = 6 + \sqrt{x}$$
$$x - 6 = \sqrt{x}$$
$$(x - 6)^2 = \left(\sqrt{x}\right)^2$$
$$x^2 - 12x + 36 = x$$
$$x^2 - 12x - x + 36 = 0$$
$$x^2 - 13x + 36 = 0$$
$$(x - 9)(x - 4) = 0$$

$$x - 9 = 0 \quad \text{or} \quad x - 4 = 0$$
$$x = 9 \quad \text{or} \qquad x = 4$$
Check: $x = 9$
$$9 = 6 + \sqrt{9}$$
$$9 = 6 + 3$$
$$9 = 9 \quad \text{True}$$
Check: $x = 4$
$$4 = 6 + \sqrt{4}$$
$$4 = 6 + 2$$
$$4 = 8 \quad \text{False}$$
The solution is 9.

**57.** $b = \sqrt{\dfrac{V}{2}}$

**a.** $b = \sqrt{\dfrac{20}{2}} \approx 3.2$

$b = \sqrt{\dfrac{200}{2}} = 10$

$b = \sqrt{\dfrac{2000}{2}} \approx 31.6$

| $V$ | 20 | 200 | 2000 |
|---|---|---|---|
| $b$ | 3.2 | 10 | 31.6 |

**b.** No; it increases by a factor of $\sqrt{10}$.

**c.** $b = \sqrt{\dfrac{V}{2}}$

$b^2 = \left( \sqrt{\dfrac{V}{2}} \right)^2$

$b^2 = \dfrac{V}{2}$

$2b^2 = V$

**59.** Answers may vary.

**61.** $\sqrt{x+1} = 2x - 3$, $y_1 = \sqrt{x+1}$, $y_2 = 2x - 3$

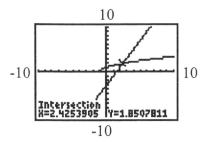

The solution is 2.43.

**63.** $-\sqrt{x+5} = -7x + 1$

$y_1 = -\sqrt{x+5}$, $y_2 = -7x + 1$

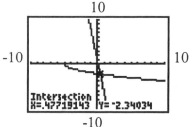

The solution is 0.48.

**Exercise Set 8.6**

**1.** $a^2 + b^2 = c^2$

$2^2 + 3^2 = c^2$

$4 + 9 = c^2$

$13 = c^2$

$\sqrt{13} = c$

The length is $\sqrt{13} \approx 3.61$.

**3.** $a^2 + b^2 = c^2$

$3^2 + b^2 = 6^2$

$9 + b^2 = 36$

$b^2 = 27$

$b = \sqrt{27}$

$b = 3\sqrt{3}$

The length is $3\sqrt{3} \approx 5.20$.

**5.** $a^2 + b^2 = c^2$

$7^2 + 24^2 = c^2$

$49 + 576 = c^2$

$625 = c^2$

$$\sqrt{625} = c$$
$$25 = c$$
The length is 25.

**7.**
$$a^2 + b^2 = c^2$$
$$a^2 + \left(\sqrt{3}\right)^2 = 5^2$$
$$a^2 + 3 = 25$$
$$a^2 = 22$$
$$a = \sqrt{22}$$
The length is $\sqrt{22} \approx 4.69$.

**9.**
$$a^2 + b^2 = c^2$$
$$4^2 + b^2 = 13^2$$
$$16 + b^2 = 169$$
$$b^2 = 153$$
$$b = \sqrt{153}$$
$$b = 3\sqrt{17}$$
The length is $3\sqrt{17} \approx 12.37$.

**11.**
$$a^2 + b^2 = c^2$$
$$4^2 + 5^2 = c^2$$
$$16 + 25 = c^2$$
$$41 = c^2$$
$$\sqrt{41} = c$$
The length is $\sqrt{41} \approx 6.40$.

**13.**
$$a^2 + b^2 = c^2$$
$$a^2 + 2^2 = 6^2$$
$$a^2 + 4 = 36$$
$$a^2 = 32$$
$$a = \sqrt{32}$$

$$a = 4\sqrt{2}$$
The length is $4\sqrt{2} \approx 5.66$.

**15.**
$$a^2 + b^2 = c^2$$
$$\left(\sqrt{10}\right)^2 + b^2 = 10^2$$
$$10 + b^2 = 100$$
$$b^2 = 90$$
$$b = \sqrt{90}$$
$$b = 3\sqrt{10}$$
The length is $3\sqrt{10} \approx 9.49$.

**17.**
$$a^2 + b^2 = c^2$$
$$40^2 + b^2 = 65^2$$
$$1600 + b^2 = 4225$$
$$b^2 = 4225 - 1600$$
$$b^2 = 2625$$
$$\sqrt{b^2} = \sqrt{2625}$$
$$b \approx 51.2 \text{ ft}$$

**19.**
$$a^2 + b^2 = c^2$$
$$5^2 + 20^2 = c^2$$
$$25 + 400 = c^2$$
$$425 = c^2$$
$$\sqrt{425} = c$$
The length is $\sqrt{425} \approx 20.6$ feet.

**21.**
$$a^2 + b^2 = c^2$$
$$6^2 + 10^2 = c^2$$
$$36 + 100 = c^2$$
$$136 = c^2$$
$$\sqrt{136} = c$$
The length is $\sqrt{136} \approx 11.7$ feet.

**23.** $(3, 6)$ and $(5, 11)$

$$d = \sqrt{(x_2 - x_1)^2 + (y_2 - y_1)^2}$$
$$= \sqrt{(5-3)^2 + (11-6)^2}$$
$$= \sqrt{2^2 + 5^2}$$
$$= \sqrt{4 + 25}$$
$$= \sqrt{29}$$

**25.** $(-3, 1)$ and $(5, -2)$

$$d = \sqrt{(x_2 - x_1)^2 + (y_2 - y_1)^2}$$
$$= \sqrt{[5 - (-3)]^2 + (-2 - 1)^2}$$
$$= \sqrt{(8)^2 + (-3)^2}$$
$$= \sqrt{64 + 9}$$
$$= \sqrt{73}$$

**27.** $(3, -2)$ and $(1, -8)$

$$d = \sqrt{(x_2 - x_1)^2 + (y_2 - y_1)^2}$$
$$= \sqrt{(1-3)^2 + [-8 - (-2)]^2}$$
$$= \sqrt{(-2)^2 + (-6)^2}$$
$$= \sqrt{4 + 36}$$
$$= \sqrt{40}$$
$$= \sqrt{4 \cdot 10}$$
$$= 2\sqrt{10}$$

**29.** $\left(\dfrac{1}{2}, 2\right)$ and $(2, -1)$

$$d = \sqrt{(x_2 - x_1)^2 + (y_2 - y_1)^2}$$
$$= \sqrt{\left(2 - \dfrac{1}{2}\right)^2 + (-1 - 2)^2}$$

$$= \sqrt{\left(\dfrac{3}{2}\right)^2 + (-3)^2}$$
$$= \sqrt{\dfrac{9}{4} + 9}$$
$$= \sqrt{\dfrac{45}{4}}$$
$$= \dfrac{\sqrt{45}}{\sqrt{4}}$$
$$= \dfrac{3\sqrt{5}}{2}$$

**31.** $(3, -2)$ and $(5, 7)$

$$d = \sqrt{(x_2 - x_1)^2 + (y_2 - y_1)^2}$$
$$= \sqrt{(5-3)^2 + [7 - (-2)]^2}$$
$$= \sqrt{2^2 + 9^2}$$
$$= \sqrt{4 + 81}$$
$$= \sqrt{85}$$

**33.** $b = \sqrt{\dfrac{3V}{h}}$

$$6 = \sqrt{\dfrac{3V}{2}}$$
$$6^2 = \left(\sqrt{\dfrac{3V}{2}}\right)^2$$
$$36 = \dfrac{3V}{2}$$
$$24 = V$$

The volume is 24 cubic feet.

**35.** $s = \sqrt{30\,fd}$

$s = \sqrt{30(0.35)(280)}$

$= \sqrt{2940}$

$\approx 54$

It was moving at 54 mph.

**37.** $v = \sqrt{2.5r}$

$v = \sqrt{2.5(300)}$

$= \sqrt{750}$

$\approx 27$

It can travel at 27 mph.

**39.** $d = 3.5\sqrt{h}$

$d = 3.5\sqrt{305.4}$

$\approx 61.2$

You can see 61.2 km.

**41.** $2^5 = 2 \cdot 2 \cdot 2 \cdot 2 \cdot 2 = 32$

**43.** $\left(-\dfrac{1}{5}\right)^2 = \left(-\dfrac{1}{5}\right)\left(-\dfrac{1}{5}\right) = \dfrac{1}{25}$

**45.** $x^2 \cdot x^3 + x^{2+3} = x^5$

**47.** $y^3 \cdot y = y^{3+1} = y^4$

**49.** Let $y$ = length of whole base and
$z$ = length of unlabeled section of base.
Find $y$ :

$y^2 + 3^2 = 7^2$

$y^2 + 9 = 49$

$y^2 = 40$

$y = \sqrt{40} = 2\sqrt{10}$

Find $z$

$z^2 + 3^2 = 5^2$

$z^2 + 9 = 25$

$z^2 = 16$

$z = \sqrt{16} = 4$

Find $x$ :

$x = y - z$

$= 2\sqrt{10} - 4$

**51.** 
$a^2 + b^2 = c^2$

$\left[60(3)\right]^2 + \left[30(3)\right]^2 = c^2$

$180^2 + 90^2 = c^2$

$32,400 + 8100 = c^2$

$40,500 = c^2$

$\sqrt{40,500} = c$

$201 \approx c$

They are about 201 miles apart.

**53.** Answers may vary.

**Exercise Set 8.7**

**1.** $8^{1/3} = \sqrt[3]{8} = 2$

**3.** $9^{1/2} = \sqrt{9} = 3$

**5.** $16^{3/4} = \left(\sqrt[4]{16}\right)^3 = 2^3 = 8$

**7.** $32^{2/5} = \left(\sqrt[5]{32}\right)^2 = 2^2 = 4$

**9.** $-16^{-1/4} = -\dfrac{1}{16^{1/4}} = -\dfrac{1}{\sqrt[4]{16}} = -\dfrac{1}{2}$

**11.** $16^{-3/2} = \dfrac{1}{16^{3/2}} = \dfrac{1}{\left(\sqrt{16}\right)^3} = \dfrac{1}{4^3} = \dfrac{1}{64}$

**39.** $9^{3/2} = \left(\sqrt{9}\right)^3 = (3)^3 = 27$

**13.** $81^{-3/2} = \dfrac{1}{81^{3/2}} = \dfrac{1}{\left(\sqrt{81}\right)^3} = \dfrac{1}{9^3} = \dfrac{1}{729}$

**41.** $64^{3/2} = \left(\sqrt{64}\right)^3 = (8)^3 = 512$

**15.** $\left(\dfrac{4}{25}\right)^{-1/2} = \dfrac{1}{\left(\frac{4}{25}\right)^{1/2}} = \dfrac{1}{\sqrt{\frac{4}{25}}} = \dfrac{1}{\frac{2}{5}} = \dfrac{5}{2}$

**43.** $-8^{2/3} = -\left(8^{2/3}\right)$

$\qquad = -\left(\sqrt[3]{8}\right)^2$

$\qquad = -(2)^2$

$\qquad = -(4)$

$\qquad = -4$

**17.** Answers may vary.

**19.** $2^{1/3} \cdot 2^{2/3} = 2^{3/3} = 2^1 = 2$

**45.** $4^{5/2} = \left(\sqrt{4}\right)^5 = (2)^5 = 32$

**21.** $\dfrac{4^{3/4}}{4^{1/4}} = 4^{3/4-1/4} = 4^{4/2} = 4^{1/2} = \sqrt{4} = 2$

**47.** $\left(\dfrac{4}{9}\right)^{3/2} = \dfrac{4^{3/2}}{9^{3/2}} = \dfrac{\left(\sqrt{4}\right)^3}{\left(\sqrt{9}\right)^3} = \dfrac{2^3}{3^3} = \dfrac{8}{27}$

**23.** $\dfrac{x^{1/6}}{x^{5/6}} = x^{\frac{1}{6}-\frac{5}{6}} = x^{-4/6} = x^{-2/3} = \dfrac{1}{x^{2/3}}$

**25.** $\left(x^{1/2}\right)^6 = x^{6/2} = x^3$

**49.** $\left(\dfrac{1}{81}\right)^{3/4} = \dfrac{1^{3/4}}{81^{3/4}} = \dfrac{\left(\sqrt[4]{1}\right)^3}{\left(\sqrt[4]{81}\right)^3} = \dfrac{1^3}{3^3} = \dfrac{1}{27}$

**28.** Answers may vary.

**29.** $81^{1/2} = \sqrt{81} = 9$

**51.** $4^{-1/2} = \dfrac{1}{4^{1/2}} = \dfrac{1}{\sqrt{4}} = \dfrac{1}{2}$

**31.** $(-8)^{1/3} = \sqrt[3]{-8} = -2$

**53.** $215^{-1/3} = \dfrac{1}{215^{1/3}} = \dfrac{1}{\sqrt[3]{215}} = \dfrac{1}{5}$

**33.** $-81^{1/4} = -\left(81^{1/4}\right) = -\left(\sqrt[4]{81}\right) = -(3) = -3$

**55.** $625^{-3/4} = \dfrac{1}{625^{3/4}} = \dfrac{1}{\left(\sqrt[4]{625}\right)^3} = \dfrac{1}{5^3} = \dfrac{1}{125}$

**35.** $\left(\dfrac{1}{81}\right)^{1/2} = \dfrac{1^{1/2}}{81^{1/2}} = \sqrt{\dfrac{1}{81}} = \dfrac{1}{9}$

**57.** $3^{4/3} \cdot 3^{2/3} = 3^{4/2+2/3} = 3^{6/3} = 3^2 = 9$

**37.** $\left(\dfrac{27}{64}\right)^{1/3} = \dfrac{27^{1/3}}{64^{1/3}} = \dfrac{\sqrt[3]{27}}{\sqrt[3]{64}} = \dfrac{3}{4}$

**59.** $\dfrac{6^{2/3}}{6^{1/3}} = 6^{2/3-1/3} = 6^{1/3}$

**61.** $\left(x^{2/3}\right)^9 = x^{\frac{2}{3}\cdot 9} = x^6$

**63.** $\dfrac{6^{1/3}}{6^{-5/3}} = 6^{1/3-(-5/3)} = 6^{6/3} = 6^2 = 36$

**65.** $\dfrac{3^{-3/5}}{3^{2/5}} = 3^{-3/5-2/5} = 3^{-5/5} = 3^{-1} = \dfrac{1}{3}$

**67.** $\left(\dfrac{x^{1/3}}{y^{3/4}}\right)^2 = \dfrac{\left(x^{1/3}\right)^2}{\left(y^{3/4}\right)^2} = \dfrac{x^{2/3}}{y^{3/2}}$

**69.** $\left(\dfrac{x^{2/5}}{y^{3/4}}\right)^8 = \dfrac{\left(x^{2/5}\right)^8}{\left(y^{3/4}\right)^8} = \dfrac{x^{16/5}}{y^6}$

**\*71.**

**73.**
$$x^2 - 4 = 3x$$
$$x^2 - 3x - 4 = 0$$
$$(x-4)(x+1) = 0$$
$$x - 4 = 0 \quad \text{or} \quad x + 1 = 0$$
$$x = 4 \qquad\qquad x = -1$$

The solutions are $-1$ and $4$.

**75.**
$$2x^2 - 5x - 3 = 3x$$
$$2x^2 - 6x + x - 3 = 0$$
$$2x(x-3) + 1(x-3) = 0$$
$$(2x+1)(x-3) = 0$$
$$2x + 1 = 0 \quad \text{or} \quad x - 3 = 0$$
$$x = -\dfrac{1}{2} \qquad\qquad x = 3$$

The solutions are $-\dfrac{1}{2}$ and $3$.

**77.** Let $N = 1.5$ and $P_o = 10,000$

$P = P_o\left(1.08\right)^N = 10,000\left(1.08\right)^{1.5} = 11,224$

The population will be 11,224 people.

**79.** $5^{3/4} = 3.344$

**81.** $18^{3/5} = 5.665$

## Chapter 8 Review

**1.** $\sqrt{81} = 9$, because $9^2 = 81$ and 9 is positive.

**2.** $-\sqrt{49} = -7$, because $\sqrt{49} = 7$.

**3.** $\sqrt[3]{27} = 3$, because $(3)^3 = 27$.

**4.** $\sqrt[4]{16} = 2$, because $2^4 = 16$.

**5.** $-\sqrt{\dfrac{9}{64}} = -\dfrac{3}{8}$, because $\sqrt{\dfrac{9}{64}} = \dfrac{3}{8}$.

**6.** $\sqrt{\dfrac{36}{81}} = \dfrac{6}{9} = \dfrac{2}{3}$, because $\left(\dfrac{6}{9}\right)^2 = \dfrac{36}{81}$.

**7.** $\sqrt[4]{-\dfrac{16}{81}}$; not a real number

**8.** $\sqrt[3]{-\dfrac{27}{64}} = -\dfrac{3}{4}$ because $\left(-\dfrac{3}{4}\right)^3 = -\dfrac{27}{64}$.

**9.** $\sqrt{76}$; irrational, 8.718

**10.** $\sqrt{576}$; rational, 24

**11.** $\sqrt{x^{12}} = x^6$, because $\left(x^6\right)^2 = x^{12}$.

**12.** $\sqrt{x^8} = x^4$, because $\left(x^4\right)^2 = x^8$.

**13.** $\sqrt{9x^6} = 3x^3$, because $\left(3x^3\right)^2 = 9x^6$.

**14.** $\sqrt{25x^4} = 5x^2$, because $\left(5x^2\right)^2 = 25x^4$.

**15.** $\sqrt{\dfrac{16}{y^{10}}} = \dfrac{4}{y^5}$ because $\left(\dfrac{4}{y^5}\right)^2 = \dfrac{16}{y^{10}}$.

**16.** $\sqrt{\dfrac{y^{12}}{49}} = \dfrac{y^6}{7}$ because $\left(\dfrac{y^6}{7}\right)^2 = \dfrac{y^{12}}{49}$.

**17.** $\sqrt{54} = \sqrt{9 \cdot 6} = \sqrt{9} \cdot \sqrt{6} = 3\sqrt{6}$

**18.** $\sqrt{88} = \sqrt{4 \cdot 22} = \sqrt{4} \cdot \sqrt{22} = 2\sqrt{22}$

**19.** $\sqrt{150x^3} = \sqrt{25x^2}\sqrt{6x} = 5x\sqrt{6x}$

**20.** $\sqrt{92y^5} = \sqrt{4 \cdot 23 \cdot y^4 \cdot y} = 2y^2\sqrt{23y}$

**21.** $\sqrt[3]{54} = \sqrt[3]{27}\sqrt[3]{2} = 3\sqrt[3]{2}$

**22.** $\sqrt[3]{88} = \sqrt[3]{8 \cdot 11} = 2\sqrt[3]{11}$

**23.** $\sqrt[4]{48} = \sqrt[4]{16}\sqrt[4]{3} = 2\sqrt[4]{3}$

**24.** $\sqrt[4]{162} = \sqrt[4]{81 \cdot 2} = 3\sqrt[4]{2}$

**25.** $\sqrt{\dfrac{18}{25}} = \dfrac{\sqrt{18}}{\sqrt{25}} = \dfrac{\sqrt{9} \cdot \sqrt{2}}{5} = \dfrac{3\sqrt{2}}{5}$

**26.** $\sqrt{\dfrac{75}{64}} = \dfrac{\sqrt{75}}{\sqrt{64}} = \dfrac{\sqrt{25} \cdot \sqrt{3}}{8} = \dfrac{5\sqrt{3}}{8}$

**27.** $\sqrt{\dfrac{45y^2}{4x^4}} = \dfrac{\sqrt{45y^2}}{\sqrt{4x^4}}$

$\qquad = \dfrac{\sqrt{9y^2}\sqrt{5}}{2x^2}$

$\qquad = \dfrac{3y\sqrt{5}}{2x^2}$

**28.** $\sqrt{\dfrac{20x^5}{9x^2}} = \sqrt{\dfrac{20x^3}{9}} = \dfrac{\sqrt{4 \cdot 5 \cdot x^2 \cdot x}}{\sqrt{9}} = \dfrac{2x\sqrt{5x}}{3}$

**29.** $\sqrt[4]{\dfrac{9}{16}} = \dfrac{\sqrt[4]{9}}{\sqrt[4]{16}} = \dfrac{\sqrt[4]{9}}{2}$

**30.** $\sqrt[3]{\dfrac{40}{27}} = \dfrac{\sqrt[3]{8 \cdot 5}}{\sqrt[3]{27}} = \dfrac{2\sqrt[3]{5}}{3}$

**31.** $\sqrt[3]{\dfrac{3}{8}} = \dfrac{\sqrt[3]{3}}{\sqrt[3]{8}} = \dfrac{\sqrt[3]{3}}{2}$

**32.** $\sqrt[4]{\dfrac{5}{81}} = \dfrac{\sqrt[4]{5}}{\sqrt[4]{81}} = \dfrac{\sqrt[4]{5}}{3}$

**33.** $3\sqrt[3]{2} + 2\sqrt[3]{3} - 4\sqrt[3]{2} = (3-4)\sqrt[3]{2} + 2\sqrt[3]{3}$
$$= -\sqrt[3]{2} + 2\sqrt[3]{2}$$

**34.** $5\sqrt{2} + 2\sqrt[3]{2} - 8\sqrt{2} = (5-8)\sqrt{2} + 2\sqrt[3]{2}$
$$= -3\sqrt{2} + 2\sqrt[3]{2}$$

**35.** $\sqrt{6} + 2\sqrt[3]{6} - 4\sqrt[3]{6} + 5\sqrt{6} = (1+5)\sqrt{6} + (2-4)\sqrt[3]{6}$
$$= 6\sqrt{6} - 2\sqrt[3]{6}$$

**36.** $3\sqrt{5} - \sqrt[3]{5} - 2\sqrt{5} + 3\sqrt[3]{5} = (3-2)\sqrt{5} + (-1+3)\sqrt[3]{5}$
$$= \sqrt{5} + 2\sqrt[3]{5}$$

**37.** $\sqrt{28x} + \sqrt{63x} + \sqrt[3]{56}$
$$= \sqrt{4} \cdot \sqrt{7x} + \sqrt{9} \cdot \sqrt{7x} + \sqrt[3]{8} \cdot \sqrt[3]{7}$$
$$= 2\sqrt{7x} + 3\sqrt{7x} + 2\sqrt[3]{7}$$
$$= 5\sqrt{7x} + 2\sqrt[3]{7}$$

**38.** $\sqrt{75y} + \sqrt{48y} - \sqrt[4]{16}$
$$= \sqrt{25} \cdot \sqrt{3y} + \sqrt{16} \cdot \sqrt{3y} - 2$$
$$= 5\sqrt{3y} + 4\sqrt{3y} - 2$$
$$= 9\sqrt{3y} - 2$$

**39.** $\sqrt{\dfrac{5}{9}} - \sqrt{\dfrac{5}{36}} = \dfrac{\sqrt{5}}{\sqrt{9}} - \dfrac{\sqrt{5}}{\sqrt{36}} = \dfrac{\sqrt{5}}{3} - \dfrac{\sqrt{5}}{6}$
$$= \dfrac{2\sqrt{5}}{6} - \dfrac{\sqrt{5}}{6} = \dfrac{2\sqrt{5} - \sqrt{5}}{6} = \dfrac{\sqrt{5}}{6}$$

**40.** $\sqrt{\dfrac{11}{25}} + \sqrt{\dfrac{11}{16}} = \dfrac{\sqrt{11}}{\sqrt{25}} + \dfrac{\sqrt{11}}{\sqrt{16}} = \dfrac{\sqrt{11}}{5} + \dfrac{\sqrt{11}}{4}$
$$= \dfrac{4\sqrt{11}}{20} + \dfrac{5\sqrt{11}}{20} = \dfrac{4\sqrt{11} + 5\sqrt{11}}{20} = \dfrac{9\sqrt{11}}{20}$$

**41.** $2\sqrt[3]{125} - 5\sqrt[3]{8} = 2(5) - 5(2)$
$$= 10 - 10$$
$$= 0$$

**42.** $3\sqrt[3]{16} - 2\sqrt[3]{2} = 3\sqrt[3]{8 \cdot 2} - 2\sqrt[3]{2}$
$$= 6\sqrt[3]{2} - 2\sqrt[3]{2}$$
$$= 4\sqrt[3]{2}$$

**43.** $3\sqrt{10} \cdot 2\sqrt{5} = 6\sqrt{50}$
$$= 6\sqrt{25}\sqrt{2}$$
$$= 6(5)\sqrt{2}$$
$$= 30\sqrt{2}$$

**44.** $2\sqrt[3]{4} \cdot 5\sqrt[3]{6} = 10\sqrt[3]{24} = 10\sqrt[3]{8 \cdot 3} = 20\sqrt[3]{3}$

**45.** $\sqrt{3}\left(2\sqrt{6} - 3\sqrt{12}\right) = 2\sqrt{18} - 3\sqrt{36}$
$$= 2\sqrt{9}\sqrt{2} - 3(6)$$
$$= 2(3)\sqrt{2} - 18$$

**46.** $4\sqrt{5}\left(2\sqrt{10} - 5\sqrt{5}\right) = 8\sqrt{50} - 20\sqrt{25}$
$$= 8\sqrt{25 \cdot 2} - 20(5)$$
$$= 40\sqrt{2} - 100$$

**47.** $\left(\sqrt{3} + 2\right)\left(\sqrt{6} - 5\right)$
$$= \sqrt{18} - 5\sqrt{3} + 2\sqrt{6} - 10$$
$$= \sqrt{9} \cdot \sqrt{2} - 5\sqrt{3} + 2\sqrt{6} - 10$$
$$= 3\sqrt{2} - 5\sqrt{3} + 2\sqrt{6} - 10$$

**48.** $\left(2\sqrt{5}+1\right)\left(4\sqrt{5}-3\right)$

$= 8\sqrt{25}-6\sqrt{5}+4\sqrt{5}-3$

$= 8\cdot 5 - 2\sqrt{5}-3$

$= 40 - 3 - 2\sqrt{5}$

$= 37 - 2\sqrt{5}$

**49.** $\dfrac{\sqrt{96}}{\sqrt{3}} = \sqrt{\dfrac{96}{3}} = \sqrt{32} = \sqrt{16}\cdot\sqrt{2} = 4\sqrt{2}$

**50.** $\dfrac{\sqrt{160}}{\sqrt{8}} = \sqrt{\dfrac{160}{8}} = \sqrt{20} = \sqrt{4}\cdot\sqrt{5} = 2\sqrt{5}$

**51.** $\dfrac{\sqrt{15x^6}}{\sqrt{12x^3}} = \sqrt{\dfrac{15x^6}{12x^3}}$

$= \sqrt{\dfrac{5x^3}{4}}$

$= \dfrac{\sqrt{5x^3}}{\sqrt{4}}$

$= \dfrac{\sqrt{x^2}\sqrt{5x}}{2}$

$= \dfrac{x\sqrt{5x}}{2}$

**52.** $\dfrac{\sqrt{50y^8}}{\sqrt{72y^3}} = \dfrac{\sqrt{25\cdot 2y^8}}{\sqrt{36\cdot 2\cdot y^2\cdot y}}$

$= \dfrac{5y^4\sqrt{2}}{6y\sqrt{2y}}$

$= \dfrac{5y^3}{6}\sqrt{\dfrac{2}{2y}}$

$= \dfrac{5y^3}{6\sqrt{y}}$

$= \dfrac{5y^3\sqrt{y}}{6y}$

$= \dfrac{5y^2\sqrt{y}}{6}$

**53.** $\sqrt{\dfrac{5}{6}} = \dfrac{\sqrt{5}}{\sqrt{6}} = \dfrac{\sqrt{5}}{\sqrt{6}}\cdot\dfrac{\sqrt{6}}{\sqrt{6}} = \dfrac{\sqrt{30}}{\sqrt{36}} = \dfrac{\sqrt{30}}{6}$

**54.** $\sqrt{\dfrac{7}{10}} = \dfrac{\sqrt{7}}{\sqrt{10}} = \dfrac{\sqrt{7}}{\sqrt{10}}\cdot\dfrac{\sqrt{10}}{\sqrt{10}} = \dfrac{\sqrt{70}}{\sqrt{100}} = \dfrac{\sqrt{70}}{10}$

**55.** $\sqrt{\dfrac{3}{2x}} = \dfrac{\sqrt{3}}{\sqrt{2x}}\cdot\dfrac{\sqrt{2x}}{\sqrt{2x}} = \dfrac{\sqrt{6x}}{2x}$

**56.** $\sqrt{\dfrac{6}{5y}} = \dfrac{\sqrt{6}}{\sqrt{5y}}\cdot\dfrac{\sqrt{5y}}{\sqrt{5y}} = \dfrac{\sqrt{30y}}{25y^2} = \dfrac{\sqrt{30y}}{5y}$

**57.** $\sqrt{\dfrac{7}{20y^2}} = \dfrac{\sqrt{7}}{\sqrt{20y^2}}$

$= \dfrac{\sqrt{7}}{\sqrt{4y^2}\sqrt{5}}$

$= \dfrac{\sqrt{7}}{2y\sqrt{5}}\cdot\dfrac{\sqrt{5}}{\sqrt{5}}$

$= \dfrac{\sqrt{35}}{2y(5)}$

$= \dfrac{\sqrt{35}}{10y}$

**58.** $\sqrt{\dfrac{5z}{12x^2}} = \dfrac{\sqrt{5z}}{\sqrt{12x^2}}\cdot\dfrac{\sqrt{3}}{\sqrt{3}} = \dfrac{\sqrt{15z}}{\sqrt{36x^2}} = \dfrac{\sqrt{15z}}{6x}$

**59.** $\sqrt[3]{\dfrac{7}{9}} = \dfrac{\sqrt[3]{7}}{\sqrt[3]{9}}\cdot\dfrac{\sqrt[3]{3}}{\sqrt[3]{3}} = \dfrac{\sqrt[3]{21}}{\sqrt[3]{27}} = \dfrac{\sqrt[3]{21}}{3}$

**60.** $\sqrt[3]{\dfrac{3}{4}} = \dfrac{\sqrt[3]{3}}{\sqrt[3]{4}} \cdot \dfrac{\sqrt[3]{2}}{\sqrt[3]{2}} = \dfrac{\sqrt[3]{6}}{\sqrt[3]{8}} = \dfrac{\sqrt[3]{6}}{2}$

**61.** $\sqrt[3]{\dfrac{3}{2}} = \dfrac{\sqrt[3]{3}}{\sqrt[3]{2}} \cdot \dfrac{\sqrt[3]{4}}{\sqrt[3]{4}} = \dfrac{\sqrt[3]{12}}{\sqrt[3]{8}} = \dfrac{\sqrt[3]{12}}{2}$

**62.** $\sqrt[3]{\dfrac{5}{4}} = \dfrac{\sqrt[3]{5}}{\sqrt[3]{4}} \cdot \dfrac{\sqrt[3]{2}}{\sqrt[3]{2}} = \dfrac{\sqrt[3]{10}}{\sqrt[3]{8}} = \dfrac{\sqrt[3]{10}}{2}$

**63.** $\dfrac{3}{\sqrt{5}-2} = \dfrac{3}{\sqrt{5}-2} \cdot \dfrac{\sqrt{5}+2}{\sqrt{5}+2} = \dfrac{3\left(\sqrt{5}+2\right)}{\left(\sqrt{5}\right)^2 - 2^2}$

$= \dfrac{3\left(\sqrt{5}+2\right)}{5-4} = \dfrac{3\left(\sqrt{5}+2\right)}{1} = 3\sqrt{5}+6$

**64.** $\dfrac{8}{\sqrt{10}-3} = \dfrac{8}{\sqrt{10}-3} \cdot \dfrac{\sqrt{10}+3}{\sqrt{10}+3} = \dfrac{8\left(\sqrt{10}+3\right)}{\left(\sqrt{10}\right)^2 - 3^2}$

$= \dfrac{8\left(\sqrt{10}+3\right)}{10-9} = \dfrac{8\left(\sqrt{10}+3\right)}{1} = 8\sqrt{10}+24$

**65.** $\dfrac{8}{\sqrt{6}+2} = \dfrac{8\left(\sqrt{6}-2\right)}{\left(\sqrt{6}+2\right)\left(\sqrt{6}-2\right)}$

$= \dfrac{8\left(\sqrt{6}-2\right)}{6-4}$

$= \dfrac{8\left(\sqrt{6}-2\right)}{2}$

$= 4\left(\sqrt{6}-2\right)$

$= 4\sqrt{6}-8$

**66.** $\dfrac{12}{\sqrt{15}-3} = \dfrac{12}{\sqrt{15}-3} \cdot \dfrac{\sqrt{15}+3}{\sqrt{15}+3}$

$= \dfrac{12\left(\sqrt{15}+3\right)}{\sqrt{225}-9}$

$= \dfrac{12\sqrt{15}+36}{15-9}$

$= \dfrac{12\sqrt{15}+36}{6}$

$= 2\sqrt{15}+6$

**67.** $\dfrac{\sqrt{2}}{4+\sqrt{2}} = \dfrac{\sqrt{2}\left(4-\sqrt{2}\right)}{\left(4+\sqrt{2}\right)\left(4-\sqrt{2}\right)}$

$= \dfrac{\sqrt{2}\left(4-\sqrt{2}\right)}{16-2}$

$= \dfrac{4\sqrt{2}-2}{14}$

$= \dfrac{2\left(2\sqrt{2}-1\right)}{14}$

$= \dfrac{2\sqrt{2}-1}{7}$

**68.** $\dfrac{\sqrt{3}}{5+\sqrt{3}} = \dfrac{\sqrt{3}}{5+\sqrt{3}} \cdot \dfrac{5-\sqrt{3}}{5-\sqrt{3}}$

$= \dfrac{\sqrt{3}\left(5-\sqrt{3}\right)}{25-\sqrt{9}}$

$= \dfrac{5\sqrt{3}-\sqrt{9}}{25-3}$

$= \dfrac{5\sqrt{3}-3}{22}$

**69.** $\dfrac{2\sqrt{3}}{\sqrt{3}-5} = \dfrac{2\sqrt{3}\left(\sqrt{3}+5\right)}{\left(\sqrt{3}-5\right)\left(\sqrt{3}+5\right)}$

$= \dfrac{2\sqrt{3}\left(\sqrt{3}+5\right)}{3-25}$

$= \dfrac{2(3)+10\sqrt{3}}{-22}$

$= \dfrac{6+10\sqrt{3}}{-22}$

$= \dfrac{2\left(3+5\sqrt{3}\right)}{-22}$

$= -\dfrac{3+5\sqrt{3}}{11}$

**70.** $\dfrac{7\sqrt{2}}{\sqrt{2}-4} = \dfrac{7\sqrt{2}}{\sqrt{2}-4} \cdot \dfrac{\sqrt{2}+4}{\sqrt{2}+4}$

$= \dfrac{7\sqrt{2}\left(\sqrt{2}+4\right)}{\sqrt{4}-16}$

$= \dfrac{7\sqrt{4}+28\sqrt{2}}{2-16}$

$= \dfrac{14+28\sqrt{2}}{-14}$

$= -1-2\sqrt{2}$

**71.** $\sqrt{2x} = 6$

$\left(\sqrt{2x}\right)^2 = 6^2$

$2x = 36$

$x = 18$

**72.** $\sqrt{x+3} = 4$

$\left(\sqrt{x+3}\right)^2 = 4^2$

$x+3 = 16$

$x = 13$

**73.** $\sqrt{x}+3 = 8$

$\sqrt{x} = 5$

$\left(\sqrt{x}\right)^2 = 5^2$

$x = 25$

**74.** $\sqrt{x}+8 = 3$

$\sqrt{x} = -5$

The square root cannot be negative, therefore there is no solution.

**75.** $\sqrt{2x+1} = x-7$

$\left(\sqrt{2x+1}\right)^2 = \left(x-7\right)^2$

$2x+1 = x^2 -14x+49$

$0 = x^2 -16x+48$

$0 = \left(x-12\right)\left(x-4\right)$

$x-12 = 0 \quad \text{or} \quad x-4 = 0$

$x = 12 \qquad\qquad x = 4$

$\qquad\qquad\qquad\qquad$ (extraneous)

**76.** $\sqrt{3x+1} = x-1$

$\left(\sqrt{3x+1}\right)^2 = \left(x-1\right)^2$

$3x+1 = x^2 -2x+1$

$0 = x^2 -5x$

$0 = x\left(x-5\right)$

$$x = 0 \quad \text{or} \quad x - 5 = 0$$
$$x = 0 \qquad x = 5$$
(extraneous)

**77.** $\sqrt{x + 3} + x = 9$

$$\sqrt{x + 3} = 9 - x$$
$$\left(\sqrt{x + 3}\right)^2 = (9 - x)^2$$
$$x + 3 = x^2 - 18x + 81$$
$$0 = x^2 - 19x + 78$$
$$0 = (x - 6)(x - 13)$$

$$x - 6 = 0 \quad \text{or} \quad x - 13 = 0$$
$$x = 6 \qquad x = 13$$
(extraneous)

**78.** $\sqrt{2x} + x = 4$

$$\sqrt{2x} = 4 - x$$
$$\left(\sqrt{2x}\right)^2 = (4 - x)^2$$
$$2x = x^2 - 8x + 16$$
$$0 = x^2 - 10x + 16$$
$$0 = (x - 2)(x - 8)$$

$$x - 2 = 0 \quad \text{or} \quad x - 8 = 0$$
$$x = 2 \qquad x = 8$$
(extraneous)

**79.** $a^2 + b^2 = c^2$

$$5^2 + b^2 = 9^2$$
$$25 + b^2 = 81$$
$$b^2 = 56$$
$$b = \sqrt{56}$$
$$b = 2\sqrt{14}$$
The length is $2\sqrt{14} \approx 7.48$.

**80.** $a^2 + b^2 = c^2$

$$6^2 + 9^2 = c^2$$
$$36 + 81 = c^2$$
$$117 = c^2$$
$$\sqrt{117} = c$$
The length is $\sqrt{117} \approx 10.82$.

**81.** $a^2 + b^2 = c^2$

$$20^2 + 12^2 = c^2$$
$$400 + 144 = c^2$$
$$544 = c^2$$
$$\sqrt{544} = c$$
$$4\sqrt{34} = c$$
They are $4\sqrt{34}$ feet apart.

**82.** $a^2 + b^2 = c^2$

$$a^2 + 5^2 = 10^2$$
$$a^2 + 25 = 100$$
$$a^2 = 75$$
$$a = \sqrt{75}$$
$$a = 5\sqrt{3}$$
The length is $5\sqrt{3}$ inches.

**83.** $(6, -2)$ and $(-3, 5)$

$$d = \sqrt{(x_2 - x_1)^2 (y_2 - y_1)^2}$$

$$d = \sqrt{(-3 - 6)^2 + [5 - (-2)]^2}$$

$$d = \sqrt{(-9)^2 + (5 + 2)^2}$$

$$d = \sqrt{81 + (7)^2}$$

$$d = \sqrt{81 + 49}$$

$$d = \sqrt{130}$$

**84.** $(2, 8)$ and $(-6, 10)$

$$\sqrt{(-6 - 2)^2 + (10 - 8)^2} = \sqrt{(-8)^2 + 2^2}$$
$$= \sqrt{64 + 4}$$
$$= \sqrt{68}$$
$$= \sqrt{4 \cdot 17}$$
$$= 2\sqrt{17}$$

**85.** $r = \sqrt{\dfrac{S}{4\pi}}$

$$r = \sqrt{\dfrac{72}{4\pi}} \approx 2.4$$

The radius is about 2.4 inches.

**86.** $r = \sqrt{\dfrac{S}{4\pi}}$

$$6 = \sqrt{\dfrac{S}{4\pi}}$$

$$6^2 = \left(\sqrt{\dfrac{S}{4\pi}}\right)^2$$

$$36 = \dfrac{S}{4\pi}$$

$$144\pi = S$$

The surface area is $144\pi$ square inches.

**87.** $\sqrt{a^5} = a^{5/2}$

**88.** $\sqrt[5]{a^3} = a^{3/5}$

**89.** $\sqrt[6]{x^{15}} = x^{15/6} = x^{5/2}$

**90.** $\sqrt[4]{x^{12}} = x^{12/4} = x^3$

**91.** $16^{1/2} = \sqrt{16} = 4$

**92.** $36^{1/2} = \sqrt{36} = 6$

**93.** $(-8)^{1/3} = \sqrt[3]{-8} = -2$

**94.** $(-32)^{1/5} = \sqrt[5]{-32} = -2$

**95.** $-64^{3/2} = -\left(64^{3/2}\right) = -\left(\sqrt{64}\right)^3 = -(8)^3 = -512$

**96.** $-8^{2/3} = -\sqrt[3]{8^2} = -\sqrt[3]{64} = -4$

**97.** $\left(\dfrac{16}{81}\right)^{3/4} = \dfrac{16^{3/4}}{81^{3/4}} = \dfrac{\left(\sqrt[4]{16}\right)^3}{\left(\sqrt[4]{81}\right)^3} = \dfrac{2^3}{3^3} = \dfrac{8}{27}$

**98.** $\left(\dfrac{9}{25}\right)^{3/2} = \left(\sqrt{\dfrac{9}{25}}\right)^3 = \left(\dfrac{3}{5}\right)^3 = \dfrac{27}{125}$

**99.** $25^{-1/2} = \dfrac{1}{25^{1/2}} = \dfrac{1}{\sqrt{25}} = \dfrac{1}{5}$

**100.** $64^{-2/3} = \dfrac{1}{64^{2/3}} = \dfrac{1}{\left(\sqrt[3]{64}\right)^2} = \dfrac{1}{(4y)^2} = \dfrac{1}{16}$

**101.** $8^{1/3} \cdot 8^{4/3} = 8^{5/3} = \left(\sqrt[3]{8}\right)^5 = 2^5 = 32$

**102.** $4^{3/2} \cdot 4^{1/2} = 4^{\frac{3}{2}+\frac{1}{2}} = 4^{4/2} = 4^2 = 16$

**103.** $\dfrac{3^{\frac{1}{6}}}{3^{\frac{5}{6}}} = 3^{\frac{1}{6}-\frac{5}{6}} = 3^{-4/6} = 3^{-2/3} = \dfrac{1}{3^{2/3}}$

**104.** $\dfrac{2^{1/4}}{2^{-3/5}} = 2^{\frac{1}{4}-\left(-\frac{3}{5}\right)} = 2^{\frac{1}{4}+\frac{3}{5}} = 2^{\frac{5+12}{20}} = 2^{17/20}$

**105.** $\left(x^{-1/3}\right)^6 = x^{-\frac{1}{3}\cdot 6} = x^{-2} = \dfrac{1}{x^2}$

**106.** $\left(\dfrac{x^{1/2}}{y^{1/3}}\right)^2 = \dfrac{x^{2/3}}{y^{2/3}} = \dfrac{x}{y^{2/3}}$

### Chapter 8 Test

**1.** $\sqrt{16} = 4$, because $4^2 = 16$ and 4 is positive.

**2.** $\sqrt[3]{125} = 5$, because $(5)^3 = 125$.

**3.** $16^{3/4} = \left(\sqrt[4]{16}\right)^3 = 2^3 = 8$

**4.** $\left(\dfrac{9}{16}\right)^{1/2} = \dfrac{9^{1/2}}{16^{1/2}} = \dfrac{\sqrt{9}}{\sqrt{16}} = \dfrac{3}{4}$

**5.** $\sqrt[4]{-81}$ is not a real number.

**6.** $27^{-2/3} = \dfrac{1}{27^{2/3}} = \dfrac{1}{\left(\sqrt[3]{27}\right)^2} = \dfrac{1}{3^2} = \dfrac{1}{9}$

**7.** $\sqrt{54} = \sqrt{9} \cdot \sqrt{6} = 3\sqrt{6}$

**8.** $\sqrt{92} = \sqrt{4} \cdot \sqrt{23} = 2\sqrt{23}$

**9.** $\sqrt{3x^6} = \sqrt{x^6}\sqrt{3} = x^3\sqrt{3}$

**10.** $\sqrt{8x^4y^7} = \sqrt{4x^4y^6}\sqrt{2y} = 2x^2y^3\sqrt{2y}$

**11.** $\sqrt{9x^9} = \sqrt{9x^8}\sqrt{x} = 3x^4\sqrt{x}$

**12.** $\sqrt[3]{8} = 2$

**13.** $\sqrt[3]{40} = \sqrt[3]{8}\sqrt[3]{5} = 2\sqrt[3]{5}$

**14.** $\sqrt{12} - 2\sqrt{75} = \sqrt{4}\sqrt{3} - 2\sqrt{25}\sqrt{3}$
$$= 2\sqrt{3} - 2(5)\sqrt{3}$$
$$= 2\sqrt{3} - 10\sqrt{3}$$
$$= -8\sqrt{3}$$

**15.** $\sqrt{2x^2} + \sqrt[3]{54} - x\sqrt{18}$
$$= \sqrt{x^2}\sqrt{2} + \sqrt[3]{27}\sqrt[3]{2} - x\sqrt{9}\sqrt{2}$$
$$= x\sqrt{2} + 3\sqrt[3]{2} - x(3)\sqrt{2}$$
$$= x\sqrt{2} + 3\sqrt[3]{2} - 3x\sqrt{2}$$
$$= -2x\sqrt{2} + 3\sqrt[3]{2}$$

**16.** $\sqrt{\dfrac{5}{16}} = \dfrac{\sqrt{5}}{\sqrt{16}} = \dfrac{\sqrt{5}}{4}$

**17.** $\sqrt[3]{\dfrac{2}{27}} = \dfrac{\sqrt[3]{2}}{\sqrt[3]{27}} = \dfrac{\sqrt[3]{2}}{3}$

**18.** $3\sqrt{8x} = 3\sqrt{4}\sqrt{2x} = 3(2)\sqrt{2x} = 6\sqrt{2x}$

**19.** $\sqrt{5}\left(\sqrt{5} + 2\sqrt{7}\right) = \sqrt{25} + 2\sqrt{35} = 5 + 2\sqrt{35}$

**20.** $\left(2\sqrt{x}+3\right)\left(2\sqrt{x}-3\right)=\left(2\sqrt{x}\right)^2-3^2=4x-9$

**21.** $\sqrt{\dfrac{2}{3}}=\dfrac{\sqrt{2}}{\sqrt{3}}=\dfrac{\sqrt{2}}{\sqrt{3}}\cdot\dfrac{\sqrt{3}}{\sqrt{3}}=\dfrac{\sqrt{6}}{\sqrt{9}}=\dfrac{\sqrt{6}}{3}$

**22.** $\sqrt[3]{\dfrac{5}{9}}=\dfrac{\sqrt[3]{5}}{\sqrt[3]{9}}=\dfrac{\sqrt[3]{5}}{\sqrt[3]{9}}\cdot\dfrac{\sqrt[3]{3}}{\sqrt[3]{3}}=\dfrac{\sqrt[3]{15}}{\sqrt[3]{27}}=\dfrac{\sqrt[3]{15}}{3}$

**23.**
$$\sqrt{\dfrac{5}{12x^2}}=\dfrac{\sqrt{5}}{\sqrt{12x^2}}$$
$$=\dfrac{\sqrt{5}}{\sqrt{4x^2\cdot3}}$$
$$=\dfrac{\sqrt{5}}{2x\sqrt{3}}$$
$$=\dfrac{\sqrt{5}}{2x\sqrt{3}}\cdot\dfrac{\sqrt{3}}{\sqrt{3}}$$
$$=\dfrac{\sqrt{15}}{2x(3)}$$
$$=\dfrac{\sqrt{15}}{6x}$$

**24.**
$$\dfrac{2\sqrt{3}}{\sqrt{3}-3}=\dfrac{2\sqrt{3}\left(\sqrt{3}+3\right)}{\left(\sqrt{3}-3\right)\left(\sqrt{3}+3\right)}$$
$$=\dfrac{2\sqrt{3}\left(\sqrt{3}+3\right)}{3-9}$$
$$=\dfrac{2(3)+6\sqrt{3}}{-6}$$
$$=\dfrac{6+6\sqrt{3}}{-6}$$
$$=\dfrac{6\left(1+\sqrt{3}\right)}{-6}$$
$$=-1\left(1+\sqrt{3}\right)$$
$$=-1-\sqrt{3}$$

**25.**
$$\sqrt{x}+8=11$$
$$\sqrt{x}=3$$
$$\left(\sqrt{x}\right)^2=3^2$$
$$x=9$$

**26.**
$$\sqrt{3x-6}=\sqrt{x+4}$$
$$\left(\sqrt{3x-6}\right)^2=\left(\sqrt{x+4}\right)^2$$
$$3x-6=x+4$$
$$2x=10$$
$$x=5$$

**27.**
$$\sqrt{2x-2}=x-5$$
$$\left(\sqrt{2x-2}\right)^2=\left(x-5\right)^2$$
$$2x-2=x^2-10x+25$$
$$0=x^2-12x+27$$
$$0=\left(x-9\right)\left(x-3\right)$$

$$x - 9 = 0 \quad \text{or} \quad x - 3 = 0$$
$$x = 9 \qquad\qquad x = 3$$
$$\text{(extraneous)}$$

**28.** $a^2 + b^2 = c^2$

$$8^2 + b^2 = 12^2$$
$$64 + b^2 = 144$$
$$b^2 = 80$$
$$b = \sqrt{80}$$
$$b = 4\sqrt{5}$$

The length is $4\sqrt{5}$ inches.

**29.** $(-3, 6)$ and $(-2, 8)$

$$d = \sqrt{(x_2 - x_1)^2 (y_2 - y_1)^2}$$
$$d = \sqrt{\left[-2 - (-3)\right]^2 + (8 - 6)^2}$$
$$d = \sqrt{(-2 + 3)^2 + (2)^2}$$
$$d = \sqrt{1^2 + 4}$$
$$d = \sqrt{1 + 4}$$
$$d = \sqrt{5}$$

**30.** $16^{-3/4} \cdot 16^{-1/4} = 16^{-4/4} = 16^{-1} = \dfrac{1}{16}$

**31.** $\left(\dfrac{x^{2/3}}{y^{2/5}}\right)^5 = \dfrac{x^{10/3}}{y^{10/5}} = \dfrac{x^{10/3}}{y^2}$

**Cumulative Review Chapter 8**

**1.** **a.** $\dfrac{(-12)(-3) + 3}{-7 - (-2)} = \dfrac{36 + 3}{-7 + 2} = \dfrac{39}{-5} = -\dfrac{39}{5}$

**b.** $\dfrac{2(-3)^2 - 20}{-5 + 4} = \dfrac{2(9) - 20}{-1} = -(18 - 20)$
$$= -(-2) = 2$$

**2.** **a.** $\dfrac{4(-3) - (-6)}{-8 + 4} = \dfrac{-12 + 6}{-4} = \dfrac{-6}{-4} = \dfrac{3}{2}$

**b.** $\dfrac{3 + (-3)(-2)^3}{-1 - (-4)} = \dfrac{3 + (-3)(-8)}{-1 + 4}$
$$= \dfrac{3 + 24}{3} = \dfrac{27}{3} = 9$$

**3.** $2x + 3x - 5 + 7 = 10x + 3 - 6x - 4$
$$5x + 2 = 4x - 1$$
$$x + 2 = -1$$
$$x = -3$$

**4.** $6y - 11 + 4 + 2y = 8 + 15y - 8y$
$$8y - 7 = 8 + 7y$$
$$y - 7 = 8$$
$$y = 15$$

**5.** $y = 3x$

$$x = -1: \ y = 3(-1) = -3$$
$$y = 0: \ 0 = 3x \Rightarrow x = 0$$
$$y = -9: \ -9 = 3x \Rightarrow x = -3$$

| $x$ | $y$ |
|-----|-----|
| $-1$ | $-3$ |
| $0$ | $0$ |
| $-3$ | $-9$ |

**6.** $2x + y = 6$

$x = 0: \; 2(0) + y = 6 \Rightarrow y = 6$

$y = -2: \; 2x + (-2) = 6 \Rightarrow 2x = 8 \Rightarrow x = 4$

$x = 3: \; 2(3) + y = 6 \Rightarrow 6 + y = 6 \Rightarrow y = 0$

| $x$ | $y$ |
|---|---|
| 0 | 6 |
| 4 | −2 |
| 3 | 0 |

**7.** $m = \dfrac{1}{4}, \; b = -3$

$y = mx + b$

$y = \dfrac{1}{4}x - 3$

**8.** $m = -2, \; b = 4$

$y = mx + b$

$y = -2x + 4$

$2x + y = 4$

**9.** $y = 5$ is horizontal so a parallel line is also horizontal. $y = c$

Point $(-2, -3)$

$y = -3$

**10.** $y = m_1 x + b_1$

$y = 2x + 4 \Rightarrow m_1 = 2$

Perpendicular line: $m_2 = -\dfrac{1}{m_1} = -\dfrac{1}{2}$

Point on line 2: $(1, 5)$

$y = mx + b$

$5 = -\dfrac{1}{2}(1) + b$

$\dfrac{10}{2} + \dfrac{1}{2} = b$

$\dfrac{11}{2} = b$

The equation is;

$y = -\dfrac{1}{2}x + \dfrac{11}{2}$

**11. a. b. c.**

**12. a. c. d.**

**13.** $\begin{cases} 2x - 3y = 6 \\ x = 2y \end{cases}$

**a.** $(12, 6)$

$\begin{array}{ll} 2x - 3y = 6 & x = 2y \\ 2(12) - 3(6) \overset{?}{=} 6 & 12 \overset{?}{=} 2(6) \\ 24 - 18 \overset{?}{=} 6 & 12 \overset{?}{=} 12 \\ 6 = 6 \;\; \text{True} & 12 = 12 \;\; \text{True} \end{array}$

$(12, 6)$ is a solution

**b.** $(0, -2)$

$\begin{array}{ll} 2x - 3y = 6 & x = 2y \\ 2(0) - 3(-2) \overset{?}{=} 6 & 0 \overset{?}{=} 2(-2) \\ 0 + 6 \overset{?}{=} 6 & 0 \overset{?}{=} -4 \\ 6 = 6 \;\; \text{True} & 0 = -4 \;\; \text{False} \end{array}$

$(0, -2)$ is not a solution

**14.** $\begin{cases} 2x + y = 4 \\ x + y = 2 \end{cases}$

   **a.** $(1, 1)$

$$2x + y = 4 \qquad\qquad x + y = 2$$
$$2(1) + (1) \overset{?}{=} 4 \qquad\qquad 1 + 1 \overset{?}{=} 2$$
$$2 + 1 \overset{?}{=} 4 \qquad\qquad 2 = 2$$
$$3 = 4 \ \text{ False} \qquad 2 = 2 \ \text{ True}$$

       $(1, 1)$ is not a solution

   **b.** $(2, 0)$

$$2x + y = 4 \qquad\qquad x + y = 2$$
$$2(2) + (0) \overset{?}{=} 4 \qquad\qquad 2 + 0 \overset{?}{=} 2$$
$$4 + 0 \overset{?}{=} 4 \qquad\qquad 2 \overset{?}{=} 2$$
$$4 = 4 \ \text{ True} \qquad 2 = 2 \ \text{ True}$$

       $(2, 0)$ is a solution

**15.** $\begin{cases} 2x + y = 10 \\ x = y + 2 \end{cases}$

Substitute $y + 2$ for $x$ in the first equation.
$$2(y + 2) + y = 10$$
$$2y + 4 + y = 10$$
$$3y = 6$$
$$y = 2$$
Let $y = 2$ in the second equation.
$$x = (2) + 2$$
$$x = 4$$
The solution of the system is $(4, 2)$.

**16.** $\begin{cases} 3y = x + 10 \\ x = 3y - 10 \\ \\ 2x + 5y = 24 \end{cases}$

Substitute $3y - 10$ for $x$ in the second equation.
$$2(3y - 10) + 5y = 24$$
$$6y - 20 + 5y = 24$$
$$11y = 44$$
$$y = 4$$
Let $y = 4$ in the first equation.
$$x = 3(4) - 10$$
$$x = 2$$
The solution of the system is $(2, 4)$.

**17.** $\begin{cases} -x - \dfrac{y}{2} = \dfrac{5}{2} \\ -\dfrac{x}{2} + \dfrac{y}{4} = 0 \end{cases}$

Multiply the first equation by 2 and the second equation by 4.
$$-2x - y = 5$$
$$\underline{-2x + y = 0}$$
$$-4x \quad\;\; = 5$$
$$x = -\dfrac{5}{4}$$

Let $x = -\dfrac{5}{4}$ in the second equation.
$$-\dfrac{-5/4}{2} + \dfrac{y}{4} = 0$$
$$\dfrac{5}{8} + \dfrac{y}{4} = 0$$

$$5 + 2y = 0$$
$$2y = -5$$
$$y = -\frac{5}{2}$$

The solution of the system is $\left(-\frac{5}{4}, -\frac{5}{2}\right)$.

**18.** $\begin{cases} \dfrac{x}{2} + y = \dfrac{5}{6} \\ 2x - y = \dfrac{5}{6} \end{cases}$

Multiply the both equations by 6.

$$3x + 6y = 5$$
$$\underline{12x - 6y = 5}$$
$$15x \phantom{aaa} = 10$$
$$x = \frac{2}{3}$$

Let $x = \dfrac{2}{3}$ in the second equation.

$$2\left(\frac{2}{3}\right) - y = \frac{5}{6}$$
$$\frac{4}{3} - y = \frac{5}{6}$$
$$8 - 6y = 5$$
$$-6y = -3$$
$$y = \frac{1}{2}$$

The solution of the system is $\left(\dfrac{2}{3}, \dfrac{1}{2}\right)$.

**19.** Let $x =$ the amount of 25% saline.

*No. of liters · Strength = Amt of Saline*

| 25% | $x$ | 0.25 | $0.25x$ |
|-----|-----|------|---------|
| 5% | $10 - x$ | 0.05 | $0.05(10 - x)$ |
| 20% | 20 | 0.2 | $0.2(10)$ |

$$0.25x + 0.05(10 - x) = 0.2(10)$$
$$0.25x + 0.5 - 0.05x = 2$$
$$0.2x + 0.5 = 2$$
$$0.2x = 1.5$$
$$x = 7.5$$
$$10 - x = 10 - 7.5 = 2.5$$

Mix 7.5 liters of 25% saline

with 2.5 liters of 5% saline.

**20.** Let $x =$ slower speed.

|  | $r$ | $\cdot$   $t$ | $=$   $d$ |
|--------|------|------|---------------|
| Slower | $x$ | 0.2 | $0.2x$ |
| Faster | $x + 15$ | 0.2 | $0.2(x + 15)$ |

$$0.2x + 0.2(x + 15) = 11$$
$$0.2x + 0.2x + 3 = 11$$
$$0.4x = 8$$
$$x = 20$$
$$x + 15 = 20 + 15 = 35$$

The slower streetcar traves at 20 mph.

The faster streetcar travels at 35 mph.

**21.** $\begin{cases} 3x \geq y \\ x + 2y \leq 8 \end{cases}$

$3x \geq y$        $x + 2y \leq 8$

Test $(0, 1)$     Test $(0, 0)$

$3(0) \overset{?}{\geq} 1$      $0 + 2(0) \overset{?}{\leq} 8$

False          True

Shade below    Shade below

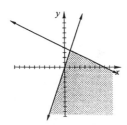

**22.** $\begin{cases} x + y \le 1 \\ 2x - y \ge 2 \end{cases}$

$x + y \le 1 \qquad 2x - y \ge 2$

Test $(0, 0)$    Test $(0, 0)$

$0 + 0 \overset{?}{\le} 1 \qquad 2(0) - 0 \overset{?}{\ge} 2$

True       False

Shade below   Shade below

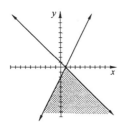

**23.** $-9x^2 + 3xy - 5y^2 + 7xy$

$= -9x^2 - 5y^2 + 10xy$

**24.** $4a^2 + 3a - 2a^2 + 7a - 5 = 2a^2 + 10a - 5$

**25.** $x^2 + 5xy + 6y^2 = (x + 2y)(x + 3y)$

**26.** $3x^2 + 15x + 18 = 3(x^2 + 5x + 6)$

$= 3(x + 2)(x + 3)$

**27.** $\dfrac{4 - x^2}{3x^2 - 5x - 2} = \dfrac{-(x^2 - 4)}{3x^2 - 5x - 2}$

$= \dfrac{-(x + 2)(x - 2)}{(3x + 1)(x - 2)}$

$= -\dfrac{x + 2}{3x + 1}$

**28.** $\dfrac{2x^2 + 7x + 3}{x^2 - 9} = \dfrac{(2x + 1)(x + 3)}{(x + 3)(x - 3)} = \dfrac{2x + 1}{x - 3}$

**29.** $\dfrac{3x^3 y^7}{40} \div \dfrac{4x^3}{y^2} = \dfrac{3x^3 y^7}{40} \cdot \dfrac{y^2}{4x^3} = \dfrac{3y^9}{160}$

**30.** $\dfrac{12x^2 y^3}{5} \div \dfrac{3y^3}{x} = \dfrac{12x^2 y^3}{5} \cdot \dfrac{x}{3y^3} = \dfrac{4x^3}{5}$

**31.** $\dfrac{2y}{2y - 7} - \dfrac{7}{2y - 7} = \dfrac{2y - 7}{2y - 7} = 1$

**32.** $\dfrac{-4x^2}{x + 1} - \dfrac{4x}{x + 1} = \dfrac{-4x^2 - 4x}{x1}$

$= -\dfrac{4x(x + 1)}{x + 1}$

$= -4x$

**33.** $\dfrac{2x}{x^2 + 2x + 1} + \dfrac{x}{x^2 - 1}$

$= \dfrac{2x}{(x + 1)(x + 1)} + \dfrac{x}{(x + 1)(x - 1)}$

$= \dfrac{2x}{(x + 1)(x + 1)} \cdot \dfrac{x - 1}{x - 1} + \dfrac{x}{(x + 1)(x - 1)} \cdot \dfrac{x + 1}{x + 1}$

$= \dfrac{2x(x - 1) + x(x + 1)}{(x + 1)(x + 1)(x - 1)}$

$$= \frac{2x^2 - 2x + x^2 + x}{(x+1)(x+1)(x-1)}$$

$$= \frac{3x^2 - x}{(x+1)(x+1)(x-1)}$$

$$= \frac{x(3x-1)}{(x+1)^2(x-1)}$$

**34.** $\dfrac{3x}{x^2+5x+6} + \dfrac{1}{x^2+2x-3}$

$$= \frac{3x}{(x+2)(x+3)} + \frac{1}{(x+3)(x-1)}$$

$$= \frac{3x}{(x+2)(x+3)} \cdot \frac{x-1}{x-1}$$

$$+ \frac{1}{(x+3)(x-1)} \cdot \frac{x+2}{x+2}$$

$$= \frac{3x(x-1) + 1(x+2)}{(x+2)(x+3)(x-1)}$$

$$= \frac{3x^2 - 3x + x + 2}{(x+2)(x+3)(x-1)}$$

$$= \frac{3x^2 - 2x + 2}{(x+2)(x+3)(x-1)}$$

$$= \frac{x(3x-1)}{(x+1)^2(x-1)}$$

**35.** $\dfrac{x}{2} + \dfrac{8}{3} = \dfrac{1}{6}$

$$6\left(\frac{x}{2} + \frac{8}{3}\right) = 6\left(\frac{1}{6}\right)$$

$$6\left(\frac{x}{2}\right) + 6\left(\frac{8}{3}\right) = 1$$

$$3x + 16 = 1$$

$$3x = -15$$

$$x = -5$$

**36.** $\dfrac{1}{21} + \dfrac{x}{7} = \dfrac{5}{3}$

$$21\left(\frac{1}{21} + \frac{x}{7}\right) = 21\left(\frac{5}{3}\right)$$

$$21\left(\frac{1}{21}\right) + 21\left(\frac{x}{7}\right) = 35$$

$$1 + 3x = 35$$

$$3x = 34$$

$$x = \frac{34}{3}$$

**37.** $\dfrac{x}{3} = \dfrac{10}{2}$

$$2x = 3(10)$$

$$2x = 30$$

$$x = 15 \text{ yards}$$

**38.** Let $x =$ the missing length

$$\frac{x}{5} = \frac{5}{2}$$

$$2x = 5(5)$$

$$2x = 25$$

$$x = \frac{25}{2}$$

**39.** $\dfrac{\dfrac{1}{z} - \dfrac{1}{2}}{\dfrac{1}{3} - \dfrac{z}{6}} = \dfrac{6z\left(\dfrac{1}{z} - \dfrac{1}{2}\right)}{6z\left(\dfrac{1}{3} - \dfrac{z}{6}\right)} = \dfrac{6z\left(\dfrac{1}{z}\right) - 6z\left(\dfrac{1}{2}\right)}{6z\left(\dfrac{1}{3}\right) - 6z\left(\dfrac{z}{6}\right)}$

$$= \frac{6 - 3z}{2z - z^2} = \frac{3(2-z)}{z(2-z)} = \frac{3}{z}$$

**40.** $\dfrac{x+3}{\dfrac{1}{x}+\dfrac{1}{3}} = \dfrac{3x(x+3)}{3x\left(\dfrac{1}{x}+\dfrac{1}{3}\right)} = \dfrac{3x(x+3)}{3x\left(\dfrac{1}{x}\right)+3x\left(\dfrac{1}{3}\right)}$

$= \dfrac{3x(x+3)}{3+x} = 3x$

**41. a.** $\sqrt{54} = \sqrt{9\cdot 6} = \sqrt{9}\cdot\sqrt{6} = 3\sqrt{6}$

    **b.** $\sqrt{12} = \sqrt{4\cdot 3} = \sqrt{4}\cdot\sqrt{3} = 2\sqrt{3}$

    **c.** $\sqrt{200} = \sqrt{100\cdot 2} = \sqrt{100}\cdot\sqrt{2} = 10\sqrt{2}$

    **d.** $\sqrt{35} = \sqrt{35}$

**42. a.** $\sqrt{40} = \sqrt{4\cdot 10} = \sqrt{4}\cdot\sqrt{10} = 2\sqrt{10}$

    **b.** $\sqrt{500} = \sqrt{100\cdot 5} = \sqrt{100}\cdot\sqrt{5} = 10\sqrt{5}$

    **c.** $\sqrt{63} = \sqrt{9\cdot 7} = \sqrt{9}\cdot\sqrt{7} = 3\sqrt{7}$

    **d.** $\sqrt{169} = 13$

**43. a.** $\left(\sqrt{5}-7\right)\left(\sqrt{5}+7\right) = \left(\sqrt{5}\right)^2 - 7^2$

$= 5 - 49$

$= -44$

    **b.** $\left(\sqrt{7x}+2\right)^2 = \left(\sqrt{7x}\right)^2 + 2\left(\sqrt{7x}\right)(2) + 2^2$

$= 7x + 4\sqrt{7x} + 4$

**44. a.** $\left(\sqrt{6}+2\right)^2 = \left(\sqrt{6}\right)^2 + 2\left(\sqrt{6}\right)(2) + 2^2$

$= 6 + 4\sqrt{6} + 4$

$= 10 + 4\sqrt{6}$

    **b.** $\left(\sqrt{x}+5\right)\left(\sqrt{x}-5\right) = \left(\sqrt{x}\right)^2 - 5^2$

$= x - 25$

**45.** $\sqrt{x}+6 = 4$

$\sqrt{x} = -2$

The square root of a real number cannot be negative. There is no solution.

**46.** $\sqrt{x+4} = \sqrt{3x-1}$

$\left(\sqrt{x+4}\right)^2 = \left(\sqrt{3x-1}\right)^2$

$x+4 = 3x-1$

$-2x+4 = -1$

$-2x = -5$

$x = \dfrac{5}{2}$

**47.** $a = 6,\ b = 8$

$c^2 = a^2 + b^2$

$c^2 = 6^2 + 8^2$

$c^2 = 36 + 64$

$c^2 = 100$

$\sqrt{c^2} = \sqrt{100}$

$c = 10$

The hypotenuse is 10 inches long

**48.** $c = 13,\ b = 9$

$a^2 + b^2 = c^2$

$a^2 + 9^2 = 13^2$

$a^2 + 81 = 169$

$a^2 = 88$

$\sqrt{c^2} = \sqrt{88}$

$c = \sqrt{4}\cdot\sqrt{22}$

$c = 2\sqrt{22}$

The other leg is $2\sqrt{22}$ inches long

**49. a.** $4^{3/2} = \left(\sqrt{4}\right)^3 = 2^3 = 8$

    **b.** $27^{2/3} = \left(\sqrt[3]{27}\right)^2 = 3^2 = 9$

    **c.** $-16^{3/4} = -\left(\sqrt[4]{16}\right)^3 = -2^3 = -8$

**50.**   **a**.  $9^{5/2} = \left(\sqrt{9}\right)^{5} = 3^{5} = 243$

     **b**.  $-81^{1/4} = -\left(\sqrt[4]{81}\right) = -3$

     **c**.  $(-64)^{2/3} = \left(\sqrt[3]{-64}\right)^{2} = (-4)^{2} = 16$

# Chapter 9

**Exercise Set 9.1**

**1.** $x^2 = 64$

$x = \sqrt{64} = 8$ or $x = -\sqrt{64} = -8$

The solutions are $\pm 8$.

**3.** $x^2 = 21$

$x = \sqrt{21}$ or $x = -\sqrt{21}$

The solutions are $\pm\sqrt{21}$.

**5.** $x^2 = \dfrac{1}{25}$

$x = \sqrt{\dfrac{1}{25}} = \dfrac{1}{5}$ or $x = -\sqrt{\dfrac{1}{25}} = -\dfrac{1}{5}$

The solutions are $\pm\dfrac{1}{5}$.

**7.** $x^2 = -4$

This equation has no real solution because $\sqrt{-4}$ is not a real number.

**9.** $3x^2 = 13$

$x^2 = \dfrac{13}{3}$

$x = \sqrt{\dfrac{13}{3}}$ or $x = -\sqrt{\dfrac{13}{3}}$

$x = \sqrt{\dfrac{13}{3}} \cdot \dfrac{\sqrt{3}}{\sqrt{3}}$ $x = -\sqrt{\dfrac{13}{3}} \cdot \dfrac{\sqrt{3}}{\sqrt{3}}$

$x = \dfrac{\sqrt{39}}{3}$ $x = -\dfrac{\sqrt{39}}{3}$

The solutions are $\pm\dfrac{\sqrt{39}}{3}$.

**11.** $7x^2 = 4$

$x^2 = \dfrac{4}{7}$

$x = \sqrt{\dfrac{4}{7}}$ or $x = -\sqrt{\dfrac{4}{7}}$

$x = \dfrac{2}{\sqrt{7}} \cdot \dfrac{\sqrt{7}}{\sqrt{7}}$ $x = -\dfrac{2}{\sqrt{7}} \cdot \dfrac{\sqrt{7}}{\sqrt{7}}$

$x = \dfrac{2\sqrt{7}}{7}$ $x = -\dfrac{2\sqrt{7}}{7}$

The solutions are $\pm\dfrac{2\sqrt{7}}{7}$.

**13.** $x^2 - 2 = 0$

$x^2 = 2$

$x = \sqrt{2}$ or $x = -\sqrt{2}$

The solutions are $\pm\sqrt{2}$.

**15.** $2x^2 - 10 = 0$

$2x^2 = 10$

$x^2 = 5$

$x = \pm\sqrt{5}$

The solutions are $\pm\sqrt{5}$.

**17.** Answers may vary.

**19.** $(x-5)^2 = 49$

$x - 5 = \sqrt{49}$ or $x - 5 = -\sqrt{49}$

$x - 5 = 7$ $x - 5 = -7$

$x = 5 + 7 = 12$ $x = 5 - 7 = -2$

The solutions are $-2$ and $12$.

366

**21.** $(x+2)^2 = 7$

$$x+2 = \sqrt{7} \quad \text{or} \quad x+2 = -\sqrt{7}$$
$$x = -2+\sqrt{7} \qquad x = -2-\sqrt{7}$$

The solutions are $-2 \pm \sqrt{7}$.

**23.** $\left(m-\dfrac{1}{2}\right)^2 = \dfrac{1}{4}$

$$m-\dfrac{1}{2} = \sqrt{\dfrac{1}{4}} \quad \text{or} \quad m-\dfrac{1}{2} = -\sqrt{\dfrac{1}{4}}$$
$$m-\dfrac{1}{2} = \dfrac{1}{2} \qquad m-\dfrac{1}{2} = -\dfrac{1}{2}$$
$$m = \dfrac{1}{2}+\dfrac{1}{2} = 1 \qquad m = \dfrac{1}{2}-\dfrac{1}{2} = 0$$

The solutions are 0 and 1.

**25.** $(p+2)^2 = 10$

$$p+2 = \sqrt{10} \quad \text{or} \quad p+2 = -\sqrt{10}$$
$$p = -2+\sqrt{10} \qquad p = -2-\sqrt{10}$$

The solutions are $-2 \pm \sqrt{10}$.

**27.** $(3y+2)^2 = 100$

$$3y+2 = \sqrt{100} \quad \text{or} \quad 3y+2 = -\sqrt{100}$$
$$3y+2 = 10 \qquad 3y+2 = -10$$
$$3y = -2+10 \qquad 3y = -2-10$$
$$y = \dfrac{-2+10}{3} \qquad y = \dfrac{-2-10}{3}$$
$$y = \dfrac{8}{3} \qquad y = -4$$

The solutions are $-4$ and $\dfrac{8}{3}$.

**29.** $(z-4)^2 = -9$

This equation has no real solution because $\sqrt{-9}$ is not a real number.

**31.** $(2x-11)^2 = 50$

$$2x-11 = \sqrt{50} \quad \text{or} \quad 2x-11 = -\sqrt{50}$$
$$2x-11 = 5\sqrt{2} \qquad 2x-11 = -5\sqrt{2}$$
$$2x = 11+5\sqrt{2} \qquad 2x = 11-5\sqrt{2}$$
$$x = \dfrac{11+5\sqrt{2}}{2} \qquad x = \dfrac{11-5\sqrt{2}}{2}$$

The solutions are $\dfrac{11 \pm 5\sqrt{2}}{2}$.

**33.** $(3x-7)^2 = 32$

$$3x-7 = \sqrt{32} \quad \text{or} \quad 3x-7 = -\sqrt{32}$$
$$3x-7 = 4\sqrt{2} \qquad 3x-7 = -4\sqrt{2}$$
$$3x = 7+4\sqrt{2} \qquad 3x = 7-4\sqrt{2}$$
$$x = \dfrac{7+4\sqrt{2}}{3} \qquad x = \dfrac{7-4\sqrt{2}}{3}$$

The solutions are $\dfrac{7 \pm 4\sqrt{2}}{3}$.

**35.** $(2p-5)^2 = 121$

$$2p-5 = \pm\sqrt{121}$$
$$2p-5 = 11 \quad \text{or} \quad 2p-5 = -11$$
$$2p = 16 \quad \text{or} \quad 2p = -6$$
$$p = 8 \quad \text{or} \quad p = -3$$

The solutions are 8 and $-3$.

**37.** Let $d = 400$

$$d = 16t^2$$

$$400 = 16t^2$$

$$\frac{400}{16} = t^2$$

$$25 = t^2$$

$$\sqrt{25} = t \quad \text{or} \quad -\sqrt{25} = t$$

$$5 = t \qquad\qquad -5 = t$$

The length of time is not a negative number so the fall lasted 5 seconds.

**39.** Let $h = 87.6$

$$h = 16t^2$$

$$87.6 = 16t^2$$

$$\frac{87.6}{16} = t^2$$

$$5.475 = t^2$$

$$\sqrt{5.475} = t \quad \text{or} \quad -\sqrt{5.475} = t$$

$$2.3 \approx t \qquad\qquad -2.3 \approx t$$

The length of the dive is not a negative number so the dive lasted approximately 2.3 seconds.

**41.** $16 \text{ mi} = 16 \text{ mi} \cdot \dfrac{5280 \text{ ft}}{1 \text{ mi}} = 84,480 \text{ ft}$

Let $h = 84,480$

$$h = 16t^2$$

$$84,480 = 16t^2$$

$$\frac{84,480}{16} = t^2$$

$$5280 = t^2$$

$$\sqrt{5280} = t \quad \text{or} \quad -\sqrt{5280} = t$$

$$72.7 \approx t \qquad\qquad -72.7 \approx t$$

The length of the fall is not a negative number so the fall lasted approximately 72.7 seconds.

**43.** Let $A = 20$

$$A = s^2$$

$$20 = s^2$$

$$\sqrt{20} = s \quad \text{or} \quad -\sqrt{20} = s$$

$$4.47 \approx s \qquad\qquad -4.47 \approx s$$

The length of a side is not a negative number so the length is approximately 4.47 inches.

**45.** Let $A = 20$

$$A = s^2$$

$$3039 = s^2$$

$$\sqrt{3039} = s \quad \text{or} \quad -\sqrt{3039} = s$$

$$55.13 \approx s \qquad\qquad -55.13 \approx s$$

The length of a side is not a negative number so the length is approximately 55.13 feet.

**47.** $x^2 + 6x + 9 = (x)^2 + 2(3)x + (3)^2$

$$= (x + 3)^2$$

**49.** $x^2 - 4x + 4 = (x)^2 - 2(2)x + (2)^2$

$$= (x - 2)^2$$

**51.** $x^2 + 4x + 4 = 16$

$$(x + 2)^2 = 16$$

$$x + 2 = \sqrt{16} \quad \text{or} \quad x + 2 = -\sqrt{16}$$

$$x + 2 = 4 \qquad\qquad x + 2 = -4$$

$$x = 2 \qquad\qquad x = -6$$

The solutions are $-6$ and $2$

**53.** $x^2 + 14x + 49 = 31$

$$(x+7)^2 = 31$$

$$x + 7 = \pm\sqrt{31}$$

$$x = -7 \pm \sqrt{31}$$

The solutions are $-7 \pm \sqrt{31}$

**55.** Let $a = x$, $b = \dfrac{3}{4}x$ and $c = 13$

$$a^2 + b^2 = c^2$$

$$x^2 + \left(\frac{3}{4}x\right)^2 = 13^2$$

$$x^2 + \frac{9}{16}x^2 = 13^2$$

$$\frac{25}{16}x^2 = 13^2$$

$$x^2 = \frac{16}{25} \cdot 13^2$$

$$x = \frac{4}{5} \cdot 13 = \frac{52}{5} = 10.4$$

$$\frac{3}{4}x = \frac{3}{4} \cdot 10.4 = 7.8$$

The sides measure 10.4 in. and 7.8 in.

**57.** $x^2 = 1.78$

$$x = \sqrt{1.78} \quad \text{or} \quad x = -\sqrt{1.78}$$

$$x = 1.33 \qquad x = -1.33$$

$$x = \pm 1.33$$

**59.** Let $y = 1451$

$$y = 2(x + 14.75)^2 + 415.875$$

$$1451 = 2(x + 14.75)^2 + 415.875$$

$$1035.125 = 2(x + 14.75)^2$$

$$517.5625 = (x + 14.75)^2$$

$$\pm\sqrt{517.5625} = x + 14.75$$

$$-14.75 \pm \sqrt{517.5625} = x$$

$$-14.75 \pm 22.75 = x$$

$$-14.75 + 22.75 = x \quad \text{or}$$

$$-14.75 - 22.75 = x$$

$$x = 8 \text{ or } x = -37.5$$

Since years cannot be negative, $x = 8$
The year will be $1998 + 8 = 2006$.

**Mental Math 9.2**

**1.** $p^2 + 8p$

$$\left(\frac{8}{2}\right)^2 = 4^2 = 16$$

**2.** $p^2 + 6p$

$$\left(\frac{6}{2}\right)^2 = 3^2 = 9$$

**3.** $x^2 + 20x$

$$\left(\frac{20}{2}\right)^2 = 10^2 = 100$$

**4.** $x^2 + 18x$

$$\left(\frac{18}{2}\right)^2 = 9^2 = 81$$

**5.** $y^2 + 14y$

$$\left(\frac{14}{2}\right)^2 = 7^2 = 49$$

**6.** $y^2 + 2y$

$$\left(\frac{2}{2}\right)^2 = 1^2 = 1$$

**Exercise Set 9.2**

**1.** $x^2 + 4x \Rightarrow \left(\frac{4}{2}\right)^2 = 2^2 = 4$

$x^2 + 4x + 4 = (x+2)^2$

**3.** $k^2 - 12k \Rightarrow \left(\frac{-12}{2}\right)^2 = 6^2 = 36$

$k^2 - 12k + 36 = (k-6)^2$

**5.** $x^2 - 3x \Rightarrow \left(\frac{-3}{2}\right)^2 = \frac{9}{4}$

$x^2 - 3x + \frac{9}{4} = \left(x - \frac{3}{2}\right)^2$

**7.** $m^2 - m \Rightarrow \left(\frac{-1}{2}\right)^2 = \frac{1}{4}$

$m^2 - m + \frac{1}{4}\left(m - \frac{1}{2}\right)^2$

**9.** $x^2 - 6x = 0$

$x^2 - 6x + 9 = 0 + 9$

$(x-3)^2 = 9$

$x - 3 = \pm\sqrt{9}$

$x = 3 \pm 3$

$x = 3 + 3$ or $x = 3 - 3$

$x = 6 \qquad x = 0$

The solutions are 0 and 6.

**11.** $x^2 + 8x = -12$

$x^2 + 8x + 16 = -12 + 16$

$(x+4)^2 = 4$

$x + 4 = \pm\sqrt{4}$

$x = -4 \pm 2$

$x = -4 + 2$ or $x = -4 - 2$

$x = -2 \qquad x = -6$

The solutions are $-6$ and $-2$.

**13.** $x^2 + 2x - 5 = 0$

$x^2 + 2x = 5$

$x^2 + 2x + 1 = 5 + 1$

$(x+1)^2 = 6$

$x + 1 = \pm\sqrt{6}$

$x = -1 \pm \sqrt{6}$

The solutions are $-1 \pm \sqrt{6}$.

**15.** $x^2 + 6x - 25 = 0$

$x^2 + 6x = 25$

$x^2 + 6x + 9 = 25 + 9$

$(x+3)^2 = 34$

$x + 3 = \pm\sqrt{34}$

$x = -3 \pm \sqrt{34}$

The solutions are $-3 \pm \sqrt{34}$.

**17.** $z^2 + 5z = 7$

$z^2 + 5z + \frac{25}{4} = 7 + \frac{25}{4}$

$\left(z + \frac{5}{2}\right)^2 = \frac{53}{4}$

$$z + \frac{5}{2} = \pm\sqrt{\frac{53}{4}}$$

$$z = -\frac{5}{2} \pm \frac{\sqrt{53}}{2}$$

$$z = \frac{-5 \pm \sqrt{53}}{2}$$

The solutions are $\dfrac{-5 \pm \sqrt{53}}{2}$.

**19.** 
$$x^2 - 2x - 1 = 0$$
$$x^2 - 2x = 1$$
$$x^2 - 2x + 1 = 1 + 1$$
$$(x-1)^2 = 2$$
$$x - 1 = \pm\sqrt{2}$$
$$x = 1 \pm \sqrt{2}$$

The solutions are $1 \pm \sqrt{2}$.

**21.** 
$$y^2 + 5y + 4 = 0$$
$$y^2 + 5y = -4$$
$$y^2 + 5y + \frac{25}{4} = -4 + \frac{25}{4}$$
$$\left(y + \frac{5}{2}\right)^2 = \frac{9}{4}$$
$$y + \frac{5}{2} = \pm\sqrt{\frac{9}{4}}$$
$$y = -\frac{5}{2} \pm \frac{3}{2}$$
$$y = -\frac{5}{2} + \frac{3}{2} \quad \text{or} \quad y = -\frac{5}{2} - \frac{3}{2}$$
$$y = -1 \qquad\qquad y = -4$$

The solutions are $-4$ and $-1$.

**23.** 
$$x(x+3) = 18$$
$$x^2 + 3x = 18$$
$$x^2 + 3x + \frac{9}{4} = 18 + \frac{9}{4}$$
$$\left(x + \frac{3}{2}\right)^2 = \frac{81}{4}$$
$$x + \frac{3}{2} = \pm\sqrt{\frac{81}{4}}$$
$$x = -\frac{3}{2} \pm \frac{9}{2}$$
$$x = -\frac{3}{2} + \frac{9}{2} \quad \text{or} \quad x = -\frac{3}{2} - \frac{9}{2}$$
$$x = 3 \qquad\qquad x = -6$$

The solutions are $-6$ and $3$.

**25.** 
$$4x^2 - 24x = 13$$
$$\frac{4x^2}{4} - \frac{24x}{4} = \frac{13}{4}$$
$$x^2 - 6x = \frac{13}{4}$$
$$x^2 - 6x + 9 = \frac{13}{4} + 9$$
$$(x-3)^2 = \frac{13}{4} + \frac{36}{4}$$
$$(x-3)^2 = \frac{49}{4}$$
$$x - 3 = \pm\sqrt{\frac{49}{4}}$$
$$x - 3 = \pm\frac{7}{2}$$
$$x = 3 \pm \frac{7}{2}$$

$$x = 3 + \frac{7}{2} \quad \text{or} \quad x = 3 - \frac{7}{2}$$

$$x = \frac{6}{2} + \frac{7}{2} \quad \text{or} \quad x = \frac{6}{2} - \frac{7}{2}$$

$$x = \frac{13}{2} \quad \text{or} \quad x = -\frac{1}{2}$$

The solutions are $\dfrac{13}{2}$ and $-\dfrac{1}{2}$.

**27.** $5x^2 + 10x + 6 = 0$

$$5x^2 + 10x = -6$$

$$x^2 + 2x = -\frac{6}{5}$$

$$x^2 + 2x + 1 = -\frac{6}{5} + 1$$

$$(x+1)^2 = -\frac{1}{5}$$

This equation has no real solution

because $\sqrt{-\dfrac{1}{5}}$ is not a real number.

**29.** $\qquad 2x^2 = 6x + 5$

$$2x^2 - 6x = 5$$

$$x^2 - 3x = \frac{5}{2}$$

$$x^2 - 3x + \frac{9}{4} = \frac{5}{2} + \frac{9}{4}$$

$$\left(x - \frac{3}{2}\right)^2 = \frac{19}{4}$$

$$x - \frac{3}{2} = \pm\sqrt{\frac{19}{4}}$$

$$x = \frac{3}{2} \pm \frac{\sqrt{19}}{2}$$

The solutions are $\dfrac{3 \pm \sqrt{19}}{2}$ .

**31.** $\qquad 3x^2 - 6x = 24$

$$x^2 - 2x = 8$$

$$x^2 - 2x + 1 = 8 + 1$$

$$(x-1)^2 = 9$$

$$x - 1 = \pm\sqrt{9}$$

$$x = 1 \pm 3$$

$$x = 1 + 3 \quad \text{or} \quad x = 1 - 3$$

$$x = 4 \qquad\qquad x = -2$$

The solutions are $-2$ and $4$.

**33.** $2y^2 + 8y + 5 = 0$

$$2y^2 + 8y = -5$$

$$y^2 + 4y = -\frac{5}{2}$$

$$y^2 + 4y + 4 = -\frac{5}{2} + 4$$

$$(y+2)^2 = \frac{3}{2}$$

$$y + 2 = \pm\sqrt{\frac{3}{2}}$$

$$y = -2 \pm \sqrt{\frac{3}{2}}$$

$$y = -2 \pm \sqrt{\frac{3}{2}} \cdot \sqrt{\frac{2}{2}}$$

$$y = -2 \pm \frac{\sqrt{6}}{2}$$

The solutions are $\dfrac{-4 \pm 2\sqrt{6}}{2}$.

**35.** $2y^2 - 3y + 1 = 0$

$$2y^2 - 3y = -1$$

$$y^2 - \frac{3}{2}y = -\frac{1}{2}$$

$$y^2 - \frac{3}{2}y + \frac{9}{16} = -\frac{1}{2} + \frac{9}{16}$$

$$\left(y - \frac{3}{4}\right)^2 = \frac{1}{16}$$

$$y - \frac{3}{4} = \pm\sqrt{\frac{1}{16}}$$

$$y = \frac{3}{4} \pm \frac{1}{4}$$

$$y = \frac{3}{4} + \frac{1}{4} \quad \text{or} \quad y = \frac{3}{4} - \frac{1}{4}$$

$$y = 1 \qquad\qquad y = \frac{1}{2}$$

The solutions are $\frac{1}{2}$ and 1.

**37.** $3y^2 - 2y - 4 = 0$

$$3y^2 - 2y = 4$$

$$\frac{3y^2}{3} - \frac{2y}{3} = \frac{4}{3}$$

$$y^2 - \frac{2}{3}y = \frac{4}{3}$$

$$y^2 - \frac{2}{3}y + \frac{1}{9} = \frac{4}{3} + \frac{1}{9}$$

$$\left(y - \frac{1}{3}\right)^2 = \frac{12}{9} + \frac{1}{9}$$

$$\left(y - \frac{1}{3}\right)^2 = \frac{13}{9}$$

$$y - \frac{1}{3} = \pm\sqrt{\frac{13}{9}}$$

$$y - \frac{1}{3} = \pm\frac{\sqrt{13}}{3}$$

$$y = \frac{1}{3} \pm \frac{\sqrt{13}}{3}$$

$$y = \frac{1 \pm \sqrt{13}}{3}$$

The solutions are $\dfrac{1 \pm \sqrt{13}}{3}$.

**39.** Answers may vary.

**41.** $\dfrac{3}{4} - \sqrt{\dfrac{25}{16}} = \dfrac{3}{4} - \dfrac{5}{4} = -\dfrac{2}{4} = -\dfrac{1}{2}$

**43.** $\dfrac{1}{2} - \sqrt{\dfrac{9}{4}} = \dfrac{1}{2} - \dfrac{3}{2} = -\dfrac{2}{2} = -1$

**45.** $\dfrac{6 + 4\sqrt{5}}{2} = \dfrac{2\left(3 + 2\sqrt{5}\right)}{2} = 3 + 2\sqrt{5}$

**47.** $\dfrac{3 - 9\sqrt{2}}{6} = \dfrac{3\left(1 - 3\sqrt{2}\right)}{3 \cdot 2} = \dfrac{1 - 3\sqrt{2}}{2}$

**49.** $x^2 + kx + 16$

$$\left(\frac{k}{2}\right)^2 = 16$$

$$\frac{k^2}{4} = 16$$

$$k^2 = 64$$

$$k = \pm\sqrt{64}$$

$$k = \pm 8$$

**51.** Let $y = 47,390$

$$y = 268x^2 + 720x + 13,390$$

$$47,390 = 268x^2 + 720x + 13,390$$

$$0 = 268x^2 + 720x - 34,000$$

$$0 = 67x^2 + 180x - 8500$$

$$0 = (67x + 850)(x - 10)$$

$$67x + 850 = 0 \quad \text{or} \quad x - 10 = 0$$

$$x = -\frac{850}{67} \qquad x = 10$$

The number of years cannot be negative so the year will be $1998 + 10 = 2008$.

**53.** $x^2 + 8x = -12$

$$y_1 = x^2 + 8x$$

$$y_2 = -12$$

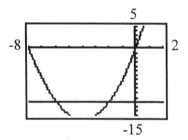

The $x$-coordinates of the intersections, $-6$ and $-2$, are the solutions.

**55.** $2x^2 = 6x + 5$

$$y_1 = 2x^2$$

$$y_2 = 6x + 5$$

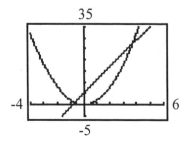

The $x$-coordinates of the intersections, $-0.68$ and $3.68$, are the solutions.

**Mental Math 9.3**

**1.** $2x^2 + 5x + 3 = 0$

$a = 2, b = 5, c = 3$

**2.** $5x^2 - 7x + 1 = 0$

$a = 5, b = -7, c = 1$

**3.** $10x^2 - 13x - 2 = 0$

$a = 10, b = -13, c = -2$

**4.** $x^2 + 3x - 7 = 0$

$a = 1, b = 3, c = -7$

**5.** $x^2 - 6 = 0$

$a = 1, b = 0, c = -6$

**6.** $9x^2 - 4 = 0$

$a = 9, b = 0, c = -4$

**Exercise Set 9.3**

**1.** $\dfrac{-1 \pm \sqrt{1^2 - 4(1)(-2)}}{2(1)} = \dfrac{-1 \pm \sqrt{1 + 8}}{2}$

$$= \dfrac{-1 \pm \sqrt{9}}{2}$$

$$= \dfrac{-1 \pm 3}{2}$$

$$= 1, -2$$

**3.** $\dfrac{-5 \pm \sqrt{5^2 - 4(1)(2)}}{2(1)} = \dfrac{-5 \pm \sqrt{25 - 8}}{2}$

$\qquad\qquad\qquad = \dfrac{-5 \pm \sqrt{17}}{2}$

**5.** $\dfrac{-(-4) \pm \sqrt{(-4)^2 - 4(2)(1)}}{2(2)} = \dfrac{4 \pm \sqrt{16 - 8}}{4}$

$\qquad\qquad\qquad\qquad = \dfrac{4 \pm \sqrt{8}}{4}$

$\qquad\qquad\qquad\qquad = \dfrac{4 \pm \sqrt{4}\sqrt{2}}{4}$

$\qquad\qquad\qquad\qquad = \dfrac{4 \pm 2\sqrt{2}}{4}$

$\qquad\qquad\qquad\qquad = \dfrac{2\left(2 \pm \sqrt{2}\right)}{4}$

$\qquad\qquad\qquad\qquad = \dfrac{2 \pm \sqrt{2}}{2}$

**7.** $x^2 - 3x + 2 = 0$

$\qquad a = 1, b = -3, \text{ and } c = 2$

$\qquad x = \dfrac{-b \pm \sqrt{b^2 - 4ac}}{2a}$

$\qquad x = \dfrac{-(-3) \pm \sqrt{(-3)^2 - 4(1)(2)}}{2(1)}$

$\qquad\quad = \dfrac{3 \pm \sqrt{9 - 8}}{2} = \dfrac{3 \pm \sqrt{1}}{2} = \dfrac{3 \pm 1}{2}$

$\qquad x = \dfrac{3 + 1}{2} = 2 \ \text{ or } \ x = \dfrac{3 - 1}{2} = 1$

The solutions are 1 and 2.

**9.** $3k^2 + 7k + 1 = 0$

$\qquad a = 3, b = 7, \text{ and } c = 1$

$\qquad k = \dfrac{-b \pm \sqrt{b^2 - 4ac}}{2a}$

$\qquad k = \dfrac{-(7) \pm \sqrt{(7)^2 - 4(3)(1)}}{2(3)}$

$\qquad\quad = \dfrac{-7 \pm \sqrt{49 - 12}}{6} = \dfrac{-7 \pm \sqrt{37}}{6}$

The solutions are $\dfrac{-7 \pm \sqrt{37}}{6}$.

**11.** $49x^2 - 4 = 0$

$\qquad a = 49, b = 0, \text{ and } c = -4$

$\qquad x = \dfrac{-b \pm \sqrt{b^2 - 4ac}}{2a}$

$\qquad x = \dfrac{-(0) \pm \sqrt{(0)^2 - 4(49)(-4)}}{2(49)}$

$\qquad\quad = \dfrac{\pm\sqrt{784}}{98} = \dfrac{\pm 28}{98} = \pm\dfrac{2}{7}$

The solutions are $\pm\dfrac{2}{7}$.

**13.** $5z^2 - 4z + 3 = 0$

$\qquad a = 5, b = -4, \text{ and } c = 3$

$\qquad z = \dfrac{-(-4) \pm \sqrt{(-4)^2 - 4(5)(3)}}{2(5)}$

$\qquad\quad = \dfrac{4 \pm \sqrt{16 - 60}}{10} = \dfrac{4 \pm \sqrt{-44}}{10}$

There is no real solution because $\sqrt{-44}$ is not a real number.

**15.** $y^2 = 7y + 30$

$y^2 - 7y - 30 = 0$

$a = 1, b = -7,$ and $c = -30$

$y = \dfrac{-(-7) \pm \sqrt{(-7)^2 - 4(1)(-30)}}{2(1)}$

$= \dfrac{7 \pm \sqrt{49 + 120}}{2} = \dfrac{7 \pm \sqrt{169}}{2}$

$= \dfrac{7 \pm 13}{2}$

$y = \dfrac{7 + 13}{2} = 10$   or   $y = \dfrac{7 - 13}{2} = -3$

The solutions are $-3$ and $10$.

**17.** $2x^2 = 10$

$2x^2 - 10 = 0$

$a = 2, b = 0,$ and $c = -10$

$x = \dfrac{-(0) \pm \sqrt{(0)^2 - 4(2)(-10)}}{2(2)}$

$= \dfrac{\pm\sqrt{80}}{4} = \dfrac{\pm 4\sqrt{5}}{4} = \pm\sqrt{5}$

The solutions are $\pm\sqrt{5}$.

**19.** $m^2 - 12 = m$

$m^2 - m - 12 = 0$

$a = 1, b = -1,$ and $c = -12$

$m = \dfrac{-(-1) \pm \sqrt{(-1)^2 - 4(1)(-12)}}{2(1)}$

$= \dfrac{1 \pm \sqrt{1 + 48}}{2} = \dfrac{1 \pm \sqrt{49}}{2} = \dfrac{1 \pm 7}{2}$

$m = \dfrac{1 + 7}{2} = 4$   or   $m = \dfrac{1 - 7}{2} = -3$

The solutions are $-3$ and $4$.

**21.** $3 - x^2 = 4x$

$-x^2 - 4x + 3 = 0$

$a = -1, b = -4,$ and $c = 3$

$x = \dfrac{-(-4) \pm \sqrt{(-4)^2 - 4(-1)(3)}}{2(-1)}$

$= \dfrac{4 \pm \sqrt{16 + 12}}{-2} = \dfrac{4 \pm \sqrt{28}}{-2}$

$= \dfrac{4 \pm 2\sqrt{7}}{-2} = -2 \pm \sqrt{7}$

The solutions are $-2 \pm \sqrt{7}$.

**23.** $2a^2 - 7a + 3 = 0$

$a = 2, b = -7, c = 3$

$a = \dfrac{7 \pm \sqrt{49 - 24}}{4}$

$a = \dfrac{7 \pm \sqrt{25}}{4}$

$a = \dfrac{7 \pm 5}{4}$

$a = \dfrac{7 + 5}{4}$   or   $a = \dfrac{7 - 5}{4}$

$a = \dfrac{12}{4} = 3$   or   $a = \dfrac{2}{4} = \dfrac{1}{2}$

The solutions are $3$ and $\dfrac{1}{2}$.

**25.** $x^2 - 5x - 2 = 0$

$a = 1, b = -5, c = -2$

$x = \dfrac{-(-5) \pm \sqrt{(--5)^2 - 4(1)(-2)}}{2(1)}$

$x = \dfrac{5 \pm \sqrt{25 + 8}}{2}$

$$x = \frac{5 \pm \sqrt{33}}{2}$$

The solutions are $\dfrac{5 \pm \sqrt{33}}{2}$.

**27.** $3x^2 - x - 14 = 0$

$a = 3,\ b = -1,\ c = -14$

$$x = \frac{-(-1) \pm \sqrt{(-1)^2 - 4(3)(-14)}}{2(3)}$$

$$x = \frac{1 \pm \sqrt{1 + 168}}{6}$$

$$x = \frac{1 \pm \sqrt{169}}{6}$$

$$x = \frac{1 \pm 13}{6}$$

$$x = \frac{1 + 13}{6} \quad \text{or} \quad x = \frac{1 - 13}{6}$$

$$x = \frac{14}{6} = \frac{7}{3} \quad \text{or} \quad x = \frac{-12}{6} = 2$$

The solutions are $\dfrac{7}{3}$ and $-2$.

**29.** $\qquad 6x^2 + 9x = 2$

$6x^2 + 9x - 2 = 0$

$a = 6,\ b = 9,\ \text{and } c = -2$

$$x = \frac{-(9) \pm \sqrt{(9)^2 - 4(6)(-2)}}{2(6)}$$

$$= \frac{-9 \pm \sqrt{81 + 48}}{12} = \frac{-9 \pm \sqrt{129}}{12}$$

The solutions are $\dfrac{-9 \pm \sqrt{129}}{12}$.

**31.** $\qquad 7p^2 + 2 = 8p$

$7p^2 - 8p + 2 = 0$

$a = 7,\ b = -8,\ \text{and } c = 2$

$$p = \frac{-(-8) \pm \sqrt{(-8)^2 - 4(7)(2)}}{2(7)}$$

$$= \frac{8 \pm \sqrt{64 - 56}}{14} = \frac{8 \pm \sqrt{8}}{14}$$

$$= \frac{8 \pm 2\sqrt{2}}{14} = \frac{4 \pm \sqrt{2}}{7}$$

The solutions are $\dfrac{4 \pm \sqrt{2}}{7}$.

**33.** $a^2 - 6a + 2 = 0$

$a = 1,\ b = -6,\ \text{and } c = 2$

$$a = \frac{-(-6) \pm \sqrt{(-6)^2 - 4(1)(2)}}{2(1)}$$

$$= \frac{6 \pm \sqrt{36 - 8}}{2} = \frac{6 \pm \sqrt{28}}{2}$$

$$= \frac{6 \pm 2\sqrt{7}}{2} = 3 \pm \sqrt{7}$$

The solutions are $3 \pm \sqrt{7}$.

**35.** $2x^2 - 6x + 3 = 0$

$a = 2,\ b = -6,\ \text{and } c = 3$

$$x = \frac{-(-6) \pm \sqrt{(-6)^2 - 4(2)(3)}}{2(2)}$$

$$= \frac{6 \pm \sqrt{36 - 24}}{4} = \frac{6 \pm \sqrt{12}}{4}$$

$$= \frac{6 \pm 2\sqrt{3}}{4} = \frac{3 \pm \sqrt{3}}{2}$$

The solutions are $\dfrac{3 \pm \sqrt{3}}{2}$.

**37.**
$$3x^2 = 1 - 2x$$
$$3x^2 + 2x - 1 = 0$$
$$a = 3, b = 2, \text{ and } c = -1$$
$$x = \frac{-(2) \pm \sqrt{(2)^2 - 4(3)(-1)}}{2(3)}$$
$$= \frac{-2 \pm \sqrt{4+12}}{6} = \frac{-2 \pm \sqrt{16}}{6} = \frac{-2 \pm 4}{6}$$
$$x = \frac{-2+4}{6} = \frac{1}{3} \quad \text{or} \quad x = \frac{-2-4}{6} = -1$$

The solutions are $-1$ and $\dfrac{1}{3}$.

**39.**
$$20y^2 = 3 - 11y$$
$$20y^2 + 11y - 3 = 0$$
$$a = 20, b = 11, \text{ and } c = -3$$
$$y = \frac{-(11) \pm \sqrt{(11)^2 - 4(20)(-3)}}{2(20)}$$
$$= \frac{-11 \pm \sqrt{121 + 240}}{40} = \frac{-11 \pm \sqrt{361}}{40}$$
$$= \frac{-11 \pm 19}{40}$$
$$y = \frac{-11+19}{40} = \frac{1}{5} \quad \text{or} \quad y = \frac{-11-19}{40} = -\frac{3}{4}$$

The solutions are $-\dfrac{3}{4}$ and $\dfrac{1}{5}$.

**41.** $x^2 + x + 1 = 0$
$$a = 1, b = 1, \text{ and } c = 1$$
$$x = \frac{-(1) \pm \sqrt{(1)^2 - 4(1)(1)}}{2(1)}$$
$$= \frac{-1 \pm \sqrt{1-4}}{2} = \frac{-1 \pm \sqrt{-3}}{2}$$

There is no real solution because $\sqrt{-3}$ is not a real number.

**43.**
$$4y^2 = 6y + 1$$
$$4y^2 - 6y - 1 = 0$$
$$a = 4, b = -6, \text{ and } c = -1$$
$$y = \frac{-(-6) \pm \sqrt{(-6)^2 - 4(4)(-1)}}{2(4)}$$
$$= \frac{6 \pm \sqrt{36 + 16}}{8} = \frac{6 \pm \sqrt{52}}{8}$$
$$= \frac{6 \pm 2\sqrt{13}}{8} = \frac{3 \pm \sqrt{13}}{4}$$

The solutions are $\dfrac{3 \pm \sqrt{13}}{4}$.

**45.** $3p^2 - \dfrac{2}{3}p + 1 = 0$
$$9p^2 - 2p + 3 = 0$$
$$a = 9, b = -2, \text{ and } c = 3$$
$$p = \frac{-(-2) \pm \sqrt{(-2)^2 - 4(9)(3)}}{2(9)}$$
$$= \frac{2 \pm \sqrt{4 - 108}}{18} = \frac{2 \pm \sqrt{-104}}{18}$$

There is no real solution because $\sqrt{-104}$ is not a real number.

**47.**
$$\frac{m^2}{2} = m + \frac{1}{2}$$
$$m^2 = 2m + 1$$
$$m^2 - 2m - 1 = 0$$
$$a = 1, b = -2, \text{ and } c = -1$$
$$m = \frac{-(-2) \pm \sqrt{(-2)^2 - 4(1)(-1)}}{2(1)}$$

$$= \frac{2 \pm \sqrt{4+4}}{2} = \frac{2 \pm \sqrt{8}}{2}$$

$$= \frac{2 \pm 2\sqrt{2}}{2} = 1 \pm \sqrt{2}$$

The solutions are $1 \pm \sqrt{2}$.

**49.**
$$4p^2 + \frac{3}{2} = -5p$$

$$8p^2 + 3 = -10p$$

$$8p^2 + 10p + 3 = 0$$

$$a = 8, b = 10, \text{ and } c = 3$$

$$p = \frac{-(10) \pm \sqrt{(10)^2 - 4(8)(3)}}{2(8)}$$

$$= \frac{-10 \pm \sqrt{100 - 96}}{16} = \frac{-10 \pm \sqrt{4}}{16}$$

$$= \frac{-10 \pm 2}{16}$$

$$p = \frac{-10 + 2}{16} = -\frac{1}{2} \quad \text{or} \quad p = \frac{-10 - 2}{16} = -\frac{3}{4}$$

The solutions are $-\frac{3}{4}$ and $-\frac{1}{2}$.

**51.**
$$5x^2 = \frac{7}{2}x + 1$$

$$10x^2 = 7x + 2$$

$$10x^2 - 7x - 2 = 0$$

$$a = 10, b = -7, \text{ and } c = -2$$

$$x = \frac{-(-7) \pm \sqrt{(-7)^2 - 4(10)(-2)}}{2(10)}$$

$$= \frac{7 \pm \sqrt{49 + 80}}{20} = \frac{7 \pm \sqrt{129}}{20}$$

The solutions are $\frac{7 \pm \sqrt{129}}{20}$.

**53.**
$$28x^2 + 5x + \frac{11}{4} = 0$$

$$112x^2 + 20x + 11 = 0$$

$$a = 112, b = 20, \text{ and } c = 11$$

$$p = \frac{-(20) \pm \sqrt{(20)^2 - 4(112)(11)}}{2(112)}$$

$$= \frac{-20 \pm \sqrt{400 - 4928}}{224} = \frac{-20 \pm \sqrt{-4528}}{224}$$

There is no real solution because $\sqrt{-4528}$ is not a real number.

**55.**
$$5z^2 - 2z = \frac{1}{5}$$

$$25z^2 - 10z = 1$$

$$25z^2 - 10z - 1 = 0$$

$$a = 25, b = -10, \text{ and } c = -1$$

$$x = \frac{-(-10) \pm \sqrt{(-10)^2 - 4(25)(-1)}}{2(25)}$$

$$= \frac{10 \pm \sqrt{100 + 100}}{50} = \frac{10 \pm \sqrt{200}}{50}$$

$$= \frac{10 \pm 10\sqrt{2}}{50} = \frac{1 \pm \sqrt{2}}{5}$$

The solutions are $\frac{1 \pm \sqrt{2}}{5}$.

**57.** $x^2 + 3\sqrt{2}\,x - 5 = 0$

$$a = 1, b = 3\sqrt{2}, \text{ and } c = -5$$

$$x = \frac{-(3\sqrt{2}) \pm \sqrt{(3\sqrt{2})^2 - 4(1)(-5)}}{2(1)}$$

$$x^2 + 3\sqrt{2}\,x - 5 = 0$$

$$a = 1, b = 3\sqrt{2}, \text{ and } c = -5$$

$$x = \frac{-\left(3\sqrt{2}\right) \pm \sqrt{\left(3\sqrt{2}\right)^2 - 4(1)(-5)}}{2(1)}$$

$$= \frac{-3\sqrt{2} \pm \sqrt{18 + 20}}{2} = \frac{-3\sqrt{2} \pm \sqrt{38}}{2}$$

The solutions are $\dfrac{-3\sqrt{2} \pm \sqrt{38}}{2}$.

**59.** $x^2 + 3x - 1 = 0$

$$a = 1, b = 3, c = -1$$

$$b^2 - 4ac = 3^2 - 4(1)(-1)$$

$$= 9 + 4$$

$$= 13$$

Since the discriminant is a positive number, this equation has two distinct real solutions.

**61.** $3x^2 + x + 5 = 0$

$$a = 3, b = 1, c = 5$$

$$b^2 - 4ac = 1^2 - 4(3)(5)$$

$$= 1 - 60$$

$$= -59$$

Since the discriminant is a negative number, this equation has no real solution.

**63.**    $4x^2 + 4x = -1$

$$4x^2 + 4x + 1 = 0$$

$$a = 4, b = 4, c = 1$$

$$b^2 - 4ac = 4^2 - 4(4)(1)$$

$$= 16 - 16$$

$$= 0$$

Since the discriminant is 0, this equation has one real solution.

**65.** $9x^2 + 2x = 0$

$$a = 9, b = 2, c = 0$$

$$b^2 - 4ac = 2^2 - 4(9)(0)$$

$$= 4 - 0$$

$$= 4$$

Since the discriminant is a positive number, this equation has two distinct real solutions.

**67.** $5x^2 + 1 = 0$

$$a = 5, b = 0, c = 1$$

$$b^2 - 4ac = 0^2 - 4(5)(1)$$

$$= 0 - 20$$

$$= -20$$

Since the discriminant is a negative number, this equation has no real solution.

**69.**    $x^2 + 36 = -12x$

$$x^2 + 12x + 36 = 0$$

$$a = 1, b = 12, c = 36$$

$$b^2 - 4ac = 12^2 - 4(1)(36)$$

$$= 144 - 144$$

$$= 0$$

Since the discriminant is 0, this equation has one real solution.

**71.**    $2x^2 - 5 = 9x$

$$2x^2 - 9x - 5 = 0$$

$$b = -9$$

The answer is d.

**73.** $\sqrt{48} = \sqrt{16 \cdot 3} = \sqrt{16} \cdot \sqrt{3} = 4\sqrt{3}$

**75.** $\sqrt{50} = \sqrt{25 \cdot 2} = \sqrt{25} \cdot \sqrt{2} = 5\sqrt{2}$

**77.** Let $x =$ the base and $4x =$ the height.

$$\frac{1}{2}bh = A$$

$$\frac{1}{2}(x)(4x) = 18$$

$$\frac{1}{2}(4x^2) = 18$$

$$2x^2 = 18$$

$$x^2 = 9$$

$$x = \pm\sqrt{9}$$

$$x = \pm 3$$

Since length can't be negative,

base $= 3$ ft. and height $= 4(3) = 12$ ft.

**79.** Let $x =$ the width then, $x + 5 =$ the length.

$A = lw$

$35 = x(x+5)$

$35 = x^2 + 5x$

$0 = x^2 + 5x - 35$

$a = 1, b = 5, c = -35$

$$x = \frac{-(5) \pm \sqrt{(5)^2 - 4(1)(-35)}}{2(1)}$$

$$= \frac{-5 \pm \sqrt{25 + 140}}{2} = \frac{-5 \pm \sqrt{165}}{2}$$

Because the width cannot be negative,

$$x = \frac{-5 + \sqrt{165}}{2} \approx 3.9$$

$x + 5 \approx 3.9 + 5 = 8.9$

Length $= 8.9$ ft., width $= 3.9$ ft.

**81.**
$$x^2 + x = 15$$

$$x^2 + x - 15 = 0$$

$$a = 1, b = 1, c = -15$$

$$x = \frac{-1 \pm \sqrt{1^2 - 4(1)(-15)}}{2(1)}$$

$$x = \frac{-1 \pm \sqrt{61}}{2}$$

$$x \approx -4.4, 3.4$$

The solutions are $-4.4$ and $3.4$.

**83.** $1.2x^2 - 5.2x - 3.9 = 0$

$a = 1, b = -5.2, c = -3.9$

$$x = \frac{-(-5.2) \pm \sqrt{(-5.2)^2 - 4(1.2)(-3.9)}}{2(1.2)}$$

$$x = \frac{5.2 \pm \sqrt{45.76}}{2.4}$$

$$x \approx -0.7, 5.0$$

**85.** Let $h = 30$

$h = -16t^2 + 120t + 80$

$30 = -16t^2 + 120t + 80$

$a = -16, b = 120, \text{ and } c = 50$

$$t = \frac{-(120) \pm \sqrt{(120)^2 - 4(-16)(50)}}{2(-16)}$$

$$= \frac{-120 \pm \sqrt{14,400 + 3200}}{-32}$$

$$= \frac{-120 \pm \sqrt{17,600}}{-32}$$

Since the time cannot be negative,

$$t = \frac{-120 - \sqrt{17,600}}{-32} \approx 7.9$$

After 7.9 seconds.

**87.** Let $y = 12,371$

$$y = -14x^2 - 257x + 14,417$$

$$12,371 = -14x^2 - 257x + 14,417$$

$$0 = -14x^2 - 257x + 2046$$

$a = -14, b = -257,$ and $c = 2046$

$$b^2 - 4ac = (-257)^2 - 4(-14)(2046)$$

$$= 66,049 + 114,576$$

$$= 180,625$$

$$x = \frac{-(-257) \pm \sqrt{180,625}}{2(-14)}$$

$$x = \frac{257 \pm 425}{-28}$$

Since time cannot be negative,

$$x = \frac{257 - 425}{-28} = 6$$

It will be the year $2000 + 6 = 2006$.

**89.**

$$y = -x^2 + 11x + 846$$

$$870 = -x^2 + 11x + 846$$

$$0 = -x^2 + 11x - 24$$

$$0 = -(x - 3)(x - 8)$$

$x - 3 = 0 \quad$ or $\quad x - 8 = 0$

$\quad x = 3 \qquad\qquad x = 8$

$1998 + 3 = 2001, \quad 1998 + 8 = 2006.$

The number of stores will be 870
in 2001 and again in 2006.

**Integrated Review-Summary on Solving Quadratic Equations**

**1.** $\quad 5x^2 - 11x + 2 = 0$

$$(5x - 1)(x - 2) = 0$$

$5x - 1 = 0 \quad$ or $\quad x - 2 = 0$

$\quad 5x = 1 \qquad\qquad x = 2$

$$x = \frac{1}{5}$$

The solutions are $\frac{1}{5}$ and 2.

**2.** $\quad 5x^2 + 13x - 6 = 0$

$$(5x - 2)(x + 3) = 0$$

$5x - 2 = 0 \quad$ or $\quad x + 3 = 0$

$\quad 5x = 2 \qquad\qquad x = -3$

$$x = \frac{2}{5}$$

The solutions are $\frac{2}{5}$ and $-3$.

**3.** $\qquad x^2 - 1 = 2x$

$$x^2 - 2x = 1$$

$$x^2 - 2x + 1 = 1 + 1$$

$$(x - 1)^2 = 2$$

$$x - 1 = \pm\sqrt{2}$$

$$x = 1 \pm \sqrt{2}$$

The solutions are $1 \pm \sqrt{2}$.

**4.** $\qquad x^2 + 7 = 6x$

$$x^2 - 6x = -7$$

$$x^2 - 6x + 9 = -7 + 9$$

$$(x - 3)^2 = 2$$

$$x - 3 = \pm\sqrt{2}$$
$$x = 3 \pm \sqrt{2}$$

The solutions are $3 \pm \sqrt{2}$.

**5.** $a^2 = 20$
$$a = \pm\sqrt{20}$$
$$= \pm 2\sqrt{5}$$

The solutions are $\pm 2\sqrt{5}$.

**6.** $a^2 = 72$
$$a = \pm\sqrt{72}$$
$$= \pm 6\sqrt{2}$$

The solutions are $\pm 6\sqrt{2}$.

**7.** $x^2 - x + 4 = 0$
$$x^2 - x = -4$$
$$x^2 - x + \frac{1}{4} = -4 + \frac{1}{4}$$
$$\left(x - \frac{1}{2}\right)^2 = -\frac{15}{4}$$

There is no real solution.

**8.** $x^2 - 2x + 7 = 0$
$$x^2 - 2x = -7$$
$$x^2 - 2x + 1 = -7 + 1$$
$$(x - 1)^2 = -6$$

There is no real solution.

**9.** $3x^2 - 12x + 12 = 0$
$$x^2 - 4x + 4 = 0$$
$$(x - 2)^2 = 0$$
$$x - 2 = 0$$

$$x = 2$$

The solution is 2.

**10.** $5x^2 - 30x + 45 = 0$
$$x^2 - 6x + 9 = 0$$
$$(x - 3)^2 = 0$$
$$x - 3 = 0$$
$$x = 3$$

The solution is 3.

**11.** $9 - 6p + p^2 = 0$
$$(p - 3)^2 = 0$$
$$p - 3 = 0$$
$$p = 3$$

The solution is 3.

**12.** $49 - 28p + 4p^2 = 0$
$$(2p - 7)^2 = 0$$
$$2p - 7 = 0$$
$$2p = 7$$
$$p = \frac{7}{2}$$

The solution is $\frac{7}{2}$.

**13.** $\qquad 4y^2 - 16 = 0$
$$4y^2 = 16$$
$$y^2 = 4$$
$$y = \pm\sqrt{4}$$
$$y = \pm 2$$

The solutions are $\pm 2$.

**14.**
$$3y^2 - 27 = 0$$
$$3y^2 = 27$$
$$y^2 = 9$$
$$y = \pm\sqrt{9}$$
$$y = \pm 3$$
The solutions are $\pm 3$.

**15.** $x^4 - 3x^3 + 2x^2 = 0$
$$x^2\left(x^2 - 3x + 2\right) = 0$$
$$x^2\left(x-1\right)\left(x-2\right) = 0$$
$$x^2 = 0 \quad \text{or} \quad x - 1 = 0 \quad \text{or} \quad x - 2 = 0$$
$$x = 0 \qquad\quad x = 1 \qquad\qquad x = 2$$
The solutions are 0, 1, and 2.

**16.** $x^3 + 7x^2 + 12x = 0$
$$x\left(x^2 + 7x + 12\right) = 0$$
$$x\left(x+4\right)\left(x+3\right) = 0$$
$$x = 0 \quad \text{or} \quad x + 4 = 0 \quad \text{or} \quad x + 3 = 0$$
$$x = 0 \qquad\quad x = -4 \qquad\qquad x = -3$$
The solutions are $-4$, $-3$, and 0.

**17.** $\left(2x+5\right)^2 = 25$
$$2x + 5 = \pm\sqrt{25}$$
$$2x = -5 \pm 5$$
$$x = \frac{-5 \pm 5}{2}$$
$$x = \frac{-5-5}{2} = -5 \quad \text{or} \quad x = \frac{-5+5}{2} = 0$$
The solutions are 0 and $-5$.

**18.** $\left(3z-4\right)^2 = 16$
$$3z - 4 = \pm\sqrt{16}$$
$$3z = 4 \pm 4$$
$$z = \frac{4 \pm 4}{3}$$
$$z = \frac{4-4}{3} = 0 \quad \text{or} \quad z = \frac{4+4}{3} = \frac{8}{3}$$
The solutions are 0 and $\dfrac{8}{3}$.

**19.**
$$30x = 25x^2 + 2$$
$$0 = 25x^2 - 30x + 2 = 0$$
$$a = 25, b = -30, \text{ and } c = 2$$
$$x = \frac{-(-30) \pm \sqrt{(-30)^2 - 4(25)(2)}}{2(25)}$$
$$= \frac{30 \pm \sqrt{900 - 200}}{50} = \frac{30 \pm \sqrt{700}}{50}$$
$$= \frac{30 \pm 10\sqrt{7}}{50} = \frac{3 \pm \sqrt{7}}{5}$$
The solutions are $\dfrac{3 \pm \sqrt{7}}{5}$.

**20.**
$$12x = 4x^2 + 4$$
$$0 = 4x^2 - 12x + 4$$
$$0 = x^2 - 3x + 1$$
$$a = 1, b = -3, \text{ and } c = 1$$
$$x = \frac{-(-3) \pm \sqrt{(-3)^2 - 4(1)(1)}}{2(1)}$$
$$= \frac{3 \pm \sqrt{9 - 4}}{2} = \frac{3 \pm \sqrt{5}}{2}$$
The solutions are $\dfrac{3 \pm \sqrt{5}}{2}$.

**21.** $\dfrac{2}{3}m^2 - \dfrac{1}{3}m - 1 = 0$

$2m^2 - m - 3 = 0$

$(2m - 3)(m + 1) = 0$

$2m - 3 = 0$   or   $m + 1 = 0$

$2m = 3$         $m = -1$

$m = \dfrac{3}{2}$

The solutions are $-1$ and $\dfrac{3}{2}$.

**22.** $\dfrac{5}{8}m^2 + m - \dfrac{1}{2} = 0$

$5m^2 + 8m - 4 = 0$

$(5m - 2)(m + 2) = 0$

$5m - 2 = 0$   or   $m + 2 = 0$

$5m = 2$         $m = -2$

$m = \dfrac{2}{5}$

The solutions are $-2$ and $\dfrac{2}{5}$.

**23.** $x^2 - \dfrac{1}{2}x - \dfrac{1}{5} = 0$

$10x^2 - 5x - 2 = 0$

$a = 10, b = -5,$ and $c = -2$

$x = \dfrac{-(-5) \pm \sqrt{(-5)^2 - 4(10)(-2)}}{2(10)}$

$= \dfrac{5 \pm \sqrt{25 + 80}}{20} = \dfrac{5 \pm \sqrt{105}}{20}$

The solutions are $\dfrac{5 \pm \sqrt{105}}{20}$.

**24.** $x^2 + \dfrac{1}{2}x - \dfrac{1}{8} = 0$

$8x^2 + 4x - 1 = 0$

$a = 8, b = 4,$ and $c = -1$

$x = \dfrac{-(4) \pm \sqrt{(4)^2 - 4(8)(-1)}}{2(8)}$

$= \dfrac{-4 \pm \sqrt{16 + 32}}{16} = \dfrac{-4 \pm \sqrt{48}}{16}$

$= \dfrac{-4 \pm 4\sqrt{3}}{16} = \dfrac{-1 \pm \sqrt{3}}{4}$

The solutions are $\dfrac{-1 \pm \sqrt{3}}{4}$.

**25.** $4x^2 - 27x + 35 = 0$

$(4x - 7)(x - 5) = 0$

$4x - 7 = 0$   or   $x - 5 = 0$

$4x = 7$         $x = 5$

$x = \dfrac{7}{4}$

The solutions are $\dfrac{7}{4}$ and 5.

**26.** $9x^2 - 16x + 7 = 0$

$(9x - 7)(x - 1) = 0$

$9x - 7 = 0$   or   $x - 1 = 0$

$9x = 7$         $x = 1$

$x = \dfrac{7}{9}$

The solutions are $\dfrac{7}{9}$ and 1.

**27.** $(7-5x)^2 = 18$

$$7-5x = \pm\sqrt{18}$$
$$7-5x = \pm 3\sqrt{2}$$
$$-5x = -7 \pm 3\sqrt{2}$$
$$\frac{-5x}{-5} = \frac{-7 \pm 3\sqrt{2}}{-5}$$
$$x = \frac{7 \pm 3\sqrt{2}}{5}$$

The solutions are $\dfrac{7 \pm 3\sqrt{2}}{5}$.

**28.** $(5-4x)^2 = 75$

$$5-4x = \pm\sqrt{75}$$
$$5-4x = \pm 5\sqrt{3}$$
$$-4x = -5 \pm 5\sqrt{3}$$
$$\frac{-4x}{-4} = \frac{-5 \pm 5\sqrt{3}}{-5}$$
$$x = \frac{5 \pm 5\sqrt{3}}{4}$$

The solutions are $\dfrac{5 \pm 5\sqrt{3}}{4}$.

**29.** $3z^2 - 7z = 12$

$$3z^2 - 7z - 12 = 0$$
$$a = 3, b = -7, \text{ and } c = -12$$
$$z = \frac{-(-7) \pm \sqrt{(-7)^2 - 4(3)(-12)}}{2(3)}$$
$$= \frac{7 \pm \sqrt{49 + 144}}{6} = \frac{7 \pm \sqrt{193}}{6}$$

The solutions are $\dfrac{7 \pm \sqrt{193}}{6}$.

**30.** $6z^2 + 7z = 6$

$$6z^2 + 7z - 6 = 0$$
$$a = 6, b = 7, \text{ and } c = -6$$
$$z = \frac{-(7) \pm \sqrt{(7)^2 - 4(6)(-6)}}{2(6)}$$
$$= \frac{-7 \pm \sqrt{49 + 144}}{12} = \frac{-7 \pm \sqrt{193}}{12}$$

The solutions are $\dfrac{-7 \pm \sqrt{193}}{12}$.

**31.** $x = x^2 - 110$

$$0 = x^2 - x - 110$$
$$0 = (x+10)(x-11)$$
$$x + 10 = 0 \quad \text{or} \quad x - 11 = 0$$
$$x = -10 \qquad\qquad x = 11$$

The solutions are $-10$ and $11$.

**32.** $\qquad\qquad x = 56 - x^2$

$$x^2 + x - 56 = 0$$
$$(x+8)(x-7) = 0$$
$$x + 8 = 0 \quad \text{or} \quad x - 7 = 0$$
$$x = -8 \qquad\qquad x = 7$$

The solutions are $-8$ and $7$.

**33.** $\qquad \dfrac{3}{4}x^2 - \dfrac{5}{2}x - 2 = 0$

$$3x^2 - 10x - 8 = 0$$
$$(3x+2)(x-4) = 0$$
$$3x + 2 = 0 \quad \text{or} \quad x - 4 = 0$$
$$3x = -2 \qquad\qquad x = 4$$
$$x = -\frac{2}{3}$$

The solutions are $-\dfrac{2}{3}$ and 4.

**34.**    $x^2 - \dfrac{6}{5}x - \dfrac{8}{5} = 0$

$5x^2 - 6x - 8 = 0$

$(5x+4)(x-2) = 0$

$5x+4 = 0$   or   $x-2 = 0$

$5x = -4 \qquad\qquad x = 2$

$x = -\dfrac{4}{5}$

The solutions are $-\dfrac{4}{5}$ and 2.

**35.**   $x^2 - 0.6x + 0.05 = 0$

$100x^2 - 60x + 5 = 0$

$20x^2 - 12x + 1 = 0$

$(10x-1)(2x-1) = 0$

$10x - 1 = 0$   or   $2x - 1 = 0$

$10x = 1 \qquad\qquad 2x = 1$

$x = \dfrac{1}{10} = 0.1 \qquad x = \dfrac{1}{2} = 0.5$

The solutions are 0.1 and 0.5.

**36.**   $x^2 - 0.1x + 0.06 = 0$

$100x^2 - 10x + 6 = 0$

$50x^2 - 5x + 3 = 0$

$(5x+1)(10x-3) = 0$

$5x+1 = 0$    or    $10x - 3 = 0$

$5x = -1 \qquad\qquad 10x = 3$

$x = -\dfrac{1}{5} = -0.2 \qquad x = \dfrac{3}{10} = 0.3$

The solutions are $-0.2$ and 0.3.

**37.**   $10x^2 - 11x + 2 = 0$

$a = 10, b = -11,$ and $c = 2$

$x = \dfrac{-(-11) \pm \sqrt{(-11)^2 - 4(10)(2)}}{2(10)}$

$= \dfrac{11 \pm \sqrt{121 - 80}}{20} = \dfrac{11 \pm \sqrt{41}}{20}$

The solutions are $\dfrac{11 \pm \sqrt{41}}{20}$.

**38.**   $20x^2 - 11x + 1 = 0$

$a = 20, b = -11,$ and $c = 1$

$x = \dfrac{-(-11) \pm \sqrt{(-11)^2 - 4(20)(1)}}{2(20)}$

$= \dfrac{11 \pm \sqrt{121 - 80}}{40} = \dfrac{11 \pm \sqrt{41}}{40}$

The solutions are $\dfrac{11 \pm \sqrt{41}}{40}$.

**39.**   $\dfrac{1}{2}z^2 - 2z + \dfrac{3}{4} = 0$

$z^2 - 4z = -\dfrac{3}{2}$

$z^2 - 4z + 4 = -\dfrac{3}{2} + 4$

$(z-2)^2 = \dfrac{5}{2}$

$z - 2 = \pm\sqrt{\dfrac{5}{2}}$

$z = 2 \pm \sqrt{\dfrac{5}{2}} = 2 \pm \dfrac{\sqrt{10}}{2}$

$= \dfrac{4 \pm \sqrt{10}}{2}$

The solutions are $\dfrac{4 \pm \sqrt{10}}{2}$.

**40.** $\dfrac{1}{5}z^2 - \dfrac{1}{2}z - 2 = 0$

$2z^2 - 5z - 20 = 0$

$a = 2, b = -5,$ and $c = -20$

$z = \dfrac{-(-5) \pm \sqrt{(-5)^2 - 4(2)(-20)}}{2(2)}$

$= \dfrac{5 \pm \sqrt{25 + 160}}{4} = \dfrac{5 \pm \sqrt{185}}{4}$

The solutions are $\dfrac{5 \pm \sqrt{185}}{4}$.

**41.** Answers may vary.

**Exercise Set 9.4**

**1.** $\sqrt{-9} = \sqrt{-1 \cdot 9} = \sqrt{-1}\sqrt{9} = i \cdot 3 = 3i$

**3.** $\sqrt{-100} = \sqrt{-1 \cdot 100} = \sqrt{-1}\sqrt{100} = i \cdot 10 = 10i$

**5.** $\sqrt{-50} = \sqrt{-1 \cdot 25 \cdot 2}$

$= \sqrt{-1}\sqrt{25}\sqrt{2}$

$= i \cdot 5\sqrt{2}$

$= 5i\sqrt{2}$

**7.** $\sqrt{-63} = \sqrt{-1 \cdot 9 \cdot 7}$

$= \sqrt{-1}\sqrt{9}\sqrt{7}$

$= i \cdot 3\sqrt{7}$

$= 3i\sqrt{7}$

**9.** $(2 - i) + (-5 + 10i) = 2 - 5 + (-i + 10i)$

$= -3 + 9i$

**11.** $(-11 + 3i) - (1 - 3i) = -11 + 3i - 1 + 3i$

$= -12 + 6i$

**13.** $(3 - 4i) - (2 - i) = 3 - 4i - 2 + i$

$= 3 - 2 + (-4i + i)$

$= 1 - 3i$

**15.** $(16 + 2i) + (-7 - 6i) = 16 + 2i - 7 - 6i$

$= 9 - 4i$

**17.** $4i(3 - 2i) = 12i - 8i^2$

$= 12i - 8(-1)$

$= 12i + 8$

$= 8 + 12i$

**19.** $(6 - 2i)(4 + i) = 6(4) + 6i - 2i(4) - 2i(i)$

$= 24 + 6i - 8i - 2i^2$

$= 24 - 2i - 2(-1)$

$= 24 - 2i + 2$

$= 26 - 2i$

**21.** $(3 + 8i)(3 - 8i) = 3^2 - (8i)^2$

$= 9 - 64i^2$

$= 9 - 64(-1)$

$= 9 + 64$

$= 73$

**23.** Answers may vary.

**25.** $\dfrac{8 - 12i}{4} = \dfrac{4(2 - 3i)}{4} = 2 - 3i$

**27.**

$$\frac{7-i}{4-3i} = \frac{(7-i)}{(4-3i)} \cdot \frac{(4+3i)}{(4+3i)}$$

$$= \frac{7(4)+7(3)-i(4)-i(3i)}{(4)^2-(3i)^2}$$

$$= \frac{28+21i-4i-3i^2}{16-9i^2}$$

$$= \frac{28+17i-3}{16-9(-1)}$$

$$= \frac{28+17i+3}{16+9}$$

$$= \frac{31+17i}{25}$$

$$= \frac{31}{25} + \frac{17}{25}i$$

**29.** $(x+1)^2 = -9$

$$x+1 = \pm\sqrt{-9}$$

$$x = -1 \pm 3i$$

The solutions are $-1 \pm 3i$.

**31.** $(2z-3)^2 = -12$

$$2z-3 = \pm\sqrt{-12}$$

$$2z-3 = \pm\sqrt{-1}\sqrt{4}\sqrt{3}$$

$$2z-3 = \pm 2i\sqrt{3}$$

$$2z = 3 \pm 2i\sqrt{3}$$

$$z = \frac{3 \pm 2i\sqrt{3}}{2}$$

The solutions are $\dfrac{3 \pm 2i\sqrt{3}}{2}$.

**33.** $y^2 + 6y + 13 = 0$

$a = 1$, $b = 6$, $c = 13$

$$y = \frac{-6 \pm \sqrt{6^2 - 4(1)(13)}}{2(1)}$$

$$x = \frac{-6 \pm \sqrt{36-52}}{2}$$

$$y = \frac{-6 \pm \sqrt{-16}}{2}$$

$$y = \frac{-6 \pm 4i}{2}$$

$$y = \frac{2(-3 \pm 2i)}{2}$$

$$y = -3 \pm 2i$$

The solutions are $-3 \pm 2i$.

**35.** $4x^2 + 7x + 4 = 0$

$a = 4$, $b = 7$, $c = 4$

$$x = \frac{-7 \pm \sqrt{7^2 - 4(4)(4)}}{2(4)}$$

$$x = \frac{-7 \pm \sqrt{49-64}}{8}$$

$$x = \frac{-7 \pm \sqrt{-15}}{8}$$

$$x = \frac{-7 \pm i\sqrt{15}}{8}$$

The solutions are $\dfrac{-7 \pm i\sqrt{15}}{8}$.

**37.** $2m^2 - 4m + 5 = 0$

$a = 2$, $b = -4$, $c = 5$

$$m = \frac{-(-4) \pm \sqrt{(-4)^2 - 4(2)(5)}}{2(2)}$$

$$m = \frac{4 \pm \sqrt{16 - 40}}{4}$$

$$m = \frac{4 \pm \sqrt{-24}}{4}$$

$$m = \frac{4 \pm \sqrt{-1 \cdot 4 \cdot 6}}{4}$$

$$m = \frac{4 \pm 2i\sqrt{6}}{4}$$

$$m = \frac{2 \pm i\sqrt{6}}{2}$$

The solutions are $\frac{2 \pm i\sqrt{6}}{2}$.

**39.** $3 + (12 - 7i) = 3 + 12 - 7i = 15 - 7i$

**41.** $-9i(5i - 7) = -45i^2 + 63i$
$$= -45(-1) + 63i$$
$$= 45 + 63i$$

**43.** $(2 - i) - (3 - 4i) = 2 - i - 3 + 4i$
$$= 2 - 3 + (-i + 4i)$$
$$= -1 + 3i$$

**45.** $\frac{15 + 10i}{5i} = \frac{(15 + 10i)}{5i} \cdot \frac{(-i)}{(-i)}$
$$= \frac{-15i - 10i^2}{-5i^2}$$
$$= \frac{-15i - 10(-1)}{-5(-1)}$$
$$= \frac{-15i + 10}{5}$$

$$= \frac{5(-3i + 2)}{5}$$
$$= -3i + 2$$
$$= 2 - 3i$$

**47.** $-5 + i - (2 + 3i) = -5 + i - 2 - 3i$
$$= -5 - 2 + (i - 3i)$$
$$= -7 - 2i$$

**49.** $(4 - 3i)(4 + 3i) = (4)^2 - (3i)^2$
$$= 16 - 9i^2$$
$$= 16 - 9(-1)$$
$$= 16 + 9$$
$$= 25$$

**51.** $\frac{4 - i}{1 + 2i} = \frac{(4 - i)(1 - 2i)}{(1 + 2i)(1 - 2i)}$
$$= \frac{4(1) + 4(-2i) - i(1) - i(-2i)}{(1)^2 - (2i)^2}$$
$$= \frac{4 - 8i + -i + 2i^2}{1 - 4i^2}$$
$$= \frac{4 - 9i + 2(-1)}{1 - 4(-1)}$$
$$= \frac{2 - 9i}{5}$$
$$= \frac{2}{5} - \frac{9}{5}i$$

**53.** $(5 + 2i)^2 = (5)^2 + 2(5)(2i) + (2i)^2$
$$= 25 + 20i + 4i^2$$
$$= 25 + 20i + 4(-1)$$
$$= 25 + 20i - 4$$
$$= 21 + 20i$$

**55.** $(y-4)^2 = -64$

$$y-4 = \pm\sqrt{-64}$$
$$y-4 = \pm 8i$$
$$x = 4 \pm 8i$$

The solutions are $4 \pm 8i$.

**57.** $4x^2 = -100$

$$x^2 = -25$$
$$x = \pm\sqrt{-25}$$
$$x = \pm 5i$$

The solutions are $\pm 5i$.

**59.** $z^2 + 6z + 10 = 0$

$a = 1,\ b = 6,\ c = 10$

$$z = \frac{-6 \pm \sqrt{6^2 - 4(1)(10)}}{2(1)}$$
$$z = \frac{-6 \pm \sqrt{36 - 40}}{2}$$
$$z = \frac{-6 \pm \sqrt{-4}}{2}$$
$$z = \frac{2(-3 \pm i)}{2}$$
$$z = -3 \pm i$$

The solutions are $-3 \pm i$.

**61.** $2a^2 - 5a + 9 = 0$

$a = 2,\ b = -5,\ c = 9$

$$a = \frac{-(-5) \pm \sqrt{(-5)^2 - 4(2)(9)}}{2(2)}$$
$$a = \frac{5 \pm \sqrt{25 - 72}}{4}$$

$$a = \frac{5 \pm \sqrt{-47}}{4}$$
$$a = \frac{5 \pm i\sqrt{47}}{4}$$

The solutions are $\dfrac{5 \pm i\sqrt{47}}{4}$.

**63.** $(2x+8)^2 = -20$

$$2x + 8 = \pm\sqrt{-20}$$
$$2x + 8 = \pm\sqrt{1}\sqrt{4}\sqrt{5}$$
$$2x + 8 = \pm 2i\sqrt{5}$$
$$2x = -8 \pm 2i\sqrt{5}$$
$$x = \frac{-8 \pm 2i\sqrt{5}}{2}$$
$$x = \frac{2(-4 \pm i\sqrt{5})}{2}$$
$$x = -4 \pm i\sqrt{5}$$

The solutions are $-4 \pm i\sqrt{5}$.

**65.** $3m^2 + 108 = 0$

$$3m^2 = -108$$
$$m^2 = -\frac{108}{3}$$
$$m^2 = -36$$
$$m = \pm\sqrt{-36}$$
$$m = \pm 6i$$

The solutions are $\pm 6i$.

**67.** $x^2 + 14x + 50 = 0$

$a = 1, b = 14, c = 50$

$$x = \frac{-14 \pm \sqrt{14^2 - 4(1)(50)}}{2(1)}$$

$$x = \frac{-14 \pm \sqrt{196 - 200}}{2}$$

$$x = \frac{-14 \pm \sqrt{-4}}{2}$$

$$x = \frac{-14 \pm 2i}{2}$$

$$x = \frac{2(-7) \pm i}{2}$$

$$x = -7 \pm i$$

The solutions are $-7 \pm i$.

**69.** $y = -3$

$y = -3$ for all values of $x$.

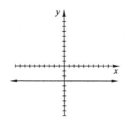

**71.** $y = 3x - 2$

| $x$ | $y$ |
|-----|-----|
| 0 | −2 |
| 3 | 7 |

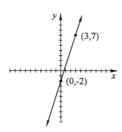

**73.** 51%

**75.** Let $x = 10$

$y = 4.8x + 42.1$

$y = 4.8(10) + 42.1$

$y = 4.8$

Expect 90.1% of the households to have computers.

**77.** False

**79.** False

**Graphing Calculator Explorations 9.5**

**1.** $x^2 - 7x - 3 = 0$

$y_1 = x^2 - 7x - 3$

$y_2 = 0$

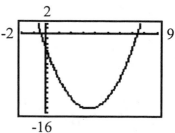

The $x$-coordinates of the intersections, $-0.41$ and $7.41$, are the solutions.

**3.** $-1.7x^2 + 5.6x - 3.7 = 0$

$y_1 = -1.7x^2 + 5.6x - 3.7$

$y_2 = 0$

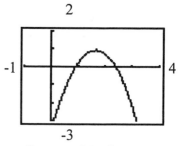

The $x$-coordinates of the intersections, $0.91$ and $2.38$, are the solutions.

**5.** $5.8x^2 - 2.6x - 1.9 = 0$

$y_1 = 5.8x^2 - 2.6x - 1.9$

$y_2 = 0$

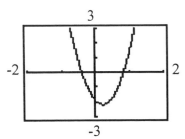

The *x*-coordinates of the intersections, $-0.39$ and $0.84$, are the solutions.

**Exercise Set 9.5**

**1.** $y = 2x^2$

| $x$ | $y$ |
|---|---|
| $-2$ | 8 |
| $-1$ | 2 |
| 0 | 0 |
| 1 | 2 |
| 2 | 8 |

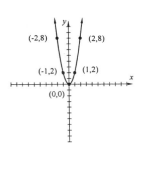

**3.** $y = -x^2$

| $x$ | $y$ |
|---|---|
| $-2$ | $-4$ |
| $-1$ | $-1$ |
| 0 | 0 |
| 1 | $-1$ |
| 2 | $-4$ |

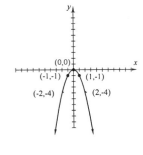

**5.**
$y = \dfrac{1}{3}x^2$

| $x$ | $y$ |
|---|---|
| $-6$ | 12 |
| $-3$ | 3 |
| 0 | 0 |
| 3 | 3 |
| 6 | 12 |

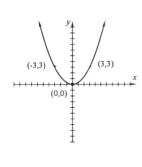

**7.** $y = x^2 - 1$

*y*-intercept: $x = 0$, $y = 0^2 - 1 = -1$, $(0, -1)$

vertex: $(0, -1)$

*x*-intercepts: $y = 0$,

$0 = x^2 - 1$

$0 = (x + 1)(x - 1)$

$x + 1 = 0$ or $x - 1 = 0$

$x = -1$ $\qquad x = 1$

$(-1, 0)$ and $(1, 0)$

| $x$ | $y$ |
|---|---|
| $-2$ | 3 |
| $-1$ | 0 |
| 0 | $-1$ |
| 1 | 0 |
| 2 | 3 |

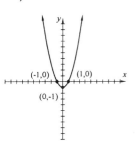

**9.** $y = x^2 + 4$

*y*-intercept: $x = 0$, $y = 0^2 + 4 = 4$, $(0, 4)$

vertex: $(0, 4)$

*x*-intercepts: $y = 0$,

$0 = x^2 + 4$

$-4 = x^2$

There are no $x$-intercepts because there is no real solution to this equation.

| $x$ | $y$ |
|-----|-----|
| $-2$ | 8 |
| $-1$ | 5 |
| 0 | 4 |
| 1 | 5 |
| 2 | 8 |

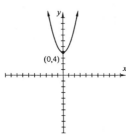

**11.** $y = x^2 + 6x$

vertex: $\left.\begin{array}{l} x = -\dfrac{b}{2a} = -\dfrac{6}{2(1)} = -3 \\ y = (-3)^2 + 6(-3) = -9 \end{array}\right\}(-3, -9)$

$y$-intercept: $x = 0$, $y = 0^2 + 6(0) = 0$,
$$(0, 0)$$

$x$-intercepts: $y = 0$,
$0 = x^2 + 6x$
$0 = x(x + 6)$
$x = -6$ or $x = 0$
$(-6, 0)$ and $(0, 0)$

| $x$ | $y$ |
|-----|-----|
| $-7$ | 7 |
| $-6$ | 0 |
| $-3$ | $-9$ |
| 0 | 0 |
| 1 | 7 |

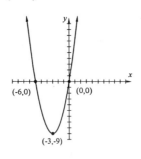

**13.** $y = x^2 + 2x - 8$

vertex: $\left.\begin{array}{l} x = -\dfrac{b}{2a} = -\dfrac{2}{2(1)} = -1 \\ y = (-1)^2 + 2(-1) - 8 = -9 \end{array}\right\}(-1, -9)$

$y$-intercept: $x = 0$, $y = 0^2 + 2(0) - 8 = -8$,
$$(0, -8)$$
$x$-intercepts: $y = 0$,
$0 = x^2 + 2x - 8$
$0 = (x + 4)(x - 2)$
$x = -4$ or $x = 2$
$(-4, 0)$ and $(2, 0)$

| $x$ | $y$ |
|-----|-----|
| $-4$ | 0 |
| $-2$ | $-8$ |
| $-1$ | $-9$ |
| 0 | $-8$ |
| 2 | 0 |

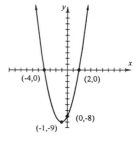

**15.** $y = -x^2 + x + 2$

vertex: $\left.\begin{array}{l} x = -\dfrac{b}{2a} = -\dfrac{1}{2(-1)} = \dfrac{1}{2} \\ y = -\left(\dfrac{1}{2}\right)^2 + \left(\dfrac{1}{2}\right) + 2 = \dfrac{9}{4} \end{array}\right\}\left(\dfrac{1}{2}, \dfrac{9}{4}\right)$

$y$-intercept: $x = 0$, $y = -0^2 + (0) + 2 = 2$,
$$(0, 2)$$
$x$-intercepts: $y = 0$,
$0 = -x^2 + x + 2$
$0 = x^2 - x - 2$
$0 = (x + 1)(x - 2)$
$x = -1$ or $x = 2$
$(-1, 0)$ and $(2, 0)$

| $x$ | $y$ |
|-----|-----|
| $-1$ | 0 |
| 0 | 2 |
| 1/2 | 9/4 |
| 1 | 2 |
| 2 | 0 |

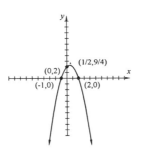

**17.** $y = x^2 + 5x + 4$

vertex: $x = -\dfrac{b}{2a} = -\dfrac{5}{2(1)} = -\dfrac{5}{2}$

$$y = \left(-\dfrac{5}{2}\right)^2 + 5\left(-\dfrac{5}{2}\right) + 4 = -\dfrac{9}{4}$$

$$\left(-\dfrac{5}{2}, -\dfrac{9}{4}\right)$$

$y$-intercept: $x = 0$, $y = 0^2 + 5(0) + 4 = 4$,

$$(0, 4)$$

$x$-intercepts: $y = 0$,

$0 = x^2 + 5x + 4$

$0 = (x + 4)(x + 1)$

$x = -4$   or   $x = -1$

$(-4, 0)$ and $(-1, 0)$

| $x$ | $y$ |
|-----|-----|
| $-5$ | 4 |
| $-4$ | 0 |
| $-5/2$ | $-9/4$ |
| $-1$ | 0 |
| 0 | 4 |

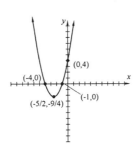

$$\left(\dfrac{1}{2}, 0\right) \text{ and } (5, 0)$$

**19.** $y = -x^2 + 4x - 3$

$$\left.\begin{array}{l} \text{vertex: } x = -\dfrac{b}{2a} = -\dfrac{4}{2(-1)} = 2 \\[2mm] y = -(2)^2 + 4(2) - 3 = 1 \end{array}\right\} (2, 1)$$

$y$-intercept: $x = 0$, $y = -0^2 + 4(0) - 3 = -3$,

$$(0, -3)$$

$x$-intercepts: $y = 0$,

$0 = -x^2 + 4x - 3$

$0 = x^2 - 4x + 3$

$0 = (x - 1)(x - 3)$

$x = 1$   or   $x = 3$

$(1, 0)$ and $(3, 0)$

| $x$ | $y$ |
|-----|-----|
| 0 | $-3$ |
| 1 | 0 |
| 2 | 1 |
| 3 | 0 |
| 4 | $-3$ |

**21.** $y = x^2 + 2x - 2$

Find the vertex.

$$x = \dfrac{-b}{2a} = \dfrac{-2}{2(1)} = -1$$

$y = (-1)^2 + 2(-1) - 2 = 1 - 2 - 2 = -3$

vertex $= (-1, -3)$

Find $y$-intercept. Let $x = 0$.

$y = 0^2 + 2(0) - 2$

$y = -2$

$y$-intercept $= (0, -2)$

Find $x$-intercepts.  Let $y = 0$.

$0 = x^2 + 2x - 2$

$x = \dfrac{-2 \pm \sqrt{2^2 - 4(1)(-2)}}{2(1)}$

$x = \dfrac{-2 \pm \sqrt{12}}{2}$

$x = \dfrac{-2 \pm 2\sqrt{3}}{2}$

$x = -1 \pm \sqrt{3}$

$x$-intercepts $= \left(-1 - \sqrt{3},\, 0\right), \left(-1 + \sqrt{3},\, 0\right)$

**23.** $y = x^2 - 3x + 1$

Find the vertex.

$x = \dfrac{-b}{2a} = \dfrac{-(-3)}{2(1)} = \dfrac{3}{2} = 1\dfrac{1}{2}$

$y = \left(\dfrac{3}{2}\right)^2 - 3\left(\dfrac{3}{2}\right) + 1 = \dfrac{9}{4} - \dfrac{9}{2} + 1 = \dfrac{9}{4} - \dfrac{18}{4}$

$= -\dfrac{5}{4} = -1\dfrac{1}{4}$

vertex $= \left(1\dfrac{1}{2},\, -1\dfrac{1}{4}\right)$

Find $y$-intercept.  Let $x = 0$.

$y = 0^2 - 3(0) + 1$

$y = 1$

$y$-intercept $= (0,\, 1)$

Find $x$-intercepts.  Let $y = 0$.

$0 = x^2 - 3x + 1$

$x = \dfrac{3 \pm \sqrt{(-3)^2 - 4(1)(1)}}{2(1)}$

$= \dfrac{3 \pm \sqrt{5}}{2}$

**25.** $\dfrac{\frac{1}{7}}{\frac{2}{5}} = \dfrac{1}{7} \cdot \dfrac{5}{2} = \dfrac{5}{14}$

**27.** $\dfrac{\frac{1}{x}}{\frac{2}{x^2}} = \dfrac{1}{x} \cdot \dfrac{x^2}{2} = \dfrac{x}{2}$

**29.** $\dfrac{2x}{1 - \frac{1}{x}} = \dfrac{2x}{\frac{x-1}{x}} = \dfrac{2x}{1} \cdot \dfrac{x}{x-1} = \dfrac{2x^2}{x-1}$

**31.** $\dfrac{\frac{a-b}{2b}}{\frac{b-a}{8b^2}} = \dfrac{a-b}{2b} \cdot \dfrac{8b^2}{b-a} = \dfrac{a-b}{2b} \cdot \dfrac{8b^2}{-1(a-b)} = -4b$

**33.** domain: all real numbers: range: $y \le 3$

**35.** domain: all real numbers: range: $y \le 1$

**37.** **a.** 256 feet

   **b.** 4 seconds

   **c.** 8 seconds

**39.** **B**

**41.** **D**

## Chapter 9 Review

**1.** $(x - 4)(5x + 3) = 0$

$5x + 3 = 0 \quad \text{or} \quad x - 4 = 0$

$x = -\dfrac{3}{5} \qquad x = 4$

The solutions are $-\dfrac{3}{5}$ and 4.

**2.** $(x+7)(3x+4)=0$

$$3x+4=0 \quad \text{or} \quad x+7=0$$

$$x=-\frac{4}{3} \qquad x=-7$$

The solutions are $-\dfrac{4}{3}$ and $-7$.

**3.** $\quad 3m^2-5m=2$

$$3m^2-5m-2=0$$

$$(m-2)(3m+1)=0$$

$$m-2=0 \quad \text{or} \quad 3m+1=0$$

$$m=2 \qquad\qquad m=-\frac{1}{3}$$

The solutions are $-\dfrac{1}{3}$ and 2.

**4.** $\quad 7m^2+2m=5$

$$7m^2+2m-5=0$$

$$(m+1)(7m-5)=0$$

$$m+1=0 \quad \text{or} \quad 7m-5=0$$

$$m=-1 \qquad\qquad m=\frac{5}{7}$$

The solutions are $\dfrac{5}{7}$ and $-1$.

**5.** $k^2=50$

$$k=\pm\sqrt{50}$$

$$=\pm 5\sqrt{2}$$

The solutions are $\pm 5\sqrt{2}$.

**6.** $k^2=45$

$$k=\pm\sqrt{45}$$

$$=\pm 3\sqrt{5}$$

The solutions are $\pm 3\sqrt{5}$.

**7.** $\qquad\qquad (x-5)(x-1)=12$

$$x(x)+x(-1)-5(x)-5(-1)=12$$

$$x^2-x-5x+5=12$$

$$x^2-6x+5-12=0$$

$$x^2-6x-7=0$$

$$(x-7)(x+1)=0$$

$$x-7=0 \quad \text{or} \quad x+1=0$$

$$x=7 \quad \text{or} \qquad x=-1$$

The solutions are 7 and $-1$.

**8.** $(x-3)(x+2)=6$

$$x^2-x-6=6$$

$$x^2-x-12=0$$

$$(x-4)(x+3)=0$$

$$x-4=0 \quad \text{or} \quad x+3=0$$

$$x=4 \quad \text{or} \qquad x=-3$$

The solutions are 4 and $-3$.

**9.** $(x-11)^2=49$

$$x-11=\pm\sqrt{49}$$

$$x-11=\pm 7$$

$$x=11\pm 7$$

$$x=11-7=4 \quad \text{or} \quad x=11+7=18$$

The solutions are 4 and 18.

**10.** $(x+3)^2=100$

$$x+3=\pm\sqrt{100}$$

$$x+3=\pm 10$$

$$x=-3\pm 10$$

$$x=-3-10=-13 \quad \text{or} \quad x=-3+10=7$$

The solutions are $-13$ and 7.

**11.**
$$6x^3 - 54x = 0$$
$$6x(x^2 - 9) = 0$$
$$6x(x + 3)(x - 3) = 0$$
$$6x = 0 \quad \text{or} \quad x + 3 = 0 \quad \text{or} \quad x - 3 = 0$$
$$x = 0 \quad \text{or} \quad x = -3 \quad \text{or} \quad x = 3$$
The solutions are $0, -3,$ and $3$.

**12.** $2x^2 - 8 = 0$
$$2x^2 = 8$$
$$x^2 = 4$$
$$x = \pm\sqrt{4}$$
The solutions are $\pm 2$.

**13.** $(4p + 2)^2 = 100$
$$4p + 2 = \pm\sqrt{100}$$
$$4p + 2 = \pm 10$$
$$4p = -2 \pm 10$$
$$p = \frac{-2 \pm 10}{4}$$
$$p = \frac{-2 - 10}{4} = -3 \quad \text{or} \quad p = \frac{-2 + 10}{4} = 2$$
The solutions are $-3$ and $2$.

**14.** $(3p + 6)^2 = 81$
$$3p + 6 = \pm\sqrt{81}$$
$$3p + 6 = \pm 9$$
$$3p = -6 \pm 9$$
$$p = \frac{-6 \pm 9}{3}$$
$$p = \frac{-6 - 9}{3} = -5 \quad \text{or} \quad p = \frac{-6 + 9}{3} = 1$$
The solutions are $-5$ and $1$.

**15.** Let $h = 100$
$$h = 16t^2$$
$$16t^2 = h$$
$$16t^2 = 100$$
$$t^2 = \frac{100}{16}$$
$$t = \pm\sqrt{\frac{100}{16}} = \pm\frac{10}{4} = \pm 2.5$$
The length of time is not a negative number so the dive lasted 2.5 seconds.

**16.** Let $h = 5 \cdot 5280 = 26,400$
$$h = 16t^2$$
$$16t^2 = h$$
$$16t^2 = 26,400$$
$$t^2 = \frac{26,400}{16} = 1650$$
$$t = \pm\sqrt{1650} = \pm 40.6$$
The length of time is not a negative number so the fall lasted 40.6 seconds.

**17.** $a^2 + 4a \Rightarrow \left(\frac{4}{2}\right)^2 = 2^2 = 4$
$$a^2 + 4a + 4 = (a + 2)^2$$

**18.** $a^2 - 12a \Rightarrow \left(\frac{-12}{2}\right)^2 = 36$
$$a^2 - 12a + 36 = (a - 6)^2$$

**19.** $m^2 - 3m \Rightarrow \left(\frac{-3}{2}\right)^2 = \frac{9}{4}$
$$m^2 - 3m + \frac{9}{4} = \left(m - \frac{3}{2}\right)^2$$

**20.** $m^2 + 5m \Rightarrow \left(\dfrac{5}{2}\right)^2 = \dfrac{25}{4}$

$m^2 + 5m + \dfrac{25}{4} = \left(m + \dfrac{5}{2}\right)^2$

**21.** $x^2 - 6x + 7 = 0$

$x^2 - 6x = -7$

$x^2 - 6x + 9 = -7 + 9$

$(x - 3)^2 = 2$

$x - 3 = \pm\sqrt{2}$

$x = 3 \pm \sqrt{2}$

The solutions are $3 \pm \sqrt{2}$.

**22.** $x^2 + 6x + 7 = 0$

$x^2 + 6x = -7$

$x^2 + 6x + 9 = -7 + 9$

$(x + 3)^2 = 2$

$x + 3 = \pm\sqrt{2}$

$x = -3 \pm \sqrt{2}$

The solutions are $-3 \pm \sqrt{2}$.

**23.** $2y^2 + y - 1 = 0$

$y^2 + \dfrac{1}{2}y - \dfrac{1}{2} = 0$

$y^2 + \dfrac{1}{2}y = \dfrac{1}{2}$

$y^2 + \dfrac{1}{2}y + \dfrac{1}{16} = \dfrac{1}{2} + \dfrac{1}{16}$

$\left(y + \dfrac{1}{4}\right)^2 = \dfrac{9}{16}$

$y + \dfrac{1}{4} = \pm\sqrt{\dfrac{9}{16}}$

$y = -\dfrac{1}{4} \pm \dfrac{3}{4}$

$y = -\dfrac{1}{4} + \dfrac{3}{4}$   or   $y = -\dfrac{1}{4} - \dfrac{3}{4}$

$y = \dfrac{1}{2}$           $y = -1$

The solutions are $\dfrac{1}{2}$ and $-1$.

**24.** $y^2 + 3y - 1 = 0$

$y^2 + 3y = 1$

$y^2 + 3y + \dfrac{9}{4} = 1 + \dfrac{9}{4}$

$\left(y + \dfrac{3}{2}\right)^2 = \dfrac{13}{4}$

$y + \dfrac{3}{2} = \pm\sqrt{\dfrac{13}{4}}$

$y = -\dfrac{3}{2} \pm \dfrac{\sqrt{13}}{2}$

$y = \dfrac{-3 \pm \sqrt{13}}{2}$

The solutions are $\dfrac{-3 \pm \sqrt{13}}{2}$.

**25.** $x^2 - 10x + 7 = 0$

$a = 1, b = -10,$ and $c = 7$

$x = \dfrac{-(-10) \pm \sqrt{(-10)^2 - 4(1)(7)}}{2(1)}$

$= \dfrac{10 \pm \sqrt{100 - 28}}{2} = \dfrac{10 \pm \sqrt{72}}{2}$

$= \dfrac{10 \pm 6\sqrt{2}}{2} = 5 \pm 3\sqrt{2}$

The solutions are $5 \pm 3\sqrt{2}$.

**26.** $x^2 + 4x - 7 = 0$

$a = 1, b = 4,$ and $c = -7$

$$x = \frac{-(4) \pm \sqrt{(4)^2 - 4(1)(-7)}}{2(1)}$$

$$= \frac{-4 \pm \sqrt{16 + 28}}{2} = \frac{-4 \pm \sqrt{44}}{2}$$

$$= \frac{-4 \pm 2\sqrt{11}}{2} = -2 \pm \sqrt{11}$$

The solutions are $-2 \pm \sqrt{11}$.

**27.** $2x^2 + x - 1 = 0$

$a = 2, b = 1,$ and $c = -1$

$$x = \frac{-(1) \pm \sqrt{(1)^2 - 4(2)(-1)}}{2(2)}$$

$$= \frac{-1 \pm \sqrt{1 + 8}}{4} = \frac{-1 \pm \sqrt{9}}{4} = \frac{-1 \pm 3}{4}$$

$$x = \frac{-1 + 3}{4} = \frac{1}{2} \quad \text{or} \quad x = \frac{-1 - 3}{4} = -1$$

The solutions are $-1$ and $\dfrac{1}{2}$.

**28.** $x^2 + 3x - 1 = 0$

$a = 1, b = 3,$ and $c = -1$

$$x = \frac{-(3) \pm \sqrt{(3)^2 - 4(1)(-1)}}{2(1)}$$

$$= \frac{-3 \pm \sqrt{9 + 4}}{2} = \frac{-3 \pm \sqrt{13}}{2}$$

The solutions are $\dfrac{-3 \pm \sqrt{13}}{2}$.

**29.** $9x^2 + 30x + 25 = 0$

$a = 9, b = 30,$ and $c = 25$

$$x = \frac{-(30) \pm \sqrt{(30)^2 - 4(9)(25)}}{2(9)}$$

$$= \frac{-30 \pm \sqrt{900 - 900}}{18} = \frac{-30 \pm \sqrt{0}}{18} = -\frac{5}{3}$$

The solution is $-\dfrac{5}{3}$.

**30.** $16x^2 - 72x + 81 = 0$

$a = 16, b = -72,$ and $c = 81$

$$x = \frac{-(-72) \pm \sqrt{(-72)^2 - 4(16)(81)}}{2(16)}$$

$$= \frac{72 \pm \sqrt{5184 - 5184}}{32} = \frac{72 \pm \sqrt{0}}{32} = \frac{9}{4}$$

The solution is $\dfrac{9}{4}$.

**31.** $15x^2 + 2 = 11x$

$15x^2 - 11x + 2 = 0$

$a = 15, b = -11,$ and $c = 2$

$$x = \frac{-(-11) \pm \sqrt{(-11)^2 - 4(15)(2)}}{2(15)}$$

$$= \frac{11 \pm \sqrt{121 - 120}}{30} = \frac{11 \pm \sqrt{1}}{30} = \frac{11 \pm 1}{30}$$

$$x = \frac{11 + 1}{30} = \frac{2}{5} \quad \text{or} \quad x = \frac{11 - 1}{30} = \frac{1}{3}$$

The solutions are $\dfrac{2}{5}$ and $\dfrac{1}{3}$.

**32.** $15x^2 + 2 = 13x$

$15x^2 - 13x + 2 = 0$

$a = 15, b = -13,$ and $c = 2$

$$x = \frac{-(-13) \pm \sqrt{(-13)^2 - 4(15)(2)}}{2(15)}$$

$$= \frac{13 \pm \sqrt{169 - 120}}{30} = \frac{13 \pm \sqrt{49}}{30} = \frac{13 \pm 7}{30}$$

$$x = \frac{13 + 7}{30} = \frac{2}{3} \quad \text{or} \quad x = \frac{13 - 7}{30} = \frac{1}{5}$$

The solutions are $\frac{2}{3}$ and $\frac{1}{5}$.

**33.** $2x^2 + x + 5 = 0$

$a = 2, b = 1,$ and $c = 5$

$$x = \frac{-(1) \pm \sqrt{(1)^2 - 4(2)(5)}}{2(2)}$$

$$= \frac{-1 \pm \sqrt{1 - 40}}{4} = \frac{-1 \pm \sqrt{-39}}{4}$$

There is no real solution because $\sqrt{-39}$ is not a real number.

**34.** $7x^2 - 3x + 1 = 0$

$a = 7, b = -3,$ and $c = 1$

$$x = \frac{-(-3) \pm \sqrt{(-3)^2 - 4(7)(1)}}{2(7)}$$

$$= \frac{3 \pm \sqrt{9 - 28}}{14} = \frac{3 \pm \sqrt{-19}}{14}$$

There is no real solution because $\sqrt{-19}$ is not a real number.

**35.** $x^2 - 7x - 1 = 0$

$a = 1, b = -7, c = -1$

$b^2 - 4ac = (-7)^2 - 4(1)(-1)$

$= 49 + 4$

$= 53$

Since the discriminant is a positive number, this equation has two distinct real solutions.

**36.** $x^2 + x + 5 = 0$

$a = 1, b = 1, c = 5$

$b^2 - 4ac = 1^2 - 4(1)(5)$

$= 1 - 20$

$= -19$

Since the discriminant is a negative number, this equation has no real solution.

**37.** $9x^2 + 1 = 6x$

$9x^2 - 6x + 1 = 0$

$a = 9, b = -6, c = 1$

$b^2 - 4ac = (-6)^2 - 4(9)(1)$

$= 36 - 36$

$= 0$

Since the discriminant is 0, this equation has one real solution.

**38.** $x^2 + 6x = 5$

$x^2 + 6x - 5 = 0$

$a = 1, b = 6, c = -5$

$b^2 - 4ac = 6^2 - 4(1)(-5)$

$= 36 + 20$

$= 56$

Since the discriminant is a positive number, this equation has two distinct real solutions.

**39.** $5x^2 + 4 = 0$

$a = 5,\ b = 0,\ c = 4$

$b^2 - 4ac = 0^2 - 4(5)(4)$

$= 0 - 80$

$= -80$

Since the discriminant is a negative number, this equation has no real solution.

**40.** $x^2 + 25 = 10x$

$x^2 - 10x + 25 = 0$

$a = 1,\ b = -10,\ c = 25$

$b^2 - 4ac = (-10)^2 - 4(1)(25)$

$= 100 - 100$

$= 0$

Since the discriminant is 0, this equation has one real solution.

**41.** $5z^2 + z - 1 = 0$

$a = 5,\ b = 1,\ c = -1$

$z = \dfrac{-1 \pm \sqrt{1^2 - 4(5)(-1)}}{2(5)}$

$z = \dfrac{-1 \pm \sqrt{1 + 20}}{10}$

$z = \dfrac{-1 \pm \sqrt{-21}}{10}$

The solutions are $\dfrac{-1 \pm \sqrt{21}}{10}$.

**42.** $4z^2 + 7z - 1 = 0$

$a = 4,\ b = 7,\ c = -1$

$z = \dfrac{-1 \pm \sqrt{7^2 - 4(4)(-1)}}{2(4)}$

$z = \dfrac{-7 \pm \sqrt{49 + 16}}{8}$

$z = \dfrac{-7 \pm \sqrt{65}}{8}$

The solutions are $\dfrac{-7 \pm \sqrt{65}}{8}$.

**43.**
$$4x^4 = x^2$$
$$4x^4 - x^2 = 0$$
$$x^2(4x^2 - 1) = 0$$
$$x^2(2x + 1)(2x - 1) = 0$$

$x^2 = 0$    or    $2x + 1 = 0$    or    $2x - 1 = 0$

$x = 0$    or    $2x = -1$    or    $2x = 1$

$x = -\dfrac{1}{2}$       $x = \dfrac{1}{2}$

The solutions are $0,\ -\dfrac{1}{2},$ and $\dfrac{1}{2}$.

**44.**
$$9x^3 = x$$
$$9x^3 - x = 0$$
$$x(9x^2 - 1) = 0$$
$$x(3x - 1)(3x + 1) = 0$$

$x = 0$    or    $3x - 1 = 0$    or    $3x + 1 = 0$

$3x = 1$    or    $3x = -1$

$x = \dfrac{1}{3}$       $x = -\dfrac{1}{3}$

The solutions are $0,\ \dfrac{1}{3},$ and $-\dfrac{1}{3}$.

**45.** $2x^2 - 15x + 7 = 0$

$(2x - 1)(x - 7) = 0$

$2x - 1 = 0$    or    $x - 7 = 0$

$2x = -1$    or      $x = 7$

$x = -\dfrac{1}{2}$

The solutions are $\dfrac{1}{2}$, 7.

**46.** $x^2 - 6x - 7 = 0$

$(x - 7)(x + 1) = 0$

$x - 7 = 0$    or    $x + 1 = 0$

$x = 7$    or      $x = -1$

The solutions are 7, −1.

**47.** $(3x - 1)^2 = 0$

$3x - 1 = \pm\sqrt{0}$

$3x - 1 = 0$

$3x = 1$

$x = \dfrac{1}{3}$

The solution is $\dfrac{1}{3}$.

**48.** $(2x - 3)^2 = 0$

$2x - 3 = \pm\sqrt{0}$

$2x - 3 = 0$

$2x = 3$

$x = \dfrac{3}{2}$

The solution is $\dfrac{3}{2}$.

**49.** $x^2 = 6x - 9$

$x^2 - 6x + 9 = 0$

$(x - 3)(x - 3) = 0$

$(x - 3)^2 = 0$

$x - 3 = \pm\sqrt{0}$

$x - 3 = 0$

$x = 3$

The solution is 3.

**50.** $x^2 = 10x - 25$

$x^2 - 10x + 25 = 0$

$(x - 5)(x - 5) = 0$

$(x - 5)^2 = 0$

$x - 5 = \pm\sqrt{0}$

$x - 5 = 0$

$x = 5$

The solution is 5.

**51.** $\left(\dfrac{1}{2}x - 3\right)^2 = 64$

$\dfrac{1}{2}x - 3 = \pm\sqrt{64}$

$\dfrac{1}{2}x - 3 = \pm 8$

$\dfrac{1}{2}x = 3 \pm 8$

$x = 2(3 \pm 8)$

$x = 2(+8)$    or    $x = 2(3 - 8)$

$x = 2(11)$    or    $x = 2(-5)$

$x = 22$      or    $x = -10$

The solutions are 22 and −10.

**52.** $\left(\dfrac{1}{3}x+1\right)^2 = 49$

$\dfrac{1}{3}x+1 = \pm\sqrt{49}$

$\dfrac{1}{3}x+1 = \pm 7$

$\dfrac{1}{3}x = -1 \pm 7$

$\dfrac{1}{3}x = -8$    or    $\dfrac{1}{3}x = 6$

$x = -24$    or    $x = 18$

The solutions are $-24$ and $18$.

**53.** $x^2 - 0.3x + 0.01 = 0$

$100\left(x^2 - 0.3x + 0.01\right) = 100(0)$

$100x^2 - 30x + 1 = 0$

$a = 100,\ b = -30,\ c = 1$

$x = \dfrac{-(-30) \pm \sqrt{(-30)^2 - 4(100)(1)}}{2(100)}$

$x = \dfrac{30 \pm \sqrt{900 - 400}}{200}$

$x = \dfrac{30 \pm \sqrt{500}}{200}$

$x = \dfrac{30 \pm \sqrt{100}\sqrt{5}}{200}$

$x = \dfrac{30 \pm 10\sqrt{5}}{200}$

$x = \dfrac{10\left(3 \pm \sqrt{5}\right)}{200}$

$x = \dfrac{3 \pm \sqrt{5}}{20}$

The solutions are $\dfrac{3 \pm \sqrt{5}}{20}$

**54.** $x^2 + 0.6x - 0.16 = 0$

$100x^2 + 60x - 16 = 0$

$a = 100,\ b = 60,\ c = -16$

$x = \dfrac{-60 \pm \sqrt{60^2 - 4(100)(-16)}}{2(100)}$

$x = \dfrac{-60 \pm \sqrt{3600 + 6400}}{200}$

$x = \dfrac{-60 \pm \sqrt{10000}}{200}$

$x = \dfrac{-60 \pm 100}{200}$

$x = \dfrac{-60 + 100}{200}$    or    $x = \dfrac{-60 - 100}{200}$

$x = \dfrac{40}{200} = \dfrac{1}{5}$    or    $x = \dfrac{-160}{200} = -\dfrac{4}{5}$

The solutions are $\dfrac{1}{5}$ and $-\dfrac{4}{5}$.

**55.** $\dfrac{1}{10}x^2 + x - \dfrac{1}{2} = 0$

$10\left(\dfrac{1}{10}x^2 + x - \dfrac{1}{2}\right) = 0$

$x^2 + 10x - 5 = 0$

$a = 1,\ b = 10,\ c = -5$

$x = \dfrac{-10 \pm \sqrt{10^2 - 4(1)(-5)}}{2(1)}$

$x = \dfrac{-10 \pm \sqrt{100 + 20}}{2}$

$$x = \frac{-10 \pm \sqrt{120}}{2}$$

$$x = \frac{-10 \pm \sqrt{4}\sqrt{30}}{2}$$

$$x = \frac{-10 \pm 2\sqrt{30}}{2}$$

$$x = \frac{2\left(-5 \pm \sqrt{30}\right)}{2}$$

$$x = -5 \pm \sqrt{30}$$

The solutions are $-5 \pm \sqrt{30}$.

**56.** $\dfrac{1}{12}x^2 - \dfrac{1}{2}x + \dfrac{1}{3} = 0$

$\quad x^2 - 6x + 4 = 0$

$\quad a = 1,\ b = -6,\ c = 4$

$$x = \frac{-(-6) \pm \sqrt{(-6)^2 - 4(1)(4)}}{2(1)}$$

$$x = \frac{6 \pm \sqrt{36 - 16}}{2}$$

$$x = \frac{6 \pm \sqrt{20}}{2}$$

$$x = \frac{6 \pm \sqrt{4 \cdot 5}}{2}$$

$$x = \frac{6 \pm 2\sqrt{5}}{2}$$

$$x = 3 \pm \sqrt{5}$$

The solutions are $3 \pm \sqrt{5}$.

**57.** Let $y = 186$

$\quad y = -8x^2 - 13x + 552$

$\quad 186 = -8x^2 - 13x + 552$

$\quad 0 = -8x^2 - 13x + 366$

$a = -8, b = -13,$ and $c = 366$

$$x = \frac{-(-13) \pm \sqrt{(-13)^2 - 4(-8)(366)}}{2(-8)}$$

$$= \frac{13 \pm \sqrt{169 + 11,712}}{-16}$$

$$= \frac{13 \pm \sqrt{11,881}}{-16} = \frac{13 \pm 109}{-16}$$

$$x = \frac{13 + 109}{-16} \approx 7.6 \quad \text{or} \quad x = \frac{13 - 109}{-16} = 6$$

The number of years cannot be negative so the year will be $1999 + 6 = 2005$.

**58.** Let $y = 331$

$\quad y = -93x^2 + 263x + 379$

$\quad 331 = -93x^2 + 263x + 379$

$\quad 0 = -93x^2 + 263x + 48$

$\quad a = -93, b = 263,$ and $c = 48$

$$x = \frac{-(263) \pm \sqrt{(263)^2 - 4(-93)(48)}}{2(-93)}$$

$$= \frac{-263 \pm \sqrt{69,169 + 17,856}}{-186}$$

$$= \frac{-263 \pm \sqrt{87,025}}{-186} = \frac{-263 \pm 295}{-186}$$

$$x = \frac{-263 - 295}{-186} = 3$$

$$\text{or} \quad x = \frac{-263 + 295}{-186} \approx -0.17$$

The number of years cannot be negative so the year will be $1999 + 3 = 2002$.

**59.** $\sqrt{-144} = \sqrt{-1 \cdot 144}$
$= \sqrt{-1} \cdot \sqrt{144}$
$= i \cdot 12$
$= 12i$

**60.** $\sqrt{-36} = \sqrt{36 \cdot i} = \sqrt{36} \cdot \sqrt{-1} = 6i$

**61.** $\sqrt{-108} = \sqrt{-1 \cdot 36 \cdot 3}$
$= \sqrt{-1} \cdot \sqrt{36} \cdot \sqrt{3}$
$= i \cdot 6\sqrt{3}$
$= 6i\sqrt{3}$

**62.** $\sqrt{-500} = \sqrt{100 \cdot 5 \cdot -1}$
$= \sqrt{100} \cdot \sqrt{-1} \cdot \sqrt{5}$
$= 10i\sqrt{5}$

**63.** $2i(3 - 5i) = (2i)(3) - (2i)(5i)$
$= 6i - 10i^2$
$= 6i - 10(-1)$
$= 10 + 6i$

**64.** $i(-7 - i) = (i)(-7) - (i)(i)$
$= -7i - i^2$
$= -7i - (-1)$
$= 1 - 7i$

**65.** $(7 - i) + (14 - 9i) = 7 - i + 14 - 9i$
$= 7 + 14 + (-i - 9i)$
$= 21 - 10i$

**66.** $(10 - 4i) + (9 - 21i) = 10 - 4i + 9 - 21i$
$= 19 - 25i$

**67.** $3 - (11 + 2i) = 3 - 11 - 2i = -8 - 2i$

**68.** $(-4 - 3i) + 5i = -4 - 3i + 5i = -4 + 2i$

**69.** $(2 - 3i)(3 - 2i) = 2(3) + 2(-2i) - 3i(3) - 3i(-2i)$
$= 6 - 4i - 9i + 6i^2$
$= 6 - 13i + 6(-1)$
$= 6 - 13i - 6$
$= -13i$

**70.** $(2 + 5i)(5 - i) = 10 - 2i + 25i - 5i^2$
$= 10 - 2i + 25i - 5(-1)$
$= 10 + 5 - 2i + 25i$
$= 15 + 23i$

**71.** $(3 - 4i)(3 + 4i) = (3)^2 - (4i)^2$
$= 9 - 16i^2$
$= 9 - 16(-1)$
$= 9 + 16$
$= 25$

**72.** $(7 - 2i)(7 - 2i) = 49 - 14i - 14i + 4i^2$
$= 49 - 4 - 28i$
$= 45 - 28i$

**73.** $\dfrac{2 - 6i}{4i} = \dfrac{(2 - 6i)(-i)}{4i(i)}$
$= \dfrac{-2i + 6i^2}{-4i^2}$
$= \dfrac{-2i + 6(-1)}{-4(-1)}$
$= \dfrac{-2i - 6}{4}$
$= \dfrac{2(-i - 3)}{4}$

$$= -\frac{i}{2} - \frac{3}{2}$$

$$= -\frac{3}{2} - \frac{1}{2}i$$

**74.** $\dfrac{5-i}{2i} = \dfrac{5-i}{2i} \cdot \dfrac{2i}{2i}$

$$= \frac{2i(5-i)}{4i^2}$$

$$= \frac{10i - 21i^2}{-4}$$

$$= \frac{2 + 10i}{-4}$$

$$= \frac{1 + 5i}{-2}$$

$$= -\frac{1}{2} - \frac{5}{2}i$$

**75.** $\dfrac{4-i}{1+2i} = \dfrac{(4-i)(1-2i)}{(1+2i)(1-2i)}$

$$= \frac{4(1) + 4(-2i) - i(1) - i(-2i)}{(1)^2 - (2i)^2}$$

$$= \frac{4 - 8i - i + 2i^2}{1 - 4i^2}$$

$$= \frac{4 - 9i + 2(-1)}{1 - 4(-1)}$$

$$= \frac{4 - 9i - 2}{1 + 4}$$

$$= \frac{2 - 9i}{5}$$

$$= \frac{2}{5} - \frac{9}{5}i$$

**76.** $\dfrac{1+3i}{2-7i} = \dfrac{1+3i}{2-7i} \cdot \dfrac{2+7i}{2+7i}$

$$= \frac{2 + 7i + 6i + 21i^2}{4 - 49i^2}$$

$$= \frac{2 - 21 + 13i}{4 + 49}$$

$$= \frac{-19 + 13i}{53}$$

$$= -\frac{19}{53} + \frac{13}{53}i$$

**77.** $3x^2 = -48$

$$x^2 = -16$$

$$x = \pm\sqrt{-16}$$

$$x = \pm\sqrt{4i}$$

The solutions are $\pm 4i$.

**78.** $5x^2 = -125$

$$\frac{5x^2}{5} = \frac{-125}{5}$$

$$x^2 = -25$$

$$x = \pm\sqrt{-25}$$

$$x = \pm\sqrt{25 \cdot -1}$$

$$x = \pm 5i$$

The solutions are $\pm 5i$.

**79.** $x^2 - 4x + 13 = 0$

$$a = 1, b = -4, c = 13$$

$$x = \frac{-(-4) \pm \sqrt{(-4)^2 - 4(1)(13)}}{2(1)}$$

$$x = \frac{4 \pm \sqrt{16 - 52}}{2}$$

$$x = \frac{4 \pm \sqrt{-36}}{2}$$

$$x = \frac{4 \pm 6i}{2}$$

$$x = \frac{2(2 \pm 3i)}{2}$$

$$x = 2 \pm 3i$$

The solutions are $2 \pm 3i$.

**80.** $x^2 + 4x + 11 = 0$

$a = 1, \; b = 4, \; c = 11$

$$x = \frac{-4 \pm \sqrt{4^2 - 4(1)(11)}}{2(1)}$$

$$x = \frac{-4 \pm \sqrt{16 - 44}}{2}$$

$$x = \frac{-4 \pm \sqrt{-28}}{2}$$

$$x = \frac{-4 \pm \sqrt{4 \cdot 7 \cdot -1}}{2}$$

$$x = \frac{-4 \pm 2i\sqrt{7}}{2}$$

$$x = -2 \pm i\sqrt{7}$$

The solutions are $-2 \pm i\sqrt{7}$.

**81.** $y = -3x^2$

$$x = \frac{-b}{2a} = \frac{-0}{2(-3)} = 0$$

$$y = -3(0)^2 = 0$$

vertex $(0, 0)$

axis of symmetry $x = 0$

Parabola opens downward because $a < 0$.

**82.** $y = -\frac{1}{2}x^2$

$$x = \frac{-b}{2a} = \frac{-0}{2(-\frac{1}{2})} = 0$$

$$y = -\frac{1}{2}(0)^2 = 0$$

vertex $(0, 0)$

axis of symmetry $x = 0$

Parabola opens downward because $a < 0$.

**83.** $y = (x - 3)^2$

$y = x^2 - 6x + 9$

$$x = \frac{-b}{2a} = \frac{-(-6)}{2(1)} = 3$$

$$y = 3^2 - 6(3) + 9 = 0$$

vertex $(3, 0)$

axis of symmetry $x = 3$

Parabola opens upward because $a > 0$.

**84.** $y = (x - 5)^2$

$y = x^2 - 10x + 25$

$$x = \frac{-b}{2a} = \frac{-(-10)}{2(1)} = 5$$

$$y = 5^2 - 10(5) + 25 = 0$$

vertex $(5, 0)$

axis of symmetry $x = 5$

Parabola opens upward because $a > 0$.

**85.** $y = 3x^2 - 7$

$$x = \frac{-b}{2a} = \frac{-0}{2(-3)} = 0$$

$$y = 3(0)^2 - 7 = -7$$

vertex $(0, -7)$

axis of symmetry $x = 0$

Parabola opens upward because $a > 0$.

| $x$ | $y$ |
|---|---|
| $-2$ | $-4$ |
| $-1$ | $-1$ |
| $0$ | $0$ |
| $1$ | $-1$ |
| $2$ | $-4$ |

**86.** $y = -2x^2 + 25$

$x = \dfrac{-b}{2a} = \dfrac{-0}{2(-2)} = 0$

$y = -2(0)^2 + 25 = 25$

vertex $(0, 25)$

axis of symmetry $x = 0$

Parabola opens downward because $a < 0$.

**87.** $y = -5(x - 72)^2 + 14$

vertex $(72, 14)$

axis of symmetry $x = 72$

Parabola opens downward.

**88.** $y = 2(x - 35)^2 - 21$

vertex $(35, -21)$

axis of symmetry $x = 35$

Parabola opens upward.

**89.** $y = -x^2$

vertex: $x = -\dfrac{b}{2a} = -\dfrac{0}{2(-1)} = 0$

$y = -(0)^2$

$(0, 0)$

$y$-intercept: $x = 0$, $y = 0$, $(0, 0)$

$x$-intercepts: $y = 0$, $x = 0$, $(0, 0)$

**90.** $y = 4x^2$

vertex: $x = -\dfrac{b}{2a} = -\dfrac{0}{2(4)} = 0$

$y = 4(0)^2 = 0$, $(0, 0)$

$y$-intercept: $x = 0$, $y = 0$, $(0, 0)$

$x$-intercepts: $y = 0$, $x = 0$, $(0, 0)$

| $x$ | $y$ |
|---|---|
| $-2$ | $16$ |
| $-1$ | $4$ |
| $0$ | $0$ |
| $1$ | $4$ |
| $2$ | $16$ |

**91.** $y = \dfrac{1}{2}x^2$

vertex: $x = -\dfrac{b}{2a} = -\dfrac{0}{2\left(\dfrac{1}{2}\right)} = 0$

$y = \dfrac{1}{2}(0)^2 = 0$, $(0, 0)$

$y$-intercept: $x = 0$, $y = 0$, $(0, 0)$

$x$-intercepts: $y = 0$, $x = 0$, $(0, 0)$

$x$-intercepts: $y = 0$, $x = 0$, $(0, 0)$

| x | y |
|---|---|
| -2 | 2 |
| -1 | 1/2 |
| 0 | 0 |
| 1 | 1/2 |
| 2 | 2 |

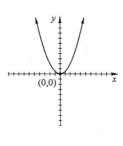

**92.** $y = \dfrac{1}{4}x^2$

vertex: $x = -\dfrac{b}{2a} = -\dfrac{0}{2\left(\dfrac{1}{4}\right)} = 0$

$y = \dfrac{1}{4}(0)^2 = 0, \ (0,0)$

$y$-intercept: $x = 0, \ y = 0, \ (0,0)$

| x | y |
|---|---|
| -2 | 1 |
| -1 | 1/4 |
| 0 | 0 |
| 1 | 1/4 |
| 2 | 1 |

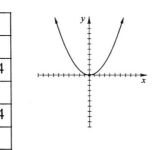

**93.** $y = x^2 + 5x + 6$

vertex: $x = -\dfrac{b}{2a} = -\dfrac{5}{2(1)} = -\dfrac{5}{2}$

$y = \left(-\dfrac{5}{2}\right)^2 + 5\left(-\dfrac{5}{2}\right) + 6 = -\dfrac{1}{4}$

$\left(-\dfrac{5}{2}, -\dfrac{1}{4}\right)$

$y$-intercept: $x = 0, \ y = 6, \ (0,6)$

$x$-intercepts: $y = 0$,

$0 = x^2 + 5x + 6$

$0 = (x+2)(x+3)$

$x = -2 \ $ or $\ x = -3$

$(-2,0) \ $ and $\ (-3,0)$

| x | y |
|---|---|
| -5 | 6 |
| -3 | 0 |
| -5/2 | -1/4 |
| -2 | 0 |
| 0 | 6 |

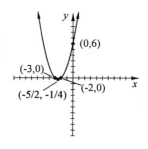

**94.** $y = x^2 - 4x - 8$

vertex: $x = -\dfrac{b}{2a} = -\dfrac{-4}{2(1)} = 2$

$y = (2)^2 - 4(2) - 8 = -12$

$(2,-12)$

$y$-intercept: $x = 0, \ y = -8, \ (0,-8)$

$x$-intercepts: $y = 0$,

$0 = x^2 - 4x - 8$

$x = \dfrac{-(-4) \pm \sqrt{(-4)^2 - 4(1)(-8)}}{2(1)}$

$x = \dfrac{4 \pm \sqrt{48}}{2} = \dfrac{4 \pm 4\sqrt{3}}{2}$

$x \approx 5.46 \ $ or $\ x \approx -1.46$

$(5.46, 0) \ $ and $\ (-1.46, 0)$

| $x$ | $y$ |
|------|------|
| $-1.46$ | $0$ |
| $0$ | $-8$ |
| $2$ | $-12$ |
| $4$ | $-8$ |
| $5.46$ | $0$ |

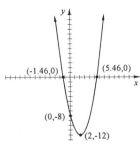

**95.** $y = 2x^2 - 11x - 6$

vertex: $x = -\dfrac{b}{2a} = -\dfrac{-11}{2(2)} = \dfrac{11}{4}$

$$y = 2\left(\dfrac{11}{4}\right)^2 - 11\left(\dfrac{11}{4}\right) - 6 = -\dfrac{169}{8}$$

$$\left(\dfrac{11}{4}, -\dfrac{169}{8}\right)$$

*y*-intercept: $x = 0,\ y = -6,\ (0, -6)$

*x*-intercepts: $y = 0,$

$0 = 2x^2 - 11x - 6$

$0 = (2x + 1)(x - 6)$

$x = -\dfrac{1}{2}$   or   $x = 6$

$\left(-\dfrac{1}{2}, 0\right)$ and $(6, 0)$

| $x$ | $y$ |
|------|------|
| $-1/2$ | $0$ |
| $0$ | $-6$ |
| $11/4$ | $-169/8$ |
| $11/2$ | $-6$ |
| $6$ | $0$ |

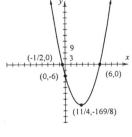

**96.** $y = 3x^2 - x - 2$

vertex: $x = -\dfrac{b}{2a} = -\dfrac{-1}{2(3)} = \dfrac{1}{6}$

$$y = 3\left(\dfrac{1}{6}\right)^2 - \left(\dfrac{1}{6}\right) - 2 = -\dfrac{25}{12}$$

$$\left(\dfrac{1}{6}, -\dfrac{25}{12}\right)$$

*y*-intercept: $x = 0,\ y = -2,\ (0, -2)$

*x*-intercepts: $y = 0,$

$0 = 3x^2 - x - 2$

$0 = (3x + 2)(x - 1)$

$x = -\dfrac{2}{3}$   or   $x = 1$

$\left(-\dfrac{2}{3}, 0\right)$ and $(1, 0)$

| $x$ | $y$ |
|------|------|
| $-2/3$ | $0$ |
| $0$ | $-2$ |
| $1/6$ | $-25/12$ |
| $1$ | $0$ |

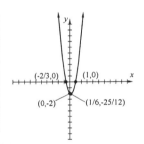

**97.** The equation has one solution because the graph intersects the *x*-axis at one point $(-2, 0)$.

**98.** The equation has two solutions because the graph intersects the *x*-axis at two points: $\left(-\dfrac{3}{2}, 0\right)$ and $(3, 0)$.

**99.** The equation has no real solution because the graph does not intersect the $x$-axis.

**100.** The equation has two solutions because the graph intersects the $x$-axis at two points: $(-2, 0)$ and $(2, 0)$.

## Chapter 9 Test

**1.**
$$5k^2 = 80$$
$$k^2 = 16$$
$$k = \pm\sqrt{16}$$
$$k = \pm 4$$

The solutions are $\pm 4$.

**2.** $(3m-5)^2 = 8$
$$3m - 5 = \pm\sqrt{8}$$
$$3m - 5 = \pm 2\sqrt{2}$$
$$3m = 5 \pm 2\sqrt{2}$$
$$m = \frac{5 \pm 2\sqrt{2}}{3}$$

The solutions are $\dfrac{5 \pm 2\sqrt{2}}{3}$.

**3.** $x^2 - 26x + 160 = 0$
$$x^2 - 26x = -160$$
$$x^2 - 26x + 169 = -160 + 169$$
$$(x-13)^2 = 9$$
$$x - 13 = \pm\sqrt{9}$$
$$x = 13 \pm 3$$
$$x = 13 - 3 \quad \text{or} \quad x = 13 + 3$$
$$x = 10 \qquad\qquad x = 16$$

The solutions are 10 and 16.

**4.** $3x^2 + 12x - 4 = 0$
$$x^2 + 4x - \frac{4}{3} = 0$$
$$x^2 + 4x = \frac{4}{3}$$
$$x^2 + 4x + 4 = \frac{4}{3} + 4$$
$$(x+2)^2 = \frac{16}{3}$$
$$x + 2 = \pm\sqrt{\frac{16}{3}}$$
$$x = -2 \pm \frac{4}{\sqrt{3}}$$
$$x = -2 \pm \frac{4\sqrt{3}}{3}$$
$$x = \frac{-6 \pm 4\sqrt{3}}{3}$$

The solutions are $\dfrac{-6 \pm 4\sqrt{3}}{3}$.

**5.** $x^2 - 3x - 10 = 0$
$$a = 1, b = -3, \text{ and } c = -10$$
$$x = \frac{-(-3) \pm \sqrt{(-3)^2 - 4(1)(-10)}}{2(1)}$$
$$= \frac{3 \pm \sqrt{9 + 40}}{2} = \frac{3 \pm \sqrt{49}}{2} = \frac{3 \pm 7}{2}$$
$$x = \frac{3 - 7}{2} = -2 \quad \text{or} \quad x = \frac{3 + 7}{2} = 5$$

The solutions are $-2$ and 5.

**6.** $p^2 - \dfrac{5}{3}p - \dfrac{1}{3} = 0$

$3p^2 - 5p - 1 = 0$

$a = 3, b = -5,$ and $c = -1$

$p = \dfrac{-(-5) \pm \sqrt{(-5)^2 - 4(3)(-1)}}{2(3)}$

$= \dfrac{5 \pm \sqrt{25 + 12}}{6} = \dfrac{5 \pm \sqrt{37}}{6}$

The solutions are $\dfrac{5 \pm \sqrt{37}}{6}$.

**7.** $(3x - 5)(x + 2) = -6$

$3x^2 + x - 10 = -6$

$3x^2 + x - 4 = 0$

$(3x + 4)(x - 1) = 0$

$3x + 4 = 0$   or   $x - 1 = 0$

$x = -\dfrac{4}{3}$      $x = 1$

The solutions are $-\dfrac{4}{3}$ and 1.

**8.** $(3x - 1)^2 = 16$

$3x - 1 = \pm\sqrt{16}$

$3x = 1 \pm 4$

$x = \dfrac{1 \pm 4}{3}$

$x = \dfrac{1 - 4}{3} = -1$   or   $x = \dfrac{1 + 4}{3} = \dfrac{5}{3}$

The solutions are $-1$ and $\dfrac{5}{3}$.

**9.** $3x^2 - 7x - 2 = 0$

$a = 3, b = -7,$ and $c = -2$

$z = \dfrac{-(-7) \pm \sqrt{(-7)^2 - 4(3)(-2)}}{2(3)}$

$= \dfrac{7 \pm \sqrt{49 + 24}}{6} = \dfrac{7 \pm \sqrt{73}}{6}$

The solutions are $\dfrac{7 \pm \sqrt{73}}{6}$.

**10.** $x^2 - 4x + 5 = 0$

$a = 1, b = -4,$ and $c = 5$

$x = \dfrac{-(-4) \pm \sqrt{(-4)^2 - 4(1)(5)}}{2(1)}$

$= \dfrac{4 \pm \sqrt{16 - 20}}{2} = \dfrac{4 \pm \sqrt{-4}}{2}$

$= \dfrac{4 \pm 2i}{2} = 2 \pm i$

The solutions are $2 \pm i$.

**11.**     $3x^2 - 7x + 2 = 0$

$(3x - 1)(x - 2) = 0$

$3x - 1 = 0$   or   $x - 2 = 0$

$x = \dfrac{1}{3}$      $x = 2$

The solutions are $\dfrac{1}{3}$ and 2.

**12.** $2x^2 - 6x + 1 = 0$

$a = 2, b = -6,$ and $c = 1$

$$x = \frac{-(-6) \pm \sqrt{(-6)^2 - 4(2)(1)}}{2(2)}$$

$$= \frac{6 \pm \sqrt{36 - 8}}{4} = \frac{6 \pm \sqrt{28}}{4}$$

$$= \frac{6 \pm 2\sqrt{7}}{4} = \frac{3 \pm \sqrt{7}}{2}$$

The solutions are $\dfrac{3 \pm \sqrt{7}}{2}$.

**13.**
$$9x^3 = x$$
$$9x^3 - x = 0$$
$$x(9x^2 - 1) = 0$$
$$x(3x + 1)(3x - 1) = 0$$
$$x = 0 \quad \text{or} \quad 3x + 1 = 0 \quad \text{or} \quad 3x - 1 = 0$$
$$x = 0 \qquad x = -\frac{1}{3} \qquad x = \frac{1}{3}$$

The solutions are $0$ and $\pm\dfrac{1}{3}$.

**14.** $\sqrt{-25} = i\sqrt{25} = 5i$

**15.** $\sqrt{-200} = i\sqrt{100 \cdot 2} = i\sqrt{100} \cdot \sqrt{2} = 10i\sqrt{2}$

**16.** $(3 + 2i) + (5 - i) = 3 + 2i + 5 - i = 8 + i$

**17.** $(3 + 2i) - (3 - 2i) = 3 + 2i - 3 + 2i = 4i$

**18.** $(3 + 2i)(3 - 2i) = (3)^2 - (2i)^2 = 9 - 4i^2$
$$= 9 - 4(-1) = 9 + 4 = 13$$

**19.** $\dfrac{3 - i}{1 + 2i} = \dfrac{3 - i}{1 + 2i} \cdot \dfrac{1 - 2i}{1 - 2i}$

$$= \frac{3 - 6i - i + 2i^2}{1^2 - (2i)^2}$$

$$= \frac{3 - 7i + 2(-1)}{1 - 4i^2}$$

$$= \frac{1 - 7i}{1 + 4}$$

$$= \frac{1}{5} - \frac{7}{5}i$$

**20.** $y = -3x^2$

vertex: $x = -\dfrac{b}{2a} = -\dfrac{0}{2(-3)} = 0$

$$y = -3(0)^2 = 0, \quad (0, 0)$$

$y$-intercept: $x = 0, \ y = 0, \ (0, 0)$

$x$-intercepts: $y = 0, \ x = 0, \ (0, 0)$

| $x$ | $y$ |
|-----|-----|
| $-2$ | $-12$ |
| $-1$ | $-3$ |
| $0$ | $0$ |
| $1$ | $-3$ |
| $2$ | $-12$ |

**21.** $y = x^2 - 7x + 10$

vertex: $x = -\dfrac{b}{2a} = -\dfrac{-7}{2(1)} = \dfrac{7}{2}$

$$y = \left(\frac{7}{2}\right)^2 - 7\left(\frac{7}{2}\right) + 10 = -\frac{9}{4}$$

$$\left(\frac{7}{2}, -\frac{9}{4}\right)$$

$y$-intercept: $x = 0, \ y = 10, \ (0, 10)$

$x$-intercepts: $y = 0$,

$0 = x^2 - 7x + 10$

$0 = (x - 2)(x - 5)$

$x = 2$  or  $x = 5$

$(2, 0)$ and $(5, 0)$

| $x$ | $y$ |
|-----|------|
| 0 | 10 |
| 2 | 0 |
| 7/2 | −9/4 |
| 5 | 0 |
| 7 | 10 |

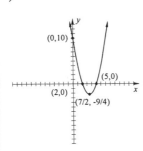

**22.** Let $h = 120.75$

$h = 16t^2$

$16t^2 = h$

$16t^2 = 120.75$

$t^2 = \dfrac{120.75}{16}$

$t = \pm\sqrt{\dfrac{120.75}{16}} \approx \pm 2.7$

The length of time is not a negative number so the dive lasted 2.7 seconds.

**Cumulative Review Chapter 9**

**1. a.** $\dfrac{x - y}{12 + x} = \dfrac{2 - (-5)}{12 + 2} = \dfrac{7}{14} = \dfrac{1}{2}$

 **b.** $x^2 - y = (2)^2 - (-5) = 4 + 5 = 9$

**2. a.** $\dfrac{x - y}{7 - x} = \dfrac{(-4) - (7)}{7 - (-4)} = \dfrac{-11}{11} = -1$

 **b.** $x^2 + 2y = (-4)^2 + 2(7) = 16 + 14 = 30$

**3. a.** $2x + 3x + 5 + 2 = 5x + 7$

 **b.** $-5a - 3 + a + 2 = -4a - 1$

 **c.** $4y - 3y^2 = 4y - 3y^2$

 **d.** $2.3x + 5x - 6 = 7.3x - 6$

 **e.** $-\dfrac{1}{2}b + b = \dfrac{1}{2}b$

**4. a.** $4x - 3 + 7 - 5x = -x + 4$

 **b.** $-6y + 3y - 8 + 8y = 5y - 8$

 **c.** $2 + 8.1a + a - 6 = 9.1a - 4$

 **d.** $2x^2 - 2x = 2x^2 - 2x$

**5. a.** $x$-intercept: $(-3, 0)$

   $y$-intercept: $(0, 2)$

 **b.** $x$-intercepts: $(-4, 0), (-1, 0)$

   $y$-intercept: $(0, 1)$

 **c.** $x$-intercept: $(0, 0)$

   $y$-intercept: $(0, 0)$

 **d.** $x$-intercept: $(2, 0)$

   $y$-intercept: none

 **e.** $x$-intercepts: $(-1, 0), (3, 0)$

   $y$-intercepts: $(0, 2), (0, -1)$

**6. a.** $x$-intercept: $(4, 0)$

   $y$-intercept: $(0, 1)$

 **b.** $x$-intercepts: $(-2, 0), (0, 0), (3, 0)$

   $y$-intercept: $(0, 0)$

 **c.** $x$-intercept: none

   $y$-intercept: $(0, -3)$

 **d.** $x$-intercepts: $(-3, 0), (3, 0)$

   $y$-intercepts: $(0, -3), (0, 3)$

**7.** $y = -\dfrac{1}{5}x + 1 : m_1 = -\dfrac{1}{5}$

$2x + 10y = 30$

$y = -\dfrac{1}{5}x + 3 : m_2 = -\dfrac{1}{5}$

$m_1 = m_2$. They are parallel

**8.** $y = 3x + 7 : m_1 = 3$

$x + 3y = -15$

$y = -\dfrac{1}{3}x - 5 : m_2 = -\dfrac{1}{3}$

$m_1 \cdot m_2 = -1$. They are perpendicular

**9.** $2x + y = 7$ $\qquad$ $2y = -4x$

| $x$ | $y$ |
|-----|-----|
| 0 | 7 |
| −2 | 3 |

| $x$ | $y$ |
|-----|-----|
| 0 | 0 |
| 6 | −3 |

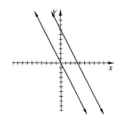

There is no solution.

**10.** $y = x + 2$ $\qquad$ $2x + y = 5$

| $x$ | $y$ |
|-----|-----|
| 0 | 2 |
| 5 | 7 |

| $x$ | $y$ |
|-----|-----|
| 0 | 5 |
| 3 | −1 |

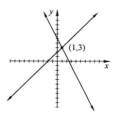

The solution.is $(1, 3)$

**11.** $\begin{cases} 7x - 3y = -14 \\ -3x + y = 6 \end{cases}$

Multiply the second equation by 3.

$\phantom{-}7x - 3y = -14$

$\underline{-9x + 3y = 18}$

$-2x \phantom{+3y} = 4$

$\phantom{-2}x = -2$

Let $x = -2$ in the second equation.

$-3(-2) + y = 6$

$6 + y = 6$

$y = 0$

The solution of the system is $(-2, 0)$.

**12.** $\begin{cases} 5x + y = 3 \\ y = -5x \end{cases}$

Substitute $-5x$ for $y$ in the first equation.

$5x + (-5x) = 3$

$0 = 3$

There is no solution.

**13.** $\begin{cases} 3x - 2y = 2 \\ -9x + 6y = -6 \end{cases}$

Multiply the first equation by 3.

$9x - 6y = 6$

$\underline{-9x + 6y = -6}$

$\phantom{-9x + 6y} 0 = 0$

The system has an infinite number of solutions.

**14.** $\begin{cases} -2x + y = 7 \\ 6x - 3y = -21 \end{cases}$

Multiply the first equation by 3.

$-6x + 3y = 21$

$\underline{6x - 3y = -21}$

$\phantom{6x - 3y} 0 = 0$

The system has an infinite number of solutions.

**15.** Let $x$ = the rate of Albert

$Rate \cdot Time = Distance$

| | | | |
|---|---|---|---|
| Albert | $x$ | 2 | $2x$ |
| Louis | $x+1$ | 2 | $2(x+1)$ |
| Total | | | 15 |

$2x + 2(x+1) = 15$

$2x + 2x + 2 = 15$

$4x + 2 = 15$

$4x = 13$

$x = 3.25$

$x + 1 = 3.25 + 1 = 4.25$

Albert: 3.25 mph; Louis: 4.25 mph.

**16.** Let $x$ = the number of dimes, and

$15 - x$ = the number of quarters.

$No. of\ Coins \cdot Value = Amt\ of\ Money$

| | | | |
|---|---|---|---|
| Dimes | $x$ | .1 | $.1x$ |
| Quarters | $15-x$ | .25 | $.25(15-x)$ |
| Total | 15 | | 2.85 |

$.1x + .25(15 - x) = 2.85$

$.1x + 3.75 - .25x = 2.85$

$-.15x + 3.75 = 2.85$

$-.15x = -.9$

$x = 6$

$15 - x = 15 - 6 = 9$

There are 6 dimes and 9 quarters in the purse.

**17.** $-3x + 4y < 12$      $x \geq 2$

Test $(0,0)$      Shade right

$-3(0) + 4(0) \overset{?}{<} 12$

True

Shade below

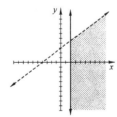

**18.** $2x - y \leq 6$      $y \geq 2$

Test $(0,0)$      Shade above

$2(0) - (0) \overset{?}{\leq} 6$

True

Shade above

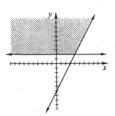

**19. a.** $\left(\dfrac{-5x^2}{y^3}\right)^2 = \dfrac{(-5)^2 x^4}{y^6} = \dfrac{25x^4}{y^6}$

**b.** $\dfrac{\left(x^3\right)^4 x}{x^7} = \dfrac{x^{12}x}{x^7} = x^{12+1-7} = x^6$

**c.** $\dfrac{(2x)^5}{x^3} = \dfrac{2^5 x^5}{x^3} = 32x^{5-3} = 32x^2$

**d.** $\dfrac{\left(a^2 b\right)^3}{a^3 b^2} = \dfrac{a^6 b^3}{a^3 b^2} = a^{6-3}b^{3-2} = a^3 b$

**20. a.** $\left(\dfrac{-6x}{y^3}\right)^3 = \dfrac{(-6)^3 x^3}{y^9} = \dfrac{-216x^3}{y^9}$

**b.** $\dfrac{a^2 b^7}{\left(2b^2\right)^5} = \dfrac{a^2 b^7}{2^5 b^{10}} = \dfrac{a^2}{32b^{10-7}} = \dfrac{a^2}{32b^3}$

**c.** $\dfrac{(3y)^2}{y^2} = \dfrac{3^2 y^2}{y^2} = 9y^{2-2} = 9$

**d.** $\dfrac{\left(x^2 y^4\right)^2}{xy^3} = \dfrac{x^4 y^8}{xy^3} = x^{4-1}y^{8-3} = x^3 y^5$

**21.** $(5x-1)\left(2x^2 + 15x + 18\right) = 0$

$(5x-1)(2x+3)(x+6) = 0$

$5x-1 = 0 \ \text{ or } \ 2x+3 = 0 \ \text{ or } \ x+6 = 0$

$x = \dfrac{1}{5} \qquad x = -\dfrac{3}{2} \qquad x = -6$

**22.** $(x+1)\left(2x^2 - 3x - 5\right) = 0$

$(x+1)(2x-5)(x+1) = 0$

$x+1 = 0 \ \text{ or } \ 2x-5 = 0$

$x = -1 \qquad\qquad x = \dfrac{5}{2}$

**23.** $\dfrac{\frac{45}{x}} {} = \dfrac{5}{7}$

$7x\left(\dfrac{45}{x}\right) = 7x\left(\dfrac{5}{7}\right)$

$315 = 5x$

$63 = x$

**24.** $\dfrac{2x+7}{3} = \dfrac{x-6}{2}$

$6\left(\dfrac{2x+7}{3}\right) = 6\left(\dfrac{x-6}{2}\right)$

$2(2x+7) = 3(x-6)$

$4x+14 = 3x-18$

$x+14 = -18$

$x = -32$

**25. a.** $\sqrt[4]{16} = \sqrt[4]{2^4} = 2$

**b.** $\sqrt[5]{-32} = \sqrt[5]{(-2)^5} = -2$

**c.** $-\sqrt[3]{8} = -\sqrt[3]{(2)^3} = -2$

**d.** $\sqrt[4]{-81}$ is not a real number.

**26. a.** $\sqrt[3]{27} = \sqrt[3]{3^3} = 3$

**b.** $-\sqrt[4]{256} = -\sqrt[4]{(4)^4} = -4$

**c.** $\sqrt[3]{-125} = \sqrt[3]{(-5)^3} = -5$

**d.** $\sqrt[5]{1} = \sqrt[5]{1^5} = 1$

418

**27.** **a.** $\sqrt{\dfrac{25}{36}} = \dfrac{\sqrt{25}}{\sqrt{36}} = \dfrac{5}{6}$

**b.** $\sqrt{\dfrac{3}{64}} = \dfrac{\sqrt{3}}{\sqrt{64}} = \dfrac{\sqrt{3}}{8}$

**c.** $\sqrt{\dfrac{40}{81}} = \dfrac{\sqrt{40}}{\sqrt{81}} = \dfrac{\sqrt{4 \cdot 10}}{9} = \dfrac{2\sqrt{10}}{9}$

**28.** **a.** $\sqrt{\dfrac{4}{25}} = \dfrac{\sqrt{4}}{\sqrt{25}} = \dfrac{2}{5}$

**b.** $\sqrt{\dfrac{16}{121}} = \dfrac{\sqrt{16}}{\sqrt{121}} = \dfrac{4}{11}$

**c.** $\sqrt{\dfrac{2}{49}} = \dfrac{\sqrt{2}}{\sqrt{49}} = \dfrac{\sqrt{2}}{7}$

**29.** **a.** $\sqrt{50} + \sqrt{8} = \sqrt{25 \cdot 2} + \sqrt{4 \cdot 2}$
$$= 5\sqrt{2} + 2\sqrt{2} = 7\sqrt{2}$$

**b.** $7\sqrt{12} - \sqrt{75} = 7\sqrt{4 \cdot 3} - \sqrt{25 \cdot 3}$
$$= 7 \cdot 2\sqrt{3} - 5\sqrt{3}$$
$$= 9\sqrt{3}$$

**c.** $\sqrt{25} - \sqrt{27} - 2\sqrt{18} - \sqrt{16}$
$$= 5 - \sqrt{9 \cdot 3} - 2\sqrt{9 \cdot 2} - 4$$
$$= 1 - 3\sqrt{3} - 2 \cdot 3\sqrt{2}$$
$$= 1 - 3\sqrt{3} - 6\sqrt{2}$$

**30.** **a.** $\sqrt{80} + \sqrt{20} = \sqrt{16 \cdot 5} + \sqrt{4 \cdot 5}$
$$= 4\sqrt{5} + 2\sqrt{5} = 6\sqrt{5}$$

**b.** $2\sqrt{98} - 2\sqrt{18} = 2\sqrt{49 \cdot 2} - 2\sqrt{9 \cdot 2}$
$$= 2 \cdot 7\sqrt{2} - 2 \cdot 3\sqrt{2}$$
$$= 8\sqrt{2}$$

**c.** $\sqrt{32} + \sqrt{121} - \sqrt{12}$
$$= \sqrt{16 \cdot 2} + 11 - \sqrt{4 \cdot 3}$$
$$= 11 + 4\sqrt{2} - 2\sqrt{3}$$

**31.** **a.** $\sqrt{7} \cdot \sqrt{3} = \sqrt{7 \cdot 3} = \sqrt{21}$

**b.** $\sqrt{3} \cdot \sqrt{15} = \sqrt{3 \cdot 15} = \sqrt{9 \cdot 5} = 3\sqrt{5}$

**c.** $2\sqrt{6} \cdot 5\sqrt{2} = 10\sqrt{6 \cdot 2} = 10\sqrt{4 \cdot 3}$
$$= 10 \cdot 2\sqrt{3} = 20\sqrt{3}$$

**d.** $\left(3\sqrt{2}\right)^2 = 9\sqrt{4} = 9 \cdot 2 = 18$

**32.** **a.** $\sqrt{2} \cdot \sqrt{5} = \sqrt{2 \cdot 5} = \sqrt{10}$

**b.** $\sqrt{56} \cdot \sqrt{7} = \sqrt{56 \cdot 7} = \sqrt{392} = \sqrt{196 \cdot 2}$
$$= 14\sqrt{2}$$

**c.** $\left(4\sqrt{3}\right)^2 = 16\sqrt{9} = 16 \cdot 3 = 48$

**d.** $3\sqrt{8} \cdot 7\sqrt{2} = 21\sqrt{8 \cdot 2} = 21\sqrt{16}$
$$= 21 \cdot 4 = 84$$

**33.** $\sqrt{x} = \sqrt{5x - 2}$
$$x = 5x - 2$$
$$0 = 4x - 2$$
$$2 = 4x$$
$$\dfrac{2}{4} = x$$
$$\dfrac{1}{2} = x$$

**34.** $\sqrt{x - 4} + 7 = 2$
$$\sqrt{x - 4} = -5$$

The square root of a real number cannot be negative. There is no solution.

**35.** $a^2 + b^2 = c^2.$

$$\left(\overline{PQ}\right)^2 + \left(\overline{QR}\right)^2 = \left(\overline{PR}\right)^2$$

$$\left(\overline{PQ}\right)^2 + (240)^2 = (320)^2$$

$$\left(\overline{PQ}\right)^2 + 57{,}600 = 102{,}400$$

$$\left(\overline{PQ}\right)^2 = 44{,}800$$

$$\overline{PQ} = \sqrt{44{,}800} \approx 212 \text{ feet}$$

**36.** $(-7, 4)$ and $(2, 5)$

$$d = \sqrt{\left(x_2 - x_1\right)^2 + \left(y_2 - y_1\right)^2}$$

$$d = \sqrt{\left(2 - (-7)\right)^2 + (5 - 4)^2}$$

$$d = \sqrt{9^2 + 1^2}$$

$$d = \sqrt{81 + 1}$$

$$d = \sqrt{82}$$

**37.** **a.** $25^{1/2} = \sqrt{25} = 5$

 **b.** $8^{1/3} = \sqrt[3]{8} = 2$

 **c.** $-16^{1/4} = -\sqrt[4]{16} = -2$

 **d.** $(-27)^{1/3} = \sqrt[3]{-27} = -3$

 **e.** $\left(\dfrac{1}{9}\right)^{1/2} = \sqrt{\dfrac{1}{9}} = \dfrac{1}{3}$

**38.** **a.** $-49^{1/2} = -\sqrt{49} = -7$

 **b.** $256^{1/4} = \sqrt[4]{256} = \sqrt[4]{(4)^4} = 4$

 **c.** $(-64)^{1/3} = \sqrt[3]{-64} = -4$

 **d.** $\left(\dfrac{25}{36}\right)^{1/2} = \sqrt{\dfrac{25}{36}} = \dfrac{5}{6}$

 **e.** $(32)^{1/5} = \sqrt[5]{32} = \sqrt[5]{(2)^5} = 2$

**39.** $2x^2 = 7$

$$x^2 = \frac{7}{2}$$

$$x = \pm\sqrt{\frac{7}{2}} = \pm\frac{\sqrt{7}}{\sqrt{2}} = \pm\frac{\sqrt{7}}{\sqrt{2}} \cdot \frac{\sqrt{2}}{\sqrt{2}} = \pm\frac{\sqrt{14}}{2}$$

**40.** $3(x - 4)^2 = 9$

$$(x - 4)^2 = 3$$

$$x - 4 = \pm\sqrt{3}$$

$$x = 4 \pm \sqrt{3}$$

**41.** $x^2 - 10x = -14$

$$x^2 - 10x + 25 = -14 + 25$$

$$(x - 5)^2 = 11$$

$$x - 5 = \pm\sqrt{11}$$

$$x = 5 \pm \sqrt{11}$$

**42.** $x^2 + 4x = 8$

$$x^2 + 4x + 4 = 8 + 4$$

$$(x + 2)^2 = 12$$

$$x + 2 = \pm\sqrt{12}$$

$$x = -2 \pm \sqrt{4 \cdot 3}$$

$$x = -2 \pm 2\sqrt{3}$$

**43.** $2x^2 - 9x = 5$

$$2x^2 - 9x - 5 = 0$$

$a = 2, b = -9,$ and $c = -5$

$$x = \frac{-(-9) \pm \sqrt{(-9)^2 - 4(2)(-5)}}{2(2)}$$

$$= \frac{9 \pm \sqrt{81 + 40}}{4} = \frac{9 \pm \sqrt{121}}{4}$$

$$= \frac{9 \pm 11}{4}$$

$$x = \frac{9+11}{4} = 5 \text{ or } x = \frac{9-11}{4} = -\frac{1}{2}$$

The solutions are 5 and $-\frac{1}{2}$.

**44.** $2x^2 + 5x = 7$

$2x^2 + 5x - 7 = 0$

$a = 2, b = 5,$ and $c = -7$

$$x = \frac{-(5) \pm \sqrt{(5)^2 - 4(2)(-7)}}{2(2)}$$

$$= \frac{-5 \pm \sqrt{25 + 56}}{4} = \frac{-5 \pm \sqrt{81}}{4}$$

$$= \frac{-5 \pm 9}{4}$$

$$x = \frac{-5+9}{4} = 1 \text{ or } x = \frac{-5-9}{4} = -\frac{7}{2}$$

The solutions are 1 and $-\frac{7}{2}$.

**45. a.** $\sqrt{-4} = i\sqrt{4} = 2i$

**b.** $\sqrt{-11} = i\sqrt{11}$

**c.** $\sqrt{-20} = i\sqrt{20} = i\sqrt{4 \cdot 5} = 2i\sqrt{5}$

**46. a.** $\sqrt{-7} = i\sqrt{7}$

**b.** $\sqrt{-16} = i\sqrt{16} = 4i$

**c.** $\sqrt{-27} = i\sqrt{27} = i\sqrt{9 \cdot 3} = 3i\sqrt{3}$

**47.** $y = x^2 - 4$

$y$-intercept: $x = 0, y = 0^2 - 4 = -4,$

$$(0, -4)$$

vertex: $(0, -4)$

$x$-intercepts: $y = 0,$

$0 = x^2 - 4$

$0 = (x+2)(x-2)$

$x + 2 = 0$ or $x - 2 = 0$

$x = -2 \qquad x = 2$

$(-2, 0)$ and $(2, 0)$

| $x$ | $y$ |
|-----|-----|
| $-2$ | $0$ |
| $-1$ | $-3$ |
| $0$ | $-4$ |
| $1$ | $-3$ |
| $2$ | $0$ |

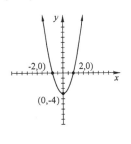

**48.** $y = x^2 + 2x + 3$

vertex: $x = -\dfrac{b}{2a} = -\dfrac{2}{2(1)} = -1 \Big\}(-1, 2)$

$\quad\quad y = (-1)^2 + 2(-1) + 3 = 2$

$y$-intercept: $x = 0, y = 0^2 + 2(0) + 3 = 3,$

$$(0, 3)$$

$x$-intercepts: $y = 0,$

$0 = x^2 + 2x + 3$

There are no real solutions to the equation, so there are no $x$-intercepts.

| $x$ | $y$ |
|-----|-----|
| $-3$ | $6$ |
| $-2$ | $3$ |
| $-1$ | $2$ |
| $0$ | $3$ |
| $1$ | $6$ |

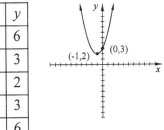

# Appendix   A

**1.**
$$9.076$$
$$+ 8.004$$
$$17.080$$

**3.**
$$27.004$$
$$- 14.200$$
$$12.804$$

**5.**
$$107.92$$
$$+ 3.04$$
$$110.96$$

**7.**
$$10.0$$
$$- 7.6$$
$$2.4$$

**9.**
$$126.32$$
$$- 97.89$$
$$28.43$$

**11.**
$$3.25$$
$$\times \quad 70$$
$$227.50$$

**13.**

$$
\begin{array}{r}
2.7 \\
3\overline{)8.1} \\
\underline{6} \\
21 \\
\underline{21} \\
0
\end{array}
$$

**15.**
$$55.4050$$
$$- \quad 6.1711$$
$$49.2339$$

**17.**

$$
\begin{array}{r}
80. \\
75.\overline{)6000.} \\
\underline{600} \\
00 \\
\underline{00}
\end{array}
$$

**19.**

$$
\begin{array}{r}
0.07612 \\
100.\overline{)7.61200} \\
\underline{700} \\
612 \\
\underline{600} \\
120 \\
\underline{100} \\
200 \\
\underline{200} \\
0
\end{array}
$$

**21.**

$$
\begin{array}{r}
4.56 \\
27.\overline{)123.12} \\
\underline{108} \\
151 \\
\underline{135} \\
162 \\
\underline{162} \\
0
\end{array}
$$

**23.**
$$
\begin{array}{r}
569.20 \\
71.25 \\
+\ \ \ 8.01 \\
\hline
648.46
\end{array}
$$

**25.**
$$
\begin{array}{r}
768.00 \\
-\ \ \ 0.17 \\
\hline
767.83
\end{array}
$$

**27.**
$$
\begin{array}{r}
12.00 \\
+\ \ 0.062 \\
\hline
12.062
\end{array}
$$

**29.**
$$
\begin{array}{r}
76.00 \\
-\ 14.52 \\
\hline
61.48
\end{array}
$$

**31.**
$$
\begin{array}{r}
7.7 \\
43.\overline{)331.1} \\
\underline{301} \\
30\ 1 \\
\underline{30\ 1} \\
0
\end{array}
$$

**33.**
$$
\begin{array}{r}
762.12 \\
89.70 \\
+\ \ 11.55 \\
\hline
863.37
\end{array}
$$

**35.**
$$
\begin{array}{r}
23.400 \\
-\ \ 0.821 \\
\hline
22.579
\end{array}
$$

**37.**
$$
\begin{array}{r}
476.12 \\
-\ 112.97 \\
\hline
363.15
\end{array}
$$

**39.**
$$
\begin{array}{r}
0.007 \\
+\ 7.000 \\
\hline
7.007
\end{array}
$$

# Appendix   B

**1.** $90° - 19° = 71°$

**3.** $90° - 70.8° = 19.2°$

**5.** $90° - 11\frac{1}{4}^° = 78\frac{3}{4}^°$

**7.** $180° - 150° = 30°$

**9.** $180° - 30.2° = 149.8°$

**11.** $180° - 79\frac{1}{2}^° = 100\frac{1}{2}^°$

**13.** $m\angle 1 = 110°$
$m\angle 2 = 180° - 110° = 70°$
$m\angle 3 = m\angle 2 = 70°$
$m\angle 4 = m\angle 2 = 70°$
$m\angle 5 = m\angle 1 = 110°$
$m\angle 6 = m\angle 4 = 70°$
$m\angle 7 = m\angle 5 = 110°$

**15.** $180° - 11° - 79° = 90°$

**17.** $180° - 25° - 65° = 90°$

**19.** $180° - 30° - 60° = 90°$

**21.** $90° - 45° = 45°$
$45°, 90°$

**23.** $90° - 17° = 73°$
$73°, 90°$

**25.** $90° - 39\frac{3}{4}^° = 50\frac{1}{4}^°$
$50\frac{1}{4}^°, 90°$

**27.** $\dfrac{12}{4} = \dfrac{18}{x}$
$4x\left(\dfrac{12}{4}\right) = 4x\left(\dfrac{18}{x}\right)$
$12x = 72$
$x = 6$

**29.** $\dfrac{6}{9} = \dfrac{3}{x}$
$9x\left(\dfrac{6}{9}\right) = 9x\left(\dfrac{3}{x}\right)$
$6x = 27$
$x = 4.5$

**31.** $a^2 + b^2 = c^2$
$6^2 + 8^2 = c^2$
$36 + 64 = c^2$
$100 = c^2$
$10 = c$

**33.** $a^2 + b^2 = c^2$
$5^2 + b^2 = 13^2$
$25 + b^2 = 169$
$b^2 = 144$
$b = 12$

# Appendix D

**1.** 21, 28, 16, 42, 38

$$\bar{x} = \frac{21+28+16+42+38}{5} = \frac{145}{5} = 29$$

16, 21, 28, 38, 42

median = 28

no mode

**3.** 7.6, 8.2, 8.2, 9.6, 5.7, 9.1

$$\bar{x} = \frac{7.6+8.2+8.2+9.6+5.7+9.1}{6}$$

$$= \frac{48.4}{6} = 8.1$$

5.7, 7.6, 8.2, 8.2, 9.1, 9.6

$$median = \frac{8.2+8.2}{2} = 8.2$$

mode = 8.2

**5.** 0.2, 0.3, 0.5, 0.6, 0.6, 0.9, 0.2, 0.7, 1.1

$$\bar{x} = \frac{0.2+0.3+0.5+0.6+0.6+0.9+0.2+0.7+1.1}{9}$$

$$= \frac{5.1}{9}$$

$$= 0.6$$

0.2, 0.2, 0.3, 0.5, 0.6, 0.6, 0.7, 0.9, 1.1

median = 0.6

mode = 0.2 and 0.6

**7.** 231, 543, 601, 293, 588, 109, 334, 268

$$\bar{x} = \frac{231+543+601+293+588+109+334+268}{8}$$

$$= \frac{2967}{8}$$

$$= 370.9$$

109, 231, 268, 293, 334, 543, 588, 601

$$median = \frac{293+334}{2} = 313.5$$

no mode

**9.** 1454, 1250, 1136, 1127, 1107

$$\bar{x} = \frac{1454+1250+1136+1127+1107}{5}$$

$$= \frac{6074}{5}$$

$$= 1214.8 \text{ feet}$$

**11.** 1454, 1250, 1136, 1127,
1107, 1046, 1023, 1002

$$median = \frac{1127+1107}{2} = 1117 \text{ feet}$$

**13.** $$\bar{x} = \frac{7.8+6.9+7.5+4.7+6.9+7.0}{6}$$

$$= \frac{40.8}{6} = 6.8 \text{ seconds}$$

**15.** 4.7, 6.9, 6.9, 7.0, 7.5, 7.8

mode = 6.9

**17.** 74, 77, 85, 86, 91, 95

$$median = \frac{85+86}{2} = 85.5$$

**19.** Sum $= 78 + 80 + 66 + 68 + 71$
$\phantom{\text{Sum} = } + 64 + 82 + 71 + 70 + 65$
$\phantom{\text{Sum} = } + 70 + 75 + 77 + 86 + 72$
$\phantom{\text{Sum}} = 1095$

$\bar{x} = \dfrac{1095}{15} = 73$

**21.** 64, 65, 66, 68, 70, 70, 71, 71, 72, 75,
77, 78, 80, 82, 86
mode = 70 and 71

**23.** 64, 65, 66, 68,
70, 70, 71, 71, 72, 75, 77, 78, 80, 82, 86
$\uparrow$
mean = 73
9 rates were lower than the mean.

**25.** _, _, 16, 18, _;
Since the mode is 21, at least two of
the missing numbers must be 21. The
mean is 20. Let the one unknown
number be $x$.

$\bar{x} = \dfrac{21 + 21 + 16 + 18 + 24}{5} = 20$

$\dfrac{76 + x}{5} = 20$

$76 + x = 100$

$x = 24$

The missing numbers are 21, 21, 24.

# Appendix F

**1.** Volume $= lwh = 6(4)(3) = 72$ cu in.

Surface area

$= 2lh + 2wh + 2lw$

$= 2(6)(3) + 2(4)(3) + 2(6)(4)$

$= 36 + 24 + 38$

$= 108$ sq in.

**3.** Volume $= s^3 = 8^3 = 512$ cu cm.

Surface area $= 6s^2 = 6(8^2) = 384$ sq cm.

**5.** Volume $= \frac{1}{3}\pi r^2 h = \frac{1}{3}\pi(2)^2(3) = 4\pi$ cu yd

$= 4\left(\frac{22}{7}\right) = 12\frac{4}{7}$ cu yd.

Surface area $= \pi r\sqrt{r^2 + h^2} + \pi r^2$

$= \pi(2)\sqrt{2^2 + 3^2} + \pi 2^2$

$= 2\sqrt{13}\pi + 4\pi$ sq yd

$= 2\sqrt{13}(3.14) + 4(3.14)$

$= 35.20$ sq yd.

**7.** Volume $= \frac{4}{3}\pi r^3 = \frac{4}{3}\pi(5)^3 = \frac{500}{3}\pi$ cu in

$= \frac{500}{3}\left(\frac{22}{7}\right) = 523\frac{17}{21}$ cu in.

Surface area $= 4\pi r^2 = 4\pi(5)^2 = 100\pi$ sq in.

$= 100\left(\frac{22}{7}\right) = 314\frac{2}{7}$ sq in.

**9.** Volume $= \frac{1}{3}s^2 h = \frac{1}{3}(6)^2(4) = 48$ cu cm

Surface area $= B + \frac{1}{2}pl$

$= 36 + \frac{1}{2}(24)(5)$

$= 96$ sq cm.

**11.** Volume $= s^3 = \left(1\frac{1}{3}\right)^3 = 2\frac{10}{27}$ cu in.

**13.** Surface area

$= 2lh + 2wh + 2lw$

$= 2(2)(1.4) + 2(3)(1.4) + 2(2)(3)$

$= 5.6 + 8.4 + 12$

$= 26$ sq ft.

**15.** Volume $= \frac{1}{3}s^2 h = \frac{1}{3}(5)^2(1.3) = 10\frac{5}{6}$ cu in

**17.** Volume $= \frac{1}{3}s^2 h = \frac{1}{3}(12)^2(20)$

$= 960$ cu cm

**19.** Surface area $= 4\pi r^2 = 4\pi(7)^2$

$= 196\pi$ sq in.

**21.** Volume $= lwh = 2\left(2\frac{1}{2}\right)\left(1\frac{1}{2}\right) = 7\frac{1}{2}$ cu ft.

**23.** Volume $= \frac{1}{3}\pi r^2 h = \frac{1}{3}\left(\frac{22}{7}\right)(2)^2(3)$

$= 12\frac{4}{7}$ cu cm